Agronomy: Advanced Techniques and Applications

Agronomy: Advanced Techniques and Applications

Editor: Harry Abbott

STATES
ACADEMIC PRESS
www.statesacademicpress.com

States Academic Press,
109 South 5th Street,
Brooklyn, NY 11249, USA

Visit us on the World Wide Web at:
www.statesacademicpress.com

ISBN: 978-1-63989-037-8 (Hardback)

Cataloging-in-Publication Data

Agronomy : advanced techniques and applications / edited by Harry Abbott.
p. cm.
Includes bibliographical references and index.
ISBN 978-1-63989-037-8
1. Agronomy. 2. Agricultural innovations. I. Abbott, Harry.
SB91 .A37 2022
630--dc23

Table of Contents

Preface

The world is advancing at a fast pace like never before. Therefore, the need is to keep up with the latest developments. This book was an idea that came to fruition when the specialists in the area realized the need to coordinate together and document essential themes in the subject. That's when I was requested to be the editor. Editing this book has been an honour as it brings together diverse authors researching on different streams of the field. The book collates essential materials contributed by veterans in the area which can be utilized by students and researchers alike.

Agronomy is the science concerned with the study of agriculture. It encompasses the study of the principles and practices of crop production and soil management. Besides crop production, this field is also vital for fruit and vegetable production. The discipline of agronomy is also concerned with soil fertility and soil conservation. Different varieties of soil are analyzed and percentage composition of macronutrients such as compounds of phosphorus, nitrogen, potassium, magnesium, calcium and sulfur are studied. Based on these compounds, strategies are devised for balancing soil nutrients for optimal plant growth. An important area of agronomy is the management of agricultural systems from an ecological and environmental perspective. This includes the practice of organic farming, sustainable agriculture, alternative cropping systems and food systems. This book outlines the processes and applications of agronomy in detail. It is a valuable compilation of topics, ranging from the basic to the most complex advancements in this field. For all readers who are interested in agronomy, the case studies included in this book will serve as an excellent guide to develop a comprehensive understanding.

Each chapter is a sole-standing publication that reflects each author´s interpretation. Thus, the book displays a multi-facetted picture of our current understanding of application, resources and aspects of the field. I would like to thank the contributors of this book and my family for their endless support.

Editor

1

Soil Quality and Water Redistribution Influences on Plant Production over Low Hillslopes on Reclaimed Mined Land

Stephen D. Merrill,[1] **Mark A. Liebig**,[1] **John D. Hendrickson**,[1] **and Abbey F. Wick**[2]

[1]*Northern Great Plains Research Laboratory, Agricultural Research Service, U.S. Department of Agriculture, P.O. Box 459, 1701 10th Ave. SW, Mandan, ND 58554, USA*
[2]*Department of Soil Science, Walster Hall, North Dakota State University, Department 7680, P.O. Box 6050, Fargo, ND 58108-6050, USA*

Correspondence should be addressed to Stephen D. Merrill; steve.merrill@ars.usda.gov

Academic Editor: Glaciela Kaschuk

Coal surface mining in northern Great Plains USA led to reclamation experiments with soil respreading. Respread soil depth (RSD) and runoff of water redistribution (WR) effects interacted in original North Dakota studies, complicating interpretations. We determined WR and soil depth/soil quality (SQ) effects on hillslope production patterns for sites with soil wedges (2%–5% slope, 50-m length) over sodic mine spoils. At Zap, cool-season forages crested wheatgrass (CWG: *Agropyron cristatum*) and Russian wildrye *(Psathyrostachys juncea)* generally decreased as RSD increased upslope. At Stanton, alfalfa *(Medicago sativa)*, native grasses (*Bouteloua* spp.), and CWG responded to RSD, increasing 70% to midslope and decreasing further. A SQ index (SQI) based on six indicator properties was highly correlated ($r > 0.7$) with RSD. Yield regressions with RSD or SQI were generally significant for Stanton forages and for spring wheat *(Triticum aestivum)* at both sites. Yield regressions with WR index (catchment area-based) indicated dominance of WR effects at Zap. Cool-season forages at Zap evidently responded to springtime runoff, while Stanton forages and spring wheat at both sites used water later in the season and responded to soil depth/SQ effects. Results suggest models for interaction of SQ and landform WR affecting productivity should include plant community composition and water-use information.

1. Introduction

A need for land reclamation was created by the advent of coal surface mining in the northern Great Plains of the USA. In portions of this region Paleocene strata can give rise to mine spoil high in exchangeable sodium and clay content making revegetation difficult [1, 2]. Reclamation experiments using soil salvage and respreading were initiated by Power and colleagues in the state of North Dakota [1, 2]. Reclaimed landforms in these experiments were hillslopes with 2% to 5% slopes and about 50-m in length, and interactions of landform water redistribution (WR) and soil depth effects complicated interpretation of productivity results. The experiments occurred before full development of current methods for analyzing soil quality (SQ) and landform WR. Thus a principal goal of this study is to quantify relative effects of soil depth SQ and WR on productivity and the role of plant community in modifying such effects.

In one experiment near Stanton ND [2], a wedge of Haplustoll subsoil material (B and C horizons, 0 to 2.4 m thick) was placed over mine spoil and covered with 0-, 20-, or 60-cm of topsoil material (A-horizon; Figure 1(a)). Production of three perennial forage species and spring wheat increased with soil depth to midslope and then declined somewhat further upslope.

In another experiment near Zap ND [3], a double wedge, consisting of three different types of Haplustoll subsoil materials 0 to 1.2 m thick, was covered with 20 cm of topsoil (Figure 1(b)). Response of two perennial forage species to soil depth was weak or nonexistent and yields generally declined as soil depth increased upslope. However, yield patterns of spring wheat at Zap were similar to those at Stanton. It

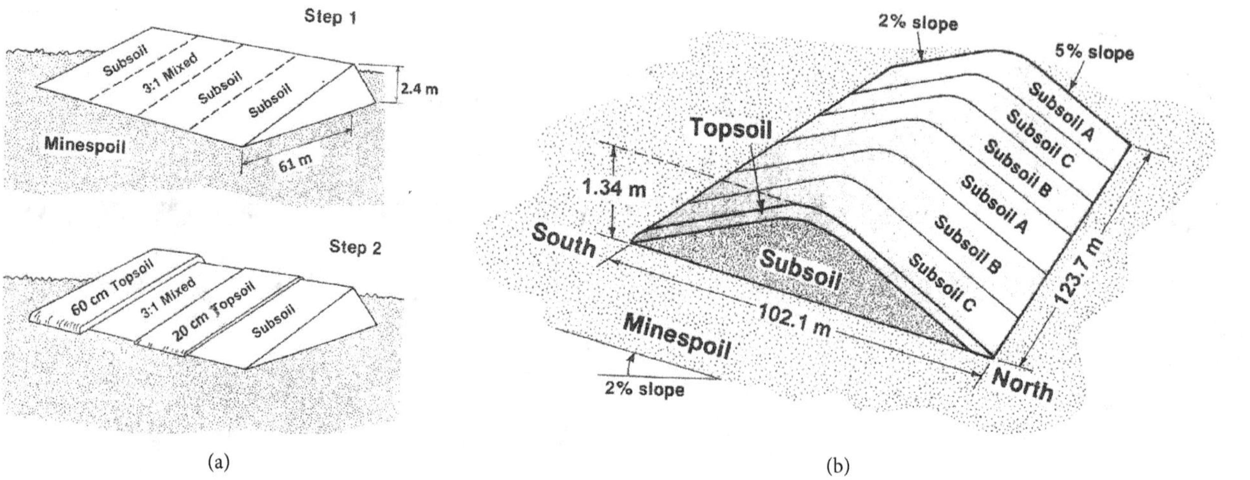

FIGURE 1: (a) One replication out of three of reconstructed soils at Stanton site. Overall length of construct was 243 m and average slope was 4%. The 3 : 1 mixed subsoil : topsoil treatment is not discussed in this paper. (b) Reconstructed soil at Zap site. Modified from Doll et al. [1].

was hypothesized [3] that water redistribution affected the hillslope patterns of forage species' yields at Zap.

The influence of soil depth where water redistribution influence should have been minimal was shown by results of Barth and Martin [4] from experiments in which they created flat-surfaced wedges of soil that were placed down into sodic mine spoils. They observed an increase of forage grass production to a soil depth of 0.7 m out of 1.5 m, with plateauing of growth at greater soil depths.

Runoff of water to lower areas of hillslopes results in higher yields downslope.

Working on pastures dominated by cool-season grasses in Iowa USA, Harmoney et al. [5] found that yields over hillslope positions ran in the order toeslope > summit > backslope. Under xeric climate in the Pacific Northwest USA, Fiez et al. [6] found that wheat yield was correlated with water use, which was generally higher at footslope or backslope positions.

Results of North Dakota-based reclamation experiments, particularly in terms of production responses to respread soil depth over mine spoils [1], have been applied to state reclamation requirements [7]. However, original interpretations of the reclamation experiments [2, 3] have left important questions unanswered that should be amenable to modern forms of soil analyses, particularly application of the soil quality (SQ) concept. Soil depth over mine spoil is a measure of soil quality (SQ). To better assure scientifically valid application of reclamation experiments featured here, the relative roles of soil depth/SQ and water redistribution in plant production responses need to be elucidated through application of current SQ assessment and landscape analysis practices.

The basic definition of SQ is the capacity of a soil to function [8]. A system for assessing SQ has been described by Andrews and coworkers [9, 10] in which a set of SQ indicator properties are chosen, the value of each indicator towards an endpoint (such as crop production) is scored, and scoring values are aggregated into a SQ index (SQI).

Indices that quantify characteristics of landforms have been reviewed [11]. Water redistribution on hillslopes has been linked to these terrain indices [12]. To best serve the specific needs of the current study, we have selected a basic terrain index based on catchment area [11, 12] for use as a hillslope water redistribution index (WRI).

The objective of this study was to determine the relative effects of SQ measures (soil depth and SQI) and apparent water redistribution (represented by WRI) upon hillslope plant production patterns on reclaimed mine soils. Other factors whose effects on plant production patterns have been studied here were plant species and structure of reclaimed soils.

2. Research Methods

2.1. Experimental Mined Land Reclamation Sites. Experimental reclamation sites were located in Mercer county, North Dakota, USA, near the towns of Zap and Stanton at 47°14′33″N-101°51′1″W and 47°15′30″N-101°20′55″W, respectively. Mean annual precipitation was 421 mm, with 270 mm occurring during the 6-mo April–September growing season, and the mean 6-mo growing season temperature was 15.6°C.

Coal surface mining at the sites produced spoil from Mesozoic strata consisting of loosely consolidated siltstone or shale material that is saline and sodic. At the Zap site, mine spoil was leveled in 1974-1975 to a 2% north-facing slope. After removal and stockpiling of nearby Haplustoll topsoil material (mostly A-horizon), three types of subsoil materials (mostly B- and C-horizon) were placed in a double wedge-shaped hillslope with 21-m wide main treatment (subsoil quality) areas with two replications (Figure 1(b); Table 1).

Topsoil (mostly A-horizon) was spread over subsoils to a depth of 20 cm. At the Stanton site, spoil was leveled in 1974 to a 0.5% slope and a single wedge of subsoil material was formed. Main treatment strips 20-m wide with three

TABLE 1: Soil properties at experimental reclamation sites. AWC, available water content; EC, electrical conductivity; SAR, sodium adsorption ratio; SOC, soil organic carbon.

	Zap reclamation site					Stanton reclamation site		
	Topsoil	Subsoil A	Subsoil B	Subsoil C	Minespoil	Topsoil	Subsoil	Minespoil
Sand, $g\,kg^{-1}$	410	82	315	679	80	526	246	74
Silt, $g\,kg^{-1}$	349	443	420	187	482	341	425	563
Clay, $g\,kg^{-1}$	241	475	265	134	438	133	329	363
Texture, USDA system	Loam	Silty clay	Clay loam	Sandy loam	Silty clay	Sandy loam	Clay loam	Silty clay loam
AWC, $kg\,kg^{-1}$	0.174	0.147	0.149	0.096	0.128	0.190	0.157	0.157
Bulk density, $g\,cm^{-3}$	1.211	1.251	1.331	1.448	1.353	1.279	1.279	1.340
$^{\$}EC_{1:1}$, $dS\,m^{-1}$	1.32	4.76	2.29	0.79	5.86	0.60	2.99	3.05
SAR, $(mol\,m^{-3})^{1/2}$	4.86	7.67	6.77	5.46	17.30	3.47	9.96	16.40
Saturation%	41.2	60.1	48.5	27.4	99.6	41.0	62.7	112.4
$^{\$}pH_{se}$	7.16	7.36	7.73	7.07	7.54	7.37	7.91	7.95
SOC, $g\,kg^{-1}$	14.3	9.7	7.1	7.2	$8.5^{\#}$	15.6	8.1	$9.5^{\#}$
Total N, $g\,kg^{-1}$	1.16	0.80	0.60	0.69	0.71	1.3	0.7	0.3
available P, $mg\,kg^{-1}$	9.2	11.4	3.6	6.7	7.1	8.7	4.0	4.9

$^{\$}$Note that EC was on 1 : 1 gravimetric basis and pH was from saturation extracts. $^{\#}$Values may be elevated to some greater or lesser degree by presence of leonardite (partially oxidized lignite).

replications were created by placement of 0-, 0.2-, and 0.6-m thicknesses of topsoil over subsoil (Figure 1(a); Table 1).

Each soil type block was divided lengthwise and four crop treatments were seeded in a randomized pattern within replications. Species seeded at Zap in 1976 were the perennial, cool-season grasses Russian wildrye (RWR; *Psathyrostachys juncea*) and crested wheatgrass (CWG; *Agropyron cristatum*) and spring wheat (Triticum aestivum). A fourth species, alfalfa (ALF; *Medicago sativa*), did not establish well and was not used for research purposes. Species seeded at the Stanton in 1975 were ALF, CWG, spring wheat, and a native, perennial, warm-season grass mix (NAT) consisting of blue grama *(Bouteloua gracilis)* and sideoats grama *(Bouteloua curtipendula)*. Before seeding, 67 and 35 $kg\,ha^{-1}$ P as triple superphosphate were broadcast on all subplots at Zap and Stanton, respectively. All subplots except those to be seeded to ALF received 55 $kg\,N\,ha^{-1}$ as ammonium nitrate every year through 1978. Spring wheat received 12 $kg\,P\,ha^{-1}$ deposited with seed.

2.2. Plant Productivity. Crops were harvested from 1976 to 1979 at the Stanton site and 1976 to 1981 at the Zap site, but, for purposes of this paper, data from 1978 and 1979 are used because they were years without drought in which both sites were represented. Vegetation was harvested from areas 4.3 m long by 1.8 m wide at 10 or 11 positions along each aspect of hillslopes, and dry weights of forage species' aboveground herbiage and spring wheat seed yields were measured. Forage grasses were harvested in late June or early July, and ALF was harvested twice, in late June or early July and again in August or early September. Spring wheat seed was harvested from the later part of July to the first part of August.

Relative yields were calculated for each seeding treatment based on the yearly average for the treatment at a site. Yield patterns of individual forage species' treatments were compared using running averages of three contiguous hillslope positions (1, 2, 3; 2, 3, 4; etc.). Applying one-way ANOVA to such averages normalized to annual treatment means, only 6 out of 132 site-position-soil treatment-year combinations indicated significant yield differences. Therefore, relative yields for a given year and site were aggregated among forage species treatments for further analyses.

2.3. Soil Measurements. Soil samples were collected by hydraulic probe in depth increments of 0–15, 15–30, 30–60, 60–90, and 90–120 cm. Samples for electrical conductivity (EC) and pH were collected at four positions from toe to summit along each hillslope aspect and those for SOC and available P at either one or two positions.

Soil chemical measurements were conducted by methods detailed in Black et al. [13] adapted to requirements of mined land reclamation research [14]. Soluble cations were determined on saturation extracts by atomic adsorption, and sodium adsorption ratio (SAR) was calculated. Electrical conductivity was determined on saturation extracts [15] and pH was determined on saturated paste [16]. Saturation extract EC (but not pH) values were converted into 1 : 1 (mass) basis using the ExtractChem model [17] for purposes of comparison with subsequent soil sampling and analyses. SOC was determined by wet oxidation [18] and available P by bicarbonate technique [19].

Details about samples used for textural analysis are given in Wick [20]. Soil texture was determined by hydrometer technique [21]. Available water content (AWC, defined as water held by soil between matric potentials of −0.33 kPa and −1.5 kPa) was calculated from textural values and organic carbon according to Gupta and Larson [22] using soil bulk density values from [20]. SOC values were multiplied by 2.0 to obtain organic carbon. Additional details about the experiments may be found in [2, 3].

2.4. Soil Quality. Soil quality assessment was carried out by calculation of a SQ index (SQI) using the SMAF tool

TABLE 2: Precipitation at experimental reclamation sites near Zap and Stanton, North Dakota, USA. Annual median precipitation near the sites at Beulah ND for 29-year period was 39.2 cm.

Month	Zap			Stanton	
	1978	1979	29-yr median 1975–2003	1978	1979
			cm		
April	4.4	4.5	3.4	4.2	3.2
May	5.8	2.0	5.7	5.2	2.7
June	6.4	6.1	7.5	8.8	7.2
July	7.8	11.4	6.1	6.2	7.9
August	1.5	2.3	3.0	0.9	12.3
September	4.1	3.2	3.2	8.8	2.5
April through September	29.9	29.5	28.9	34.1	35.8

developed by Andrews et al. [10]. In this method of SQ assessment, the first step is selection of a minimum number of properties related to an end point of interest, such as crop production. We have used six available properties, SAR, EC, pH, SOC, AWC, and available P. For each SQ indicator property, a scoring value from 0 to 100 was calculated according to the property's capacity to support crop production. Scoring functions [10] were all monotonic over ranges of interest here and were either "less is better" (SAR, EC, and pH) or "more is better" (SOC, AWC, and available P). Scoring values were averaged with equal weights [9] to yield a SQI value. Values were calculated for individual soil depths and then integrated over 0–90 cm. This depth range produced the highest average R^2 values for yield versus soil property regressions (data not shown).

In order to calculate a SQI value for each of the 8 or 10 positions along hillslopes at which plant yields were measured, soil property values were estimated from measured values. For the chemical properties EC, SAR, and pH, measured at four locations, this was done by interpolation on treatment averages for each sampling depth increment. For the properties SOC, available P, and AWC, treatment averages for reconstructed soil horizons (topsoil, subsoil, or mine spoil) and soil depths were used to derive suitable values.

2.5. Water Redistribution. Water content measurements were examined for indications of downslope water redistribution. Neutron moisture meters had been used in steel access tubes, which had been installed two per subplot at a lower midslope and a near-summit position. Although measurements were made over multiple years, only 1979 data had been analyzed by individual depth increments in predecessor studies, making them suitable for indicating downslope water redistribution. Out of measurements made five times during the year, only those in late April-early May showed substantial evidence of water redistribution in the 0–30 cm soil depth. (The 0–30 cm depth was the only increment where water content differences between the upper and lower hillslope positions had not been affected by systematic variation in respread soil depth.) ANOVA was done across soil treatments within seeding treatments. Averaged across soil treatments, 0–30 cm soil water at Zap under CWG was 14% greater at the lower hillslope position (water content = $0.21\,m^3\,m^{-3}$, p of greater $F = 0.023$). Lower versus upper position differences averaged 9% for ALF and RWR treatments (ns). At Stanton, average soil water under CWG was 4% greater at the lower hillslope position (water content = $0.23\,m^3\,m^{-3}$, p of greater $F = 0.054$). At Stanton, almost all indication of greater water content at the lower position was for the 0 cm topsoil treatment, with an average increase of 7%.

Moore et al. [11] have reviewed terrain attribute indicators based on catchment area, aspect, slope, and curvature. Western et al. [12] have examined relationships between landscape soil water distributions and various terrain indices. Given the simple landform of the low hillslopes involved here and the differences in aspect between the two sites, catchment area has been used here as a water redistribution index (WRI):

$$\mathrm{WRI} = \ln(d), \qquad (1)$$

where d is distance (in m) downslope from the summit.

2.6. Statistical Analyses. One-way ANOVA of individual seeding treatments' yields used SPC for Excel (BPI Consulting; Cypress, Texas, USA). Soil water data was analyzed within site, depth, and seeding treatment by ANOVA with hillslope location as split plot within soil treatment using SAS 79 (SAS Enterprises Inc., Cary, North Carolina, USA). Correlations among soil properties and indices and regressions of plant production versus WRI and SQ metrics used SPC for Excel. Regressions of plant production versus hillslope position used Sigma Plot (Systat Software, Inc., San Jose, California, USA).

3. Results and Discussion

3.1. Patterns of Plant Production. Growing season precipitation at Zap and Stanton during the years of the study was average or above average (Table 2). However, precipitation in May 1979 was half or less of average at both sites.

Except for ALF, yields of perennial forages and spring wheat at both sites were considerably higher in 1978 than in

TABLE 3: Biomass yields of perennial forage species or seed yield of spring wheat.

Site	Year	Crested wheatgrass	Native grass mix	Russian wildrye	Alfalfa	Spring wheat
		kg ha^{-1}				
Zap	1978	3103		2134		811
	1979	1418		1149		
Stanton	1978	4023	1462		2194	1639
	1979	1599	861		2327	

1979, reflecting a better distribution of spring precipitation during 1978 (Table 3). At Zap, yields of perennials were not significantly different among subsoil treatments in 1978 but about 25% greater on subsoil C than on A or B in 1979 (data not shown). Spring wheat seed yield in 1978 was about 50% greater on subsoil A than on subsoil C, where it was least. For perennials and spring wheat at Stanton, the two topsoiled treatments had 20% to 30% higher yields than the treatment with subsoil alone.

Three out of six 1978-79 regressions of aggregated CWG and RWR yields versus hillslope position on the Zap 5% north-facing slope had about 20% to 45% increases with greater soil depths and elevations to 10 m or more from the toe of the soil wedge (Figure 2). However, decreases in yield further upslope with increasing soil depth and elevation were greater over the majority of the length at Zap north-facing slope. Five out of six regressions had $0.07 > p > 0.001$. These growth patterns were indicative of apparent water redistribution effects. The soil treatment with subsoil C (sandy loam, nonsaline, Table 1) showed the largest increases and decreases of yield. Only two out of six regressions were significant on the 1.5% south-facing slope at Zap.

On the 4% south-facing slope at Stanton, 1978-79 relative yields of the three (ALF, CWG, and NAT) forage species aggregated showed up to about 60% increases of production with increasing soil depth and elevation to midslope and then smaller decreases of yield further upslope (Figure 3). Five of six regressions were highly significant and $p = 0.06$ for one. These patterns at Stanton were indicative of a greater response to soil depth and a lesser response to apparent water redistribution compared with patterns at the Zap site. Figure 3 indicates the consistently higher yields of the 20 cm and 60 cm topsoil treatments compared with the no topsoil treatment.

Hillslope patterns of perennial forages' yields have been aggregated across soil treatments and compared with those of spring wheat seed yields for the year 1978 in Figure 4. At Zap 5% north-facing slope, the spring wheat production pattern was quite different from that of forage species, approximately doubling from toe to midslope, then decreasing somewhat with greater soil depth towards the summit. Spring wheat production also differed greatly from the yield pattern of forages at the 1.5% Zap south-facing slope, increasing by well over 100% with soil depth from toeslope to near summit. The hillslope pattern of spring wheat production at Stanton was very similar to that of perennial forages.

Hillslope water redistribution generated by runoff is sensitive to slope [11, 12]. This is illustrated by comparison of forage production patterns for soil treatments on Zap north-facing versus south-facing slopes (5% versus 1.5%, resp.) in Figure 2, and by the same north/south comparison of Zap forages with spring wheat in Figure 4. In particular, the pattern of spring wheat production on the low south-facing slope at Zap appears to be free of any apparent water redistribution effect.

3.2. Analysis of Plant Production over Hillslopes. To understand results of regressions of plant productivity with measures of SQ and water redistribution, we examined correlations among various soil properties and indices (Table 4). Given the basic wedge-shaped design of the experiments, the two measures of SQ used here, RSD and SQI, were highly and positively correlated with distance from toe of each hillslope, and WRI was negatively correlated with distance (a). SQ measures RSD and SQI were strongly related, $R^2 = 0.890$ and 0.689 for Zap and Stanton sites, respectively (b). And WRI was negatively correlated to SQ measures. Chemical soil properties SAR, EC, and pH were negatively correlated with SQI at both sites, reflecting "more is less" SQ scoring functions (c). Soil properties AWC, SOC, and P were positively correlated with SQI at Stanton, but only P was correlated with SQI at Zap.

At Zap, subsoil B and C treatments had positive regressions of forage species production with water redistribution index (WRI); the subsoil A regression was near significant (Table 5). Water redistribution is sensitive to slope, and this was shown by comparison of yield versus WRI regressions for Zap north-facing slope versus south-facing, with R^2 values of 0.228 and 0.045, respectively (data not shown). All regressions between yield and SQ measures RSD and SQI at Zap had negative regression coefficients, evidently a consequence of negative correlations between WRI and SQ measures (see Table 4).

Multilinear regressions combining RSD or SQI with WRI also had negative coefficients for the SQ measure part of the equations. Positive regressions with WRI and negative coefficients for regressions with SQ measures reflected evident dominance of water redistribution effects on hillslope yield patterns.

In contrast to Zap, only multilinear regressions of yield versus RSD and WRI or yield versus SQI and WRI were significant at Stanton and then only for the two treatments with topsoil (Table 5). Plant yield versus hillslope position functions at Stanton were nonlinear to an extent that made linear regressions of yield versus WRI or SQ measures

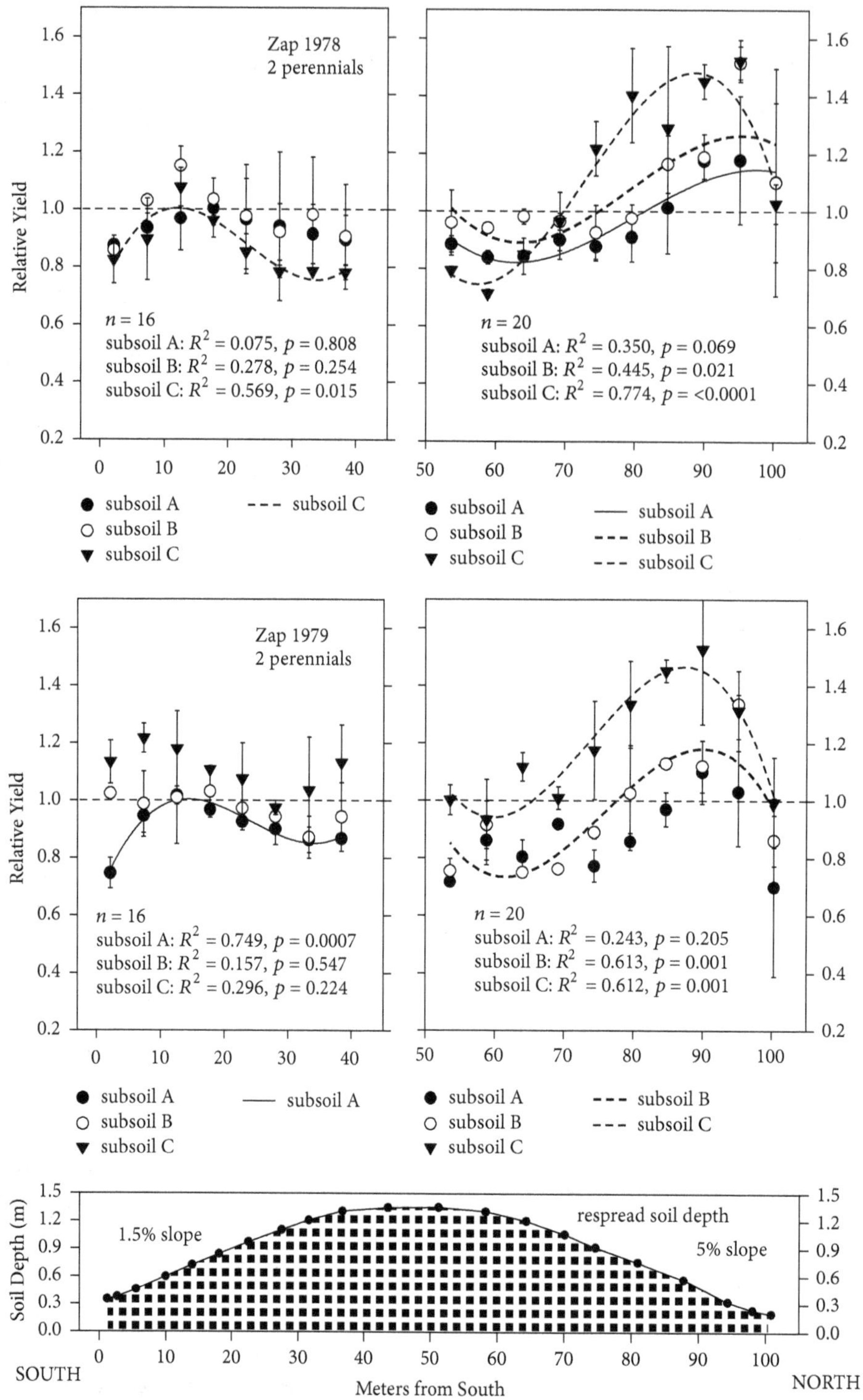

FIGURE 2: Relative yields of perennial forage crops versus hillslope positions for soil treatments at Zap site during 1978 and 1979 showing means, standard errors, and cubic regressions. Yields of crested wheatgrass *(Agropyron cristatum)* and Russian wildrye *(Psathyrostachys juncea)* have been aggregated.

nonsignificant (see Figure 3). All regression coefficients in significant equations shown for Stanton were positive.

Regressions of yield versus WRI and SQ measures comparing 1978 spring wheat seed and perennial forage biomass indicated that spring wheat at Zap displayed a pattern of results similar to those of both spring wheat and perennials at Stanton (regressions aggregated across soil treatments; Table 6). Spring wheat at both sites and perennials at Stanton had positive linear and multilinear regressions with SQ measures in 1978, while perennials at Zap had negative

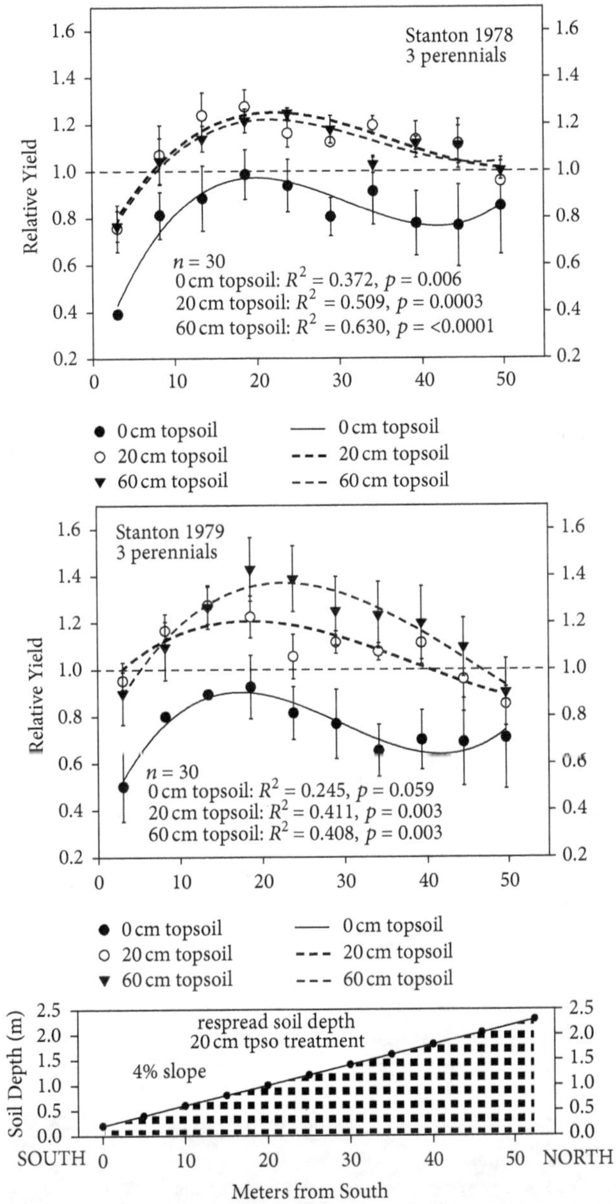

FIGURE 3: Relative yields of perennial forage crops versus hillslope positions for soil treatments at Stanton site during 1978 and 1979 showing means, standard errors, and cubic regressions. Yields of alfalfa *(Medicago sativa)*, crested wheatgrass *(Agropyron cristatum)*, and native mix (blue grama, *Bouteloua gracilis*, and sideoats grama, *Bouteloua curtipendula*) have been aggregated.

coefficients as was also shown for regressions based on individual soil treatments in Table 5.

Differences in hillslope production patterns and regression results between forage species and spring wheat at Zap appear to be related to when the species are most actively using soil water and growing during the spring and summer. Perennial grasses at Zap, CWG, and RWR are both cool-season species, actively growing in earlier spring. Under North Dakota USA conditions, CWG reaches boot stage by late May to about June 1 [23], whereas, in the same area, spring wheat reaches boot stage around the last week of June to about the beginning of July and accumulates seed mass up to the end of July [24]. In a semiarid area, water redistribution in springtime should affect cool-season forages. Later water use and growth by spring wheat would rely more upon stored soil water, making growth more responsive to soil depth/SQ effects.

Two out of three forage species at Stanton were either warm-season (NAT mix) or full season (twice-cut ALF). As noted above, there were very few differences among hillslope growth patterns of individual forage species, and this applied to cool-season CWG compared to the other forage species at Stanton. Thus, the principal reason for differences in hillslope growth patterns between perennials at Zap on one hand and

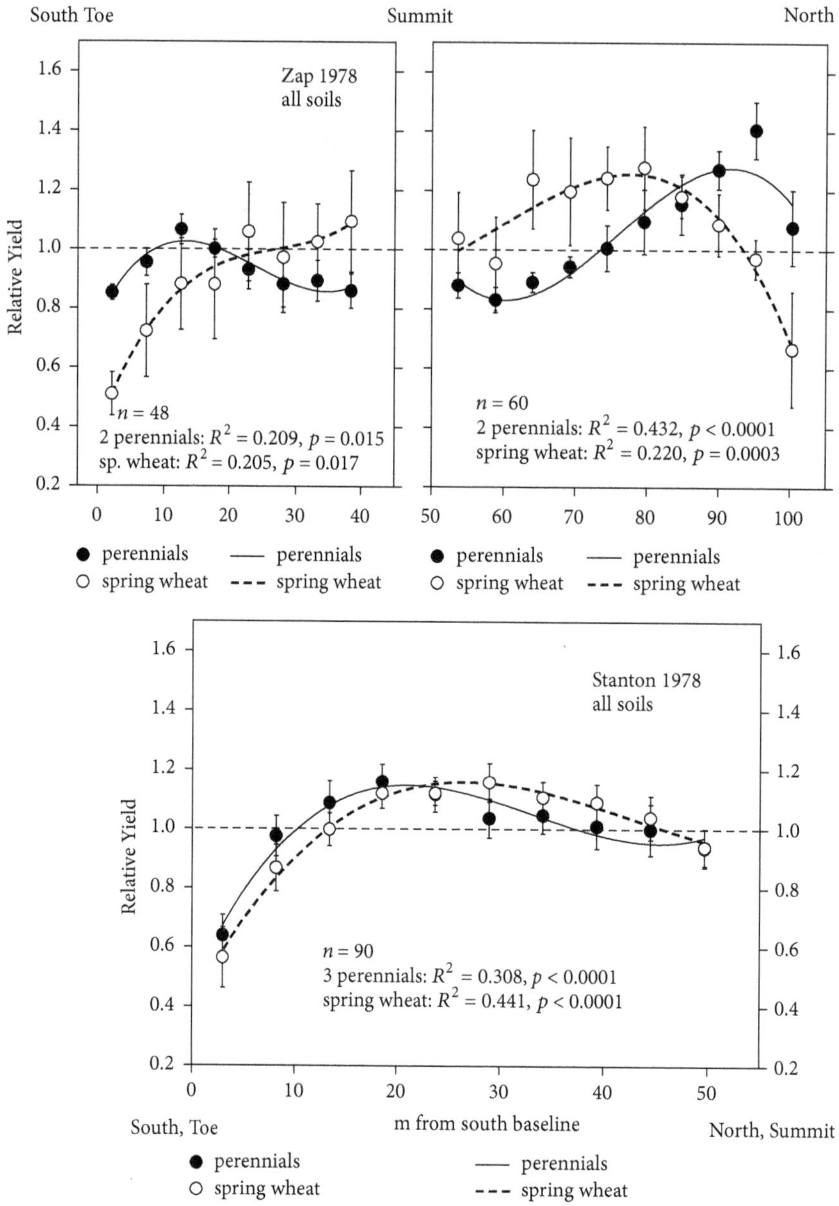

FIGURE 4: Relative yields versus hillslope positions comparing perennial forage crops' biomass and spring wheat seed for 1978 at Zap and Stanton sites showing means, standard errors, and cubic regressions. Soil treatments have been aggregated.

perennials at Stanton plus spring wheat at both sites appears to involve seasonality of growth and water use of the plant species at the sites.

Aspect is another factor that could have been involved in the differences in hillslope yield patterns between the sites. The Stanton site had a single 4% south-facing slope, making it less responsive to earlier-season water redistribution effects on plant growth than the 5% north-facing slope at Zap. Examples exist in literature showing significantly greater soil water at north aspect positions compared to south aspect on Great Plains USA agricultural lands [25].

3.3. Application of Results. Current SQ assessment practice emphasizes management effects by examination of the most dynamic part of the profile near the soil surface, with emphasis placed on biological properties [26]. An earlier conception of SQ assessment included consideration of the entire rooting zone [27]. This study and its predecessors [2, 3] have shown growth responses to respread soil depths of 1 m or greater, indicating the value of a whole-rootzone approach to SQ assessment. A study of water use and movement at the Stanton site from 1978 to 1980 by Merrill et al. revealed that root growth occurred to at least a depth of 0.9 m [28]. Furthermore, this study showed that root growth and water use were occurring in mine spoil where soil depths were 0.9 m or less, an indication of macroporosity development in this material which initially has extremely low hydraulic conductivity. As noted above, we have used the 0–90 cm

TABLE 4: (a) Correlations between distance from bottom (toe) of hillslopes and water redistribution index (WRI), respread soil depth (RSD), and soil quality index (SQI). (b) Correlation matrices for WRI, RSD, and SQI. (c) Correlations between SQI and soil properties. All r values given have $p \leq 0.001$. AWC, available water content; EC, electrical conductivity; P, available phosphorus; SAR, sodium adsorption ratio; SOC, soil organic carbon.

(a)

Site		WRI	RSD	SQI
Zap $n = 96$	Meters from south toe	−0.925	0.979	0.846
Zap $n = 120$	Meters from north toe	−0.978	0.985	0.934
Stanton $n = 180$	Meters from south toe	−0.981	0.930	0.458

(b)

Zap, $n = 216$	WRI	RSD	SQI	Stanton, $n = 180$	WRI	RSD	SQI
WRI	1	−0.762	−0.704	WRI	1	−0.903	−0.444
RSD	- - -	1	0.890	RSD	- -	1	0.689
SQI	- -	- -	1	SQI	- -	- -	1

(c)

site		SAR	EC	pH	AWC	SOC	P
Zap $n = 216$	SQI	−0.904	−0.734	−0.464	ns	ns	0.221
Stanton $n = 180$	SQI	−0.861	−0.915	−0.859	0.719	0.579	0.376

TABLE 5: Linear and multilinear regressions of perennial forages' relative yields in 1978-79 versus water redistribution index (WRI), respread soil depth (RSD), and soil quality index (SQI) for soil treatments at Zap and Stanton sites. The third column in each group indicates the sign of regression coefficient(s). For entries with $p > 0.10$, R^2 and other information are not given. At Zap and Stanton sites, $n = 72$ and 60, respectively. ns means multilinear regression coefficient has $p > 0.10$.

X-variable(s)	R^2	p	Coefficient sign	R^2	p	Coefficient sign	R^2	p	Coefficient sign
		Zap 1978-79 Subsoil A			Zap 1978-79 Subsoil B			Zap 1978-79 Subsoil C	
WRI		0.103		0.123	0.003	P†	0.217	<0.001	P
RSD	0.074	0.021	N	0.197	<0.001	N	0.188	<0.001	N
RSD + WRI	0.074	0.069	N; ns	0.197	<0.001	N; ns	0.225	<0.001	ns; P
SQI	0.075	0.020	N	0.187	<0.001	N	0.221	<0.001	N
SQI + WRI	0.076	0.066	N: ns	0.195	<0.001	N; ns	0.258	<0.001	N; P
		Stanton 1978-79 0 cm topsoil			Stanton 1978-79 20 cm topsoil			Stanton 1978-79 60 cm topsoil	
WRI		0.830			0.109			0.583	
RSD		0.418			0.479			0.639	
RSD + WRI		0.113		0.250	<0.001	P; P	0.303	<0.001	P; P
SQI		0.386			0.243		0.284	<0.001	P
SQI + WRI		0.103		0.418	<0.001	P; P	0.402	<0.001	P; P

†P = positive; N = negative.

depth for SQ indexing and analyses. What would appear to be most appropriate for disturbed lands reclamation practice is a synthesis of the near-surface and whole-rootzone approaches to SQ assessment.

Especially in subhumid or semiarid areas, use of soil water deeper in the rootzone can become important under limited precipitation. Evidence for the benefit of considering the whole profile in SQ assessment was given in Merrill et al. [29], where two taxonomically similar North Dakota USA Haplustoll soils with similar SQI values were compared. One had subsoil with higher hydraulic conductivity and good root penetrability and exhibited greater maize production under

TABLE 6: Linear and multilinear regressions of 1978 spring wheat and perennial forages' relative yields versus water redistribution index (WRI), respread soil depth (RSD), and soil quality index (SQI) aggregated across soil treatments at Zap and Stanton sites. The third column in each group indicates the sign of regression coefficient(s). For entries with $p > 0.10$, R^2 and other information are not given. ns means multilinear regression coefficient has $p > 0.10$.

X-variable(s)	Zap 1978 all soils *Spring wheat*, $n = 108$			Zap 1978 all soils *Perennials*, $n = 108$			Stanton 1978 all soils *Spring wheat*, $n = 90$			Stanton 1978 all soils *Perennials*, $n = 90$		
	R^2	p	coefficient sign	R^2	p	coefficient sign	R^2	p	coefficient sign	R^2	p	coefficient sign
WRI		0.349		0.186	<0.001	P[†]	0.057	0.021	N		0.492	
RSD	0.081	0.003	P	0.200	<0.001	N	0.204	<0.001	P	0.092	0.004	P
RSD + WRI	0.119	0.001	P; P	0.220	<0.001	N; ns	0.351	<0.001	P; P	0.311	<0.001	P; P
SQI	0.047	0.024	P	0.257	<0.001	N	0.285	<0.001	P	0.314	<0.001	P
SQI + WRI	0.055	0.052	P; ns	0.268	<0.001	N; ns	0.285	<0.001	P; ns	0.352	<0.001	P; P

[†]P = positive; N = negative.

soil water limitation compared with the other soil, which had subsoil with higher clay content and apparently low hydraulic conductivity.

Much of the disturbed land in the northern Great Plains region, such as that used for research here, has been or will be reclaimed to rangeland/pastureland usage. According to Herrick et al. [30], SQ assessments of rangeland should include measures of plant community composition and structure along with soil properties to best gauge soil and land stability. They recommend aggregate stability of surface soil as a key indicator of resistance to erosion.

Our results have shown that interpretation of reclamation experiments by soil salvage and respreading requires consideration of both soil depth through SQ analysis and basic hillslope water redistribution process. Analysis of hillslope growth patterns and regression analyses have shown qualitative differences between the sites and among plant species types. These differences may be summarized: more dominance of apparent water redistribution effects for cool-season grasses at Zap site and dominance or more balanced influence of soil depth and associated SQ factors for spring wheat at both sites and forage species at Stanton.

To obtain more quantitative analyses of reclamation experiment results such as those studied here, we suggest an analysis model be implemented that combines hydrologic processes on the hillslope scale with plant growth responses to soil factors at the point scale. For example, hydrologists have described and implemented a distributed hydrological model [31] that divides the hillslope scale into space increments and uses mass balance to derive hydrologic components at each increment, such as runoff, infiltration, transpiration, and drainage to subsoil. Plant ecological and water-use effects within the context of semiarid rangelands have been simulated with this model. The plant growth component of the Erosion Prediction Impact Calculator (EPIC) model [32] could be used for the growth response part of a hybrid hydrological-edaphic/ecological analysis model. The EPIC model accounts for plant water relations and root growth by soil depth increments. Our results have shown that plant species and associated phenology, such as seasonality of water use and growth, can affect hillslope growth pattern. McMaster et al. [33] have described a modern, standalone version of the plant growth portion of EPIC and discuss earlier versions used in other publically supported agroecosystem models in the USA.

4. Conclusions

We studied mined land reclamation experiments in the state of North Dakota, USA, in which salvaged topsoil and subsoil materials were respread on high-sodium mine spoils, forming low hillslopes. At the Zap site, hillslope yield patterns of cool-season forage grasses CWG and RWR were more dominated by apparent water redistribution effects (yield increasing downslope) than by growth responses to soil depth increasing upslope. Hillslope yield patterns of forage crops (ALF, CWG, and NAT mix) at Stanton site and of spring wheat at both sites were more dominated by response to increasing soil depths from toeslope to midslope areas, with lesser response to apparent water redistribution from midslope to shoulder slope areas.

A soil quality index (SQI) based on six soil indicator properties was highly correlated ($r > 0.7$) with respread soil depth. Regressions between forage species' yields and water redistribution index (WRI) were significant for two out of three soil treatments at Zap. Multilinear regressions of forage yield at Stanton with measures of SQ combined with WRI captured curvilinear hillslope yield patterns, and both linear and multilinear regressions of spring wheat yield with SQ measures were significant for the two sites.

The dominance of water redistribution effects on hillslope yield patterns at Zap, supported by regression analyses, appeared to be due to response of cool-season forage species to springtime runoff. Forages at Stanton and spring wheat at both sites are known to use soil water later in the season, and this was associated with hillslope yield patterns that were dominated by soil depth and SQ effects.

Our results show that assessment models for disturbed lands reclamation by soil respreading need to include information about plant community composition and phenology of growth and water use along with landform hydrology and SQ-soil structure information.

Acknowledgments

The authors wish to acknowledge the leadership of the late Dr. J. F. Power in the mined land reclamation experiments upon which this paper is based.

References

[1] E. C. Doll, S. D. Merrill, and G. A. Halvorson, "Soil replacement for reclamation of strip mined lands in North Dakota," North Dakota Agricultural Experiment Station, Bulletin 514, Fargo, North Dakota, USA, 24 pp., 1984.

[2] J. F. Power, F. M. Sandoval, R. E. Ries, and S. D. Merrill, "Effects of Topsoil and Subsoil Thickness on Soil Water Content and Crop Production on a Disturbed Soill," *Soil Science Society of America Journal*, vol. 45, no. 1, p. 124, 1981.

[3] S. D. Merrill, R. E. Ries, and J. F. Power, "Subsoil characteristics and landscape position affect productivity of reconstructed mine soils," *Soil Science Society of America Journal*, vol. 62, no. 1, pp. 263–271, 1998.

[4] R. C. Barth and B. K. Martin, "Soil depth requirements for revegetation of surface-mined areas in Wyoming, Montana, and North Dekota," *Journal of Environmental Quality*, vol. 13, no. 3, pp. 399–404, 1984.

[5] K. R. Harmoney, K. J. Moore, E. C. Brummer, C. Lee Burras, and J. R. George, "Spatial legume composition and diversity across seeded landscapes," *Agronomy Journal*, vol. 93, no. 5, pp. 992–1000, 2001.

[6] T. E. Fiez, B. C. Miller, and W. L. Pan, "Winter wheat yield and grain protein across varied landscape positions," *Agronomy Journal*, vol. 86, no. 6, pp. 1026–1032, 1994.

[7] NDAC, "North Dakota Administrative Code, Title 69, Bismarck, North Dakota, USA, 2017," http://www.legis.nd.gov/agency-rules/north-dakota-administrative-code.

[8] D. L. Karlen, M. J. Mausbach, J. W. Doran, R. G. Cline, R. F. Harris, and G. E. Schuman, "Soil quality: a concept, definition, and framework for evaluation," *Soil Science Society of America Journal*, vol. 61, no. 1, pp. 4–10, 1997.

[9] S. S. Andrews, D. L. Karlen, and J. P. Mitchel, "A comparison of soil quality indexing methods for vegetable production systems in northern California Agriculture," *Ecosystems Environment*, vol. 90, no. 1, pp. 25–45, 2002.

[10] S. S. Andrews, D. L. Karlen, and C. A. Cambardella, "The soil management assessment framework: a quantitative soil quality evaluation method," *Soil Science Society of America Journal*, vol. 68, no. 6, pp. 1945–1962, 2004.

[11] I. D. Moore, R. B. Grayson, and A. R. Ladson, "Digital terrain modelling: a review of hydrological, geomorphological, and biological applications," *Hydrological Processes*, vol. 5, no. 1, pp. 3–30, 1991.

[12] A. W. Western, R. B. Grayson, G. Blöschl, G. R. Willgoose, and T. A. McMahon, "Observed spatial organization of soil moisture and its relation to terrain indices," *Water Resources Research*, vol. 35, no. 3, pp. 797–810, 1999.

[13] C. A. Black, D. D. Evans, J. L. White, L. E. Ensminger, and F. E. Clark, *Methods of Soil Analysis Part 2, Chemical and Microbiological Properties, Agronomy Monograph 9.2*, C. A. Black, D. D. Evans, J. L. White, L. E. Ensminger, and F. E. Clark, Eds., American Society of Agronomy, Madison, Wisconsin, USA, 1965, 1550 pp. (approx.).

[14] F. M. Sandoval and J. F. Power, *Laboratory methods recommended for chemical analysis of mined land spoils and overburden in western United States, USDA-ARS Handbook 525*, U.S. Government Printing Office, Washington, District of Columbia, USA, 1978, pp. 31.

[15] C. A. Bower and L. V. Wilcox, C. A. Black, D. D. Evans, J. L. White, L. E. Ensminger, and F. E. Clark, Eds. American Society of Agronomy, Madison, Wisconsin, USA, 1965, pp. 933–951.

[16] M. Peech, *Hydrogen-ion activity, in Methods of Soil Analysis Part 2, Chemical and Microbiological Properties, Agronomy Monograph 9.2*, C. A. Black, D. D. Evans, J. L. White, L. E. Ensminger, and F. E. Clark, Eds., American Society of Agronomy, Madison, Wisconsin, USA, 1965, pp. 914–926.

[17] U.S. Salinity Laboratory, ExtractChem (software), U.S. Dept. Agriculture, Agricultural Research Service, Riverside, California, USA, 2017, https://www.ars.usda.gov/research/software/.

[18] L. E. Allison, *Organic carbon, in Methods of Soil Analysis Part 2, Chemical and Microbiological Properties, Agronomy Monograph 9.2*, C. A. Black, D. D. Evans, J. L. White, L. E. Ensminger, and F. E. Clark, Eds., American Society of Agronomy, Madison, Wisconsin, USA, 1965, pp. 1367–1378.

[19] S. R. Olsen and L. A. Dean, *Phosphorus , in Methods of Soil Analysis Part 2, Chemical and Microbiological Properties, Agronomy Monograph 9.2*, C. A. Black, D. D. Evans, J. L. White, and F. E. Clark, Eds., American Society of Agronomy, Madison, Wisconsin, USA, 1965, pp. 1035–1049.

[20] A. F. Wick, *The effects of soil depth and soil characteristics on plant community development in North Dakota*, University of Denver, Denver, Colorado, USA, 2004, pp., 204.

[21] G. W. Gee and J. W. Bauder, "Particle size analysis," in *Methods of Soil Analysis Part 1, Physical and Mineralogical Methods, SSSA Book*, A. Klute, Ed., 5, pp. 383–411, Soil Science Society of America and American Society of Agronomy, Madison, Wisconsin, USA, 2nd edition, 1986.

[22] S. C. Gupta and W. E. Larson, "Estimating soil water retention characteristics from particle size distribution, organic matter percent, and bulk density," *Water Resources Research*, vol. 15, no. 6, pp. 1633–1635, 1979.

[23] A. B. Frank and A. Bauer, "Rooting Activity and Water Use during Vegatative Development of Crested and Western Wheatgrass," *Agronomy Journal*, vol. 83, no. 5, p. 906, 1991.

[24] S. D. Merrill, A. L. Black, and A. Bauer, "Conservation tillage affects root growth of dryland spring wheat under drought," *Soil Science Society of America Journal*, vol. 60, no. 2, pp. 575–583, 1996.

[25] A. Y. Hanna, P. W. Harlan, and D. T. Lewis, "Soil Available Water as Influenced by Landscape Position and Aspectl," *Agronomy Journal*, vol. 74, no. 6, p. 999, 1982.

[26] D. E. Stott, D. L. Karlen, C. A. Cambardella, and R. Daren Harmel, "A soil quality and metabolic activity assessment after fifty-seven years of agricultural management," *Soil Science Society of America Journal*, vol. 77, no. 3, pp. 903–913, 2013.

[27] W. E. Larson and F. J. Pierce, "The dynamics of soil quality as a measure of sustainable management," in *Defining Soil Quality for a Sustainable Environment, Chap. 3., SSSA Special Pub. No. 35*, J. W. Doran, D. C. Coleman, D. F. Bezdicek, and B. A. Stewart, Eds., pp. 37–51, Soil Science Society of America and American Society of Agronomy, Madison, Wisconsin, USA, 1994.

[28] S. D. Merrill, S. J. Smith, and J. F. Power, "Effect of disturbed soil thickness on soil water use and movement under perennial grass.," *Soil Science Society of America Journal*, vol. 49, no. 1, pp. 196–202, 1985.

[29] S. D. Merrill, M. A. Liebig, D. L. Tanaka, J. M. Krupinsky, and J. D. Hanson, "Comparison of soil quality and productivity at two sites differing in profile structure and topsoil properties," *Agriculture, Ecosystems & Environment*, vol. 179, pp. 53–61, 2013.

[30] J. E. Herrick, J. R. Brown, A. J. Tugel, P. L. Shaver, and K. M. Havstad, "Application of soil quality to monitoring and management: Paradigms from rangeland ecology," *Agronomy Journal*, vol. 94, no. 1, pp. 3–11, 2002.

[31] E. R. Vivoni, D. Entekhabi, R. L. Bras, and V. Y. Ivanov, "Controls on runoff generation and scale-dependence in a distributed hydrologic model," *Hydrology and Earth System Sciences*, vol. 11, no. 5, pp. 1683–1701, 2007.

[32] J. R. Williams, C. A. Jones, J. R. Kiniry, and D. A. Spanel, "EPIC crop growth model," *Transactions of the ASAE*, vol. 32, no. 2, pp. 497–511, 1989.

[33] G. S. McMaster, J. C. Ascough, D. A. Edmunds et al., "Simulating Unstressed Crop Development and Growth Using the Unified Plant Growth Model (UPGM)," *Environmental Modeling & Assessment*, vol. 19, no. 5, pp. 407–424, 2014.

2

Evaluation of Sensor-Based Nitrogen Rates and Sources in Wheat

Olga S. Walsh,[1] Sanaz Shafian,[1] and Robin J. Christiaens[2]

[1]*Department of Plant Sciences, Southwest Research and Extension Center, University of Idaho, Parma, ID 83660, USA*
[2]*Private Enterprise, Roundup, MT 59072, USA*

Correspondence should be addressed to Olga S. Walsh; owalsh@uidaho.edu

Academic Editor: Yuanhu Xuan

Nitrogen (N) is one of the most essential nutrients needed to reach maximum grain yield in all environments. Nitrogen fertilizers represent an important production cost, in both monetary and environmental terms. The aim of this study was to assess the effect of preplant nitrogen (N) rate and topdress N source on spring wheat (*Triticum aestivum* L.) grain yield and quality. Study was conducted in North-Central and Western Montana from 2011 to 2013 (total of 6 site-years). Six different preplant nitrogen (N) rates (0, 220, 22, 44, 67, and 90 N rate, kg ha^{-1}) followed by two topdress N sources (urea, 46-0-0, and urea ammonium nitrate (UAN), 32-0-0) were applied to spring wheat (*Triticum aestivum* L.). The results showed that there were no significant differences in grain yield, protein content, or protein yield, associated with topdress N source.

1. Introduction

Wheat is one of the most important crops which provide nutrition (proteins, energy, and minerals) to most of the world population. United States Department of Agriculture (USDA) has in its World Agricultural Supply and Demand Estimates (WASDE) forecasted global wheat production to reach 743 million metric tons (MMT), and the demand for wheat in developing countries is projected to increase 60% by 2050 (https://www.proag.com/news/world-wheat-consumption-increases-spring-wheat-in-short-supply/). Wheat ranks third among the US field crops in planted acreage, production, and gross farm receipts, behind corn and soybeans (https://www.ers.usda.gov/topics/crops/wheat/). Excellent growing conditions for much of the US, especially throughout the Great Plains states such as Kansas, Montana, North Dakota, Nebraska, and Oklahoma, contribute to excellent US wheat production levels. In 2016-17, the US farmers produced a total of 24.6 million US tons of winter, spring, and durum wheat on 50.2 million acres of cropland (https://www.ers.usda.gov/topics/crops/wheat/).

Wheat grain yield and quality depend on multiple genes interacting with environmental factors, such as nitrogen (N) availability, water, and temperature [1, 2]. Nitrogen is generally the most limiting nutrient factor for wheat production by influencing on chlorophyll production, photosynthesis process, and grain yield and quality [3]. Increasing N supply to a wheat crop can increase photosynthesis rate and consequently increase canopy biomass and grain yield. However, excessive N fertilization in cultivated land has profound environmental impacts such as nitrate leaching, soil denitrification, ammonia volatilization, and nitrous oxide emissions, which contaminate water and air and aggravate the climate change [4, 5]. While N losses cannot be avoided completely, they can be substantially reduced with sustainable agricultural practice such as adjusting N rate, timing and using the most appropriate fertilizer source, and adopting precision nutrient management technologies [6].

Crop canopy sensors have been developed to help growers to assess crop N status and make the best management decision based on actual crop condition and increase crop NUE [7]. Crop canopy sensors are designed to measure crop spectral reflectance at specific wavelength and generate vegetation indices (VIs) such as Normalized Difference Vegetation Index (NDVI) to assess specific crop characteristics of interest such as N status [8, 9]. These VIs can be used to generate field maps that show spatial variability of crop condition. These VIs have also been integrated into different algorithms to estimate crop N content, make fertilizer N recommendations, and increase economic return for producers [10]. Over 30 sensor-based algorithms have been developed since early 1990s. These algorithms are crop, region, and sensor specific and need to be calibrated for new crop in new region and new sensor used. For example, spring wheat (US, Canada, and Mexico) algorithm was developed as a collaborative effort of Oklahoma State University, Agriculture and Agri-Food Canada, and International Maize and Wheat Improvement Center (CYMMIT). This algorithm is a modified version of the spring wheat algorithm developed by the Oklahoma State University [11]. This algorithm is based on the relationship between sensor data collected early in the growing season and the yield attained across years and locations. The maximum wheat yield potential was set at 8000 kg ha^{-1} and the NUE at 35%. [12]. This algorithm is based on one single N application, immediately after sensing event, which is recommended to occur at Feekes 5-6 while the timing of applications and the source of N can affect the amount of N recovered or the nitrogen use efficiency (NUE) in a given year.

In the Great Plains, wheat growers can choose to apply all N before planting or split N fertilizer between preplant starter and the midseason topdress. Several N fertilizer sources—dry and liquid— are available to producers. The growers often chose N source based on current price per unit of N as well as currently available application equipment. Contrasting results in comparing dry with liquid N products in wheat have been reported. Several studies demonstrated that there are no differences in NUE between dry and liquid fertilizers [13, 14]. Some studies showed that NUE might be lower with liquid N sources due to significant ammonia loss [15]. Other researchers found that liquid N products may be superior in terms of crop yield and quality, as well as being more environmentally friendly, due to greater plant availability and thus more efficient uptake [16]. For example, in a long-term winter wheat study in Oklahoma (total of 80 site-years), liquid N products resulted in an almost 20% advantage in NUE compared to dry granular N source. Knowing the relative NUE of preplant and topdress and what type of N products is more efficient, dry or liquid one, are important for calibrating the developed algorithms, estimating crop N needed with higher accuracy to reach the potential yield plateau.

The objective of this study was to determine whether dry or liquid N source would result in significantly superior results in spring wheat production. Specifically, we assess the effect of preplant N rate and topdress N source on spring wheat grain yield and quality.

2. Methods and Materials

2.1. Study Areas. This study was conducted for three consecutive growing seasons (2011–2013) at two experimental locations: one dryland site, at Western Triangle Agricultural Center (WTARC), Conrad, MT (48.309794 and −111.924684), and one irrigated site at Western Agricultural Research Center (WARC), Corvallis, MT (46.328179 and −114.089873). Initial soil test results (2011) for each location are detailed in Table 1. The experimental design was a randomized complete block design (RCBD) with four replications. The main plot treatments included six preplant N rates (0, 22, 44, 67, 90, and 220 kg N ha^{-1}) applied at planting by side-banding granular urea (46-0-0) approximately 2.5 cm below the seed. The subplot treatments were two topdress N sources (dry granular urea and liquid urea ammonium nitrate (UAN)) (28-0-0) (Table 2). Treatment 1 (0 kg N ha^{-1}) was established as an unfertilized check plot and treatment 2 (220 kg N ha^{-1}) served as a nonlimiting N-rich reference. In total, there were 40 plots at each experimental location. Hard red spring wheat (*Triticum aestivum* cv. Choteau) was planted with a small plot drill with Conserva Pak™ openers manufactured by Swift Machining (Washougal, WA) at a density of approximately 1.8 million plants per hectare.

Table 1: Initial preplant soil test results (0–60 cm), Conrad and Corvallis, 2011.

	Conrad	Corvallis
Soil series	Scobey clay loam	Burnt Fork silt loam
pH	7.7	7.8
O.M.	3.7	1.5
NO_3-N	50	57
P (ppm)	20	18
K (ppm)	272	221
EC (mmhos/cm)	0.48	0.35

Table 2: Treatment, preplant N rate, and topdress N source for each study site.

Trt	*Preplant N rate, kg ha^{-1}	**Topdress N source
1	0	-
2	220	urea
3	22	urea
4	44	urea
5	67	urea
6	90	urea
7	22	UAN
8	44	UAN
9	67	UAN
10	90	UAN

2.2. Field Data Collection and Generating Topdress N Prescriptions. Within each study plot, Normalized Difference Vegetative Index (NDVI), was measured using a GreenSeeker hand-held optical sensor (Trimble Navigation Ltd., Sunnyvale, CA) at Feekes 5 of growing stage (early jointing, beginning of stem elongation, prior to first visible node). Then, the Spring Wheat (US, Canada, Mexico) algorithm (http://www.soiltesting.okstate.edu/SBNRC/SBNRC.php) was used to calculate topdress N rates for each treatment. The algorithm uses maximum NDVI (measured NDVI in treatment 2, nonlimiting N reference), measured NDVI for each treatment, seeding date, date of sensing (NDVI measurement date), and yield goal (average yield goal for the area) as input parameters to calculate topdress N rates. In some cases, treatment 2 did

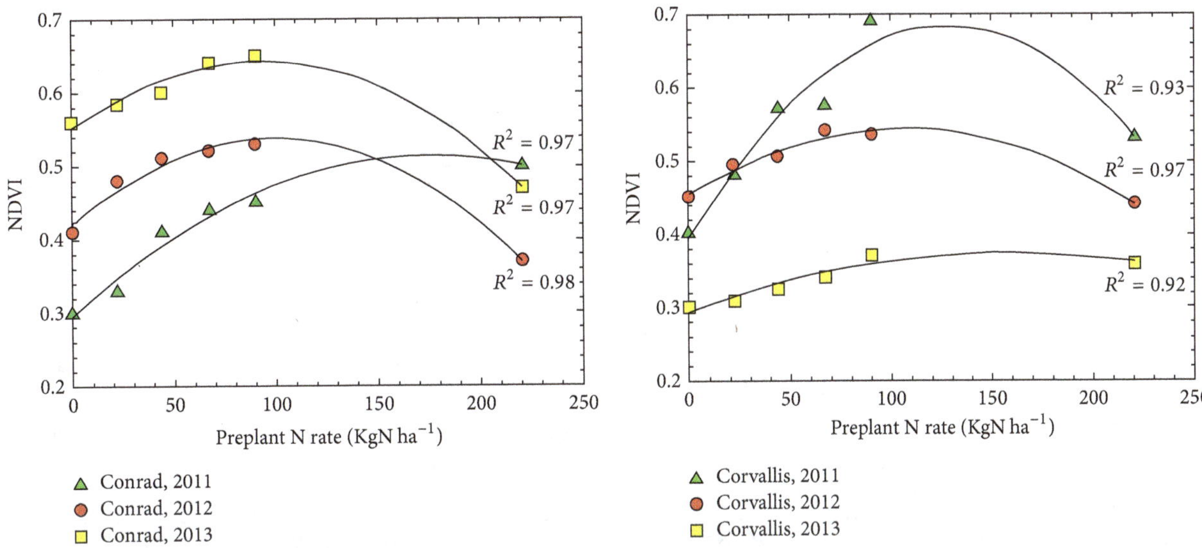

FIGURE 1: Relationships between Normalized Difference Vegetation Index (NDVI) and preplant N application rate for 2011, 2012, and 2013 at (a) Conrad and (b) Corvallis. Each data point represents the mean of four replications and was regressed against preplant N application rate.

not result in the highest NDVI for the location, possibly, due to the negative impact of a high N rate applied at planting affecting the seed. Then, a treatment with the highest NDVI values was chosen and used as a reference. The topdress N fertilizer was applied as urea (as dry prill, broadcasted manually) or as UAN (as a foliar spray, using a battery-operated backpack sprayer with a fan nozzle).

At maturity, wheat was harvested with a self-propelled Wintersteiger Classic Combine (Wintersteiger Inc., Salt Lake City, USA). At harvest, plot grain yield was recorded for each experimental plot using a Harvest Master GrainGage, by Wintersteiger. The harvested wheat grain was dried in the drying room for 14 days at the temperature of 35°C; then, the dried samples were weighed to determine the accurate by-plot grain yield, which was adjusted to 12% moisture. The by-plot subsamples were analyzed for total N content using near infrared reflectance spectroscopy (NIR) with a Perten DA 7250 NIR analyzer (Perten Instruments, Inc., Springfield, IL) at Agvise Laboratories (Northwood, ND).

2.3. Statistical Analysis. Grain N uptake for each plot was calculated by multiplying grain yield by total N concentration. The analysis of variance was conducted using the PROC GLM procedure in SAS v9.4 (SAS Institute, Inc., Cary, NC) and the mean separation was performed using the Orthogonal Contrasts method at a significance level of 0.05.

3. Results and Discussions

3.1. Relationship between Preplant N Rate and NDVI. Responses of NDVI to preplant N application rate for 2011, 2012 and 2013 at each study site are presented in Figure 1. Generally, there was a significant positive correlation between NDVI and preplant N application rate for all site-years.

The relationships between NDVI and preplant N application rate were linear or quadratic in nature (Figure 1). Positive quadratic relationships were more frequently observed between preplant N application rate and NDVI when wheat NDVI values were relatively high.

At Conrad site, the average NDVI values were 0.40, 0.47, and 0.58 in 2011, 2012, and 2013, respectively. Conrad was a dryland site, so differences in NDVI values for same preplant N application rate in different years could be related to varied precipitations and, consequently, differences in soil moisture content. In 2011 season, the amount and the distribution of rainfall from May to July, from planting time to the time the NDVI was measured, were higher than in 2012 and 2013. Moreover, the daily rainfall distribution was erratic; at times, heavy rain has caused significant runoff and leaching in 2011 compared to 2012 and 2013. This may explain lower average NDVI value at Conrad in 2011 compared to other two growing seasons. In 2013 Conrad study site received almost 20 cm of snowfall in April, before wheat planting, which increased the soil moisture content considerably. The higher soil moisture amounts available at wheat planting has likely resulted in higher NDVI values in 2013 compared to 2011 and 2012.

At Corvallis study site, the average NDVI values were 0.54, 0.49, and 0.33 in 2011, 2012, and 2013, respectively. At Corvallis, plots received 15.24 cm of irrigation water in three equally space periods between planting and soft dough stage, annually. Precipitation data can explain the differences in average NDVI values in different years. This field received 11.25 cm, 8.08 cm, and 7.76 cm of precipitations in 2011, 2012, and 2013, respectively, from planting to the date of the NDVI measurement. The higher rainfall in 2011 apparently may have resulted in higher NDVI compare to other years.

For Conrad in 2012 and 2013 and for Corvallis in 2011 and 2012, treatment 2, 220 kg N ha^{-1}, did not reach the highest NDVI value. This was possibly because the 220 kg N ha^{-1} preplant application rate might have been damaging to seeds and/or seedlings and produced relatively poor stand. The

TABLE 3: Preplant N rate, topdress N rate, topdress N source, total N rate applied, and spring wheat grain yield, Conrad and Corvallis, MT, 2011.

Trt	Conrad			Corvallis		
	Topdress N rate, kg ha^{-1}	Total N rate applied, kg ha^{-1}	Grain yield, kg ha^{-1}	Topdress N rate, kg ha^{-1}	Total N rate applied, kg ha^{-1}	Grain yield, kg ha^{-1}
1	-	-	942 f	-	-	2018 f
2	22	242	2690 a	24	244	3699 abc
3	22	44	1547 e	32	54	2757 d
4	22	66	1547 e	8	52	3430 bc
5	22	89	1883 cd	17	84	3833 abc
6	11	101	2152 b	24	114	3968 a
7	33	55	1480 e	32	54	3228 cd
8	22	66	1614 de	8	52	3497 abc
9	11	78	1950 bc	8	75	3363 bc
10	11	101	2152 b	17	107	3564 abc

*Preplant fertilizer N will be applied as urea. **Topdress fertilizer N rates were determined based on the NDVI values obtained using GreenSeeker, as prescribed by USA, Canada, Mexico Algorithm. Means within each column followed by the same letter are not significantly different at $p < 0.05$, as determined by Duncan's multiple range test.

TABLE 4: Preplant N rate, topdress N rate, topdress N source, total N rate applied, and spring wheat grain yield, Conrad and Corvallis, MT, 2012.

Trt	Conrad			Corvallis		
	Topdress N rate, kg ha^{-1}	Total N rate applied, kg ha^{-1}	Grain yield, kg ha^{-1}	Topdress N rate, kg ha^{-1}	Total N rate applied, kg ha^{-1}	Grain yield, kg ha^{-1}
1	-	-	5851 d	-	-	3901 f
2	78	298	6187 d	110	330	6456 d
3	16	38	6658 c	124	146	6725 cd
4	16	60	6994 abc	124	168	6927 bcd
5	16	83	7061 abc	124	191	7465 ab
6	30	120	7263 a	124	214	6860 bcd
7	25	47	6658 c	124	146	7196 abcd
8	16	60	6725 bc	110	154	7398 abcd
9	21	88	6927 abc	124	191	7599 a
10	21	111	7129 ab	110	200	7667 a

*Preplant fertilizer N will be applied as urea. **Topdress fertilizer N rates were determined based on the NDVI values obtained using GreenSeeker, as prescribed by USA, Canada, Mexico Algorithm. Means within each column followed by the same letter are not significantly different at $p < 0.05$, as determined by Duncan's multiple range test.

decision was made to use a treatment with the highest NDVI value as an N-rich reference for those particular site-years.

3.2. Relationship between NDVI and Prescribed Topdress N Rates. The algorithm has prescribed topdress N rates ranging from 0 to 137 kg N ha^{-1} depending on the yield goal for the location and the obtained NDVI values (Tables 3–5). As it mentioned in previous section, in several instances that treatment 2 did not reach to the highest NDVI value the decision was made to use a treatment with the highest NDVI value as an N-rich reference for that particular site-year.

In some cases, the algorithm prescribed higher topdress N rate for treatment with lower NDVI values in order to boost yield potential. For example, at Conrad, in 2012, measured NDVI showed higher value in treatment 6 compared to treatment 2 (Table 6). Prescribed 30 kg N ha^{-1} topdress N for treatment 6 resulted in a total of 120 kg N ha^{-1} applied, while 78 kg N ha^{-1} topdress N prescribed for treatment 2 resulted in a total of 298 kg N ha^{-1} applied. However, treatment 6 yielded over 1000 kg ha^{-1} more compared to treatment 2. This suggests that prescribed topdress N rate for treatment 2 was excessive and did not help to optimize yield.

In some instances, the prescribed N rates did not allow to reach optimum yield (Conrad, 2011 (Table 6)). Although higher topdress N rate of 33 kg N ha^{-1} was prescribed to treatment 7 compared to 11 kg N ha^{-1} recommended for treatment 6, it was not adequate to optimize yield. In fact, significantly lower yield was obtained with treatment 7 (1480 kg ha^{-1}), which received only 22 kg N ha^{-1} at seeding, compared to treatment 6 (2152 kg N ha^{-1}) to which 90 kg N ha^{-1} was applied at seeding. Moreover, the highest grain yield for

TABLE 5: Preplant N rate, topdress N rate, topdress N source, total N rate applied, and spring wheat grain yield, Conrad and Corvallis, MT, 2013.

Trt	Conrad			Corvallis		
	Topdress N rate, kg ha^{-1}	Total N rate applied, kg ha^{-1}	Grain yield, kg ha^{-1}	Topdress N rate, kg ha^{-1}	Total N rate applied, kg ha^{-1}	Grain yield, kg ha^{-1}
1	-	-	4304 ab	-	-	3430 a
2	91	311	4102 b	90	310	3968 a
3	54	76	4237 ab	90	112	3968 a
4	54	98	4304 ab	90	134	4035 a
5	54	121	4573 ab	137	204	3968 a
6	54	144	4708 ab	90	180	4035 a
7	54	76	4439 ab	137	159	3430 a
8	54	98	4506 ab	90	134	3430 a
9	54	121	4842 a	90	157	3430 a
10	104	194	4573 ab	137	227	3430 a

*Preplant fertilizer N will be applied as urea. **Topdress fertilizer N rates were determined based on the NDVI values obtained using GreenSeeker, as prescribed by USA, Canada, Mexico Algorithm. Means within each column followed by the same letter are not significantly different at $p < 0.05$, as determined by Duncan's multiple range test.

TABLE 6: Three cases illustrating topdress N rate recommendations prescribed by US-Canada-Mexico Algorithm and grain yield results obtained following the application of prescribed N rates.

Case	Site-year	Trt	Preplant N rate, kg ha^{-1}	Topdress N rate, kg ha^{-1}	Total N rate, kg ha^{-1}	N rate difference, kg ha^{-1}	*Grain yield, kg ha^{-1}	Yield gain, kg ha^{-1}
1	Conrad, 2012	2	220	78	298	−178	6187 d	+1076
		6	90	30	120		7263 a	
2	Conrad, 2011	6	90	11	101	−46	2152 b	+672
		7	22	33	55		1480 e	
3	Conrad, 2012	3	22	16	38	+98	6658 c	+605
		6	90	30	120		7263 a	

*Means followed by the same letter are not significantly different at $p < 0.05$, as determined by Duncan's multiple range test.

that site-year was 2690 kg N ha^{-1} (treatment 2, 220 kg N ha^{-1} applied at seeding), which suggests that topdress N rates should have been higher for both treatment 6 and treatment 7.

In some cases, the topdress N rates seemed appropriate. For instance, at Conrad, in 2012, higher preplant N rate for treatment 6 (90 kg N ha^{-1}) resulted in better plant stand, reflected by the greater NDVI value compared to treatment 3 (22 kg N ha^{-1} preplant). This prompted higher N rate recommendation of 25 kg N ha^{-1} for treatment 6 compared to only 15 kg N ha^{-1} for treatment 3. A difference of 98 kg N ha^{-1} total N applied has resulted in a surplus of 605 kg ha^{-1} of wheat grain yield.

These results show that, at all site-years, N fertilizer rates recommended by the USA/Canada/Mexico Algorithm were not appropriate for grain yield optimization. For example, much higher topdress N rates were prescribed for Corvallis (the irrigated site) compared to the Conrad (dryland sites). This makes sense since the expected yield potential (YP) at the irrigated site was much greater. On the other hand, grain yields obtained at Conrad were just as high as at Corvallis, indicating that the YP was either overestimated at Corvallis or underestimated at Conrad. This puts forward a question of whether there is a need for two separate algorithms, one developed for dryland spring wheat and the other for irrigated spring wheat production systems.

3.3. Relationship between N Rate and Source and Wheat Yield and Quality. Based on different N application rates and sources, a wide range of yields ranging from 942 kg ha^{-1} to 7667 kg ha^{-1} were obtained (Tables 3–5). In 2011 and 2012 growing seasons, grain yield response to preplant N application rate was significant at all locations (Table 7). In Conrad, the highest mean grain yield of 7263 kg ha^{-1} was obtained in 2012 when 90 kg N ha^{-1} preplan followed by 30 kg N ha^{-1} topdress was applied as urea. In Corvallis, the highest mean yield of 7667 kg N ha^{-1} was obtained in 2012 when 90 kg ha^{-1} preplant followed by 110 kg ha^{-1} topdress was applied as UAN. In contrast, the lowest grain yield, except the control, was recorded when 22 kg N ha^{-1} preplant followed by 33 kg N ha^{-1} of topdress in Conrad in 2011. These results showed that, in most cases, the higher amount of N applied resulted in higher grain yield, which in consistent with results from other studies [17, 18].

Table 7: Effect of preplant N rate and topdress N source on spring wheat grain yield (GY) and grain protein content (GP) for 6 site-years in Montana.

Effects	Growing season 2011				Growing season 2012				Growing season 2013			
	Conrad		Corvallis		Conrad		Corvallis		Conrad		Corvallis	
	GY, kg ha^{-1}	GP, %	GY, kg ha^{-1}	GP, %	GY, kg ha^{-1}	GP, %	GY, kg ha^{-1}	GP, %	GY, kg ha^{-1}	GP, %	GY, kg ha^{-1}	GP, %
Preplant N rate												
0	942	9.5	2018	14.0	5851	9.6	3901	11.0	4304	12.4	3430	14.3
22	1480	9.5	3026	15.0	6658	10.5	6994	13.0	4304	14.0	3699	17.3
44	1614	9.6	3430	13.0	6860	11.1	7129	14.0	4371	14.6	3699	16.6
67	1950	9.5	3632	15.0	6994	11.8	7532	14.0	4708	15.2	3699	17.1
90	2152	9.6	3766	15.0	7196	14.3	7263	14.0	4640	15.7	3699	17.2
220	2690	9.7	3699	16.0	6187	15.4	6456	15.0	4102	17.2	3968	17.4
F test	**	ns	***	**	***	ns	***	***	ns	***	ns	***
Topdress N source												
Urea	1749	9.6	3497	14.5	5716	11.5	5985	14.1	4439	15.2	3968	17.2
UAN	1816	9.5	3430	14.4	5582	11.6	6456	13.6	4573	14.6	3430	16.9
F test	ns	ns	ns	ns	ns	ns	***	***	ns	***	***	ns

** and *** denote significant at 0.05 and 0.01 levels, respectively; ns denotes not significant.

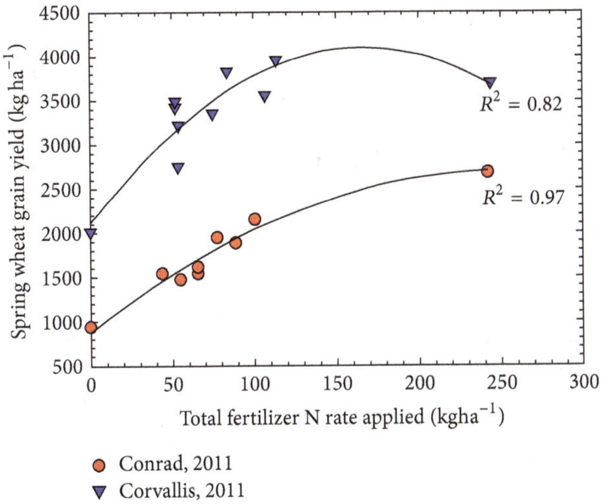

FIGURE 2: Spring wheat grain yield as affected by total amount of applied N fertilizer, Conrad and Corvallis, MT, 2011.

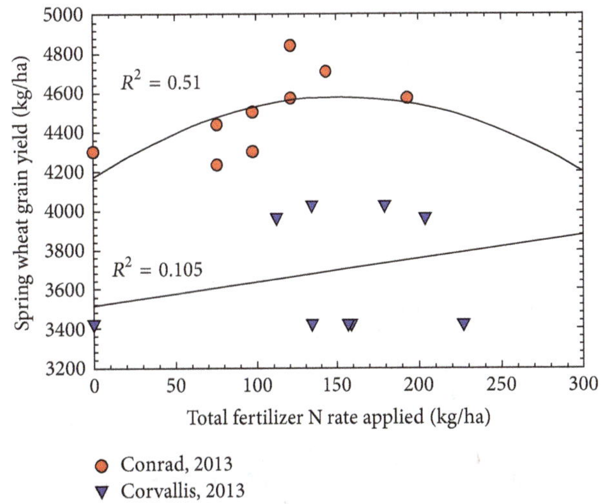

FIGURE 4: Spring wheat grain yield as affected by total amount of applied N fertilizer, Conrad and Corvallis, MT, 2013.

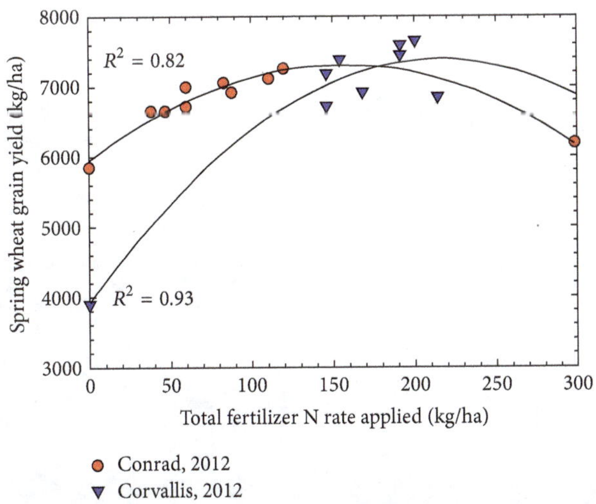

FIGURE 3: Spring wheat grain yield as affected by total amount of applied N fertilizer, Conrad and Corvallis, MT, 2012.

Although the application of urea resulted in slightly higher yields compared to UAN, with 1 site-year being virtually equal in yield for both N sources, the differences were no statistically significant. In addition, the results showed there were no significant differences in grain protein content values associated with topdress fertilizer N source (urea versus UAN), although slightly higher (but not statistically significant) grain protein content values were noted with urea topdress application compared to UAN (Table 7).

The relationship between total amount of N applied and spring wheat grain yield for each site-year is shown in Figures 2–4. In all cases, there was a positive correlation between total amount of applied N fertilizer and grain yield. In 2011 and 2012 for both fields, this correlation was strong with R^2 ranging from 0.82 to 0.95. In 2013, R^2 decreased for both locations. These results may indicate that the model should be calibrated for each location frequently. In 2011, at Conrad, wheat yield increased to 242 kg N ha^{-1} and at Corvallis to 114 kg N ha^{-1} (Figure 2). In 2012, at Conrad, yield was maximized at 120 kg N ha^{-1} and at Corvallis at 200 kg N ha^{-1} (Figure 3). In 2013, at Conrad, grain yield was maximized at 121 kg N ha^{-1} and at Corvallis at 134 kg N ha^{-1} (Figure 4).

4. Conclusions

Crop canopy sensors are convenient tools for assessment of crop N status and provide assistance to growers when making the best management decision based on actual crop condition. In this study, NDVI from a GreenSeeker sensor was integrated with spring wheat (US, Canada, and Mexico) algorithm to estimate crop N content and make fertilizer N recommendations.

Overall, no significant differences in wheat grain yield or grain protein content associated with topdress N fertilizer source were observed at any of the site-years. This suggests that topdress N fertilizer rates do not need to be adjusted based of fertilizer sources used, that is, the same N rates should be prescribed whether urea or UAN is applied. At the time of this writing, the cost of N sources per unit of N are volatile, with urea currently costing \$0.45 more per kg of N. Thus, growers are strongly advised to pay attention to fertilizer prices per unit of N when making decisions on what N source to use in any particular growing season, especially when there is no clear consistent advantage from using one source versus the other.

Results indicated that the three assessed algorithms developed in other regions did not provide the appropriate topdress N rate recommendations for spring wheat. Our case study emphasizes the importance of (1) calibrating the crop sensors for the local crop varieties and growing conditions and (2) developing N fertilization algorithms based on locally established YP prediction studies.

Acknowledgments

The project was funded in part by the Montana Fertilizer Tax Advisory Board.

References

[1] C. L. Da Silva, G. Benin, E. Bornhofen, E. Beche, M. H. Todeschini, and A. S. Milioli, "Nitrogen use efficiency is associated with chlorophyll content in Brazilian spring wheat," *Australian Journal of Crop Science*, vol. 8, no. 6, pp. 957–964, 2014.

[2] X. Xu, Z. Wu, Y. Dong, Z. Zhou, and Z. Xiong, "Effects of nitrogen and biochar amendment on soil methane concentration profiles and diffusion in a rice-wheat annual rotation system," *Scientific reports*, vol. 6, Article ID 38688, 2016.

[3] D. K. Biswas and B.-L. Ma, "Effect of nitrogen rate and fertilizer nitrogen source on physiology, yield, grain quality, and nitrogen use efficiency in corn," *Canadian Journal of Plant Science*, vol. 96, no. 3, pp. 392–403, 2016.

[4] A. R. Ravishankara, J. S. Daniel, and R. W. Portmann, "Nitrous oxide (N_2O): the dominant ozone-depleting substance emitted in the 21st century," *Science*, vol. 326, no. 5949, pp. 123–125, 2009.

[5] D. S. Reay, E. A. Davidson, K. A. Smith et al., "Global agriculture and nitrous oxide emissions," *Nature Climate Change*, vol. 2, no. 6, pp. 410–416, 2012.

[6] N. Gupta, A. K. Gupta, V. S. Gaur, and A. Kumar, "Relationship of nitrogen use efficiency with the activities of enzymes involved in nitrogen uptake and assimilation of finger millet genotypes grown under different nitrogen inputs," *The Scientific World Journal*, vol. 2012, Article ID 625731, 10 pages, 2012.

[7] L. J. Thompson, R. B. Ferguson, N. Kitchen et al., "Model and sensor-based recommendation approaches for in-season nitrogen management in corn," *Agronomy Journal*, vol. 107, no. 6, pp. 2020–2030, 2015.

[8] J. Crain, I. Ortiz-Monasterio, and B. Raun, "Evaluation of a reduced cost active NDVI sensor for crop nutrient management," *Journal of Sensors*, vol. 2012, Article ID 582028, 10 pages, 2012.

[9] M. Huang, W. Zhang, L. Jiang, and Y. Zou, "Impact of temperature changes on early-rice productivity in a subtropical environment of China," *Field Crops Research*, vol. 146, pp. 10–15, 2013.

[10] J. T. Bushong, J. L. Mullock, E. C. Miller, W. R. Raun, and D. Brian Arnall, "Evaluation of mid-season sensor based nitrogen fertilizer recommendations for winter wheat using different estimates of yield potential," *Precision Agriculture*, vol. 17, no. 4, pp. 470–487, 2016.

[11] W. R. Raun, J. B. Solie, M. L. Stone et al., "Optical sensor-based algorithm for crop nitrogen fertilization," *Communications in Soil Science and Plant Analysis*, vol. 36, no. 19-20, pp. 2759–2781, 2005.

[12] J. I. Ortiz-Monasterio and W. Raun, "Reduced nitrogen and improved farm income for irrigated spring wheat in the Yaqui Valley, Mexico, using sensor based nitrogen management," *Journal of Agricultural Science*, vol. 145, no. 3, pp. 215–222, 2007.

[13] G. Silva, *All Fertilizers are not Created Equal*, 2016, http://msue.anr.msu.edu/news/all_fertilizers_are_not_created_equal.

[14] The Mosaic Company, *Fluid and Dry Fertilizers, Fluids and Solids are Equal Agronomically*, 2013, http://www.cropnutrition.com/efu-fluid-dry-fertilizers#overview.

[15] C. J. Watson, R. J. Stevens, R. J. Laughlin, and P. Poland, "Volatilization of ammonia from solid and liquid urea surface-applied to perennial ryegrass," *The Journal of Agricultural Science*, vol. 119, no. 2, pp. 223–226, 1992.

[16] E. Lombi, M. J. McLaughlin, C. Johnston, R. D. Armstrong, and R. E. Holloway, "Mobility and lability of phosphorus from granular and fluid monoammonium phosphate differs in a calcareous soil," *Soil Science Society of America Journal*, vol. 68, no. 2, pp. 682–689, 2004.

[17] Z. Abebe and H. Feyisa, "Effects of nitrogen rates and time of application on yield of maize: rainfall variability influenced time of N application," *International Journal of Agronomy*, 2017.

[18] O. S. Walsh, A. Pandey, and R. Christiaens, "Sensor-based technologies for nitrogen management in spring wheat," in *Proceedings of the Western Nutrient Management Conference*, vol. 11, Reno, NV, USA, 2015, http://www.ipni.net/ipniweb/conference/wnmc.nsf/e0f085ed5f091b1b852579000057902e/4be3031d1d87927a85257e37004fa7a8/$FILE/WNMC2015%20Walsh%20pg163.pdf.

3

Estimates of Combining Ability and Heterosis for Yield and its Related Traits in Pearl Millet Inbred Lines under Downy Mildew Prevalent Areas of Senegal

Ghislain Kanfany,[1,2,3] Amadou Fofana,[3] Pangirayi Tongoona,[1] Agyemang Danquah,[1] Samuel Offei,[1] Eric Danquah,[1] and Ndiaga Cisse[2]

[1] *West Africa Centre for Crop Improvement, University of Ghana, PMB LG 30, Legon, Ghana*
[2] *Centre d'Etudes Régional pour l'Amélioration de l'Adaptation à la Sècheresse, BP 3320, Thies, Senegal*
[3] *Centre National de Recherches Agronomiques de Bambey, BP 211, Bambey, Senegal*

Correspondence should be addressed to Ghislain Kanfany; gkanfany@wacci.edu.gh

Academic Editor: Iskender Tiryaki

Pearl millet is an important cereal crop for smallholder farmers' food security in West and Central Africa. However, its production has stagnated due to several factors such as the continuous use of local populations. A set of 17 inbred lines was crossed with Sosat C 88 and Souna 3 following a line × tester mating design. The F_1 hybrids, their parents, and a check were evaluated in Bambey and Nioro research stations during the rainy season of 2017. Data on downy mildew incidence, plant height, flowering time, panicle length and diameter, productive tillers, thousand-grain weight, panicle, and grain yield were recorded. GCA and SCA mean squares were significant for most of the traits indicating that both additive and nonadditive gene effects were involved in the control of the inheritance of these traits. However, the contribution of GCA to total mean squares was higher than that of SCA for all the traits, providing that additive gene action was more important in their inheritance. The top-cross hybrid IBL155-2-1 × Sosat C 88 exhibited negative and significant SCA effects for downy mildew incidence, flowering time, and plant height. Lines IBL003-B-1, IBL091-1-1, IBL095-4-1, IBL110-B-1, and IBL 206-1-1 had positive GCA effects for grain yield and negative GCA effects for downy mildew, flowering time, and plant height. These lines can be used as parents to create synthetic varieties or hybrids.

1. Introduction

Pearl millet (*Pennisetum glaucum* (L.) R. Brown) is the 4th most important tropical cereal after rice, maize, and sorghum [1]. In 2014, global grain pearl millet production was estimated at 28 million tons, harvested from 32 million ha in Asia, Africa, and the Americas. The average yield is about 900 kg ha^{-1}. India and Africa are the most important producers with more than 85% of the total production in 2016 [1]. It is an important dual purpose cereal crop in Africa and Asia where it is considered as a staple food and source of fodder and feed for livestock for smallholders farmers [2]. The grain of pearl millet is a very rich source of protein, vitamins, and minerals in comparison with other cereals and is used for human consumption in diverse ways [3]. The pearl millet stover is used as fuel, material for building and fencing, and also a soil additive to enhance soil fertility [4]. It is also being experimented as a grain and forage crop in the USA, Canada, Mexico, India, and West and North Africa [5].

As the world population is continuously increasing and is projected to reach nine billion by 2050, pearl millet is expected to play an important role for achieving food security in West and Central African countries, which have the highest population growth rates in the world [6]. However, in this part of the world, yields of pearl millet are very low compared to yields in India and yet African farmers, particularly in Senegal, have not adopted improved varieties in a large scale. In pearl millet growing areas in Africa, adoption rate of improved OPVs varied from 5 to 37% [7]. In contrast, most Indian farmers are using improved varieties, particularly hybrids since the 1960s. Indeed, in India hybrids had 25–30% grain yield advantage over OPVs, leading to the rapid adoption of the hybrids whose yield increased from 305 kg ha^{-1} during 1951–1955 to 998 kg ha^{-1} during 2008–2012 [5]. Thus, enhancement of pearl millet production and productivity in Africa, which is a high priority, can be achieved through the identification of elite parent materials which can be used as parents to develop hybrid varieties.

ICRISAT has developed pearl millet inbred lines derived from landraces originating from West and Central Africa which can be useful in developing high yielding pearl millet hybrids and synthetic varieties with considerable adaptation to this pearl millet growing environment. These lines were screened for pearl millet downy mildew resistance in Senegal and some of them showed good agronomic traits and resistance to the pearl millet downy mildew. However, the per se performance of these pearl millet inbred lines does not predict the performance of hybrids for disease resistance and agronomic traits [8]. Therefore, to make effective use of these pearl millet inbred lines, their combining abilities need to be elucidated [9]. This genetic information can be obtained by different mating design including line × tester [10]. GCA and SCA estimates of pearl millet inbred parents or landraces for different traits such as micronutrients [11], grain quality [12], and fodder yield [13] were reported to be important. Additive genetic action was also reported to be important in controlling traits such as grain yield, flowering time, and panicle length [14]. There is scanty published information on the combining ability of the pearl millet inbred lines derived from landraces collected in West and Central Africa, known as the origin of the crop, for disease resistance and agronomic traits. The objectives of this study were to estimate combining ability and heterosis of pearl millet inbred lines for downy mildew, yield, and other agronomic traits under downy mildew infested fields and identify superior pearl millet hybrids for yield, yield components, and resistance to downy mildew.

2. Materials and Methods

2.1. Plant Material and Mating Design. Seventeen inbred lines were used as females and crossed each to two OPVs used as males according to the line × tester mating design [10] to generate 34 F_1 hybrids. The OPVs varieties were considered as testers and the inbred lines as lines. The two testers named Souna 3 and Sosat C 88 are popular varieties adapted to the groundnut agroecological zone. The pearl millet inbred lines used for the study were selected from a pool of pearl millet landraces from West and Central Africa converted to inbred lines through successive selfing up to S6 [15]. These inbred lines were selected through a downy mildew phenotypic evaluation conducted at Bambey and Nioro research stations during the rainy season 2016. They showed less than 10% DMI and were classified as resistant varieties. The seeds of the male parents were planted in 4 different dates in order to synchronise the flowering time of these male parents with the ones of the female parents. Thus, from January 2017, the sowing of the male parents was done every week and seeds of each of the male parents were sown in 5 rows of 15 hills per row. All the female parents were sown in one time during the second sowing date of the male parents in a one row-plot of 15 hills. At the booting stage, at least plant heads of 4 panicles per plant of the male and female parents were covered in order to avoid undesirable pollination. At flowering, each covered panicle of female plant was pollinated with bulk pollen collected from at least 20 different plants of the male parent.

At maturity stage, F_1 panicles of the female parents were harvested and the lower and upper parts of each panicle were cut before threshing to minimize outcrossing from unknown plants or selfing. Indeed, because of the protogynous nature of the crop the stigmata of a plant are receptive before the shedding of pollen and the flowering starts from the upper to the lower part. Then, the upper part of the panicle may be pollinated by unknown plants if not covered on the right time and the lower part of the panicle may be pollinated by the pollen from the same plant.

After threshing, F_1 seeds from the same female parent were bulked and used as F_1 hybrids. The 34 F_1 hybrids along with the 17 inbred lines, the two testers, and an OPV named Thialack II as check, providing 54 genotypes, were used for the evaluation (Table 1).

2.2. Study Sites, Experimental Design, and Field Management. The 34 hybrids, together with their parents and the OPV check, were evaluated under rainfed conditions during the rainy season of 2017 at two locations in Senegal. The study sites were Bambey (13°49′12″ North, 13°55′12″ West) and Nioro (13°45′0″ North, 15°48′0″ West) research stations. Both locations are in the groundnut agroecological zone, the main pearl millet growing area in Senegal, and were characterized as hotspots for downy mildew in the previous study. The genotypes were arranged in 9 × 6 alpha lattice design with three replications at each site. Each block was surrounded by a downy mildew infector row consisting of a downy mildew susceptible line, 7042 S, sown 3 weeks before the tested materials. Each plot consisted of one row of 8.1 m length with a spacing of 0.9 m between rows and between plants within a row. At least 10 seeds were planted per hole and later thinned to two plants two weeks after sowing. The fields were weeded two times after sowing. The trials received the recommended 15N-15P-15K basal fertilizer at a rate of 150 kg ha^{-1} just before sowing. During the crop development a top dressing using area at a rate of 100 kg ha^{-1} was done in two fractions: 50 kg ha^{-1} after thinning and 50 kg ha^{-1} after the second weeding.

2.3. Data Collection. The recorded data were collected according to the method described by Drabo [14]. Flowering (FWT) was recorded by counting the total number of days from sowing to the time when 50% of plants in a plot flowered. Downy mildew incidence (DMI) was obtained by dividing the total number of infected plants, 30 DAS from a plot, by the total number of plants. Panicles harvested in a plot were weighed to determine panicle yield (PY) and then threshed. Grains obtained in each plot were weighed and used

TABLE 1: List of parental lines and check used in the study.

Number	Genotype	Source	Role in crosses	Status	Response to downy mildew
1	IBL 001-4-1	ICRISAT	Line	Inbred line	Resistant
2	IBL 003-B-1	ICRISAT	Line	Inbred line	Resistant
3	IBL 011-4-1	ICRISAT	Line	Inbred line	Resistant
4	IBL 021-3-1	ICRISAT	Line	Inbred line	Resistant
5	IBL 055-4-1	ICRISAT	Line	Inbred line	Resistant
6	IBL 091-1-1	ICRISAT	Line	Inbred line	Resistant
7	IBL 095-4-1	ICRISAT	Line	Inbred line	Resistant
8	IBL 098-3-1	ICRISAT	Line	Inbred line	Resistant
9	IBL 106-B-1	ICRISAT	Line	Inbred line	Resistant
10	IBL 110-B-1	ICRISAT	Line	Inbred line	Resistant
11	IBL 114-6-1	ICRISAT	Line	Inbred line	Resistant
12	IBL 119-B-1	ICRISAT	Line	Inbred line	Resistant
13	IBL 155-2-1	ICRISAT	Line	Inbred line	Resistant
14	IBL 165-1-1	ICRISAT	Line	Inbred line	Resistant
15	IBL 179-2-1	ICRISAT	Line	Inbred line	Resistant
16	IBL 179-3-1	ICRISAT	Line	Inbred line	Resistant
17	IBL 206-1-1	ICRISAT	Line	Inbred line	Resistant
18	Souna 3	Senegal	Tester	OPV (improved)	Susceptible
19	Sosat C 88	Senegal	Tester	OPV (improved)	Resistant
20	Thialack II	Senegal	Check	OPV (improved)	Susceptible

to calculate grain yield (GY) in $\mathrm{kg\,ha^{-1}}$ using the following formula:

$$\mathrm{GY}\ \left(\mathrm{kg\,ha^{-1}}\right) = \left[\frac{\text{grain weight}\ \left(\mathrm{kg\,plot^{-1}}\right) \times 10{,}000}{\text{plot size m}^2}\right]. \quad (1)$$

Five random plants were selected in each plot to measure the plant height (PH) from the base of the plant to the upper part of the panicle, number of productive tillers (PT) by counting the number of tillers per plant which produce productive panicles, panicle length (PL), and panicle diameter (PDIA). Five random samples of 1000 grains for each plot were weighed using a sensitive balance to determine the 1000-grain weight (TGW).

2.4. Data Analysis. Analysis of variance for each experimental site as well as for combined data after the homogeneity test of variance across the two experimental sites was performed using the general linear model (GLM) procedure in SAS version 9.4 (SAS Institute, Cary, NC). The following mathematical linear model was used:

$$Y_{ijk} = \mu + G_i + L_j + \mathrm{GL}_{ij} + r_{jk} + b_{jk} + e_{ijk}, \quad (2)$$

where

Y_{ijk} is the observed value of the variable for the ith entry in the jth location within kth replication;

μ is the overall general mean;

G_i is the effect of the ith genotype;

L_j is the effect of the jth location;

GL_{ij} is the interaction effect of the ith entry and the jth location;

r_{jk} is the effect of the kth replication within the jth location;

b_{jk} is the effect of the lth block of the kth replication in the jth location;

e_{ijkr} is the experimental pooled error.

For the combining ability, analysis of variance was performed for traits that showed significant differences among hybrids using SAS software version 9.4 (SAS Institute, Cary, NC). Thus, the sum of squares of hybrids was partitioned into various variations due to lines, testers, and their interactions based on the following statistical model described by Singh and Chaudhary (1977):

$$Y_{ijk} = \mu + M_i + F_j + \mathrm{MF}_{ij} + e_{ijk}, \quad (3)$$

where

Y_{ijk} is kth observation on the ith and jth progeny;

μ is the overall general mean;

M_i is the effect of the ith male;

F_j is the effect of the jth female;

MF_{ij} is interaction effect;

e_{ijk} is error associated with each observation.

Table 2: Mean squares for studied traits across locations.

Source of variation	d.f.	DMI	FWT	PH	PL	PDIA	PT	TGW	PY	GY
Rep (site)	4	232.4**	28.4**	746.4	28.5	0.4	2.2	4.9*	2499694*	1258762**
Block (Rep × site)	48	85.5	8.2	359.9	32.9	0.14	1.5	2.3	817310	324630
Site	1	84.6	248.9***	0.1	444.1**	1.6**	29.4**	36.2***	5279671*	3168503**
Genotype	53	156.4***	82.3***	3193.5***	278.6***	0.46***	5.1***	10.4***	3975982***	1157443***
Genotype × site	53	72.7	13.62**	488.1	45.1	0.19	2.0	2.4*	1418904*	603031**
Error	164	60.04	7.9	474.7	31.6	0.17	1.9	1.6	997156	297217

DMI: downy mildew incidence; FWT: flowering time; PH: plant height; PL: panicle length; PDIA: panicle diameter; PT: productive tillers; TGW: 1000-grain weight; PY: panicle yield; GY: grain yield; d.f.: degree of freedom; Rep: replication. *,**,*** Significant at 0.05 and 0.01 and 0.001 probability levels, respectively.

The values of the general combining ability for both male and female and the specific combining ability effects for all the studied traits were estimated as follows:

$$\begin{aligned} \text{GCA}_{\text{Line}} &= \text{Line mean } (X_i) \\ &\quad - \text{Overall mean } (X\ldots) \\ \text{GCA}_{\text{tester}} &= \text{Tester mean } (X_j) \\ &\quad - \text{Overall mean } (X\ldots) \\ \text{SCA}_{\text{line}\times\text{tester}} &= \text{Cross mean } (X_{ij}) - \text{Line mean } (X_i) \\ &\quad - \text{Tester mean } (X_j) \\ &\quad + \text{Overall mean } (X\ldots), \end{aligned} \tag{4}$$

where

$X\ldots$ is overall mean;

X_i is mean of all the hybrids containing an ith line average over all replications, sites, and males;

X_j is mean of all the hybrids containing a jth tester average over all replications, sites, and females;

X_{ij} is mean of the cross between ith line and jth tester across all replications and sites.

The significance of the GCA effects was tested using the formula described by Cox and Frey (1984):

$$\begin{aligned} t_{\text{cal}} &= \frac{\text{GCA}}{\text{SE}_{\text{gca (male)}}}, \quad \text{where, S.E}_{\text{(gca male)}} = \sqrt{\frac{\text{Me}}{rts}} \\ t_{\text{cal}} &= \frac{\text{GCA}}{\text{SE}_{\text{gca (female)}}}, \quad \text{where, S.E}_{\text{(gca female)}} = \sqrt{\frac{\text{Me}}{rts}} \\ t_{\text{cal}} &= \frac{\text{SCA}}{\text{SE}_{\text{SCA (line}\times\text{tester)}}}, \quad \text{where, S.E (sca)} = \sqrt{\frac{\text{Me}}{rs}}, \end{aligned} \tag{5}$$

where

Me is the error mean sum of squares;

r, t, l, s are numbers of replications, testers lines, and sites, respectively;

SE is standard error.

Standard, mid-parent, and better parent heterosis for grain yield were also calculated for each cross across locations following Hallauer et al. (2010):

$$\begin{aligned} \text{Standard heterosis (SH)} &= \frac{(F_1 - \text{Check})}{\text{Check}} \times 100 \\ \text{Mid-parent heterosis (MPH)} &= \frac{(F_1 - \text{MP})}{\text{MP}} \times 100 \\ \text{Better parent heterosis (BPH)} &= \frac{(F_1 - \text{HP})}{\text{HP}} \times 100, \end{aligned} \tag{6}$$

where F_1 denotes the mean performance of the hybrid averaged over the two locations. The mean value of the OPV check was used to calculate the standard heterosis. The parent with the highest mean value was used as better parent in the calculation of high-parent heterosis while the average between the two parents was used for the mid-parent heterosis.

3. Results

3.1. Performance of Hybrids and Parents across Locations. Combined analysis of variance across locations showed highly significant ($P < 0.01$) genotype effect for all measured traits (Table 2). Site effect was also significant for all the traits, except for DMI and PH. However, interaction genotype × site effect was only significant for FWT, TGW, PY, and GY.

All genotypes were resistant to downy mildew with a mean DMI of 4%, except for IBL 155-2-1 and its progeny with Souna 3 which displayed both 22% DMI (Table 3). Days from sowing to 50% flowering (DAS) of genotypes across the two sites ranged from 50 to 69 DAS with an average of 56 DAS. The genotypes were tall with plant height ranging from 2 to 3.2 m. Panicle length of the pearl millet genotypes varied from 27 to 58 cm with an average of 44 cm while their diameter ranged from 1.2 to 2.7 cm with a mean diameter of 2.1 cm. The number of productive tillers ranged from 2 to 6 tillers per plant with a mean value of 4 productive tillers per plant. The 1000 seeds weight varied from 5 to 12 g with a mean of 9 g. The panicle yield of genotypes across the two sites varied from 376 kg ha^{-1} for IBL 119-B-1 to 4190 kg ha^{-1} for IBL 110-B-1 × Souna 3 and their grain yield varied from 92 kg ha^{-1} for IBL 119-B-1 to 2024 kg ha^{-1} for IBL 206-1-1 × Souna 3.

As expected, the F_1 hybrids were generally more productive compared to the inbred lines and OPVs. The top five

TABLE 3: Performance of tested genotypes for studied traits across sites.

Genotype	DMI	FWT	PHIG	PLEN	PDIA	PT	TGW	PY	GY
IBL 001-4-1	2	60	255	34	2.1	2	10	1104	585
IBL 003-B-1	4	51	231	31	2.5	5	11	1872	1340
IBL 011-4-1	14	59	237	44	1.9	3	7	1164	428
IBL 021-3-1	0	63	223	32	2.1	2	8	1178	419
IBL 055-4-1	6	65	210	37	1.6	2	6	517	286
IBL 091-1-1	0	60	246	47	2.2	4	8	2328	1136
IBL 095-4-1	0	50	243	37	1.9	5	10	2015	947
IBL 098-3-1	11	51	239	31	1.8	4	10	1946	1188
IBL 106-B-1	6	54	278	39	1.8	3	10	1340	710
IBL 110-B-1	0	62	240	45	1.6	2	6	997	404
IBL 114-6-1	0	62	199	27	1.8	2	8	750	324
IBL 119-B-1	0	69	260	41	1.2	2	5	376	92
IBL 155-2-1	22	57	200	35	1.3	3	7	1099	466
IBL 165-1-1	2	64	238	39	1.6	4	7	1349	489
IBL 179-2-1	0	60	272	43	1.9	2	8	850	394
IBL 179-3-1	8	54	233	35	2.0	5	8	1497	886
IBL 206-1-1	0	56	228	38	1.8	4	8	2102	1126
IBL 001-4-1 × Souna 3	17	53	273	43	2.0	5	11	3475	1923
IBL 003-B-1 × Souna 3	8	52	259	44	2.1	4	9	3218	1646
IBL 011-4-1 × Souna 3	5	52	255	50	1.8	4	8	2368	942
IBL 021-3-1 × Souna 3	4	57	210	44	1.9	3	7	1772	852
IBL 055-4-1 × Souna 3	0	58	241	38	2.7	3	8	2232	1397
IBL 091-1-1 × Souna 3	0	54	250	54	2.0	4	8	2369	1211
IBL 095-4-1 × Souna 3	0	52	262	58	2.0	4	9	2975	1641
IBL 098-3-1 × Souna 3	19	54	277	46	1.9	4	9	2326	1289
IBL 106-B-1 × Souna 3	6	56	256	52	1.6	3	9	2462	1351
IBL 110-B-1 × Souna 3	0	56	274	52	2.6	5	7	4190	1493
IBL 114-6-1 × Souna 3	4	57	270	47	2.0	3	8	2991	1181
IBL 119-B-1 × Souna 3	0	56	321	49	2.0	5	9	3586	1530
IBL 155-2-1 × Souna 3	22	56	287	46	1.9	5	9	2562	1101
IBL 165-1-1 × Souna 3	0	59	289	45	2.1	4	8	3420	1612
IBL 179-2-1 × Souna 3	0	59	298	44	2.0	3	9	1906	735
IBL 179-3-1 × Souna 3	2	55	275	46	2.2	4	8	2943	1402
IBL 206-1-1 × Souna 3	2	54	245	44	2.1	5	9	3671	2024
IBL 001-4-1 × Sosat C 88	9	56	284	36	2.4	4	9	2997	1579
IBL 003-B-1 × Sosat C 88	2	50	234	28	2.6	4	12	3182	1883
IBL 011-4-1 × Sosat C 88	0	50	257	39	2.1	5	10	2468	1331
IBL 021-3-1 × Sosat C 88	0	54	284	37	2.3	4	10	3053	1654
IBL 055-4-1 × Sosat C 88	0	57	298	40	2.0	3	9	1824	1131
IBL 091-1-1 × Sosat C 88	0	50	260	41	2.3	4	10	2912	2019
IBL 095-4-1 × Sosat C 88	2	55	281	39	2.3	4	10	2426	1368
IBL 098-3-1 × Sosat C 88	6	57	285	43	2.2	3	9	2722	1381
IBL 106-B-1 × Sosat C 88	0	57	255	37	2.2	3	9	1933	949
IBL 110-B-1 × Sosat C 88	2	55	256	38	2.6	4	9	2827	1433
IBL 114-6-1 × Sosat C 88	0	52	264	31	2.4	5	9	2944	1416
IBL 119-B-1 × Sosat C 88	2	60	298	41	2.0	4	8	1993	871
IBL 155-2-1 × Sosat C 88	0	51	260	35	1.9	6	10	2414	1215
IBL 165-1-1 × Sosat C 88	0	60	287	39	2.5	4	10	2785	1114
IBL 179-2-1 × Sosat C 88	0	54	277	32	2.4	6	9	2920	1586
IBL 179-3-1 × Sosat C 88	0	56	260	29	2.4	4	9	2375	1247
IBL 206-1-1 × Sosat C 8	0	53	242	34	2.4	5	9	3325	1988
Souna 3	5	56	273	55	2.0	4	8	2918	1268

Table 3: Continued.

Genotype	DMI	FWT	PHIG	PLEN	PDIA	PT	TGW	PY	GY
Sosat C 88	2	51	251	33	2.3	4	10	2568	1548
Thialack II	7	53	292	54	1.9	4	8	3380	1694
Mean	4	56	259	41	2.1	4	9	2313	1171
Range	0–22	50–69	200–321	27–58	1.2–2.7	2–6	5–12	376–4190	92–2024
Standard deviation	5.6	4.1	25.8	7.3	0.3	0.9	1.4	868.8	483.1

DMI: downy mildew incidence; FWT: flowering, PH: plant height; PL: panicle length; PDIA: panicle diameter; PT: productive tillers; TGW: 1000-grain weight; PY: panicle yield; GY: grain yield.

genotypes across sites were hybrids IBL 206-1-1 × Souna 3 (2024 kg ha^{-1}); IBL 091-1-1 × Sosat C 88 (2019 kg ha^{-1}); IBL 206-1-1 × Sosat C 88 (1988 kg ha^{-1}); IBL 001-4-1 × Souna 3 (1923 kg ha^{-1}); and IBL 003-B-1 × Sosat C 88 (1883 kg ha^{-1}). Among these top hybrids, two involved the inbred line IBL 206-1-1 as parent. The hybrid IBL 179-2-1 × Souna 3 (735 kg ha^{-1}) was the lowest yielding among the tested hybrids. The check, Thialack II, was the most productive OPV with an average grain yield of 1694 kg ha^{-1} and ranked among the ten best genotypes. The best inbred line was IBL 003-B-1 (1340 kg ha^{-1}).

Genotypes flowered 2 days earlier in Nioro (55 DAS) compared to Bambey (57 DAS) (Table 4). The average TWG in Nioro was 8 g while in Bambey it was 9 g. The panicle yield of genotypes under Nioro conditions ranged from 461 kg ha^{-1} for the inbred line IBL 110-B-1 to 4647 kg ha^{-1} for hybrid IBL 110-B-1 × Souna 3 while under Bambey conditions it varied from 251 kg ha^{-1} for the inbred IBL 119-B-1 to 4660 kg ha^{-1} for the hybrid IBL 165-1-1 × Souna 3. Grain yield of genotypes under Bambey environment ranged from 61 kg ha^{-1} for inbred IBL 119-B-1 to 2162 kg ha^{-1} for hybrid IBL 165-1-1 × Souna 3. In Nioro, the grain yield varied from 87 kg ha^{-1} for inbred line IBL 110-B-1 to 2966 kg ha^{-1} for the hybrid IBL 206-1-1 × Sosat C 88.

Based on grain yield, the ten best genotypes in Nioro were only hybrids while in Bambey the three OPVs were among the top ten genotypes. The hybrids IBL 091-1-1 × Sosat C 88 and IBL 206-1-1 × Souna 3 performed well under both locations and were among the best ten genotypes across the two environments.

3.2. Combining Ability Analysis across Locations. The total variation due to crosses was partitioned into line, tester, and line × tester interaction (Table 5). The mean squares due to hybrids were significant for all the traits except for PY and GY. Line mean squares across the two locations were also significant for all the traits except for PDIA while tester mean squares were not significant for PT, PY, and GY. Line × tester mean squares were significant for most traits except PDIA, PY, and GY. The mean squares due to line × site were significant for TGW, PY, and GY whereas the mean squares due to tester × site interaction were significant for FWT and DMI. However, the mean squares due to site × line × tester interaction were not significant for all the traits across the two locations.

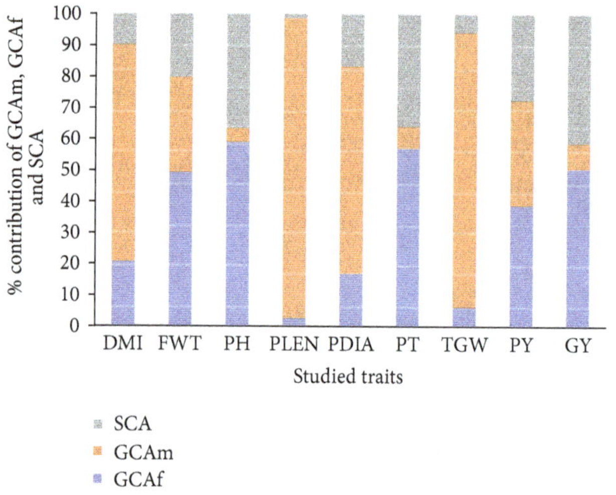

Figure 1: Proportion of total mean squares of studied traits attributable to GCA_m, GCA_f, and SCA across locations. DMI: downy mildew incidence; FWT: flowering time; PH: plant height; PL: panicle length; PDIA: panicle diameter; PT: productive tillers; TGW: 1000-grain weight; PY: panicle yield; GY: grain yield; GCA_m: general combining ability for male parent; GCA_f: general combining ability for female parent; SCA: specific combining ability.

3.3. Relative Contributions of Mean Squares to Additive and Nonadditive Effects. Across the two locations, the relative importance of mean squares, for additive effect (GCA_m + GCA_f), was higher for all the traits compared to the dominance effect (SCA) (Figure 1). GCA effects accounted for most of the variation observed for most of the traits with more than 80% of the total genotypic variation among hybrids except for PH, PT, PY, and GY. The overall contribution of GCA sums of squares to the total mean squares across the two locations varied from 58% for GY to 99% for PLEN while SCA varied from 1% for PLEN to 42% for grain yield. The contribution of GCA_m was higher than GCA_f and SCA for DMI, PLEN, PDIA, and TGW while GCA_f was larger than GCAm and SCA mean square for FWT, PH, PT, PY, and GY. The contribution of GCA_f (50%) was slightly higher than SCA (42%) to grain yield.

3.4. Estimation of General Combining Ability Effects. The contribution of lines and testers to crosses for traits studied across the two locations is presented in Table 6. For female lines, significant GCA effects were observed for most of

Table 4: Mean flowering time, yield, and related traits of genotypes per site.

Genotype	FWT (das)		TGW (g)		PY (kg ha^{-1})		GY (kg ha^{-1})	
	Bambey	Nioro	Bambey	Nioro	Bambey	Nioro	Bambey	Nioro
IBL 001-4-1	60	61	9	10	1284	924	768	402
IBL 003-B-1	51	52	11	12	1856	1888	1512	1168
IBL 011-4-1	63	55	7	7	1195	1132	437	419
IBL 021-3-1	62	63	9	7	1423	933	718	119
IBL 055-4-1	68	63	5	7	321	712	139	432
IBL 091-1-1	63	56	8	8	2557	2099	1140	1132
IBL 095-4-1	49	50	10	11	1658	2372	924	969
IBL 098-3-1	49	53	9	11	1687	2206	1057	1318
IBL 106-B-1	55	54	10	10	780	1900	428	992
IBL 110-B-1	61	62	7	5	1533	461	721	87
IBL 114-6-1	65	58	6	9	628	873	267	382
IBL 119-B-1	73	64	3	6	251	500	61	122
IBL 155-2-1	55	58	7	6	1235	962	649	284
IBL 165-1-1	64	63	7	8	1970	727	792	186
IBL 179-2-1	59	60	8	7	1078	622	559	228
IBL 179-3-1	54	53	8	8	1330	1664	746	1025
IBL 206-1-1	60	52	8	8	920	3283	564	1687
IBL 001-4-1 × Souna 3	54	51	10	12	2625	4325	1223	2624
IBL 003-B-1 × Souna 3	52	51	9	10	2758	3677	1011	2281
IBL 011-4-1 × Souna 3	54	50	9	8	2151	2585	923	961
IBL 021-3-1 × Souna 3	58	56	7	7	1672	1872	1113	591
IBL 055-4-1 × Souna 3	59	57	9	8	2140	2324	1609	1185
IBL 091-1-1 × Souna 3	55	53	8	7	2375	2363	1295	1127
IBL 095-4-1 × Souna 3	53	50	9	8	2847	3102	1513	1768
IBL 098-3-1 × Souna 3	56	53	7	10	1983	2669	1018	1559
IBL 106-B-1 × Souna 3	57	55	8	9	1908	3015	1005	1697
IBL 110-B-1 × Souna 3	59	53	7	7	3733	4647	934	2051
IBL 114-6-1 × Souna 3	57	56	7	10	3154	2828	1436	926
IBL 119-B-1 × Souna 3	59	53	10	8	4175	2996	1829	1231
IBL 155-2-1 × Souna 3	59	54	8	9	1913	3211	907	1294
IBL 165-1-1 × Souna 3	60	57	8	8	4660	2180	2162	1061
IBL 179-2-1 × Souna 3	60	58	9	8	1728	2084	617	854
IBL 179-3-1 × Souna 3	56	54	7	9	2031	3855	822	1982
IBL 206-1-1 × Souna 3	56	53	8	11	3285	4058	1609	2438
IBL 001-4-1 × Sosat C 88	57	54	9	10	2503	3490	1298	1860
IBL 003-B-1 × Sosat C 88	48	51	12	12	3562	2802	1959	1806
IBL 011-4-1 × Sosat C 88	49	51	10	10	2380	2555	1259	1403
IBL 021-3-1 × Sosat C 88	53	55	9	10	2526	3580	1304	2005
IBL 055-4-1 × Sosat C 88	58	56	9	9	1630	2017	1007	1255
IBL 091-1-1 × Sosat C 88	48	51	11	10	2563	3262	1976	2061
IBL 095-4-1 × Sosat C 88	56	53	9	10	1963	2889	1080	1657
IBL 098-3-1 × Sosat C 88	59	55	8	11	1260	4184	404	2357
IBL 106-B-1 × Sosat C 88	59	54	8	11	2133	1733	944	953
IBL 110-B-1 × Sosat C 88	56	53	9	10	2367	3288	1032	1835
IBL 114-6-1 × Sosat C 88	53	51	9	10	2917	2971	1662	1171
IBL 119-B-1 × Sosat C 88	59	60	9	7	2992	994	1261	482
IBL 155-2-1 × Sosat C 88	51	51	9	11	1939	2889	801	1630
IBL 165-1-1 × Sosat C 88	60	59	11	9	3801	1770	1450	778
IBL 179-2-1 × Sosat C 88	53	55	9	10	2449	3390	1103	2070
IBL 179-3-1 × Sosat C 88	58	54	9	10	1644	3107	727	1767

TABLE 4: Continued.

Genotype	FWT (das)		TGW (g)		PY (kg ha^{-1})		GY (kg ha^{-1})	
	Bambey	Nioro	Bambey	Nioro	Bambey	Nioro	Bambey	Nioro
IBL 206-1-1 × Sosat C 88	54	52	8	9	2411	4240	1010	2966
Souna 3	55	56	8	8	3465	2372	1731	805
Sosat C 88	50	52	9	10	2956	2180	1707	1389
Thialack II	54	52	7	8	3716	3044	1659	1729
Mean	57	55	8	9	2186	2441	1072	1270
SD	5	4	2	2	955	1089	476	703

FWT: flowering time; TGW: 1000-grain weight; PY: panicle yield; GY: grain yield.

TABLE 5: Mean squares for combining ability for studied traits across locations.

Source of variation	d.f.	DMI	FWT	PH	PLEN	PDIA	PT	TGW	PY	GY
Rep (site)	2	55.5	12.7	300.9	1.2	0.7	3.1	5.0	316965	25012
Site	1	158.8*	178.8***	711.6	280.0**	0.2	12.2*	26.1**	9629606**	6762757***
Hybrid	33	188.1***	45.7***	2852.4***	309.4***	0.5*	4.3**	6.2***	1962646	688597
Line (GCA)	16	229.6***	64.9***	3607.8***	160.6***	0.4	5.4**	4.4**	2290926*	775061*
Tester (GCA)	1	776.5***	40.5**	270.7	6226.1***	1.6*	0.7	64.8***	1998612	123285
Line × tester (SCA)	16	110.0**	26.8***	2258.6***	88.5***	0.4	3.4*	4.4***	1632117	637466
Site × line	16	52.4	7.5	311.4	18.0	0.3	2.3	3.9**	3292948**	1351732***
Site × tester	1	184.4*	54.0**	79.1	26.1	1.6	0.9	0.04	95074	444360
Site × line × tester	16	64.1	6.0	173.9	28.2	0.3	1.2	1.2	703213	424419
Error	134	39.7	5.3	281.7	25.4	0.3	1.9	1.8	1244846	414329

DMI: downy mildew incidence; FWT: flowering time; PH: plant height; PL: panicle length; PDIA: panicle diameter; PT: productive tillers; TGW: 1000-grain weight; PY: panicle yield; GY: grain yield; GCA: general combining ability; SCA: specific combining ability; d.f.: degree of freedom; Rep: replication. *,**,***Significant at 0.05 and 0.01 and 0.001 probability levels, respectively.

TABLE 6: Estimates of GCA effects of lines and testers evaluated across the two sites.

Lines	DMI (%)	FWT (das)	PH (cm)	PLEN (cm)	PDIA (cm)	PT	TGW (g)	PY (kg ha^{-1})	GY (kg ha^{-1})
IBL 001-4-1	9.7***	−0.8	10.1	−2.5	0.1	0.6	1.1	483.9	354.4
IBL 003-B-1	1.6	−4.2***	−21.7***	−5.7***	0.3	0.2	1.7*	447.9	367.4
IBL 011-4-1	−0.4	−3.8***	−12.0*	3.0*	−0.3	0.3	0.0	−334.1	−260.4
IBL 021-3-1	−1.1	0.8	−21.3***	−1.2	−0.2	−0.8	−0.5	−339.1	−143.8
IBL 055-4-1	−3.2	2.6**	0.9	−2.9*	0.3	−1.2**	−0.4	−724.2	−132.9
IBL 091-1-1	−3.2	−3.0**	−13.2*	5.5***	−0.1	−0.2	0.0	−111.2	217.9
IBL 095-4-1	−2.4	−1.7*	3.0	6.7***	−0.2	−0.2	0.0	−51.7	107.5
IBL 098-3-1	9.0***	0.8	12.5*	2.8*	0.0	−0.6	−0.1	−227.9	−62.2
IBL 106-B-1	−0.4	1.4	−12.5*	2.3	−0.2	−1.1*	0.1	−554.6	−247.0
IBL 110-B-1	−2.3	0.6	−3.4	3.0*	0.4	0.4	−0.9	756.7	66.2
IBL 114-6-1	−1.4	−0.5	−1.7	−2.7*	0.0	−0.2	−0.3	215.7	−98.2
IBL 119-B-1	−2.4	2.9**	41.2***	3.3*	−0.2	0.3	−0.4	37.4	−196.4
IBL 155-2-1	7.6**	−1.0	5.3	−1.2	−0.2	1.3**	0.2	−263.9	−239.0
IBL 165-1-1	−3.2	4.2***	19.8***	0.3	0.2	0.0	−0.1	350.7	−34.2
IBL 179-2-1	−3.2	1.6	19.4***	−3.7**	0.0	0.3	−0.1	−339.1	−236.0
IBL 179-3-1	−2.4	0.9	−1.1	−4.3**	0.1	−0.2	−0.3	−92.9	−72.4
IBL 206-1-1	−2.3	−0.9	−25.2***	−2.7*	0.0	0.9*	0.0	746.5	609.0
SE	*2.0*	*0.8*	*4.9*	*1.2*	*0.2*	*0.4*	*0.6*	*508.2*	*325.6*
Testers									
Souna 3	2.0	0.4	−1.2	5.5	−0.1	−0.1	−0.6*	99.0	−24.6
Sosat C 88	−2.0	−0.4	1.2	−5.5	0.1	0.1	0.6*	−99.0	24.6
SE	*1.0*	*0.5*	*0.6*	*0.4*	*0.1*	*0.1*	*0.0*	*21.6*	*46.7*

DMI: downy mildew incidence; FWT: flowering time; PH: plant height; PL: panicle length; PDIA: panicle diameter; PT: productive tillers; TGW: 1000-grain weight; PY: panicle yield; GY: grain yield. *,**,***Significant at 0.05 and 0.01 and 0.001 probability levels, respectively.

the traits while for male lines significant GCA effects were recorded only for TGW. For DMI, the GCA effects varied from −3.2 for IBL 091-1-1 to 9.7 for IBL 001-4-1. Positive and significant GCA effects for DMI were observed on parental lines IBL 001-4-1, IBL 098-3-1, and IBL 155-2-1. For FWT, GCA effects ranged from −4.2 for IBL 003-B-1 to 4.2 for IBL 165-1-1 and both positive and negative significant GCA effects were observed. Estimates of GCA effects for PH ranged from −25.2 for IBL 206-1-1 to 19.4 for IBL 179-2-1. Out of the 19 parental lines, six showed negative and significant effects whereas four lines exhibited positive and significant effects for PH. GCA effects for PLEN varied from −5.7 for IBL 003-B-1 to 6.7 for IBL 095-4-1 with both positive and negative significant effects whereas GCA effects for PDIA ranged from −0.3 for IBL 011-4-1 to 0.4 for IBL 110-B-1 with no significant effects. The GCA effects due to parental lines for PT across locations varied from −1.2 to 1.3 for IBL 055-4-1 and IBL 155-2-1, respectively. Significant positive GCA effects for PT were observed in lines IBL 155-2-1 and IBL 206-1-1 while significant negative GCA effects were observed in lines IBL 055-4-1 and IBL 106-B-1. Across research stations, the GCA for TGW ranged from −0.9 for IBL 110-B-1 to 1.7 for IBL 003-B-1. The tester Sosat C 88 and the inbred line IBL 003-B-1 had significant positive GCA effects while the tester Souna 3 showed significant negative GCA effects for TGW. For PY and GY traits, no significant GCA effects were showed. However, among parental lines, inbred lines IBL 206-1-1, IBL 003-B-1, IBL 001-4-1, IBL 091-1-1, IBL 095-4-1, and IBL 110-B-1 manifested desirable positive GCA effects for GY and most other studied traits for the two research stations. In contrast, inbred lines IBL 011-4-1, IBL 106-B-1, IBL 155-2-1, and IBL 179-2-1 ranked among the worst lines for GY with negative GCA effects.

3.5. Estimation of Specific Combining Ability Effects. Significant positive and negative SCA effects were recorded for all the observed traits (Table 7). The top-cross hybrid IBL 155-2-1 × Sosat C 88 was the only one which exhibited negative and significant SCA effects for DMI. In addition, its SCA effects for FWT and PH were negative and significant while its SCA effects for PT were significant and positive. Among 34 top-cross hybrids, six top-cross hybrids had significant SCA effects, of which three were positive. All the significant and positive SCA effects for PY and GY were recorded in the crosses among Sosat C 88 with the inbred lines IBL 179-2-1, IBL 091-1-1, and IBL 021-3-1.

3.6. Estimation of Standard, Best, and Mid-Parent Heterosis for Grain Yield across Locations. The estimates of best parent, mid-parent, and standard heterosis for grain yield are summarized in Table 8. The best parent heterosis for grain yield across the two locations varied from −44 to 60% and 17 hybrids displayed positive best parent heterosis. IBL 206-1-1 × Souna 3, followed by IBL 001-4-1 × Souna 3, had the largest best parent heterosis for grain yield and was among the best five hybrids while IBL 119-B-1 × Sosat C 88 had the least best parent heterosis value. The mid-parent heterosis varied from −16% for IBL 106-B-1 × Sosat C 88 to 125% for IBL 119-B-1 × Souna 3 which was not among the ten best hybrids. All the crosses displayed positive mid-parent heterosis for grain yield except IBL 106-B-1 × Sosat C 88 (−16%) and IBL 179-2-1 × Souna 3 (−12%). The standard heterosis values for grain yield across the experimental sites varied from −57% for IBL 179-2-1 × Souna 3 to 20% for IBL 206-1-1 × Souna 3. The crosses IBL 206-1-1 × Souna 3, IBL 091-1-1 × Sosat C 88, IBL 206-1-1 × Sosat C 88, IBL 001-4-1 × Souna 3, and IBL 003-B-1 × Sosat C 88 exhibited positive standard heterosis for grain yield. These hybrids were the top best five and displayed also both positive better and mid-parent heterosis values for grain yield.

4. Discussion

The significant differences observed among the genotypes for all the characters studied indicated the presence of large amount of genetic variability among the inbred lines, the OPVs, and their crosses, which is a prerequisite in the establishment of a successful breeding programme. Genetic variability for downy mildew disease and several agronomic traits has been also reported in many studies conducted in West and Central Africa [14, 16–19]. The results indicated also the influence of the environment on the performance of the genotypes for FWT, TGW, PY, and GY traits as their genotype × location interaction effect was significant. The environment effect in the performance of genotypes for flowering time was also reported in Burkina Faso [14]. The mean grain yield at Nioro research station was higher compared to Bambey research station. This could be explained by rainfall pattern and soil texture variability existing between the two locations where the experiments were established. Bambey research station is located in the northern part of the groundnut basin in the Sudano-Sahelian area and the soil texture is sandy while Nioro research station, located in the southern part of the groundnut basin in the Sudanese zone, has sandy-clay soil texture. However, despite the site effect on grain yield and yield related traits, some of the genotypes such as IBL 091-1-1, IBL 091-1-1 × Sosat C 88, and Thialack II have performed well under the two environments.

Besides the existence of useful variability, the establishment of a successful breeding programme depends on a deep understanding of the underlying gene action of the traits of interest. Indeed, this genetic information will guide breeders on which breeding methods and lines to use for the development of improved varieties [9]. In this study, GCA and SCA mean squares were significant for all the traits studied except for the SCA of PDIA, PY, and GY traits indicating that both additive and nonadditive gene actions were important for the inheritance of these traits across the two locations. This result is contrary to the findings of [16, 17] that reported only significant GCA effects for agronomic traits such as flowering time, downy mildew incidence, plant height, and panicle length. However, in the present study, the larger proportion of GCA over SCA mean squares for most of the traits such as DMI, FWT, PL, PDIA, and TGW indicated the preponderance of additive gene action over nonadditive gene action. This would imply that recurrent selection could be effectively used for improvement of these traits. The result of this study is consistent with that of [14] that reported additive gene action to be more important that

TABLE 7: Estimates of SCA effects for hybrids evaluated across the two sites.

Hybrid	DMI (%)	FWT (das)	PH (cm)	PLEN (cm)	PDIA (cm)	PT	TGW (g)	PY (kg ha^{-1})	GY (kg ha^{-1})
IBL 001-4-1 × Souna 3	2.5	−1.9*	−4.7	−1.8	0	0.6	1.2***	140.3	196.8
IBL 003-B-1 × Souna 3	1.1	0.5	13.8***	2.6	−0.3**	0.2	−0.9**	−81.1	−93.6
IBL 011-4-1 × Souna 3	0.8	0.7	0.2	0.2	−0.1	−0.1	−0.1	−148.8	−170.1
IBL 021-3-1 × Souna 3	0.1	1	−35.4***	−2.3	−0.1	−0.4	−0.8*	−739.7**	−376.5*
IBL 055-4-1 × Souna 3	−2	0.2	−27.2***	−6***	0.5***	0.4	0.1	105.2	157.6
IBL 091-1-1 × Souna 3	−2	1.8*	−3.7	0.9	0	−0.4	−0.9**	−370.6*	−379.3*
IBL 095-4-1 × Souna 3	−2.8	−2**	−8.3*	4.2**	0.1	0.4	0.1	175.4	160.8
IBL 098-3-1 × Souna 3	4.4	−1.7*	−2.9	−4*	−0.1	0.5	0.3	−296.6	−21.5
IBL 106-B-1 × Souna 3	0.9	−0.8	1.4	2.1	−0.1	−0.2	0.1	165.6	226
IBL 110-B-1 × Souna 3	−2.9	0.3	10.3**	1.9	0.2	0.6	−0.5	582.3*	54.2
IBL 114-6-1 × Souna 3	−0.2	1.8*	4.2	2.2	−0.1	−0.6	0	−75.6	−92.9
IBL 119-B-1 × Souna 3	−2.8	−2.3**	12.4**	−1	0.1	0.4	0.9**	697.4**	354.1
IBL 155-2-1 × Souna 3	8.8***	2.1**	14.5***	0.3	0.1	−0.8*	0	−25	−32.8
IBL 165-1-1 × Souna 3	−2	−0.9	2.3	−2.6	−0.2	0.4	−0.5	218.4	273.5
IBL 179-2-1 × Souna 3	−2	2.1**	11.8**	0.5	0.1	−1.2***	0.3	−605.7*	−400.8*
IBL 179-3-1 × Souna 3	−1.1	−0.9	8.5*	3.2*	−0.2	0.3	0	184.9	102.3
IBL 206-1-1 × Souna 3	−1	0.1	2.7	−0.4	0.1	−0.2	0.7*	73.9	42.3
IBL 001-4-1 × Sosat C 88	−2.5	1.9*	4.7	1.8	0	−0.6	−1.2***	−140.3	−196.8
IBL 003-B-1 × Sosat C 88	−1.1	−0.5	−13.8***	−2.6	0.3**	−0.2	0.9***	81.1	93.6
IBL 011-4-1 × Sosat C 88	−0.8	−0.7	−0.2	−0.2	0.1	0.1	0.1	148.8	170.1
IBL 021-3-1 × Sosat C 88	−0.1	−1	35.4***	2.3	0.1	0.4	0.8*	739.7**	376.5*
IBL 055-4-1 × Sosat C 88	2	−0.2	27.2***	6***	−0.5***	−0.4	−0.1	−105.2	−157.6
IBL 091-1-1 × Sosat C 88	2	−1.8*	3.7	−0.9	0	0.4	0.9**	370.6*	379.3*
IBL 095-4-1 × Sosat C 88	2.8	2**	8.3*	−4.2**	−0.1	−0.4	−0.1	−175.4	−160.8
IBL 098-3-1 × Sosat C 88	−4.4	1.7*	2.9	4*	0.1	−0.5	−0.3	296.6	21.5
IBL 106-B-1 × Sosat C 88	−0.9	0.8	−1.4	−2.1	0.1	0.2	−0.1	−165.6	−226
IBL 110-B-1 × Sosat C 88	2.9	−0.3	−10.3**	−1.9	−0.2	−0.6	0.5	−582.3*	−54.2
IBL 114-6-1 × Sosat C 88	0.2	−1.8*	−4.2	−2.2	0.1	0.6	0	75.6	92.9
IBL 119-B-1 × Sosat C 88	2.8	2.3**	−12.4**	1	−0.1	−0.4	−0.9**	−697.4**	−354.1
IBL 155-2-1 × Sosat C 88	−8.8***	−2.1**	−14.5***	−0.3	−0.1	0.8*	0	25	32.8
IBL 165-1-1 × Sosat C 88	2	0.9	−2.3	2.6	0.2	−0.4	0.5	−218.4	−273.5
IBL 179-2-1 × Sosat C 88	2	−2.1**	−11.8**	−0.5	−0.1	1.2***	−0.3	605.7*	400.8*
IBL 179-3-1 × Sosat C 88	1.1	0.9	−8.5*	−3.2*	0.2	−0.3	0	−184.9	−102.3
IBL 206-1-1 × Sosat C 8	1	−0.1	−2.7	0.4	−0.1	0.2	−0.7*	−73.9	−42.3
SE	*2.2*	*0.7*	*3.7*	*1.5*	*0.1*	*0.3*	*0.3*	*234.8*	*182.5*

DMI: downy mildew incidence; FWT: flowering time; PH: plant height; PL: panicle length; PDIA: panicle diameter; PT: productive tillers; TGW: 1000-grain weight; PY: panicle yield; GY: grain yield. *,**,***Significant at 0.05 and 0.01 and 0.001 probability levels, respectively.

nonadditive gene action in controlling agronomic traits such as grain yield, flowering time, and panicle length. Similarly, [20] reported the importance of additive gene action over nonadditive gene action in the expression of panicle length and diameter. The additive gene action was also reported for other traits in pearl millet such as Fe and Zn densities [11]. For grain yield, the significance of GCA_f and the lack of significance for SCA suggest that grain yield is controlled by additive gene effects as reported by several authors [14, 16]. However, the slight difference of their mean squares suggests that nonadditive gene action is also important in the inheritance of grain yield trait. This study has also provided information on parental effects in controlling the traits studied. The larger GCA_m mean squares over GCA_f mean squares for DMI, PL, PDIA, and TGW display the role of paternal effects in the control of these traits while the larger GCA_f mean squares over GCAm mean squares for FWT, PH, PT, and GY suggest the role of maternal effects in the control of these traits across the two locations. Similarly, [14] found a paternal effect in controlling PDIA and a maternal effect for FWT and PH under different locations in Burkina Faso. The best performing cross for high grain yield and resistance to downy mildew disease may be produced by crossing the male parents resistant to the disease with female parents having good yield potential.

Inbred lines IBL 001-4-1, IBL 003-B-1, IBL 091-1-1, IBL 095-4-1, IBL 110-B-1, and IBL 206-1-1 had positive GCA effects for grain yield indicating that these lines contributed

TABLE 8: Mean grain yield and best and mid-parent heterosis of pearl millet hybrid across locations.

Cross	GY (kg ha^{-1})	BPH	MPH	SH
IBL 001-4-1 × Souna 3	1923	52	108	14
IBL 003-B-1 × Souna 3	1646	23	26	−3
IBL 011-4-1 × Souna 3	942	−26	11	−44
IBL 021-3-1 × Souna 3	852	−33	1	−50
IBL 055-4-1 × Souna 3	1397	10	80	−18
IBL 091-1-1 × Souna 3	1211	−4	1	−29
IBL 095-4-1 × Souna 3	1641	29	48	−3
IBL 098-3-1 × Souna 3	1289	2	5	−24
IBL 106-B-1 × Souna 3	1351	7	37	−20
IBL 110-B-1 × Souna 3	1493	18	79	−12
IBL 114-6-1 × Souna 3	1181	−7	48	−30
IBL 119-B-1 × Souna 3	1530	21	125	−10
IBL 155-2-1 × Souna 3	1101	−13	27	−35
IBL 165-1-1 × Souna 3	1612	27	83	−5
IBL 179-2-1 × Souna 3	735	−42	−12	−57
IBL 179-3-1 × Souna 3	1402	11	30	−17
IBL 206-1-1 × Souna 3	2024	60	69	20
IBL 001-4-1 × Sosat C 88	1579	2	48	−7
IBL 003-B-1 × Sosat C 88	1883	22	30	11
IBL 011-4-1 × Sosat C 88	1331	−14	35	−21
IBL 021-3-1 × Sosat C 88	1654	7	68	−2
IBL 055-4-1 × Sosat C 88	1131	−27	23	−33
IBL 091-1-1 × Sosat C 88	2019	30	50	19
IBL 095-4-1 × Sosat C 88	1368	−12	10	−19
IBL 098-3-1 × Sosat C 88	1381	−11	1	−18
IBL 106-B-1 × Sosat C 88	949	−39	−16	−44
IBL 110-B-1 × Sosat C 88	1433	−7	47	−15
IBL 114-6-1 × Sosat C 88	1416	−9	51	−16
IBL 119-B-1 × Sosat C 88	871	−44	6	−49
IBL 155-2-1 × Sosat C 88	1215	−22	21	−28
IBL 165-1-1 × Sosat C 88	1114	−28	9	−34
IBL 179-2-1 × Sosat C 88	1586	2	63	−6
IBL 179-3-1 × Sosat C 88	1247	−19	2	−26
IBL 206-1-1 × Sosat C 8	1988	28	49	17

GY: grain yield; BPH: best parent heterosis; MPH: mid-parent heterosis; SH: standard heterosis.

favorable alleles for grain yield. They produced hybrids that were among the best 15 across the two locations. Thus, such lines could be used as parents to create high yielding synthetic or F_1 hybrid varieties. However, IBL 001-4-1 unlike the other five inbred lines had positive and significant GCA effect for downy mildew and produced hybrids with a certain level of disease incidence. The other lines showed negative GCA effects and would be good sources of resistance for downy mildew under Senegalese growing conditions. In addition, they had negative GCA effects for flowering time and plant height. Thus, their cross is expected to produce a medium plant height and early maturing synthetic pearl millet varieties, tolerant to the downy mildew disease with improved grain yield.

In this study, the top-cross hybrids performed better than the inbred lines and OPVs. The top five genotypes across the two locations were hybrids, showing evidence of heterosis for grain yield in pearl millet which has been also reported previously [5, 14, 16]. Grain yield showed a mid-parent heterosis ranging from −16% to 125% and most of the hybrids except IBL 106-B-1 × Sosat C 88 and IBL 179-2-1 × Souna 3 exceeded the parental lines. This finding is consistent with [18] that reported mid-parent heterosis ranging from 1.9 to 98% for top-crosses evaluated under low P conditions. Information about the performance of hybrids compared to the standard check is needed for the farmer to determine the benefit of growing hybrid. In this study, a maximum standard heterosis of 20% for grain yield was observed providing advantage of growing hybrids compared to the local cultivars. Similar standard heterosis for grain yield was also reported in Burkina Faso [14]. The higher mean performance of the crosses compared to their parents and the control check indicate great potential for hybrid pearl millet breeding. Therefore, this technology can be a good strategy to increase pearl millet production like in India where more than 70% of the pearl millet cultivated area is sown with F_1 hybrids [5]. However, a strong hybrid pearl millet breeding programme needs to be established.

5. Conclusion

The present study revealed that the crosses IBL 206-1-1 × Souna 3, IBL 091-1-1 × Sosat C 88, IBL 206-1-1 × Sosat C 88, IBL 001-4-1 × Souna 3, and IBL 003-B-1 × Sosat C 88 were the top five hybrids and exhibited positive best parent, mid-parent, and standard heterosis for grain yield. Furthermore, both additive and nonadditive gene action were involved in the inheritance of almost all the traits studied. However, the contribution of the additive gene action was higher than that of nonadditive gene action for all the traits. Inbred lines IBL 003-B-1, IBL 091-1-1, IBL 095-4-1, IBL 110-B-1, and IBL 206-1-1 exhibited positive GCA effects for grain yield and negative GCA effects for flowering time, downy mildew disease, and plant height. These lines can be used as parents for breeding high yielding synthetic varieties or hybrids F_1 adapted to West and Central African countries.

Acknowledgments

The authors are thankful to the West Africa Agricultural Productivity Program (WAAPP). This work could not have been done without the funding received from the WAAPP.

References

[1] FAO, *FAO Database for agriculture statistics*, 2015, http://faostat.fao.org.

[2] V. Rajaram, T. Nepolean, S. Senthilvel et al., "Pearl millet [Pennisetum glaucum (L.) R. Br.] consensus linkage map constructed using four RIL mapping populations and newly developed EST-SSRs," *BMC Genomics*, vol. 14, no. 1, pp. 1–15, 2013.

[3] V. S. Nambiar, J. J. Dhaduk, N. Sareen, T. Shahu, and R. Desai, "Potential functional implications of pearl millet (Pennisetum glaucum) in health and disease," *Journal of Applied Pharmaceutical Science*, vol. 1, no. 10, pp. 62–67, 2011.

[4] Y. Camara, M. C. S. Bantilan, and J. Ndjeunga, *Impacts of Sorghum and Millet Research in West And Central Africa (WCA): A Synthesis and Lessons Learnt*, International Crops Research Institute for the Semi-Arid Tropics, 2006.

[5] O. P. Yadav and K. N. Rai, "Genetic Improvement of Pearl Millet in India," *Agricultural Research*, vol. 2, no. 4, pp. 275–292, 2013.

[6] S. L. Tan, "Cassava silently, the tuber fills: the lowly cassava, regarded as a poor mans crop, may help save the Euphytica world from the curse of plastic pollution," *Utar Agriculture Science Journal*, vol. 1, pp. 12–24, 2015.

[7] B. A. Christinck, M. Diarra, and G. Horneber, *Innovations in Seed Systems, Lessons from the CCRP Funded Project Sustaining farmer-managed Seed Initiatives in Mali, Niger and Burkina Faso*, International Crops Research Institute for the Semi-Arid Tropics, 2014.

[8] A. R. Hallauer, M. J. Carena, and J. B. Miranda-Filho, *Quantitative Genetics in Maize Breeding*, Springer, New York, NY, USA, 2010.

[9] D. S. Falconer and T. F. C. Mackay, *Introduction to Quantitative Genetics*, Longman, New York, NY, USA, 1996.

[10] O. Kempthorne, *An introduction to genetic statistics*, John Wiley and Sons, Inc., New York, NY, USA and London, UK, 1957.

[11] M. Govindaraj, K. N. Rai, P. Shanmugasundaram et al., "Combining ability and heterosis for grain iron and zinc densities in pearl millet," *Crop Science*, vol. 53, no. 2, pp. 507–517, 2013.

[12] R. S. Parmar, G. S. Vala, V. N. Gohil, and A. S. Dudhat, "Studies on combining ability for development of new hybrids in pearl millet [Pennisetum gaucum (L.) R. BR.]," *International Journal of Plant Science*, vol. 8, no. 2, pp. 405–409, 2013.

[13] V. P. Chaudhary, K. K. Dhedhi, H. J. Joshi, and D. R. Mehta, "Combining ability studies in line x tester crosses of pearl millet [Pennisetum glaucum (L.) R. Br.]," *Research on Crops*, vol. 13, no. 3, pp. 1094–1097, 2012.

[14] I. Drabo, *Breeding pearl millet (Pennisetum glaucum (L) R. BR.) for downy mildew resistance and improved yield in Burkina Faso [Ph.D. thesis]*, University of Ghana, 2016.

[15] D. C. Gemenet, W. L. Leiser, R. G. Zangre et al., "Association analysis of low-phosphorus tolerance in West African pearl millet using DArT markers," *Molecular Breeding*, vol. 35, no. 8, pp. 1–20, 2015.

[16] B. Ouendeba, G. Ejeta, W. E. Nyquist, W. W. Hanna, and A. Kumar, "Heterosis and Combining Ability among African Pearl Millet Landraces," *Crop Science*, vol. 33, no. 4, pp. 735–739, 1993.

[17] A. Issaka, *Development of Downy Mildew Resistant F1 pearl millet Hybrids in Niger [Ph.D. thesis]*, University of Ghana, 2012.

[18] D. C. Gemenet, C. T. Tom, O. Sy et al., "Pearl millet inbred and testcross performance under low phosphorus in West Africa," *Crop Science*, vol. 54, no. 6, pp. 2574–2585, 2014.

[19] A. Pucher, O. Sy, M. D. Sanogo et al., "Combining ability patterns among West African pearl millet landraces and prospects for pearl millet hybrid breeding," *Field Crops Research*, vol. 195, pp. 9–20, 2016.

[20] A. S. Jethva, L. Raval, R. B. Madriya, D. R. Mehta, and C. Mandavia, "Combing ability over environments for grain yield and its related traits in pearl millet," *Crop Improvement*, vol. 38, no. 1, pp. 92–96, 2011.

4

Gas Exchange and Productivity in Temperate and Droughty Years of Four Eastern, Elite Loblolly Pine Genotypes Grown in the Western Gulf Region

Michael A. Blazier,[1] Michael C. Tyree,[2,3] Mary Anne Sword Sayer,[4] Dipesh KC,[1] Wilson G. Hood,[3] and Bradley S. Osbon[1]

[1]*Louisiana State University AgCenter, Hill Farm Research Station, Homer, LA 71040, USA*
[2]*Department of Biology, Indiana University of Pennsylvania, Indiana, PA 15705, USA*
[3]*Work performed while at: School of Forestry, Louisiana Tech University, P.O. Box 10138, Ruston, LA 71272, USA*
[4]*USDA Forest Service, Southern Research Station, Pineville, LA 71360, USA*

Correspondence should be addressed to Michael A. Blazier; mblazier@agcenter.lsu.edu

Guest Editor: Arun Lahiri Majumder

Loblolly pine plantations in the western portion of the species' range are sometimes planted with genotypes from the eastern portion of its range to improve plantation productivity. Advances in loblolly pine breeding have led to the development of clonally propagated genotypes with higher potential growth rates and better form than more commonly planted half-sib genotypes. At a site in the western portion of the loblolly pine range, four genotypes from the eastern portion of the loblolly pine range were established. Two genotypes (HS756 and HS8103) were half-sib, and two genotypes (V9 and V93) were varieties. The V93 genotype was propagated from the HS756 genotype. The objective of this study was to determine the effects of genotype on seasonal trends in gas exchange parameters at the leaf and crown levels, growth, and biomass allocation patterns. During the two-year study, one year had precipitation and temperature trends similar to the long-term average and one year had extreme drought, with record heat. The HS756, V9, and V93 genotypes had the highest height growth throughout the study. The V93 genotype was sensitive to the drought; its leaf- and crown-level A_{sat} and g_s, declined during the drought more markedly than those of the other genotypes. Although its A_{sat} and g_s were affected by drought, height growth productivity of V93 may have been sustained during the drought by its biomass partitioning pattern of allocating higher proportions of its root biomass to small and fine roots and its aboveground biomass to foliage. These results suggest that a variety such as V93 could be more susceptible to changes in C fixation and water uptake with recurrent drought.

1. Introduction

Productivity of loblolly pine (*Pinus taeda* L.) plantations in the southeastern United States has improved with widespread planting of genetically improved seedlings. Up to 25% tree volume gains have been reached by half-sib, open-pollinated loblolly pine families [1]; and it is estimated that tree volume gains of more than 50% can be achieved by loblolly pine varieties when paired with intensive silviculture to optimize growth [2–4]. However, the greater genetic uniformity of varieties relative to more conventionally-planted half-sib genotypes may lead to greater likelihood of genotype × environment (G × E) interactions in growth and the physiological processes that govern growth patterns [4].

For decades, G × E considerations of productivity and drought hardiness have affected loblolly pine plantation decision-making in the western portion of the loblolly pine range. Numerous genotype studies demonstrated that when planted in common environments in the western portion of the loblolly pine range, genotypes from the eastern portion of its range are taller and faster growing than local genotypes

[5–9]. Eastern families have also shown tendencies to produce trees with straighter stems [8] and allocate stand-level volume to fewer trees per hectare [10] relative to western families. Stands of eastern families thus often produce larger, more economically valuable trees than those of western families [6, 11]. Consequently, loblolly pine from eastern genotypes has been commonly planted in the western portion of loblolly pine's range in favor of local genotypes since the 1980s. However, an offsetting factor for volume gains achievable for eastern genotypes planted in the region is drought injury and mortality. Eastern genotypes are adapted to a climate with greater summer rainfall and lower potential evapotranspiration than that of the western portion of the loblolly pine range, so forest managers have generally avoided planting eastern genotypes on soils with poor moisture-holding capacity to lower risks of drought-related mortality [12]. As such, genotype selection in the region has been guided by compromises between potential gains in productivity and risk of drought-related mortality. The breadth of knowledge for eastern versus western genotype selection in the western portion of the loblolly pine range was overwhelmingly gained through studies of half-sib genotypes; G × E considerations for varietal genotypes of loblolly pine in the region may be more pronounced and are relatively understudied due to the more recent development of industrial-scale production of varietal loblolly pine. The substantially higher cost of varietal relative to half-sib seedlings (nearly 10 times greater cost per seedling) also contributes to the need to better understand G × E factors that affect productivity.

Leaf-level gas exchange varies among loblolly pine genotypes, and it has been implicated as a potential measure of genotype success [13–15]. Differences in leaf photosynthetic rate, respiration rate, stomatal conductance, and water-use efficiency among genotypes can affect stand-level carbon gain and water use [16, 17]. Genotype differences in leaf photosynthetic rates are also correlated with growth trends under environmental stresses such as drought; water stress differences have been inferred as a major factor governing G × E interactions in tree growth trends [18]. Leaf-level gas exchange parameters differed among tree genotypes in some studies [14, 19], whereas others found no differences [17, 20, 21]. Environmental factors such as air temperature, irradiance, ambient CO_2 concentrations, and absolute humidity deficit also affect leaf-level gas exchange [20]. Perhaps because of these variable genetic and environmental effects on leaf-level gas exchange parameters, G × E effects on them have varied in presence [14] and absence [13, 22, 23] in prior studies. Aspinwall et al. [17] suggested that studies of genetic differences in leaf-level gas exchange parameters are more robust by taking repeated measurements over the course of the season to account for seasonal variability in environmental factors and tree size. Gas exchange parameters can also be more strongly related to genotype growth trends by scaling up to the crown-level using total leaf area estimates [18, 24, 25].

Although loblolly pine varieties from the eastern portion of the range may be relatively productive in growth and form when grown in the western portion of the loblolly range, the physiological parameters that govern their growth have been relatively understudied in the western portion of the range. Gas exchange parameters of loblolly pine varieties have been predominately studied in the eastern portion of the range, for example, North Carolina [17], Virginia [14], and South Carolina [15]. The Comparing Clonal Lines on Experimental Sites (CCLONES) study of the Forest Biology Research Cooperative is the only previously reported loblolly pine variety trial to the authors' knowledge that includes sites in the western portion of the range (with sites in southwestern Louisiana and Southeastern Texas) [26], but no gas exchange examinations have yet been published from the CCLONES study. The objective of this study was to determine the effects of genotype on seasonal trends in gas exchange parameters at the leaf and crown levels for loblolly pine varieties and half-sib families from the eastern portion of the loblolly pine range grown on a site in the western portion of the range (northwestern Louisiana). During the study, the region experienced an extreme drought in summer as characterized by the U.S. drought monitor scale [27], which offered an opportunity to explore differences in these parameters under droughty conditions of concern for genotype selection practices in the western portion of the loblolly pine range.

2. Materials and Methods

2.1. Study Site and Experimental Design. The study site was established in 2004 at the Louisiana State University Agricultural Center Hill Farm Research Station near Homer, LA (32.749025 N, −93.04111111 W). Following the removal of the previous 50-year-old loblolly pine stand, the study site was burned and subsoiled to a 60 cm depth in September and October 2004, respectively. The study site had two soil series: (1) well-drained Wolfpen soil series (loamy, siliceous, semi-active, thermic Arenic Paleudalfs) loamy fine sand with a deep (69 cm) argillic layer, and (2) moderately well-drained Sacul soil (fine, mixed, active, thermic Aquic Hapludults) very fine sandy loam with a shallow (23 cm) argillic layer (USDA SCS 1989). Wolfpen series have moderate extent through the Southern Coastal Plain of Louisiana, Arkansas, and Texas; Sacul series have broad extent through the Western Coastal Plain of Louisiana, Arkansas, and Texas and the Southern Coastal Plain of Alabama and Mississippi [28, 29]. Average precipitation for the study's region is 133 $cm \cdot yr^{-1}$; drought conditions are common due to relatively low late-summer precipitation [30]. Temperature and precipitation patterns within this study were obtained from data collected from the National Oceanic and Atmospheric Administration (NOAA) weather station located nearest (5.9 km) the study site. Long-term (30-year) averages of temperature and precipitation for the region in which the study site was located (Figure 1) were obtained from the Southern Climate Impacts Planning Program (SCIPP) of NOAA; the study site location was within the North Central Louisiana region within the SCIPP Average Monthly Temperature and Precipitation Tool [31]. In 2010, conditions were somewhat droughty by the date of the first measurements of the year, so volumetric soil moisture was measured concurrently with gas exchange measurements to further

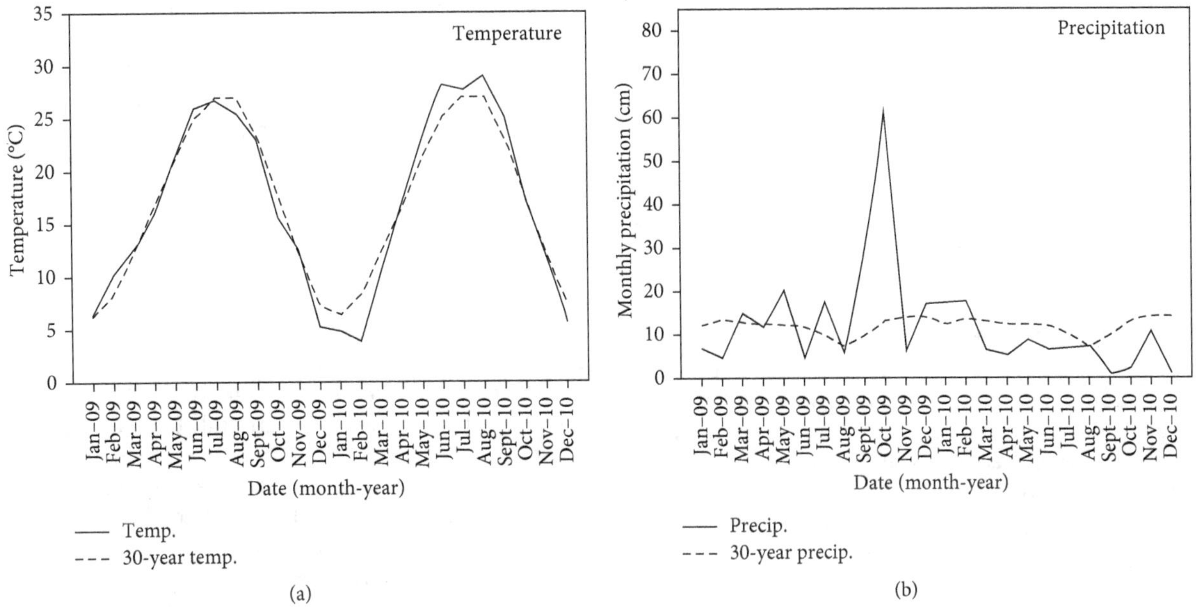

FIGURE 1: Monthly total precipitation, average monthly temperatures, and 30-year averages of precipitation and temperatures for a study site in North Central Louisiana.

characterize environmental conditions. Soil moisture was measured to a 15-cm depth at a point within 1 m of trees on which gas exchange measurements were taken using a TDR 200 time-domain reflectometry probe (Spectrum Technologies, Inc., Paxinos, PA, USA).

Containerized seedlings of four genotypes of loblolly pine from the eastern portion of the loblolly pine range were planted by hand in January 2005. Seedlings were planted in 0.1-ha plots on a 1.8 m × 4.9 m spacing. Plots were separated by at least 4 m on all sides to ensure independence of treatments. Genotypes of this study consisted of two half-sib genotypes (HS756 and HS8103) and two varieties (V9 and V93). The HS756 genotype was included in this study because of its ubiquity in operational planting and breeding programs in the southeastern United States; HS756 has a compact crown and fast growth rate. The V93 genotype was propagated from HS756 and shares its compact crown and fast growth rate. The HS8103 and V9 also have fast growth potential but they are characterized by relatively large crowns. Genotypes were replicated 12 times; the genotype treatment was applied as a randomized complete block design. There were three blocks for the experiment, with each genotype replicated four times per block. Soil type and depth to the argillic horizon within the Wolfpen soil type were the blocking factors.

After planting, applications of sulfometuron methyl (Oust XP) were applied in a 1.8 m band over seedlings in March and September 2005 to enhance establishment success. In April 2006, August 2006, and April 2007 the site was rotary-mowed to eliminate volunteer loblolly pine seedlings. Brush sawing was also conducted within rows each June from 2005 through 2007 to eliminate volunteer loblolly pine seedlings. In July 2005 and July 2007, all hardwood saplings within rows were suppressed with a basal bark application of triclopyr (Garlon 4) herbicide.

2.2. Leaf-Level Gas Exchange. Leaf-level gas exchange parameters of light-saturated photosynthetic rate (A_{sat}) and conductance (g_s) were measured using a Li-6400 portable photosynthesis system with a 6-cm^2 cuvette having a blue-red LED light source (LiCor, Inc., Lincoln, NE, USA) in April, June, August, and November 2009 and in May, June, August, and September 2010. Gas exchange parameters were measured on one tree per plot; measurement trees were selected based on representative size, crown form, and absence of deformities (forking and crown damage) and damage from insects and disease. Gas exchange measurements were taken on a branch in the upper-middle portion of the crown on the first fully elongated flushes of previous-year and current-year foliage. Saturating light (photosynthetically active radiation of 2000 μmol·m^{-2}·s^{-1}) and constant CO_2 concentrations (400 μmol·mol^{-1}) were maintained during measurements. Measurements were conducted between 800 and 1700 Eastern Standard Time, and measurements were stopped when ambient temperatures exceeded 27°C to avoid reductions in photosynthetic capacity associated with temperature-induced stomatal closure [9]. Intrinsic water-use efficiency (*iWUE*; μmol·mmol^{-1}) was calculated by dividing saturated net photosynthesis (A_{sat}·μmol·CO_2·m^{-2}·s^{-1}) by stomatal conductance (g_s; mmol·m^{-2}·s^{-1}) [32]. Following gas exchange measurements, needles were immediately removed from the cuvette and fascicle diameter was measured to the nearest 0.01 mm using digital calipers (Mitutoyo America Corporation, City of Industry, CA). The leaf area per fascicle was calculated using the following equation [33]:

$$A = 2RL(N + \pi), \tag{1}$$

where A is the fascicle area (cm^2), R is the radius of needles (cm), L is the length of the fascicle (cm), and N is the number of needles per fascicle.

2.3. Net CO_2 Assimilation-Light Response Curves. In mid-August through mid-September 2009, net CO_2 assimilation-light response curves were measured. This period received approximately two times more precipitation than the 30-year average, so it represented a best-case scenario for plant metabolism and growth (Figure 1). By graphically comparing net CO_2 assimilation within a range of light levels, light-saturated net photosynthesis (A_{sat}), light compensation point (LCP), mitochondrial respiration rate (R_M), and apparent quantum yield (Q) were determined. Measurements were taken with the Li-6400 photosynthesis system at ten light levels: 2000, 1400, 1000, 600, 400, 200, 100, 50, 20, and 0 μmol·m^{-2}·s^{-1} photosynthetic photon flux densities (PPFDs). Parameters for the net CO_2-light response curves were calculated by fitting response curves to a non-rectangular hyperbola via nonlinear least squares regression [34] using the following equation:

$$A = A_{sat} \times \left(\left[1 - \left(\frac{1 - R_M}{A_{sat}} \right) \right]^{(1-\text{PPFD/LCP})} \right), \quad (2)$$

where A is the net CO_2 assimilation at a given light level, A_{sat} is light-saturated net CO_2 assimilation, R_M is the mitochondrial respiration taken during the day, PPFD is the photosynthetic photon flux density, and LCP is the light compensation point. Apparent quantum yield (Q) was calculated using the first derivative of the above equation:

$$Q = \left(\frac{A_{sat}}{R_M} \right) \times \left(\frac{1 - \text{LCP}}{A_{sat}} \right) \times \ln \left(\frac{1 - \text{LCP}}{A_{sat}} \right). \quad (3)$$

2.4. Crown-Level Biomass and Gas Exchange. Trees were destructively harvested in September of 2009 and 2010 to develop regression equations to estimate crown biomass. For both harvests, six trees (two per block) of each genotype were selected based on the same factors as the measurement trees described above. Prior to felling trees, their total height and diameter at breast height (1.4 m, DBH) were measured. After felling, foliage was removed from branches and separated by year of production. Foliage was then oven-dried to constant weight at 40°C, and dry weights were measured. Equations were developed from these data to estimate crown foliage weight of measurement trees using their height and DBH.

Before foliage samples from the destructive harvests were dried, four subsamples of 50 fascicles each were collected from each foliage flush per tree for specific leaf area (SLA) determination. The leaf per fascicle was calculated using equation (1) (above). All measurements required for the use of equation (1) were collected using the procedures outlined by Bingham [33]. The leaf area values calculated for each fascicle within the 50-fascicle samples were then summed to yield the leaf area for the entire sample. All samples were oven-dried at 40°C to constant weight, and the leaf area of each sample was divided by its oven-dry weight to provide the SLA of each 50-fascicle sample. These procedures yielded average SLA values for each genotype.

The SLA values were used as multipliers for the crown foliage weights of all measurement trees estimated by the regression equations. From that multiplication, the crown leaf area of each measurement tree was obtained. All leaf-level parameters (A_{sat}, g_s, and $iWUE$) were multiplied by the crown leaf area to scale up to the crown level. In 2009, tree height and DBH were measured as crown model inputs in August concurrently with gas exchange measurements, which allowed for scaling of gas exchange parameters to the crown level for August. In 2010, tree height and DBH were measured concurrent with all gas exchange measurements, which allowed scaling of gas exchange parameters to the crown level for May, June, August, and September.

2.5. Biomass Partitioning. As part of the 2009 destructive harvest, above- and belowground biomass partitioning patterns were measured. Diameters and length were measured on all branches of each harvested tree. From the foliage (total of all flushes), branch, and stem weights of the harvested tree, ratios of foliage weight to branch weight and branch weight to stem weight were derived. After aboveground biomass was harvested, root systems within the 1 m^3 volume of soil around the stump were excavated. Root systems were washed and separated into fine (<2 mm diameter), small (2–5 mm diameter), large (>5 mm diameter), and taproot. After the root components were separated, they were dried to constant weight at 40°C, and then dry weights were measured. A relatively negligible amount of root biomass was found below 50 cm within the soil profile, particularly for fine and small roots; weights of all root types reported are within the upper 50 cm of soil.

2.6. Tree Growth. In January–February 2006, 2008, 2009, 2010, and 2011, tree height and DBH were measured. In 2006 and 2008, tree height was measured by a height pole. In the succeeding years, tree height was measured by a hypsometer (Vertex 3, Haglöf Sweden AB, Långsele, Västernorrland, Sverige). Calipers were used to measure DBH. Measurements were taken within the central 0.03 ha of each plot to avoid edge effects on tree growth.

2.7. Data Analysis. Analysis of variance (ANOVA) was conducted at $\alpha = 0.05$ for all parameters using the GLIMMIX procedure of the SAS System (SAS Institute, Inc., Cary, NC). When significant main effects were found in ANOVA, an F-protected least significant difference (LSD) means separation was performed by the LINES option of the LSMEANS statement. When ANOVA determined significant interactions, the SLICEBY, TUKEY, and LINES options of the SLICE statement were used to conduct Tukey means separations within the interaction. The COVTEST statement was used to carry out a restricted likelihood test of homogeneity of variance for variables within the ANOVA model. Parameters for which significant heterogeneity of variance were found were log-transformed for ANOVA (Table 1).

For leaf- and crown-level (2010 only), A_{sat}, g_s, and $iWUE$, ANOVA were performed separately for each foliage flush and year of observation. The model for these

TABLE 1: Tests of fixed effects of leaf-level and crown-level (Crn) photosynthetic capacity (A_{sat}), conductance (g_s), and intrinsic use efficiency (*iWUE*) in previous-year (PY) and current-year (CY) foliage flushes as well as foliage area (FA) and biomass partitioning parameters (foliage : branch mass, branch : stem mass, foliage mass, branch mass, stem mass, branch diameter, branch length, branch angle, taproot weight, fine: total root mass, and small : total root mass) of four loblolly pine genotypes in North Central Louisiana, USA.

	Genotype		Date/flush		Genotype × date/ flush	
Parameter	df	P	df	P	df	P
PY-A_{sat} 2009	3	0.439	2	<0.0001	6	0.005
CY-A_{sat} 2009	3	0.031	2	<0.0001	6	0.012
PY-A_{sat} 2010	3	0.802	3	<0.0001	9	0.044
CY-A_{sat} 2010	3	0.553	2	<0.0001	6	0.046
PY-g_s 2009	3	0.255	2	<0.0001	6	0.047
CY-g_s 2009	3	0.438	2	<0.0001	6	0.098
PY-g_s 2010	3	0.897	3	<0.0001	9	0.041
CY-g_s 2010	3	0.2007	2	<0.0001	6	0.010
PY-*iWUE* 2009	3	0.994	2	<0.0001	6	0.999
CY-*iWUE* 2009	3	0.332	2	<0.0001	6	0.244
PY-*iWUE* 2010	3	0.815	2	<0.0001	6	0.598
CY-*iWUE* 2010	3	0.337	2	<0.0001	6	0.268
Crn FA 2009	3	0.0002	1	<0.0001	3	0.010
Crn-A_{sat} 2009	3	0.974	1	0.912	3	0.001
Crn-g_s 2009	3	0.216	1	0.0001	3	0.001
Crn-*iWUE* 2009	3	0.086	1	0.104	3	0.045
PY-CrnA_{sat} 2010	3	0.013	3	<0.0001	9	0.859
CY-CrnA_{sat} 2010	3	0.019	2	<0.0001	6	0.806
PY-Crng_s 2010	3	<0.0001	3	<0.0001	9	0.097
CY-Crng_s 2010	3	<0.0001	2	<0.0001	6	0.155
PY-Crn*iWUE* 2010	3	0.493	3	<0.0001	9	0.986
CY-Crn*iWUE* 2010	3	0.897	2	<0.0001	6	0.642
Foliage: branch	3	0.020				
Branch: stem	3	0.020				
Foliage mass	3	0.063				
Branch mass	3	0.030				
Stem mass	3	0.187				
Branch dia.	3	0.021				
Branch angle	3	0.012				
Branch length	3	0.151				
Taproot wt.	3	0.045				
Fine: total	3	0.040				
Small: total	3	0.004				

For each parameter, degrees of freedom (df) and probability of a greater F (P) are provided. Fixed effects for leaf-level parameters are genotype, date, and genotype × date; for crown-level parameters in 2009 fixed effects are genotype, foliage flush, and genotype × flush; for biomass partitioning parameters, the fixed effect is genotype.

parameters was a repeated measures model with an autoregressive correlation structure that included genotype, month, and the family × month interaction as fixed effects and block and its interaction with genotype as random effects. The same model was used for ANOVA of tree height and DBH, with year as an independent variable instead of month. Analyses of variance for light response parameters (LCP, R_M, and Q), crown-level gas exchange parameters (A_{sat}, R_M, g_s, and *iWUE*) of 2009, and biomass partitioning parameters (biomass component weight ratios, branch dimensions, and root dimensions) were conducted with a model that included genotype as a fixed effect and block and its interaction with genotype as random effects. Analysis of variance for crown-level gas exchange parameters of 2009 was performed for each foliage flush (Table 1).

Procedures for developing the equations used to estimate foliage weight were consistent with those of Blazier et al. [35]. Equation development consisted of two phases, both conducted with the SAS system: (1) stepwise regression with the REG procedure to identify variables that significantly ($\alpha = 0.15$) affected foliage biomass and (2) nonlinear regression with the NLIN procedure of foliage weight with variables identified as significant in stepwise regression. Equation procedures were performed for each foliage flush and genotype.

3. Results

3.1. Temperature and Precipitation. There was a contrast in temperature and precipitation patterns in the two years of this study (Figure 1). Total annual precipitation of 2009 was 2.5 times greater than the 30-year average precipitation, with early fall having substantially greater rainfall than the long-term average. Temperatures in 2009 were similar to or lower than the long-term average throughout the year, with late summer and fall being 6% lower in mean monthly temperatures than the long-term average. Total annual precipitation of 2010 was 36% lower than the 30-year average; this annual precipitation was the third-lowest annual precipitation level for North Central Louisiana from 1895 through 2010 [36]. Monthly mean temperatures of 2010 exceeded the 30-year average in spring and summer. Summer temperatures of 2010 were the highest on record for North Central Louisiana from 1895 through 2010 [36]. Soil moisture continually declined throughout 2010, going from 11.2% in May to 0.38, 0.27, and 0.01% in June, August, and September, respectively (data not shown).

3.2. Leaf-Level Gas Exchange and Light Response Curve Parameters. The V93 genotype differed from the other genotypes in A_{sat} and g_s on several dates (significant genotype × date interaction). In April 2009, V93 had lower A_{sat} than HS756 in previous-year foliage (Table 2). In September 2010, V93 had lower A_{sat} than HS8103 in current-year foliage. In August 2009, V93 had lower g_s than HS756 in current-year foliage (Table 3). In the current-year foliage flush in September 2010, g_s of V93 was lower than that of all other genotypes (Table 3).

The HS8103 genotype had relatively high A_{sat} in June 2009 and September 2010 (Table 2). In June 2009, HS8103 had greater A_{sat} than HS756 and V9 in previous- and current-year foliage. In September 2010, HS8103 had greater A_{sat} than the HS756 and V93 genotypes in the previous-year and current-year foliage flushes, respectively.

For some leaf-level parameters, no differences were found among genotypes (data not shown). No differences in *iWUE* were found in either year. No differences were found in R_M, LCP, and Q.

TABLE 2: Light-saturated photosynthesis (A_{sat}; μmol·m^{-2}·s^{-1}) of previous- and current-year foliage of loblolly pine genotypes at a site in North Central Louisiana in 2009 (temperate year) and 2010 (droughty year).

	2009				2010			
	April	June	August	November	May	June	August	September
Previous year								
HS756	5.132^{a}	3.725^{b}	5.918^{a}	—	4.792^{a}	1.625^{a}	0.596^{a}	0.206^{b}
HS8103	4.723^{ab}	4.788^{a}	5.507^{a}	—	5.182^{a}	1.404^{a}	0.554^{a}	0.562^{a}
V9	4.940^{ab}	3.482^{b}	5.627^{a}	—	4.682^{a}	1.776^{a}	0.671^{a}	0.325^{ab}
V93	3.888^{b}	4.226^{ab}	5.965^{a}	—	4.387^{a}	1.856^{a}	0.614^{a}	0.252^{ab}
Current year								
HS756	—	4.567^{b}	6.503^{a}	5.436^{a}	—	3.127^{a}	1.744^{a}	0.371^{ab}
HS8103	—	5.620^{a}	6.059^{a}	5.697^{a}	—	3.001^{a}	1.650^{a}	0.625^{a}
V9	—	4.471^{b}	6.205^{a}	5.158^{a}	—	3.228^{a}	1.567^{a}	0.382^{ab}
V93	—	5.454^{ab}	6.203^{a}	5.415^{a}	—	3.401^{a}	1.549^{a}	0.227^{b}

For each foliage flush, means within a column followed by different letters are different at $P < 0.05$.

TABLE 3: Stomatal conductance (g_s; mmol·m^{-2}·s^{-1}) of previous- and current-year foliage of loblolly pine genotypes at a site in North Central Louisiana in 2009 (temperate year) and 2010 (droughty year).

	2009				2010			
	April	June	August	November	May	June	August	September
Previous year								
HS756	0.054^{a}	0.070^{a}	0.073^{a}	—	0.063^{a}	0.015^{a}	0.006^{a}	0.003^{a}
HS8103	0.040^{ab}	0.082^{a}	0.063^{a}	—	0.072^{a}	0.011^{a}	0.005^{a}	0.004^{a}
V9	0.043^{ab}	0.062^{a}	0.057^{a}	—	0.058^{a}	0.014^{a}	0.006^{a}	0.004^{a}
V93	0.032^{b}	0.067^{a}	0.078^{a}	—	0.062^{a}	0.016^{a}	0.006^{a}	0.002^{a}
Current year								
HS756	—	0.083^{a}	0.082^{a}	0.043^{a}	—	0.033^{a}	0.013^{a}	0.003^{a}
HS8103	—	0.096^{a}	0.070^{ab}	0.050^{a}	—	0.030^{a}	0.012^{a}	0.005^{a}
V9	—	0.076^{a}	0.062^{ab}	0.044^{a}	—	0.032^{a}	0.013^{a}	0.005^{a}
V93	—	0.094^{a}	0.058^{b}	0.052^{a}	—	0.033^{a}	0.011^{a}	0.002^{b}

For each year and foliage flush, means within a column followed by different letters are different at $P < 0.05$.

3.3. Crown Area and Gas Exchange. Genotype differences (significant genotype main effect) in the crown foliage area (Crn-FA) and crown-level A_{sat} and g_s (Crn-A_{sat}, Crn-g_s, respectively) were found in both years of the study (Table 4). In 2009, V93 and HS756 had greater Crn-FA than the other two genotypes in both foliage flushes. In 2010, Crn-FA of HS756 and V9 had greater Crn-FA than the other two genotypes in both foliage flushes. The V93 genotype had greater Crn-g_s than the HS8103 and V9 genotypes in both foliage flushes in 2009. In current-year foliage in 2009, Crn-g_s of HS756 exceeded that of all other genotypes. In 2010, Crn-g_s of HS756 was greater than that of the HS8103 and V93 genotypes in both foliage flushes. The V93 genotype also had lower Crn-g_s than V9 in previous-year foliage.

The HS756 genotype had relatively high Crn-A_{sat} in both years of the study (Table 4). In 2009, HS756 had greater Crn-A_{sat} than HS8103 and V9 in previous- and current-year foliage. In 2010, HS756 and V9 had higher Crn-A_{sat} than V93 in both foliage flushes.

Crown-level *iWUE* (Crn-*iWUE*) differed in current-year foliage in August 2009, but there were no differences among genotypes in Crn-*iWUE* in 2010. In August 2009, Crn-*iWUE* of V93 was greater than that of HS756 in current-year foliage (Figure 2). Crn-*iWUE* means observed in 2010 (averaged for all measurement dates) were 96.2, 95.8, 98.6, and 88.2 μmol·mmol^{-1} for HS756, HS8103, V9, and V93, respectively, in previous-year foliage. Current-year Crn-*iWUE* in 2010 averaged for all measurement dates were 97.6, 100.4, 101.0, and 103.8 μmol·mmol^{-1} for HS756, HS8103, V9, and V93, respectively (data not shown).

3.4. Tree Growth and Biomass Partitioning. There was a significant genotype × date interaction for tree height (Figure 3); the HS8103 genotype had relatively low height in all observed years after 2006. In 2008, HS756 and V93 had greater height than HS8103. In 2009 through 2011, HS8103 had lower height than all other genotypes. There was no difference among genotypes in DBH in any year (data not shown).

The genotypes differed in their aboveground biomass partitioning, with the greatest contrasts occurring between HS756 and V93 (Table 5). The V93 genotype had lower branch diameter, higher foliage : branch weight, and lower branch : stem weight than HS756. Taproot weight of HS756 exceeded that of the V9 and V93 genotypes. There were no differences among the genotypes in average branch length (data not shown). Foliage mass, which did not differ among

Table 4: Crown foliage area (m^2), light-saturated photosynthesis ($\mu mol \cdot m^{-2}$), conductance ($mmol \cdot s^{-1}$), and transpiration ($mmol \cdot s^{-1}$) of previous-year and current-year foliage of loblolly pine genotypes at a site in North Central Louisiana in August 2009 and in 2010 (averaged for May, June, August, and September).

	Previous-year foliage				Current-year foliage			
	HS756	HS8103	V9	V93	HS756	HS8103	V9	V93
Foliage area								
August 2009	13.81^a	3.11^b	4.50^b	13.37^a	28.82^a	9.72^b	9.79^b	23.24^a
May–September average 2010	77.57^a	38.52^b	57.44^a	26.28^c	169.85^a	41.66^b	139.30^a	30.43^b
Photosynthesis								
August 2009	81.41^a	18.00^b	23.35^b	80.46^a	183.10^a	57.70^b	59.10^b	145.54^a
May–September average 2010	59.43^a	17.35^{ab}	51.71^a	8.81^b	100.66^a	30.44^{ab}	116.10^a	11.80^b
Conductance								
August 2009	1.09^a	0.20^b	0.26^b	1.11^a	2.55^a	0.67^c	0.67^c	1.41^b
May–September average 2010	0.88^a	0.44^{bc}	0.71^{ab}	0.29^c	1.93^a	0.42^b	1.67^a	0.27^b

For each parameter, date, and foliage flush, means within a row followed by different letters are different at $P < 0.05$.

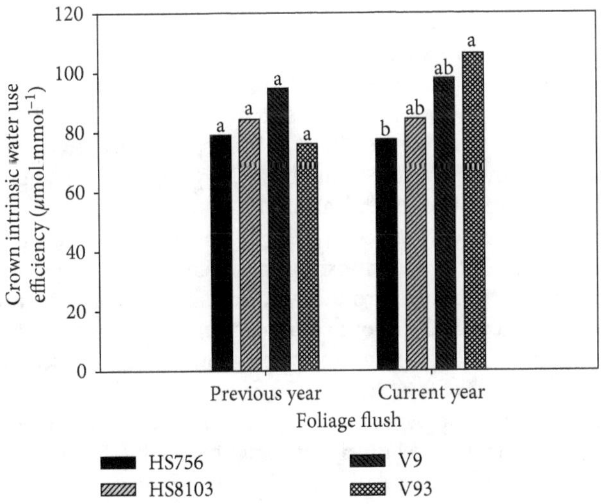

Figure 2: Crown intrinsic water use efficiency of previous- and current-year foliage flushes of loblolly pine genotypes from the eastern portion of the species' range grown at a site in North Central Louisiana in 2009.

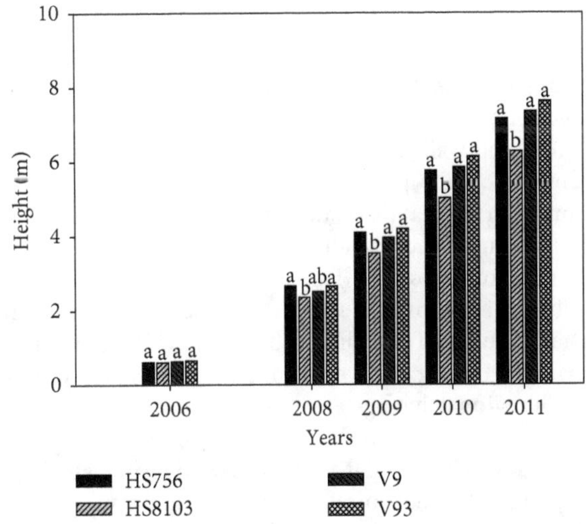

Figure 3: Tree height of loblolly pine genotypes from the eastern portion of the species' range grown at a site in North Central Louisiana.

genotypes, was 7.8, 5.1, 4.8, and 3.3 kg for HS756, HS8103, V9, and V93, respectively. V93 had lower branch mass (2.0 kg) than that of HS756 (6.9 kg); HS8103 and V9 had stem mass of 4.7 and 3.6 kg, respectively. There were no differences in stem mass among genotypes; HS756, HS8103, V9, and V93, respectively, had stem mass of 8.5, 7.6, 5.8, and 6.2 kg, respectively.

No differences were observed among genotypes in most of the types of roots in the upper 50 cm of soil, but they differed in root biomass partitioning patterns. Fine, small, large, and total root biomass were similar for the genotypes. Taproot biomass of HS756 (2000 g) was greater than that of V9 (1000 g) and V93 (1000 g) (data not shown). The HS756 genotype had lower fine : total and small : total root weight than V93; small : total root weight of HS756 was also lower than that of HS8103 (Table 5).

4. Discussion

At this Western Gulf site and under contrasting precipitation patterns, the eastern genotypes had short-term differences in leaf-level gas exchange parameters. Crown-level gas exchange and light response parameters also differed among the genotypes, and the pattern in genotype differences in crown gas exchange parameters varied in the two years observed. These genotype differences in leaf- and crown-level gas exchange parameters did not completely align with genotype variations in growth, suggesting factors other than these parameters also influenced growth patterns.

The leaf-level gas exchange parameters measured in this study in 2009 were within the ranges found in similar studies of loblolly pine, but values in 2010 were lower than in previous studies. In studies of genotypes of loblolly pine at juvenile ages conducted in the eastern portion of the loblolly

TABLE 5: Above- and belowground biomass partitioning patterns of loblolly pine genotypes at a site in North Central Louisiana in 2009.

Genotype	Foliage: branch mass	Branch: stem mass	Branch diameter (cm)	Branch angle (°)	Taproot weight (g)	Fine: total root mass	Small: total root mass
HS756	1.18^{b}	0.78^{a}	1.42^{a}	42.7^{ab}	2400^{a}	6.41^{b}	0.92^{b}
HS8103	1.23^{ab}	0.55^{ab}	1.24^{ab}	39.7^{b}	2100^{ab}	7.19^{ab}	1.87^{a}
V9	1.56^{ab}	0.59^{ab}	1.22^{ab}	48.1^{ab}	1500^{b}	8.87^{ab}	1.23^{ab}
V93	1.66^{a}	0.33^{b}	1.02^{b}	51.1^{a}	1700^{b}	12.14^{a}	1.78^{a}

For each parameter, means within a column followed by different letters are different at $P < 0.05$.

pine range, A_{sat} ranges of 1.5 to 8.0 μmol·m^{-2}·s^{-1} and g_s values of 0.03 to 0.15 mmol·m^{-2}·s^{-1} were reported [14, 17, 20]. The A_{sat} and g_s values of this study in September 2010, which was a period of extreme drought, were substantially below those of the prior studies. These findings suggest that the drought of 2010 reduced the water uptake and photosynthetic capacity at this site. The A_{max} and g_s processes of the V93 genotype appeared more sensitive to droughty conditions than the other genotypes; its A_{max} was relatively low among genotypes in September 2010, and its g_s was lower than all other genotypes in September 2010.

Our findings of genotype differences in leaf-level gas exchange parameters contrast with prior studies of loblolly pine genotypes that found no differences in these parameters [17, 20]. Aspinwall et al. [17] found no differences in A_{sat} and g_s among loblolly pine genotypes in North Carolina; the authors inferred that their study had lower propensity to find such differences because all genotypes tested were native to the region of the study. However, Yang et al. [20] tested loblolly pine genotypes from disparate regions (Atlantic Coastal Plain and Texas) in North Carolina, conducted their experiment during one year that had greater precipitation than the regional average and one with lower-than-average precipitation, and found no differences in A_{sat}, g_s, and *iWUE*. Our results relative to those of Yang et al. [20] could indicate that moving eastern genotypes to the western portion of the loblolly pine range have more potential to differentiate genotype leaf-level gas exchange parameters than moving genotypes to the eastern portion of the range. Further study with a greater array of sites and genotypes is necessary to more comprehensively understand this issue.

Crown foliage area had a dominant influence on crown-level gas exchange parameters, but the droughty conditions of 2010 also affected genotype differences in crown gas exchange parameters. Genotype differences in Crn-A_{sat} were identical to those of Crn-FA in 2009, and genotype differences in Crn-g_s were highly similar to those of Crn-FA in the same year. The two half-sib genotypes of this study were consistent in their differences in Crn-FA and Crn-g_s in both years and for both foliage flushes HS756 exceeded HS8103 in each of these parameters, with the exception of Crn-A_{sat} in current-year foliage, in 2010. Foliage area (and correspondingly Crn-g_s) of the varieties followed different trajectories between 2009 and 2010. In 2009, V93 equaled HS756 in having the highest Crn-FA and Crn-g_s, but in 2010, these parameters for V93 ranked among the lowest of the genotypes. Opposite trends occurred for V9. These results are similar to those of Yáñez et al. [37], in which varieties of loblolly pine had differences in crown characteristics (width and height to live crown in their study) over different site and silvicultural conditions. Yáñez et al. [37] inferred that the crown development of varieties is plastic, which facilitates their responding differently to varied environmental and silvicultural conditions. Our study provided evidence that there was plasticity within the genotypes tested in crown development, occurring in response to drought.

The leaf- and crown-level gas exchange trends did not align with growth trends for all genotypes. The HS8103 genotype consistently had the lowest height growth among genotypes at this site, but none of its leaf-level gas exchange parameters were lower than those of any other genotype. The V93 genotype has been consistently one of the fastest-growing in terms of height at this site, with its height relative to other genotypes unaffected by the 2010 drought. However, V93 had lower leaf-level A_{sat} and g_s in some of the months observed, particularly during the droughty month of September 2010. Relationships between leaf-level gas exchange and genotype productivity have been inconsistent in prior studies [3, 20]. Some studies found relationships between gas exchange parameters and growth trends by scaling with estimates of total leaf area as done in this study [24, 25]. In this study, the persistently lower Crn-g_s of HS8103 were consistent with its relatively low height growth. Those trends were driven by its relatively low crown foliage area, and crown leaf area has been strongly linked with tree growth [23, 38]. Yang et al. [20] found that leaf area was most strongly related to growth of the fastest and slowest loblolly pine genotypes of their study, but in this study crown foliage area patterns were inconsistent with growth differences between our slowest genotype (HS8103) and all other genotypes that were significantly faster growing in height. Crown foliage area of HS8103 was consistently among the lowest of the genotypes, but it was similar to V9, comparable to V93, and greater than V93 in some flushes and observation dates. Crown foliage area and gas exchange parameters of V93 were often inconsistent with its height growth. Despite having height growth among the highest of the genotypes, it had relatively low Crn-FA, Crn-A_{sat}, and Crn-g_s in 2010. These results indicate that although V93 showed signs of stress in C fixation and water uptake at this site, other factors facilitated its relatively high productivity in height growth at this site.

The growth and gas exchange trends of V93 and HS756 were illustrative of the variability in these characteristics possible within related genotypes. Their height growth was similar throughout the study, but they had dissimilar gas exchange patterns. King et al. [14] likewise found that loblolly pine varieties with shared parents vary in their gas exchange and growth responses to forest management practices and site fertility; they suggested investigating biomass allocation patterns (not measured in their study) to better understand these differences. Crn-FA, Crn-A_{sat}, and Crn-g_s of V93 appeared more sensitive to drought than that of HS756, but it may have maintained its height growth through its biomass allocation patterns. The V93 genotype allocated a significantly lower proportion of biomass to branch biomass (via smaller branches) and a higher proportion to foliage biomass than HS756. The V93 genotype allocated higher proportions of belowground biomass to fine and small roots than HS756, which could foster better nutrient acquisition. This relatively efficient foliage biomass allocation and greater allocation of biomass to fine and small roots may have contributed to the higher Crn-*iWUE* of V93 relative to HS756 in the temperate year of 2009. However, the larger taproot of HS756 could be important for sustaining water uptake during droughty periods as in 2010.

5. Conclusions

Growth of these genotypes from the eastern portion of the loblolly pine range at this site in the western part of the range has differentiated, and crown-level gas exchange and biomass allocation patterns have governed the genotype growth differences. The most substantial trends in gas exchange parameters were associated with the V93 genotype's response to drought. In the droughty year, V93 had lower foliage area, A_{sat} and g_s relative to the other faster growing genotypes (HS756 and V9). The lower values of these parameters are particularly of merit relative to HS756, from which V93 was propagated. While the height growth of V93 did not change relative to other genotypes after the drought, its apparent increased sensitivity to drought highlights the potential increased risk associated with establishing loblolly pine plantations of genetically homogenous genotypes of eastern origin in the western portion of the species' range. Drought risk of planting such a genotype could increase if the region experiences more recurrent drought. However, the relatively high growth of V93 and its efficient aboveground biomass allocation of less branch biomass to stem (thereby increasing its value for forest products such as sawtimber) demonstrates the growth and value potential of such a genotype. Further studies of this nature with a greater array of genotypes over more diverse edaphic and climatic conditions will continue to refine knowledge about matching genotypes to sites in a manner that minimize drought risks.

Acknowledgments

This research was funded by the LSU Ag Center and McIntire-Stennis. The authors thank Dylan Dillaway, Kenny Kidd, Brian Finch, Jessica LaGrone, Damien Reynolds, Michael Guilbeaux, Abby Sant, Dani Kennedy, and Chelsea Stringfield for their help with field measurements and the destructive harvest.

References

[1] R. W. Whetten and R. Kellison, "Research gap analysis for application of biotechnology to sustaining U.S. forests," *Journal of Forestry*, vol. 108, pp. 193–201, 2010.

[2] H. L. Allen, T. R. Fox, and R. G. Campbell, "What is ahead for intensive pine plantation silviculture in the south?," *Southern Journal of Applied Forestry*, vol. 29, pp. 62–69, 2005.

[3] T. A. Martin, P. M. Dougherty, and S. E. McKeand, "Strategies and case studies for incorporating ecophysiology into southern pine tree improvement programs," *Southern Journal of Applied Forestry*, vol. 29, no. 2, pp. 70–79, 2005.

[4] S. E. McKeand, E. J. Jokela, D. A. Huber et al., "Performance of improved genotypes of loblolly pine across different soils, climates, and silvicultural inputs," *Forest Ecology and Management*, vol. 227, no. 1-2, pp. 178–184, 2006.

[5] O. O. Wells and P. C. Wakeley, "Genotypic variation in survival, growth, and fusiform-rust infection of planted loblolly pine," *Forest Science Monographs*, vol. 11, p. 40, 1966.

[6] O. O. Wells and C. C. Lambeth, "Loblolly pine provenance test in southern Arkansas-25th year results," *Southern Journal of Applied Forestry*, vol. 7, pp. 71–75, 1983.

[7] C. A. Harrington, "Retrospective shoot growth analysis for three seed sources of loblolly pine," *Canadian Journal of Forest Research*, vol. 21, no. 3, pp. 306–317, 1991.

[8] S. D. Douglass, C. G. Williams, C. C. Lambeth, D. A. Huber, and L. C. Burris, "Family × environment interaction for sweep in local and nonlocal seed sources of loblolly pine," in *Proceedings of the 22nd Southern Forest Tree Improvement Conference*, pp. 318–326, Atlanta, GA, USA, June 1993.

[9] M. A. Blazier, T. C. Hennessey, T. B. Lynch, R. F. Wittwer, and M. E. Payton, "Productivity, crown architechture, and gas exchange of North Carolina and Oklahoma/Arkansas loblolly pine families growing on a droughty site in southeastern Oklahoma," *Forest Ecology and Management*, vol. 194, no. 1–3, pp. 83–94, 2004.

[10] C. B. Talbert and M. R. Strub, "Dynamics of stand growth and yield over 29 years in a loblolly pine source trial in Arkansas," in *Proceedings of the 19th Southern Forest Tree Improvement Conference*, pp. 30–38, College Station, TX, USA, June 1987.

[11] C. G. Tauer and J. A. Loo-Dinkins, "Seed source variation in specific gravity of loblolly pine grown in a common environment in Arkansas," *Forest Science*, vol. 36, pp. 1133–1145, 1990.

[12] C. C. Lambeth, P. M. Dougherty, W. T. Gladstone, R. B. McCullough, and O. O. Wells, "Large-scale planting of North Carolina loblolly pine in Arkansas and Oklahoma: a case of gain versus risk," *Journal of Forestry*, vol. 82, pp. 736–741, 1984.

[13] L. J. Samuelson, "Effects of nitrogen on leaf physiology and growth of different families of loblolly and slash pine," *New Forests*, vol. 19, no. 1, pp. 95–107, 2000.

[14] N. T. King, J. R. Seiler, T. R. Fox, and K. H. Johnsen, "Post-fertilization loblolly pine clone physiology and growth performance," *Tree Physiology*, vol. 28, no. 5, pp. 703–711, 2008.

[15] M. C. Tyree, J. R. Seiler, C. A. Maier, and K. H. Johnsen, "*Pinus taeda* clones and soil nutrient availability: effects of soil organic matter incorporation and fertilization on biomass partitioning and leaf physiology," *Tree Physiology*, vol. 29, no. 9, pp. 1117–1131, 2009.

[16] R. O. Tesky, B. C. Bongarten, B. M. Cregg, P. M. Dougherty, and T. C. Hennessey, "Physiology and genetics of tree growth response to moisture and temperature stress: an examination of the characteristics of loblolly pine (*Pinus taeda* L.)," *Tree Physiology*, vol. 3, pp. 41–61, 1987.

[17] M. J. Aspinwall, J. S. King, S. E. McKeand, and J. Domec, "Leaf-level gas-exchange uniformity and photosynthetic capacity among loblolly pine (*Pinus taeda* L.) genotypes of contrasting inherent genetic variation," *Tree Physiology*, vol. 31, no. 1, pp. 78–91, 2011.

[18] J. E. Major and K. H. Johnsen, "Family variation in photosynthesis of 22-year-old black spruce: a test of two models of physiological response to water stress," *Canadian Journal of Forest Research*, vol. 26, no. 11, pp. 1922–1933, 1996.

[19] D. J. Weston and W. L. Bauerle, "Inhibition and acclimation of C3 photosynthesis to moderate heat: a perspective from thermally contrasting genotypes of *Acer rubrum* (red maple)," *Tree Physiology*, vol. 27, no. 8, pp. 1083–1092, 2007.

[20] W. Q. Yang, R. Murthy, P. King, and M. A. Topa, "Diurnal changes in gas exchange and carbon partitioning in needles of fast- and slow-growing families of loblolly pine (*Pinus taeda*)," *Tree Physiology*, vol. 22, no. 7, pp. 489–498, 2002.

[21] J. R. Seiler and J. D. Johnson, "Physiological and morphological responses of three half-sib families of loblolly pine to water-stress conditioning," *Forest Science*, vol. 34, pp. 487–495, 1988.

[22] H. E. Bown, M. S. Watt, P. W. Clinton, E. G. Mason, and B. Richardson, "Partitioning concurrent influences of nitrogen and phosphorus supply on photosynthetic model parameters of *Pinus radiata*," *Tree Physiology*, vol. 27, no. 3, pp. 335–344, 2007.

[23] D. J. Chmura and M. G. Tjoelker, "Leaf traits in relation to crown development, light interception, and growth of elite families of loblolly pine and slash pine," *Tree Physiology*, vol. 28, no. 5, pp. 729–742, 2008.

[24] D. A. Michael, D. I. Dickmann, J. G. Isebrands, and N. D. Nelson, "Photosynthesis patterns during the establishment year within two *Populus* clones with contrasting morphology and phenology," *Tree Physiology*, vol. 6, no. 1, pp. 11–27, 1990.

[25] T. Wang, P. M. A. Tigerstedt, and A. Vihera-Aarnio, "Photosynthesis and canopy characteristics in genetically defined families of silver birch (*Betula pendula*)," *Tree Physiology*, vol. 15, no. 10, pp. 665–671, 1995.

[26] B. Baltunis, D. Huber, and T. Wite, *Genetic Analysis of Early Field Growth of Loblolly Pine Clones and Seedlings from the Same Full-Sib Families*, University of Florida, Gainsville, FL, USA, 2006.

[27] Forwarn, "U.S. Forest change assessment viewer. Drought assesements maps," 2017, https://forwarn.forestthreats.org/fcav2/.

[28] National Cooperative Soil Survey, "Wolfpen series. National cooperative soil survey characterization database," 2003, https://soilseries.sc.egov.usda.gov/OSD_Docs/W/WOLFPEN.html.

[29] National Cooperative Soil Survey, "Sacul series. National cooperative soil survey characterization database," 2015, https://soilseries.sc.egov.usda.gov/OSD_Docs/S/SACUL.html.

[30] United States Department of Agriculture, *Soil Conservation Service, Soil Survey of Claiborne Parish, Louisiana*, United States Department of Agriculture, Washington, DC, USA, 1989.

[31] Southern Climate Impacts Planning Program (SCIPP), "Average monthly temperature and precipitation tool," June 2018, http://charts.srcc.lsu.edu/.

[32] B. M. Cregg, J. M. Olivas-Garcia, and T. C. Hennessey, "Provenance variation in carbon isotope discrimination of mature ponderosa pine trees at two locations in the Great Plains," *Canadian Journal of Forest Research*, vol. 30, no. 3, pp. 428–439, 2000.

[33] G. E. Bingham, "Leaf area measurement of pine needles," in *LI-1600 Steady State Porometer Instruction Manual*, Li-Cor, Inc. Publication No. 8107 01R2, Lincoln, Nebraska, 1983.

[34] P. J. Hanson, R. E. McRoberts, J. G. Isebrands, and R. K. Dixon, "An optimal sampling strategy for determining CO_2 exchange-rate as a function of photosynthetic photon flux-density," *Photosynthetica*, vol. 21, pp. 98–101, 1987.

[35] M. A. Blazier, T. C. Hennessey, T. B. Lynch, and R. F. Wittwer, "Comparison of branch biomass relationships for North Carolina and Oklahoma/Arkansas loblolly pine seed sources growing in southeastern Oklahoma," *Forest Ecology and Management*, vol. 159, no. 3, pp. 241–248, 2002.

[36] Southern Climate Impacts Planning Program (SCIPP), "Historical climate trends tool," June 2018, http://charts.srcc.lsu.edu/trends/.

[37] M. A. Yáñez, T. R. Fox, and J. R. Seiler, "Early growth responses of loblolly pine varieties and families to silvicultural intensity," *Forest Ecology and Management*, vol. 356, pp. 204–215, 2015.

[38] R. L. McCrady and E. J. Jokela, "Canopy dynamics, light interception, and radiation use efficiencyt of selected loblolly pine families," *Forest Science*, vol. 44, pp. 64–72, 1998.

5

Identification of Sources of Resistance for Peanut *Aspergillus flavus* Colonization and Aflatoxin Contamination

Richard Moise Alansou Dieme,[1] Issa Faye,[2] Yedomon Ange Bovys Zoclanclounon,[3] Daniel Fonceka,[3] Ousmane Ndoye,[4] and Papa Madiallacke Diedhiou[1]

[1]*Université Gaston Berger, BP 211 Saint-Louis, Senegal*
[2]*Centre National de Recherches Agronomiques de Bambey, Institut Sénégalais de Recherches Agricoles, BP 211 Bambey, Senegal*
[3]*Centre d'Etudes Régional pour l'Amélioration de l'Adaptation à la Sécheresse, Route de Khombole, BP 3320 Thiès, Senegal*
[4]*Conseil Ouest et centre Africain pour la Recherche et le Développement Agricoles, Avenue Bourguiba, BP 48 Dakar RP, Senegal*

Correspondence should be addressed to Richard Moise Alansou Dieme; alansou7dieme@gmail.com

Academic Editor: Glaciela Kaschuk

Peanut aflatoxin contamination caused by *Aspergillus flavus* is a serious constraint for food safety and human health in Senegal. The present study aimed to identify sources of resistance for *A. flavus* colonization and aflatoxin contamination. Thus, seeds from 67 peanut genotypes were tested under laboratory conditions. Aqueous conidial suspension of an aflatoxinogenic strain of *A. flavus* was used for inoculation in Petri dishes containing ten seeds of each genotype, and data on incidence and severity were recorded. Total aflatoxin concentration in seeds was determined on 15th day after inoculation using mReader® method. Results showed a significant ($p < 0.001$) variation of aflatoxin, incidence and severity among the tested peanut genotypes. Incidence ranged from 0 to 70% with a mean of 20.36 ± 0.8%. Out of the 67 genotypes, eight showed incidence less than 10%. Severity ranged from 0 to 44% with a mean value of 8.82 ± 0.45%. The genotype 12CS_104 showed aflatoxin concentration level in conformity with the European standard (4 ppb). Out of three clusters revealed by hierarchical classification based on disease incidence and severity, the cluster 1 contained 33 genotypes characterised by low incidence and severity values. These genotypes can be tested under field conditions to confirm their resistance to *A flavus.*

1. Introduction

Peanut (*Arachis hypogaea* L.) is an important staple crop in Senegal. The national peanut production was estimated at 1,050,042 tons during the rainy season of 2016 [1]. This crop is mainly produced in Fatick, Kaolack, Kaffrine, Louga, and Thies regions, with more than 60% of the national peanut production [1]. Peanut seeds are widely used for food consumption and play a significant economic role for small-scale farmers and food industries in Senegal [2]. However, pre- and postharvest aflatoxin contamination in peanut is a serious threat for food safety and human health in Senegal [3]. It is one of the major constraints limiting sustainable and good quality seed production in the world [4]. Aflatoxin contamination is due to *Aspergillus flavus* (Link ex Fries, Teleomorph: *Petromyces flavus*) [5]. Damages caused by this facultative plant pathogen in maize, peanut, and sesame were reported in Senegal [6]. Considerable economic losses caused by this bacterium are mainly due to crop quality value and international trade restrictions on food stuffs charged in aflatoxin [7].

Aflatoxin is the name of a group of toxin known as G1, G2, B1, B2, M1, and M2 that produced the plant pathogen [8]. These toxins occur naturally and have been found in a wide range of commodities, including peanuts used for animal and human consumption [9]. Aflatoxins are toxic, mutagenic, and carcinogenic compounds [10]. Depending on their levels, toxins can severely affect the liver and induce immune-suppressing effects [9].

To handle this issue, a wide range of preharvest aflatoxin contamination management methods were developed. Application of atoxinogenic isolates of *A. flavus* [11] and host genetic resistance were tested [12]. In Senegal, previous studies reported that varieties 55-437 and 73-3 were resistant to *A. flavus* [13]. Identification of new sources of resistance merits to be investigated for efficient peanut breeding program. First step of host genetic resistance is the seed colonization test. Therefore, the present study was undertaken to identify promising peanut genotypes under laboratory conditions.

2. Materials and Methods

2.1. Plant Materials. The plant material consisted of 67 genotypes including 58 chromosomal substitutions lines [14] and nine national released varieties. The chromosomal substitution lines belong to a cross between Fleur 11 and a synthetic amphidiploid parent (Table 1).

2.2. Isolation of Aspergillus flavus, Sporangial Suspension Preparation, and Inoculation. Aflatoxinogenic strain provided from peanut seeds were purified by successive cultures on 5/2 agar medium. The aflatoxin concentration level was checked using the Reveal® Q$^+$ Aflatoxin test kit (accesso peanut enterprise corporation, USA). The spore suspension of *A. flavus* was obtained by soaking colonized seeds in 50 ml of sterile distilled water. Then, one drop of Tween 20 was added to the solution and thoroughly mixed for 10 minutes. Inoculation was carried out by introducing 100 μl of the supernatant of the spore suspension into each Petri dish.

2.3. Seed Colonization Test. The seed colonization test was conducted following a modified Mehan and McDonald procedure. For each genotype, 50 seeds were sterilized and rinsed properly in sterile distilled water. Then, the seeds were hydrated to about 20% moisture content. The 50 seeds of each genotype were placed in 5 Petri dishes containing 10 seeds, and each Petri dish was considered as a replication. The seeds were inoculated with a conidial suspension (60 μL containing approximately 1×10^8 mL^{-1} conidia of the aflatoxigenic strain of *A. flavus*). This preparation was kept at laboratory conditions (25 ± 0.12°C and 82 ± 0.42% relative humidity) for fifteen days.

2.4. Data Collection. The seeds' colonization was observed during two weeks, and aflatoxin concentration was measured using the Reveal® Q$^+$ Aflatoxin test kit (accesso peanut enterprise corporation, USA). The incidence was calculated using the following formula:

$$\text{incidence}\,(\%) = \frac{\text{number of seeds showing pathogen colonizaton}}{\text{total number of seeds}} \times 100. \quad (1)$$

Table 1: Peanut material used in this study.

No	Genotypes	Description	Country of origin
1	12CS_001	CSL*	Senegal
2	12CS_004	CSL	Senegal
3	12CS_006	CSL	Senegal
4	12CS_007	CSL	Senegal
5	12CS_008	CSL	Senegal
6	12CS_009	CSL	Senegal
7	12CS_010	CSL	Senegal
8	12CS_011	CSL	Senegal
9	12CS_012	CSL	Senegal
10	12CS_016	CSL	Senegal
11	12CS_018	CSL	Senegal
12	12CS_020	CSL	Senegal
13	12CS_021	CSL	Senegal
14	12CS_022	CSL	Senegal
15	12CS_023	CSL	Senegal
16	12CS_024	CSL	Senegal
17	12CS_027	CSL	Senegal
18	12CS_028	CSL	Senegal
19	12CS_031	CSL	Senegal
20	12CS_032	CSL	Senegal
21	12CS_033	CSL	Senegal
22	12CS_034	CSL	Senegal
23	12CS_036	CSL	Senegal
24	12CS_037	CSL	Senegal
25	12CS_039	CSL	Senegal
26	12CS_041	CSL	Senegal
27	12CS_042	CSL	Senegal
28	12CS_047	CSL	Senegal
29	12CS_048	CSL	Senegal
30	12CS_050	CSL	Senegal
31	12CS_051	CSL	Senegal
32	12CS_052	CSL	Senegal
33	12CS_053	CSL	Senegal
34	12CS_054	CSL	Senegal
35	12CS_055	CSL	Senegal
36	12CS_059	CSL	Senegal
37	12CS_060	CSL	Senegal
38	12CS_061	CSL	Senegal
39	12CS_062	CSL	Senegal
40	12CS_063	CSL	Senegal
41	12CS_066	CSL	Senegal
42	12CS_070	CSL	Senegal
43	12CS_072	CSL	Senegal
44	12CS_075	CSL	Senegal
45	12CS_076	CSL	Senegal
46	12CS_078	CSL	Senegal
47	12CS_079	CSL	Senegal
48	12CS_084	CSL	Senegal
49	12CS_085	CSL	Senegal
50	12CS_090	CSL	Senegal
51	12CS_091	CSL	Senegal
52	12CS_095	CSL	Senegal
54	12CS_096	CSL	Senegal
55	12CS_100	CSL	Senegal
56	12CS_111	CSL	Senegal
57	12CS_112	CSL	Senegal
58	12CS_118	CSL	Senegal
59	12CS_119	CSL	Senegal
60	55-33	Variety	Senegal
61	55-437	Resistant control	Senegal

Table 1: Continued.

No	Genotypes	Description	Country of origin
62	73-30	CSL	Senegal
63	73-33	Resistant control	Senegal
64	78-936	Variety	Senegal
65	Fleur11	Susceptible control	Senegal
66	GC-8-35	Variety	Senegal
67	L27	Variety	Senegal

*Chromosomal substitution lines.

The severity scale of aflatoxin on seeds was estimated using a modified Tonapi et al. [15] scale. It was defined as follows: 0, noninfected seeds; 1, seeds whose surface covered by the fungus is less than 20%; 2, 20%–40% seed surface covered by the fungus; 3, 40%–60% seed surface covered by the fungus; 4, 60%–80% seed surface covered by the fungus; and 5, 80%–100% seed surface covered by the fungus. The severity calculation based on Tonapi et al. [15] formula was as follows:

$$\text{severity}\,(\%) = \frac{\sum_{i=1}^{n} (N_i \times i)}{\text{total of seeds} \times (n-1)}, \qquad (2)$$

where $p < 0.001$ i is severity scale from 0 to 5 and N_i is the number of seed corresponding to scale i of severity.

2.5. Data Analysis. Data analysis was performed with the open-source statistical software R version 3.4.5 [16]. Descriptive statistics of recorded data were generated with pastecs package [17]. In order to find out variability of incidence and severity according to tested genotypes, data were subjected to Poisson regression analysis using glm (generalized linear model) function of package stats implemented in the R. Spearman's rank correlation test was performed to highlight relationship between incidence, severity, and aflatoxin concentration levels using correlation test function of package stats. Identification of different groups of genotypes based on incidence and severity was performed based on a principal component analysis and a hierarchical clustering with the functions PCA and HCPC of package FactoMineR [18], respectively. The Euclidean distance and Ward classification method were used to classify tested genotypes. The function fviz_pca_biplot [19] was used to plot the principal components analysis biplot in different clusters based on hierarchical classification.

3. Results

3.1. Reaction of Peanut Genotypes to Aspergillus flavus. Analysis of variance revealed highly significant ($p < 0.001$) variation of aflatoxin incidence and severity among the tested peanut genotypes (Table 2).

The severity ranged between 0 and 44%, respectively, with a mean of 8.82 ± 0.45%. The recorded incidence ranged from 0 to 70% with an average value of 20.36 ± 0.80% (Table 3).

One genotype (12CS_104) showed aflatoxin concentration level less than 4 ppb. A total of 34 genotypes presented aflatoxin concentration level up to 2000 ppb (Figure 1).

Table 2: Deviance values from the Poisson regression model on incidence and severity.

Source of variation	Degree of freedom	Incidence	Severity
Replication	4	33.9 (ns)	2.04 (ns)
Genotype	66	7242.2***	190.08***

***Significant chi-squared test at 0.001 level of probability; ns = not significant.

Out of the 67 genotypes, eight showed incidence less than 10% while 33 showed incidences between 10 and 20% and 16 with incidences ranged from 20 to 30% (Figure 2).

3.2. Correlation between Incidence, Severity, and Aflatoxin Concentration Level. Spearman's rank correlation test revealed a strong relationship ($r = 0.93$, $p < 0.001$) between incidence and severity of peanut genotypes. Positive and significant correlations were detected between aflatoxin concentration levels and disease incidence ($r = 0.28$, $p < 0.01$) and aflatoxin concentration levels and disease severity ($r = 0.35$, $p < 0.05$) (Table 4).

3.3. Classification of the Tested Genotypes according to Sensibility and Aflatoxin Concentration Level. The factorial axes 1 and 2 explained 60.5 and 39.5% of overall variability, respectively (Figure 3). Hierarchical classification performed on principal component analysis revealed three clusters of genotypes based on disease incidence and aflatoxin concentration levels (Figure 3). The clusters 1, 2, and 3 grouped 33, 20, and 14 genotypes, respectively. The incidence and aflatoxin concentration are significantly ($p < 0.001$) associated to cluster 1 (Table 4).

Mean values of these two variables in this cluster are less than the overall mean. Therefore, cluster 1 is characterized by desirable genotypes which combine low incidence values and aflatoxin concentration levels. Cluster 2 is significantly ($p < 0.001$) related to the aflatoxin concentration level (Table 5).

The mean value of aflatoxin concentration in cluster 2 (4075.5 ppb) is 190% which is higher than the overall mean (2143.8 ppb). Thus, this second cluster is characterized by genotypes with high level of aflatoxin. Incidence is linked to cluster 3 (Table 5). Mean value of this variable (35%) in cluster 3 is superior to overall mean (20.35%). Thus, the cluster 3 encompasses the most susceptible genotypes to *A. flavus*.

Based on the closest distance between each genotype and the respective cluster centres, 12CS_039, 12CS_010, and 12CS_050 were the first representative genotypes (paragon) of cluster 1, 2, and 3, respectively (Table 4). Based on the farthest distance from a genotype projected point in a cluster to the centres of the two others, clustering revealed that cluster 1, 2, and 3 were characterised by the genotypes 12CS_104, 78-936, and 12CS_021, respectively (Figure 3, Table 5). Based on results, out of 67 genotypes, 33 promising genotypes (cluster 1) were noted (Figure 3).

Table 3: Means of incidence and severity of the tested lines.

Lines	Incidence		Severity	
	Mean	Std deviation	Mean	Std deviation
12CS_001	44	13.42	1.02	0.44
12CS_004	30	12.25	0.74	0.23
12CS_006	30	15.81	0.46	0.29
12CS_007	18	8.37	0.34	0.21
12CS_008	12	16.43	0.22	0.27
12CS_009	24	8.94	0.58	0.33
12CS_010	18	8.37	0.4	0.27
12CS_011	18	14.83	0.42	0.48
12CS_012	32	13.04	0.82	0.43
12CS_016	34	8.94	0.82	0.54
12CS_018	38	13.04	0.66	0.33
12CS_020	22	10.95	0.5	0.34
12CS_021	52	10.95	1.64	0.49
12CS_022	16	15.17	0.36	0.43
12CS_023	18	13.04	0.28	0.24
12CS_024	28	10.95	0.5	0.21
12CS_027	18	8.37	0.26	0.19
12CS_028	18	8.37	0.3	0.20
12CS_031	16	5.48	0.4	0.28
12CS_032	14	11.40	0.16	0.11
12CS_033	10	12.25	0.2	0.23
12CS_034	14	15.17	0.28	0.34
12CS_036	14	5.48	0.26	0.15
12CS_037	16	5.48	0.32	0.19
12CS_039	14	8.94	0.22	0.23
12CS_041	24	15.17	0.64	0.38
12CS_042	24	18.17	0.42	0.42
12CS_047	18	13.04	0.4	0.22
12CS_048	6	8.94	0.08	0.13
12CS_050	34	11.40	0.76	0.29
12CS_051	34	20.74	0.72	0.36
12CS_052	30	15.81	0.76	0.67
12CS_053	36	15.17	0.98	0.59
12CS_054	26	11.40	0.68	0.44
12CS_055	18	8.37	0.48	0.54
12CS_059	24	16.73	0.48	0.48
12CS_060	12	13.04	0.4	0.39
12CS_061	14	11.40	0.24	0.19
12CS_062	18	8.37	0.26	0.15
12CS_063	18	14.83	0.42	0.53
12CS_066	32	22.80	0.8	0.76
12CS_070	22	19.24	0.56	0.53
12CS_072	14	13.42	0.26	0.28
12CS_075	28	13.04	0.66	0.36
12CS_076	12	4.47	0.18	0.08
12CS_078	10	12.25	0.16	0.18
12CS_079	18	16.43	0.34	0.34
12CS_084	14	5.48	0.3	0.21
12CS_085	22	8.37	0.5	0.32
12CS_090	14	13.42	0.34	0.50
12CS_091	26	11.40	0.64	0.48
12CS_095	12	8.37	0.26	0.27
12CS_096	16	8.94	0.48	0.58
12CS_100	40	12.25	0.96	0.34
12CS_104	6	8.94	0.14	0.26
12CS_111	6	8.94	0.1	0.14
12CS_112	18	16.43	0.26	0.32
12CS_118	20	10.00	0.3	0.20
12CS_119	20	7.07	0.42	0.28

Table 3: Continued.

Lines	Incidence		Severity	
	Mean	Std deviation	Mean	Std deviation
55-33	4	5.48	0.04	0.05
55-437	12	10.95	0.18	0.19
73-30	6	5.48	0.06	0.05
73-33	16	11.40	0.28	0.19
78-936	12	4.47	0.24	0.31
Fleur_11	28	16.43	0.5	0.17
GC-8-35	22	10.95	0.6	0.24
L27	10	0.00	0.16	0.13
Mean	20.36		8.82	
Standard error	0.80		0.45	

Table 4: Spearman's rank matrix correlation performed on incidence, severity, and aflatoxin concentration levels data.

	Incidence	Severity	Aflatoxin concentration levels
Incidence	1.00		
Severity	0.93***	1.00	
Aflatoxin concentration levels	0.28*	0.35**	1.00

*Significant Spearman's rank correlation test at 0.05 level of probability.
**Significant Spearman's rank correlation test at 0.01 level of probability.
***Significant Spearman's rank correlation test at 0.001 level of probability.

4. Discussion

In the present study, a wide phenotypic variation was observed among the tested genotypes for incidence, severity, and aflatoxin concentrations. This variation can be explained by the variability of seed coat structure of the tested genotypes. In fact, the seed coat can constitute a barrier to *A. flavus* seed invasion depending on its thickness and/or permeability [20], and Zhou and Liang [21] studies showed that genotypes seed coat with smaller hilum, more compact arrangement and thicker testa showed more resistance to *A. flavus*. In addition, implication of wax and cutin layers of seed coat was demonstrated to be related to genotypes resistance [22]. Another explanation of this wide variation in incidence, severity, and aflatoxin rate can be biochemical compounds' differential variability in the tested seeds. Lindsey and Turner [23] demonstrated that the presence of polyphenol compounds, specifically, tannins in seed can have inhibitor effect against *A. flavus*. Amaya et al. [24] and Liang et al. [25] showed the difference among seed coat biochemical compounds to determine sensibility to *A. flavus*. Liang [22] demonstrated that the presence of trypsin in seeds can also be related to resistance to *A. flavus*. Turner et al. [26] isolated and identified the 5,7-dimethoxyisoflavone as an inhibitor for *A. flavus* invasion in peanut seed.

12CS_104 was the most resistant genotype to aflatoxin contamination with an aflatoxin level lower than the European Union standards (4 ppb). However, except 12CS_104, all the genotypes have their aflatoxin concentration level higher than the Chinese (20 ppb) standards.

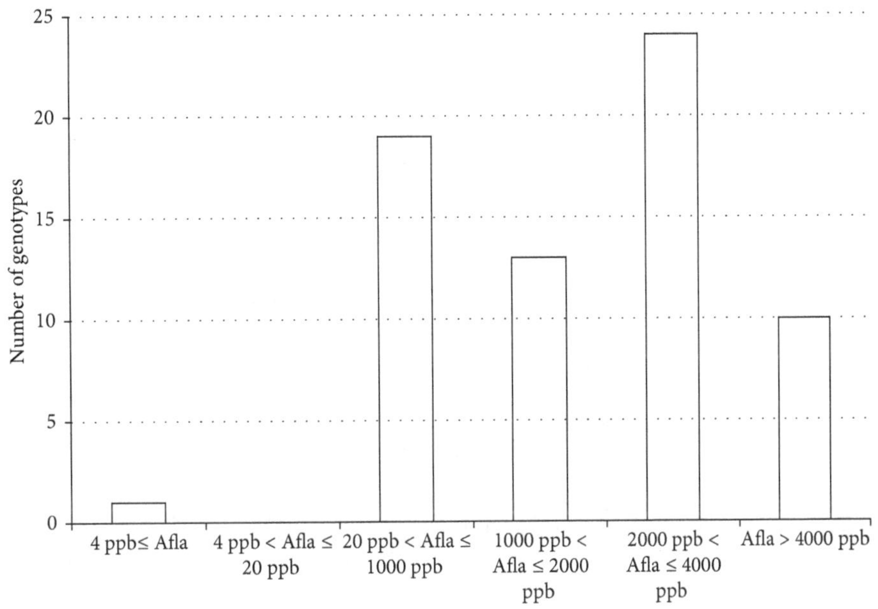

Figure 1: Number of peanut genotypes according to aflatoxin concentration level interval values.

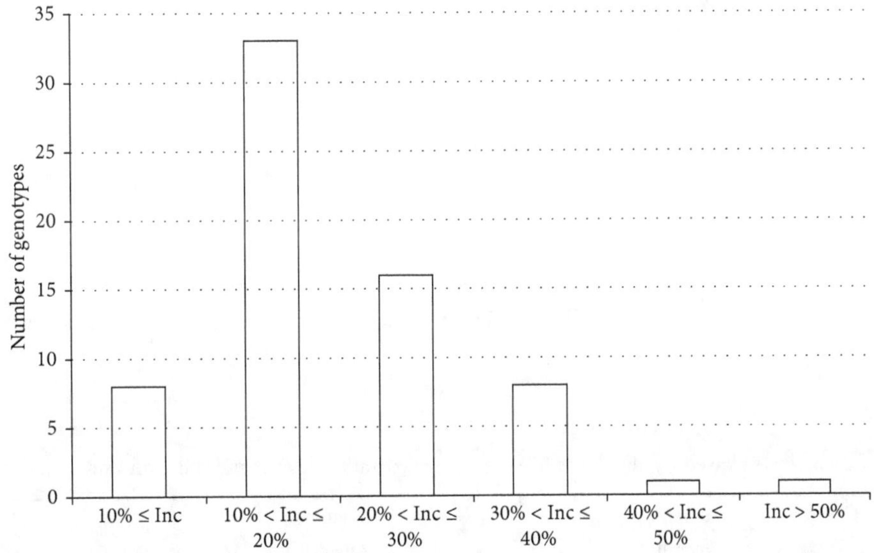

Figure 2: Number of peanut genotypes showing aflatoxin presence according to incidence intervals.

Indeed, the highest aflatoxin concentration level was observed with genotype 78-936. The contrasting genotypes observed in this study can be used as positive and negative checks, respectively, for accurate field experiment. Furthermore, these contrasted genotypes can be used to develop mapping population for genetic study such as inheritance of aflatoxin and identification of quantitative trait loci (QTL). The varieties 55-437 and 73-30 showed incidence less than 15% as reported by the previous study realized 30 years ago by Zambettakis et al. [13], but their aflatoxin concentration levels were largely up to the European Union standards.

The correlation test showed a positive relationship between *A. flavus* colonization and aflatoxin contamination. This confirmed that the presence of *A. flavus* induced aflatoxin production in seeds. Hierarchical classification highlighted three clusters according to incidence, severity, and aflatoxin concentration levels. The relatively low values of incidence observed on the 33 genotypes belonged to cluster 1 should be confirmed under field conditions. These genotypes can be evaluated in different locations on infested fields.

5. Conclusion

This study uncovered that the lines 12CS_104 exhibited low values of incidence and severity. Furthermore, its aflatoxin

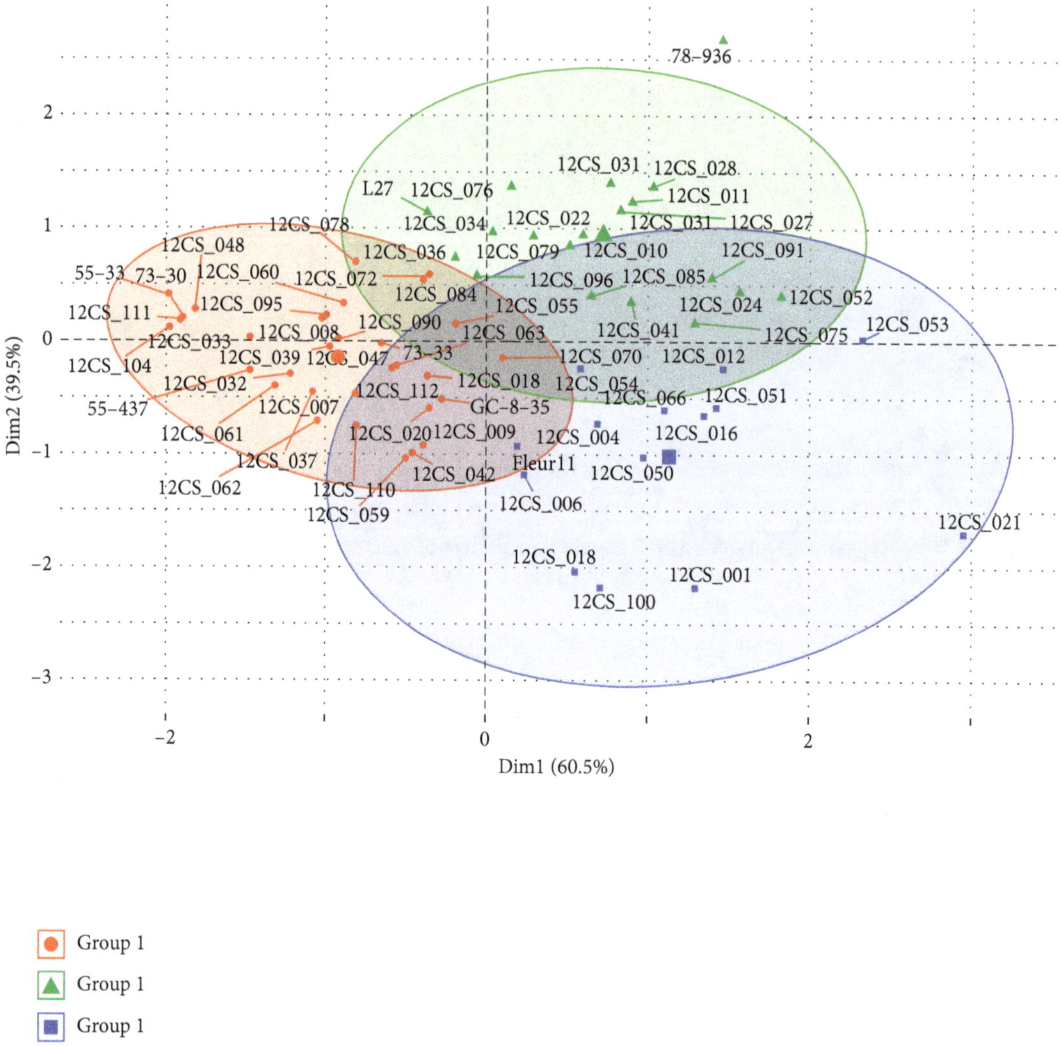

Figure 3: Principal component analysis biplot showing peanut genotypes clustering based on their severity and incidence.

Table 5: Description of each cluster based on incidence, aflatoxin concentration levels, and paragon and singular genotype per cluster.

	Cluster 1			Cluster 2			Cluster 3		
	Mean	Overall mean	p value[†]	Mean	Overall mean	p value	Mean	Overall mean	p value
Description of clusters by variables									
Incidence	15.09	20.35	$<10^{-3}$	18.8	20.35	>0.05	35.00	20.35	$<10^{-3}$
Aflatoxin	915.65	2143.89	$<10^{-3}$	4075.50	2143.89	$<10^{-3}$	2279.57	2143.89	>0.05
Description of clusters by distances									
Top 3 representative genotypes[††]	12CS_039 (0.11)	12CS_090 (0.16)	73-33 (0.29)	12CS_010 (0.12)	12CS_079 (0.22)	12CS_027 (0.24)	12CS_050 (0.15)	12CS_066 (0.41)	12CS_016 (0.42)
Top 3 characteristic genotypes[†††]	12CS_104 (2.82)	55-33 (2.75)	12CS_111 (2.74)	78-936 (3.68)	12CS_028 (2.39)	12CS_031 (2.28)	12CS_021 (3.45)	12CS_001 (3.00)	12CS_100 (2.59)

[†]Correlation signification of variables with a cluster. [††] Based on the closest distance between each genotype and the respective cluster centres. [†††] Based on the farthest distance from a genotype projected point in a cluster to the centres of the two others.

concentration level was smaller than standards. This genotype represents a relevant tool for the breeding program for resistance to *A. flavus* as a potentially resistant gene donor.

Acknowledgments

The authors sincerely thank the West Africa Agricultural Productivity Program (WAAPP) for financial support and facilities help.

References

[1] ANSD, *Bulletin Mensuel des Statistiques Économiques*, ANSD, Dakar, Sénégal, 2017.

[2] D. Clavel, A. Da Sylva, O. Ndoye, and A. Mayeux, "Amélioration de la qualité sanitaire de l'arachide au Sénégal: un challenge pour une opération de recherche dévelopement participative," *Cahiers Agricultures*, vol. 22, no. 13, pp. 174–81, 2013.

[3] P. M. Diedhiou, F. Ba, A. Kane, and N. Mbaye, "Effect of different cooking methods on aflatoxin fate in peanut products," *African Journal of Food Science and Technology*, vol. 3, no. 12, pp. 53–58, 2012.

[4] W. A. Korani, Y. Chu, C. Holbrook, J. Clevenger, and P. Osias Akins, "Genotypic regulation of aflatoxin accumulation but not Aspergillus fungal growth upon post-harvest infection of peanut (*Arachis hypogaea* L.) seeds," *Toxins*, vol. 9, no. 7, p. E218, 2017.

[5] J. C. Fountain, J. Koh, L. Yang et al., "Proteome analysis of *Aspergillus flavus* isoate-specific responses to oxidative stress in relationship to aflatoxin production capability," *Scientic Reports*, vol. 8, no. 1, p. 3430, 2018.

[6] P. M. Diedhiou, R. Bandyopadhyay, J. Atehnkeng, and P. S. Ojiambo, "Aspergillus colonization and aflatoxin contamination of maize and sesame kernels in two agro-ecoloical zones in Senegal," *Journal of Phytopathology*, vol. 159, no. 4, pp. 268–275, 2011.

[7] X. Q. Liang, M. Luo, and B. Z. Guo, "Resistance mechanisms to *Aspergillus flavus* infection and aflatoxin contamination in peanut (*Arachis hypogea* L.)," *Plant Pathology Journal*, vol. 5, no. 11, pp. 115–124, 2006.

[8] S. Amaike and N. P. Keller, "*Aspergillus flavus*," *Annual Review of Phytopathology*, vol. 49, no. 1, pp. 107–133, 2011.

[9] J. H. Williams, T. D. Phillips, P. Jolly, J. K. Styles, C. M. Jolly, and D. Aggarwal, "Human aflatoxicosis in developing countries: a review of toxicology, exposure, potential health consequences, and interventions," *American Journal of Clinical Nutrition*, vol. 80, no. 5, pp. 1106–1122, 2004.

[10] Y. C. Chen, C. D. Liao, H. Y. Lin, L. C. Chiueh, and D. Y. C. Shih, "Survey of aflatoxin contamination in peanut proucts in Taiwan from 1997 to 2011," *Journal of Food and Drug Analysis*, vol. 21, no. 3, pp. 247–252, 2013.

[11] R. Y. Kelley, W. P. Williams, J. E. Mylroie et al., "Identification of maize genes associated with host plant resistance or susceptibility to *Aspergillus flavus* infection and aflatoxin accumulation," *PLoS ONE*, vol. 7, no. 5, Article ID e36892, 2012.

[12] H. Q. Xue, T. G. Isleib, G. A. Payne, and G. OBrian, "Evaluation of post-harvest aflatoxin production in peanut germplasm with resistance to seed colonization and pre-harvest aflatoxin contamination," *Peanut Science*, vol. 31, no. 2, pp. 124–134, 2004.

[13] C. Zambettakis, F. Waliyar, A. Bockelee-Morvan, and O. de Pins, "Results of four years of research on resistance of groundnut varieties to *Aspergillus flavus*," *Oleagineux*, vol. 36, pp. 377–385, 1981.

[14] D. Fonceka, H. A. Tossim, R. Rivallan et al., "Construction of chromosome segment substitution lines in peanut (*Arachis hypogea* L.) using a wild synthetic and QTL mapping for plant morphology," *Plos ONE*, vol. 7, no. 11, Article ID e48642, 2012.

[15] V. K. Mehan and D. McDonald, *Screening for Resistance to Aspergillus Invasion and Aflatoxin Production in Groundnuts*, ICRISAT Groundnut Improvement Program Occasional Paper2, Patancheru, India, 1980.

[16] V. A. Tonapi, R. R. Mundada, S. S. Navi et al., "Effect of temperature and humidity regimes on grain mold sporulation and seed quality in sorghum (*Sorghum bicolor* (L.) Moench)," *Archives of Phytopathology and Plant Protection*, vol. 40, no. 2, pp. 113-127, 2007.

[17] R Core Team, *A Language and Environment for Statistical Computing*, R Foundation for Statistical Computing, Vienna, Austria, 2018.

[18] P. Grosjean and F. Ibanez, *PASTECS: Package for Analysis of Space Time Ecological Series, R package version 1.3-18*, 2014.

[19] S. Lê, J. Josse, and F. Husson, "FactoMineR: an R package for multivariate analysis," *Journal of Statistical Software*, vol. 25, no. 1, pp. 1–18, 2008.

[20] A. Kassambara and F. Mundt, *factoextra: Extract and visualize the results of multivariate data analyses. R package version 1.0.4*, 2017.

[21] R. A. Taber, R. E. Pettit, C. R. Benedict, J. W. Dieckert, and D. L. Kertrin, "Comparison of *Aspergillus flavus* tolerant and susceptible lines. I. Light microscope investigation," in *Proceedings of the American Peanut Research Education Association*, vol. 5, pp. 206-207, Oklahoma, USA, 1973.

[22] G. Y. Zhou and X. Q. Liang, "Studies on the ultramicroscopic structure of seed coats between resistant and susceptible to *Aspergillus flavus* invasion in peanut," *Chinese Journal of Oil Crop Sciences*, vol. 20, pp. 32–35, 1999.

[23] X. Q. Liang, G. Y. Zhou, and R. Z. Pan, "Study on the relationship of wax and cutin layers in peanut seeds and resistance to invasion and aflatoxin production by *Aspergillus flavus*," *Journal Of Tropical And Subtropical Botany*, vol. 11, pp. 11–14, 2003.

[24] D. L. Lindsey and R. B. Turner, "Inhibition of growth of *Aspergillus flavus* and Trichoderma viride by peanut embryos," *Mycopathologia*, vol. 55, no. 3, pp. 149–152, 1975.

[25] F. J. Amaya, C. T. Young, A. J. Norden, and A. C. Mixon, "Chemical screening for *Aspergillus flavus* resistance in peanut," *Oleagineux*, vol. 35, no. 5, pp. 255–259, 1980.

[26] X. Q. Liang, G. Y. Zhou, and R. Z. Pan, "Changes of some biochemical substances in peanut seeds under infection of *Aspergillus flavus* and their role in resistance to seed invasion," *Chinese Journal of Oil Crop Sciences*, vol. 23, no. 2, pp. 26–31, 2001.

6

Performance of Two Rape (*Brassica napus*) Cultivars under Different Fertilizer Management Levels in the Smallholder Sector of Zimbabwe

S. Ganya,[1] E. Svotwa,[2] and R. D. Katsaruware[3]

[1]*Ministry of Agriculture, Mashonaland West Province, Zimbabwe*
[2]*Senior Lecturer, Department of Crop Science and Postharvest Technology, Chinhoyi University of Technology, Chinhoyi, Zimbabwe*
[3]*Senior Lecturer, Department of Agriculture, Zimbabwe Open University, Mashonaland West Province, Zimbabwe*

Correspondence should be addressed to E. Svotwa; esvotwa2@gmail.com

Academic Editor: David Clay

Crop response to fertilizer application depends on the physiological and morphological characteristics of the cultivars, thus causing cultivar differences in growth rate and final yield. A study was carried out in the Zvimba smallholder farming area of Zimbabwe, from November 2016 to February 2017, to investigate the performance of two rape (*Brassica napus*) cultivars under different nitrogen fertilizer management levels. An experiment was set up in a randomized complete block design, replicated three times, with two rape varieties, Hobson and English Giant, and 5 ammonium nitrate (AN) (34.5% N) fertilizer application rates (200 kg/ha, 300 kg/ha, 350 kg/ha, and 400 kg/ha) as factors. Nitrogen fertilizer rates had a significant ($p < 0.05$) effect on leaf length at 6 WAP and 7 WAP. Nitrogen fertilizer application levels had significant effects on both in-season and total fresh leaf yield. There were no varietal effects on the leaf length ($p > 0.05$), in-season yield and total leaf yield ($p > 0.05$), and there were also no variety ∗ fertilizer application level interaction effects on leaf length, in-season yield and total leaf yield ($p > 0.05$). From the results of this study, English Giant rape would maintain a longer leaf length than Hobson up to the end of the season, but the two cultivars have similar yield response to fertilizers' application rate. Both the English Giant rape and Hobson rape cultivars can, therefore, be recommended for production in the Mhondoro smallholder farming area of Zimbabwe and other areas with similar climatic and soil characteristics.

1. Introduction

The current food production in Southern Africa and specifically in Zimbabwe is inadequate. Zimbabwe is currently importing much of its food, and most of the basic commodities and agriculture produce are imported and yet there are vast tracts of land in farms and small suitable pockets in the smallholder sector where food production can be practiced. There is need to find new efficient ways to meet these growing food needs without undermining our futures. With the population of 17 million, expected to double in 2050, the demand for food will more than double as a result of urbanization [1]. There are, however, opportunities in increasing agricultural output and profitability through promoting the use of improved and adaptable hybrids and through the application of site-specific rates of nutrient sources. Such gains are likely to catalyze economic growth and subsequent improvement in livelihoods, food security, and employment and thus, meet the Sustainable Development Goals which aim to end poverty and hunger, among other objectives.

Vegetable production in Zimbabwe is a fast expanding enterprise because of the increasing demand resulting from the rapid increasing urban populations [2]. Diversification in crop production by small holder farmers in Southern Africa has the capability of creating employment and increase in income. This, in turn, allows for the purchase of food to increase food security [3]. Vegetables provide nutrients such as vitamins, minerals, and roughage which are essential for a balanced diet [4]. *Brassica napus* and other leaf vegetables are a profitable agricultural enterprise in both local and export markets [2, 5].

Rape is one of the primary vegetable crops in Zimbabwe, grown for its leaves [4], which are rich in vitamin A, thiamine, and ascorbic acid. Rape seed oil is used for baking bread, illumination, lubrication, and the manufacture of soaps [6]. The vegetable is grown all year round in Zimbabwe and is used as a source of income through the selling of the leaf produce [7].

Rape (*Brassica napus*) has a high demand for water due to extensive leaf area and thus needs regular intervals of irrigation. The critical period for rape is the stem-elongation stage when the crop builds the branching structure and strong stems, then produce high yields [8]. Rape is sown for later transplanting at the age of 4–6 weeks after sowing, but the crop can also be grown in situ [6]. The crop has a short life span as compared to other leafy vegetables, making it the most preferred leaf vegetable grown by smallholder farmers in Zimbabwe [9].

According to Musara and Chitamba [10], there are many rape cultivars available in Zimbabwe but the most popular ones are English Giant rape and Hobson cultivars. The characteristics of the two cultivars were summarized by Muchecheti et al. [9]. The English Giant rape cultivar has dark green leaves and medium branches and has an immense growth rate. The English Giant rape has large broad leaves, with a yield potential of the cultivar of 25–40 tons/ha. The cultivar has large broad leaves and is normally preferred for its hardness. The cultivar cooks quicker than its competitor, Hobson. The English Giant rape germination takes 5–10 days and also takes 90 days to mature [11].

Another variety is Hobson rape, which is an improvement of the English Giant rape variety. This variety can be grown throughout the year and is reported to be a high yielding, with the potential of producing 25–40 tons/ha during its production period [12]. The Hobson rape cultivar has a rapid growth and responds well to regular cuttings [13]. The cultivar has pale green leaves and is a very palatable vegetable. It is an excellent variety with a wide sowing window.

Both Hobson and English Giant rape varieties are heavy feeders and respond well to applications of farm yard manure or mature compost, which should, however, be supplemented with either a compound or straight fertilizer [6, 14]. A recommended good practice is to collect soil samples from the land and send for analysis before planting in order to apply the exact amount of fertilizer and manure required by the crop [15]. The condition of the crop will dictate whether further top dressings are needed [11].

The production of leaf vegetables all year round requires land use intensification, and the enterprise is only viable and profitable when nutrients from the soils which are depleted during crop production are replenished [16, 17]. Continuous cropping in most soils in sub-Saharan Africa has led to the deficiency in nitrogen (N), phosphorus (P), or both [18]. The negative nutrient balances have been attributed by the removal of crop residues from the fields, coupled with lower rates of macronutrients applications compared to losses [19].

Rape grows on a wide range of soils, provided the inherent drainage and fertility problems associated with clays and sandy soils are alleviated [11, 20]. The improvement and maintenance of soil fertility is the major constraint to sustainable crop production in the communal smallholder farming areas of Zimbabwe. From a regional assessment carried out in the Chinamhora and Mhondoro communal areas of Zimbabwe, it was revealed that over 40% of the soils had phosphorus deficiency, while 82% had very low nitrogen [21]. In addition, communal area fields show wide ranges in the values of soil properties [21], and this coupled with continuous cropping has led to the deficiency in nitrogen (N), phosphorus (P), or both [18]. The implications are that sustainable crop yield can only be achieved through determination of site-specific variety and nutrient requirements instead of the generalized crop and fertilizer recommendations, an approach that is currently used in Zimbabwe.

The rate of fertilizer use by famers generally depends on various factors, including arable land size, household income, and distance to input purchasing point, fertilizer price, and transport cost [22, 23]. Findings by Murwira [24] were that smallholder communal farmers in Zimbabwe applied an average of 18 kg/ha of Compound D (8:14:7) fertilizer, which was below the recommended rate of 300 kg/ha. Such application of lower than the required amount of fertilizer by the smallholder farmers could be attributable to the blanket recommendations that made by research and extension services on fertilizer application rates [25]. The cost of fertilizer is prohibitively high, difficult to transport, and often in short supply, forcing many farmers to apply less than the recommended rates which eventually leads to low yields [26]. Smallholder farmers usually apply these nutrients, not at site-specific rates but using the general recommendations from agronomic guidelines [25, 27, 28]. Most countries in the region's mineral recommendation rates are geared towards large-scale commercial farmers and thus overlooked the spectrum of farming objectives and returns and investment that typifies smallholder farming systems [27]. Such imperfect fertilizer management has always resulted in the inconsistent and inappropriate application of fertilizer in agricultural production, with consequent environment risks [29].

The generally recommended fertilizer application rates for both English Giant rape and Hobson varieties 600 kg/ha Compound C fertilizer (6N $15P_2O_5$12 K_2O) applied as basal before or during transplanting. Ammonium nitrate (AN) is applied as a topdressing at the rate of 200 kg/ha (NTS, 2010). Since the crop is harvested continuously, topdressing is usually split with the first application at three weeks after transplanting and every three weeks, thereafter, as long as picking is continuous [30].

If appropriate fertilizer inputs are combined with suitable varieties, larger increases in crop yields could be achieved on a sustainable basis. It is hypothesized in this study that the two rape varieties, English Giant and Hobson, have similar growth and yield response to varying AN fertilizer application levels. The results of the experiment are useful in minimizing the cost of fertilizers through application of site-specific requirements, instead of using the general recommendation in the general rape agronomy guide.

2. Methods

2.1. Description of the Study Area. The experiment was carried out from November 2016 to February 2017 in a communal area, in Ward 4 of Chegutu District in the Mashonaland West Province of Zimbabwe, where the main economic activity is agriculture. Ward 4 of Chegutu District is located 95 km southwest of the city of Harare, about 60 km southeast of Chegutu, at an altitude of 11280 m [31].

The experimental site lies in the natural Agro-Ecological Region 2b of Zimbabwe, where the climate is characterized by an annual rainfall that ranges from 750 to 1000 mm in the months of November to March and a mean annual rainfall of 819 mm [31]. The study area has unreliable rainfall pattern which is not evenly distributed and hence is characterized by seasonal dry spells.

The area records minimum and maximum temperatures for the site area at 2°C and 32°C, respectively. The coldest and the hottest months for the area are July and October, respectively [32]. The main agricultural system in the research area is mainly subsistence farming [33]. Local people rely mainly on agriculture to make their living, cultivating subsistence crops chiefly maize, groundnuts, pulses, tobacco, and horticultural crops and keeping cattle. The soils are mainly granitic sandy soils. The predominant soils in the research area are granitic sand soils which are high in quartz and are light colored. The soils have a low inherent fertility. The soils have low available water-holding capacity and poor structure; this makes them to have high leaching losses as when compared to other soils [33].

2.2. Experimental Design. A two-factor experiment was set up in a randomized complete block design and was replicated 3 times. Factor one was rape variety (English Giant and Hobson varieties), and the second factor was ammonium nitrate (34.5% N) fertilizer application rate (200 kg/ha, 300 kg/ha, 350 kg/ha, and 400 kg/ha). Each plot consisted of three 1 by 10 m beds (30 m^2), and experimental assessments were done in the central bed (10 m^2). The experimental treatments were allocated randomly to the treatment plots.

The seedlings for the experiment were raised from the seed that was purchased from the National Tested Seeds, a company that sells seed for horticultural crops in Zimbabwe. Basal compound (NPK) and topdressing fertilizer as well as lime were applied in accordance with the recommendations made when the soil was sent for analysis (Figure 1). The basal fertilizer used in the experiment was Compound C (6% N; 15% P; 12% K). The Compound C fertilizer was applied during transplanting at the rate of 600 kg/ha. Ammonium nitrate (AN) (34.5% N) was used as a nitrogen source, and it was applied as a top dressing. Ammonium nitrate fertilizer was applied in two equal parts, with the first application done at 2 weeks after transplanting, while the second was done at 5 weeks after transplanting. The standard requirement was 200 kg AN/ha, as recommended by the result of soil analysis (Figure 1).

The experimental plots were hand weeded regularly. Insect pests were controlled to minimize economic losses. During the course of the experiment, the plots were watered 2 times a week at the rate of 30 mm, and this depended on the crop requirements and weather conditions.

2.3. Harvesting. Harvesting commenced on the first week after the second topdressing application. Two parameters, the longest leaf length and the fresh leaf mass, were measured at 5, 7, 9, and 11 weeks after transplanting (WAP). For each treatment, data were collected from 10 plants systematically selected from the central bed and tagged for repeated measurements. At every harvesting period, the longest leaf was measured, and the harvested leaves from the sampled 10 plants were weighed on a digital scale.

2.4. Data Analysis. The statistical package, GenStat Discovery 14th Edition, was used to analyze the data. All data were subjected to an analysis of variance (ANOVA), and treatment means were separated using the least significance difference (LSD) test at $p < 0.05$.

3. Results

There were significant ($p < 0.05$) AN fertilizer application rate effects on the longest leaf length at 5, 7, 9, and 11 weeks after transplanting (Table 1). All the tested AN application rates (300 kg/ha, 350 kg/ha, and 400 kg/ha) had a significant greater ($p < 0.05$) longest leaf length than the standard (200 kg/ha) at all the assessment times (5, 7, and 9 WAP). Generally, longest leaf dimensions increased with the increasing levels of AN fertilizer from 200 to 400 kg/ha. There were no significant ($p > 0.05$) longest leaf length differences between the 350 kg/ha and 400 kg/ha treated plants at 5, 7, and 11 WAP. The two rape varieties, Hobson and English Giant, had no significant ($p > 0.05$) effect on the longest leaf length differences at 7 and 9 WAP. The English Giant rape had a significantly higher ($p < 0.05$) longest leaf at 9 and 11 WAP. There were no significant variety × fertilizer application level interaction effects ($p > 0.05$) on the longest leaf length at all the assessment times (5, 7, 9, and 11 WAP).

The AN fertilizer application rate effects on the rape fresh leaf yield were significant ($p < 0.05$) at all the assessment times (5, 7, 9, and 11 WAP) (Table 2). All the tested AN application levels (300 kg/ha, 350 kg/ha, and 400 kg/ha) had a significantly greater ($p < 0.05$) in-season fresh yield than standard treatment (200 kg/ha) at all the assessment times (5, 7, 9, and 11 WAP). Generally, in-season fresh leaf yield increased with the increasing levels of fertilizer from 200 to 400 kg AN/ha. In-season fresh leaf yield for 300 kg and 350 kg/ha AN-treated plants was statistically similar at 5, 7, and 9 WAP, but lower than that from the 400 kg/ha AN-treated plants. The 200 kg/ha, 300 kg/ha, 350 kg/ha, and 400 kg/ha AN/ha treated

ALYTICAL CHEMISTRY SERVICES DIVISION
TEST REPORT NO: 165165

CLIENT ADDRESS: AGRITEX
BOX 130
MUBAIRA

Attention: GANYA SEBASTIAN

LIME & FERTILISER GUIDELINES

				SUGGESTED FERTILISERS					
LAB REF	YOUR REF	CROP	FERTILIZER	N	P_3O_3	K_3O	RATE (kg/ha)	APPLICATION TIME	LIME (kg/ha) SEE NOTES
165165	SAMPLE 1	RAPE	COMPOUND C	6	15	12	600	AT PLANTING	600
			AMMONIUM NITRATE	34.5	0	0	200	3 AND 6 WAP	

WAP········WEEKS AFTER PLANTING

Figure 1: Site-specific results of soil analysis.

Table 1: Effect of variety and AN fertilizer application rate on rape longest leaf lengths at 6, 7, 8, and 9 weeks after transplanting.

AN fertilizer application rate	5 WAP	7 WAP	9 WAP	11 WAP
0 kg/ha	37.49a	32.79a	36.4a	20.85a
200 kg/ha	47.16b	43.79b	46.4b	35.37b
300 kg/ha	50.99c	47.01c	49c	41.27c
350 kg/ha	50.44c	46.15bc	49.6c	42.28cd
400 kg/ha	52.52c	48.11c	51.6d	44.92d
p-fert	<0.001	<0.001	$p < 0.001$	<0.001
English Giant	48.59	49.63	44.2b	47.5b
Hobson	46.85	47.6	42.94a	45.8a
Pvar	0.07	0.195	0.011	0.003
pvar * fert	0.223	0.848	0.714	0.279
Lsd-fert	2.999	3.122	1.97	3.328
Lsd-var			1.25	2.105

Table 2: Effect of variety and AN fertilizer application rape fresh leaf yield (g/plant) at 6, 7, 8, and 9 weeks after planting.

AN fertilizer application rate	5 WAP	7 WAP	9 WAP	11 WAP
0 kg/ha	84.2a	84.2a	61.7a	38a
200 kg/ha	176.7b	176.7b	158.3b	194b
300 kg/ha	182.5b	182.5b	163.3b	150b
350 kg/ha	208.3b	208.3b	191.7b	172b
400 kg/ha	248.3c	248.3c	231.7c	205b
p-fert	<0.001	<0.001	<0.001	<0.001
English Giant	449.3	386.3	163.3	162
Hobson	433.7	381.7	159.3	142
Pvar	0.785	0.505	0.679	0.382
pvar * fert	0.833	0.991	0.991	0.497
Lsd-fert	31.04	32.52	31.64	72.8

plants had statistically similar in-season fresh leaf yield at 11 WAP.

There were no significant ($p > 0.05$) in-season fresh leaf mass differences between the two rape varieties, Hobson and English Giant, and there were also no significance ($p > 0.05$) on the variety x fertilizer level interaction effect on in-season fresh leaf yield all the assessment times (5, 7, 9, and 11 WAP).

There was a significance ($p < 0.05$) of AN fertilizer application rate effects on the total rape fresh leaf yield (Figure 2). All the AN application rates (200 kg/ha, 300 kg/ha, 350 kg/ha, and 400 kg/ha) had a significant greater ($p < 0.05$) total fresh yield than the negative control treatment (0 g Nm^{-2}). Total fresh leaf yield for 300 kg/ha and 350 kg/ha AN-treated plants had statistically similar fresh leaf yield with the standard (200 kg/ha) (Lsd = 2.643), while fresh leaf yield from the 400 kg/ha AN-treated plants was significantly greater than all the tested AN fertilizer rates (Lsd = 2.646). There were no significant ($p > 0.05$) total fresh leaf mass differences between the two rape varieties, Hobson and English Giant, and there were also no variety x fertilizer level interaction effects on total fresh leaf yield.

4. Discussion

Rape growth increased as levels of increased nitrogen concentration was increasing with the increase in the nitrogen rate from 0 to 400 kg/ha in spite of the time of application showing that there are additional yield benefits that could be realized if higher levels of N are applied to both varieties. According to Wang and Li [34], vegetable yields are increased continuously to a certain threshold rate, and any excess input of fertilizer more or less reduced plant growth, leading to yield decline harvests. As argued by Kant et al. [35], the increases in rape leaf yield with nitrogen availability showed that nitrogen favoured more vegetative growth producing plants with a higher leaf count and subsequently higher total green leaf area per plant. Low nitrogen fertilizer of ammonium nitrate treatments might have negatively affected the photosynthesis process leading to reduced growth rate and consequently yield [36], while high rate of nitrogen increasing leaf area development increases overall crop assimilation [37]. while high rate of nitrogen increasing leaf area development increases overall crop assimilation [38] According to Öztürk [38],

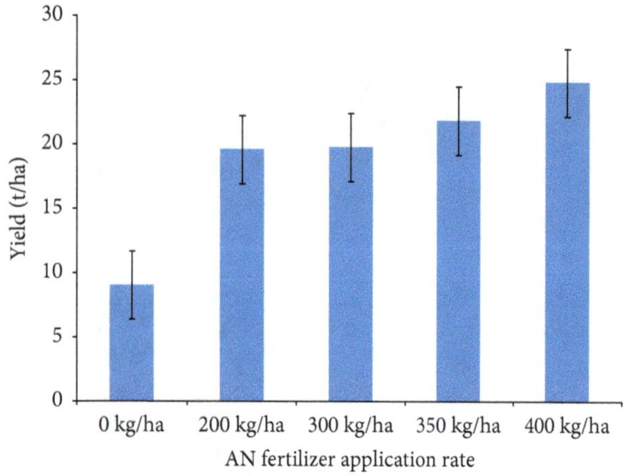

Figure 2: Average yield (t/ha) for the different fertilizer application levels.

ammonium nitrate at optimal rates of 100 to 150 kg/ha was more appropriate on rape than comparative rates using ammonium sulphate and urea, thus highlighting the practical importance of adequate fertilizer application and even suitable source in both seed and leaf yield. However, the accumulation of nitrates is one of the major disadvantages of high-level fertilizer application in rape [39]. However, NO_3–N concentration increased with supply under drought but was unaffected by supply under irrigation [40], making the practice of optimum water application a necessity.

The Giant rape and Hobson rape had similar leaf length response to AN application rate under the subtropical Zvimba District conditions. As argued by Bacha et al. [41], the genotype × environment × E interaction effects make it difficult to select the best performing as well as the most stable genotypes, and so, its efficient interpretation is an important issue in plant improvement. This similarity of the two rape cultivars to application rate could be an indication of similarity in the breeding history, may be with a common parentage.

The general decline trend in rape leaf yield in all treatments from the first to the last harvest may be attributed by changes in B napus developmental stages and declining in nutrient supply [4]. The general yield potential of both varieties is 24–40 tons/ha (National Tested Seeds, Zimbabwe). However, the average yield produced from the research was 19.3 tons/ha far below the expected yield probably to the excessiveness of 920 mm of rainfall and the above 22°C received during the nine weeks of the experiment. The excessive rains received could have caused the leaching of nitrogen fertilizers; as a result, this reduced the yield of rape [42].

5. Conclusion and Recommendations

5.1. Conclusion. The results showed that the highest vegetative growth rates, as measured by the longest leaf, as well as the highest fresh leaf yield were obtained from the 400 kg/ha AN-treated plots. This drives home the point that the higher growth rate and yield are obtained when nitrogen fertilizer is applied in large quantities. Also, the results showed no cultivar effect on growth and yield of rape vegetable. However, English Giant rape growth rate started on higher threshold levels and then subsequently declined as the harvest progressed, whereas Hobson rape showed the opposite results, it started at low levels and then increased as the harvesting progressed.

5.2. Recommendations. It is recommended that farmers should be encouraged to grow either of the two varieties as they showed that they have no effect on the growth rate and yield of rape. Both varieties indicated that they perform very well under high nitrogen levels. It also recommended that further research should be done to ascertain the hazards of nitrates in vegetables to human health and their effect to the environment. Also, another research to assess the chemical content and the taste of rape vegetables after application of higher nitrogen fertilizer levels needs to be investigated.

References

[1] IndexMundi, "Zimbabwe Population growth rate," 2018, https://www.indexmundi.com/Zimbabwe.

[2] E. Kuntashula, P. L. Mafongoya, G. Sileshi, and S. Lungu, "Potential of biomass transfer technologies in sustainable vegetable production in the wetlands (DAMBOS) of eastern Zambia," *Experimental Agriculture*, vol. 40, no. 1, pp. 37–51, 2004.

[3] P. Drechsel and D. Kunze, *Waste Composting for Urban and Peri-Urban Agriculture, Closing the Rural-Urban Nutrient Cycle in Sub-Saharan Africa*, International Water Management Institute, Food and Agriculture Organization and CABI Publishing, Wallingford, UK, 2001.

[4] I. W. Nyakudya, L. Jimu, M. Marashe, and C. A. T. Katsvanga, "Comparative growth and yield responses of rape (*Brassica napus L.*) to different soil fertility management amendments," *Electronic Journal of Environmental, Agricultural and Food Chemistry*, vol. 9, no. 1, pp. 207–214, 2010.

[5] B. Loehr, A. A. Seif, and B. Nyambo, "Vegetable IPM in Africa: current status and future prospects with emphasis on eastern and southern Africa," *Acta Horticulturae*, no. 513, pp. 99–104, 1998.

[6] G. W. Ware and J. P. McCollum, *Producing Vegetable Crops*, The Interstate Printers and Publishers, Danville, IL, USA, 1975.

[7] H. C. Thompson and W. C. Kelly, *Vegetable Crops*, McGraw-Hill Book Co., New York, NY, USA, 4th edition, 1957.

[8] G. U. Xiao-bo, L. I. Yuan-nong, D. U. Ya-dan, and Y. I. N. Min-hua, "Ridge-furrow rainwater harvesting with supplemental irrigation to improve seed yield and water use efficiency of winter oilseed rape (*Brassica napus L.*)," *Journal of Integrative Agriculture*, vol. 16, no. 5, pp. 1162–1172, 2016.

[9] F. Muchecheti, I. C. Madakadze, and P. Soundy, "Production of rape (*Brassica napus L.*) on soils amended with leguminous tree prunings: yield responses in relation to the chemical composition of the tree prunings," *African Journal of Agricultural Research*, vol. 7, no. 24, pp. 3541–3549, 2012.

[10] C. Musara and J. Chitamba, "Growth rate and yield of *Brassica napus* in response to *Acacia angustissima* leaf biomass application," *Journal of Animal & Plant Sciences*, vol. 25, no. 2, pp. 510–518, 2015.

[11] J. M. Tweedie and A. J. Mullins, *Kale-Rape, Agric Handbook*, Department of conservation and extension, Harare, Rhodesia, 1962.

[12] C. H. Oldham, *Brassica Crops and Allied Cruciferous Crops*, Biotech Books, Delhi, India, 1999.

[13] ICRAF: 2006, World Agroforestry Centre, http://www.worldagroforestrycentre.org/sea.

[14] S. P. Baliyan, P. S. Baliyan, and K. S. Madhava Rao, "Response of different combinations of manure and fertilizers to the yield of rape crop," *International Journal of Agricultural Research, Innovation and Technology*, vol. 2, no. 2, pp. 44–47, 2012.

[15] Washington University Extension (WSUE), *A Home Gardener's Guide to Soils and Fertilizers*, Washington State University Extension EM063E, Pullman, WC, USA, 2014.

[16] P. M. Kormawa, A. Y. Kamara, S. C. Jutzi, and N. Sanginga, "Economic evaluation of using mulch from multi-purpose trees in maize-based production systems in Southern-Western, Nigeria," *Experimental Agriculture*, vol. 35, no. 1, pp. 101–109, 1999.

[17] T. Losak, J. Hluseki, S. Kracmar, and L. Varga, "The effects of nitrogen and sulphur fertilization on yield and quality of Kohlrabi (*Brassica olarracea*)," *Revista Brasileira de Ciência do Solo*, vol. 32, no. 2, pp. 697–703, 2008.

[18] P. L. Mafongoya, K. E. Giller, and C. A. Palm, "Decomposition and nitrogen release patterns of tree prunings and litter," *Agroforestry Systems*, vol. 38, pp. 77–97, 1998.

[19] W. Bedada, *Compost and fertilizer—alternatives or complementary? Management feasibility and long-term effects on soil fertility in an ethiopian village. Faculty of natural resources and agricultural sciences department of soil and environment*, Ph.D. thesis, Swedish University of Agricultural Sciences, Uppsala, Sweden, 2015.

[20] A. Shumba, "Integrated soil fertility, residue and tillage management for improved soil quality and maize (Zea mays, L.) production in urban agriculture in Harare and Chitungwiza," *Master of Philosophy Degree in Agriculture (MPhil) Thesis, Soil and Agricultural Engineering, Department, Faculty of Agriculture*, University of Zimbabwe, Zimbabwe, 2018.

[21] J. Nyamangara, L. M. Mugwira, and S. E. Mpofu, "Soil fertility status in the communal areas of Zimbabwe in relation to sustainable crop production," *Journal of Sustainable Agriculture*, vol. 16, no. 2, pp. 15–29, 2000.

[22] E. Bhondayi, "An investigation of the determinants of fertilizer use by communal farmers in drought prone areas of Zimbabwe, a study of Buhera District," Research Project Report, Department of Agricultural Economics and Extension, University of Zimbabwe, Harare, Zimbabwe, 2004.

[23] F. Zinyama, "A comparative analysis of the use of pit-stored manure on vegetables and field crops: the case of Mangwende Communal Area in Murehwa District," Research Project Report, Department of Agricultural Economics and Extension, University of Zimbabwe, Harare, Zimbabwe, 2003.

[24] H. K. Murwira, "Ammonia losses from Zimbabwean cattle manure before and after incorporation into soil," *Tropical Agriculture*, vol. 72, pp. 269–276, 1995.

[25] J. Dimes, L. Muza, G. Malunga, and S. Snapp, "Trade- offs between investments in nitrogen and weeding. On farm experimentation and simulation analysis in Malawi and Zimbabwe," in *Integrated approaches to higher maize productivity in the New Millennium; Proceedings of the Seventh Eastern and Southern Africa Regional Maize Conference, (February 2001)*, D. K. Friesen and A. F. E. Palmer, Eds., pp. 452–456, CIMMTY (International Maize and Wheat Improvement Center) and KARI (Kenya Agricultural Research Institute), Nairobi, Kenya, 2004.

[26] O. Svubure, *Fertilizer Management for Optimum Maize Yields in Sandy Soils of Zimbabwe*, Agritex, Harare, Zimbabwe, 2000.

[27] D. Hikwa and L. Mukurumbira, "Highlights of previous, current, and proposed soil fertility research by the Department of Research and Specialist Services (DRSS) in Zimbabwe," in *Report on the First Meeting of the network Working Group. Soil Fertility Research Network for Maize Based Farming Systems in selected Countries of Southern Africa. Rockefeller Foundation Southern Africa Agricultural Science Program and the International Maize and Wheat Improvement Centre Maize Program, Lilongwe and Harare*, S. R. Waddington, Ed., pp. 1-2, 1995.

[28] S. F. M. Kazombo-Phiri, "Main cause of declining soil productivity in Malawi: effects on agriculture and management technologies employment to alleviate the problem," in *Sustainable Agriculture Systems for the Drylands, Proceedings of the International Symposium for Sustainable Dryland Agriculture systems, (December 2003)*, O. Go and D. Pastemak, Eds., p. 336, Niamey, Niger, 2005.

[29] X. T. Ju, C. L. Kou, F. S. Zhang, and P. Christie, "Nitrogen balance and groundwater nitrate contamination: comparison among three intensive cropping systems on the North China Plain," *Environmental Pollution*, vol. 143, no. 1, pp. 117–125, 2006.

[30] Department of Meteorological Services, *Climate Handbook of Zimbabwe*, Department of Meteorological Services, Harare, Zimbabwe, 1981.

[31] A. Turner and O. Chivinge, *Production and Marketing Of Horticultural Crops in Zimbabwe: A survey of Smallholder Farmers in the Mashonaland East province*, CIIFAD, NY, USA, 1999.

[32] Surveyor General, *The Climate of Zimbabwe*, Government of Zimbabwe Printers, Harare, Zimbabwe, 1995.

[33] AGRITEX, *Natural Regions and Farming Areas, Agritex planning Branch*, Government Printers, Harare Zimbabwe, 1983.

[34] Z. Wang and S. Li, "Effects of nitrogen and phosphorus fertilization on plant growth and nitrate accumulation in vegetables," *Journal of Plant Nutrition*, vol. 27, no. 3, pp. 539–556, 2007.

[35] S. Kant, Y. M. Bi, and S. J. Rothstein, "Understanding plant response to nitrogen limitation for the improvement of crop nitrogen use efficiency," *Journal of Experimental Botany*, vol. 62, no. 4, pp. 1499–1509, 2011.

[36] R. Fox, *The sewage solution. Practical Hydroponics and Greenhouse Magazine*, Casper Publications, 2008, http://www.hydroponics.com.au/back-issues/issue20.html.

[37] J. K. Yayock, G. Lombin, and J. J. Owonubi, *Crop Science and Production in Warm Climates*, Macmillan Publishers Ltd., London, UK, 1988.

[38] Ö Öztürk, "Effects of source and rate of nitrogen fertilizer on yield, yield components and quality of winter rapeseed (*Brassica napus L.*)," *Chilean Journal of Agricultural Research*, vol. 70, no. 1, pp. 132–141, 2010.

[39] E. Chakwizira, P. Johnstone, A. L. Fletcher, E. D. Meenken, J. M. de Ruiter, and H. E. Brown, "Effects of nitrogen rate on

nitrate–nitrogen accumulation in forage kale and rape crops," *Grass and Forage Science*, vol. 70, no. 2, pp. 268–282, 2015.

[40] C. Fallovo, M. Schreiner, D. Schwarz, G. Colla, and A. Krumbein, "Phytochemical changes induced by different nitrogen supply forms and radiation levels in two leafy *Brassica* species," *Journal of Agricultural and Food Chemistry*, vol. 59, no. 8, pp. 4198–4207, 2011.

[41] T. Bacha, S. Alemerew, and Z. Tadesse, "Genotype X environment interaction and yield stability of bread wheat (*Triticum aestivum L.*) genotype in Ethiopia using the ammi analysis," *Journal of Biology, Agriculture and Healthcare*, vol. 5, no. 11, 2015.

[42] O. Balodis and Z. Gaile, "Influence of agro ecological factors on winter oil-seed rape (*Brassica napus L.*) autumn growth," in *Research for Rural Development*, Z. Gaile, K. Špoģis, A. Kaķītis et al., Eds., pp. 36–44, International Scientific Conference Proceedings, Jelgava Latvia, 2009.

[43] P. Brady, *Soils and their Improvement*, Agrobios, Jodhpur, Rajasthan, 1990.

[44] H. A. Elwell and M. A. Stocking, "Loss of soil nutrients by sheet erosion in a major hidden farming cost," *Zimbabwe Science News*, vol. 22, no. 7-8, pp. 79–82, 1988.

[45] Food and Agriculture Organization (FAO), *Fertiliser Use by Crops*, Food and Agriculture Organization of the United Nations, Rome, Italy, 2006.

[46] K. E. Giller, M. H. Beare, P. Lavelle, A. M. N. Izac, and M. J. Swift, "Agricultural intensification, soil biodiversity and agroecosystem function," *Applied Soil Ecology*, vol. 6, no. 1, pp. 3–16, 1997.

[47] P. L. Mafongoya, A. Bationo, J. Kilharaj, and B. S. Waswa, "Appropriate technologies to replenish soil fertility in Southern Africa," *Nutrient Cycling in Agroecosystems*, vol. 76, no. 2-3, pp. 137–151, 2006.

[48] K. Webster and R. R. Wilson, *Tropical Agriculture*, CAB International, Tucson, AZ, USA, 1998.

7

Evaluation of Quizalofop-Resistant Rice for Arkansas Rice Production Systems

Zachary D. Lancaster,[1] Jason K. Norsworthy,[1] and Robert C. Scott[2]

[1]*Department of Crop, Soil, and Environmental Sciences, University of Arkansas, Fayetteville, AR 72701, USA*
[2]*Department of Crop, Soil, and Environmental Sciences, University of Arkansas, Lonoke, AR 72086, USA*

Correspondence should be addressed to Zachary D. Lancaster; zdlancas@email.uark.edu

Academic Editor: Rodomiro Ortiz

Due to the ongoing evolution of herbicide-resistant weeds, new technologies are needed to maintain effective levels of control. A new rice variety that will be resistant to quizalofop, an acetyl coenzyme A carboxylase- (ACCase-) inhibiting herbicide, is currently under development. With the anticipated launch of this technology in 2018, multiple experiments were conducted to determine effectiveness of the quizalofop-resistant rice system for common grass weed species found in Arkansas rice production. One hundred and twenty-six barnyardgrass populations were collected across Arkansas and treated with quizalofop at 80 g ai ha^{-1} to determine a baseline of response. All populations evaluated were effectively controlled (≥92%) by quizalofop, with only 13 populations resulting in lower than 98% control. A greenhouse and field trial were conducted to compare efficacy of quizalofop to currently labeled rice graminicides for control of common rice grass weeds. Results from the greenhouse experiment showed that quizalofop treatments resulted in greater efficacy of common grass weeds compared to cyhalofop or fenoxaprop. This was especially apparent at the larger grass growth stages. A field experiment conducted compared season-long weed control programs of quizalofop to fenoxaprop and cyhalofop. The quizalofop-containing treatments were no better than fenoxaprop and cyhalofop for barnyardgrass and broadleaf signalgrass control. Barnyardgrass and broadleaf signalgrass control were greater than 96% for all herbicide treatments. An additional field experiment was conducted to determine the best rate structure for sequential applications of quizalofop in rice. Sequential applications of quizalofop at 120 g ha^{-1} followed by 120 g ha^{-1} two weeks later resulted in the highest barnyardgrass and broadleaf signalgrass control. Likewise, applying the full seasonal use rate of 240 g ha^{-1} of quizalofop resulted in greater control compared to 200 and 160 g ha^{-1}. Results from this research indicate a strong benefit from quizalofop use in rice.

1. Introduction

Arkansas is the top rice producing state, contributing over half of the United States rice production [1]. Rice production in Arkansas is generally located in the eastern half of the state, in the Mississippi River Delta region [2]. Weed control is a major obstacle for Arkansas rice production. Most of the rice grown is produced in a drill seeded, delayed flooding system, with only around 5% annually produced using a water-seeded system [3]. Hence, an effective weed control program in Arkansas begins with a preemergence residual herbicide followed by postemergence herbicide applications [4]. One of the main challenges to rice production is the ever increasing herbicide resistance found in multiple common rice weeds.

Two of the most problematic weeds to Arkansas rice production are barnyardgrass and red rice (*Oryza sativa* L.). While already difficult to control in rice, both of these species have evolved resistance to commonly used rice herbicides, making effective control even more difficult. Barnyardgrass is a principle rice weed globally, along with other closely related *Echinochloa* species, and thrives in the flooded rice production system [5]. Barnyardgrass has evolved resistance to many common rice herbicides, including propanil [6], quinclorac [7], and clomazone [8]. Red rice has long been difficult to control due to physiological similarities between it and commercial rice varieties [9]. Thus, to selectively control red rice as well as other common rice weeds, imidazolinone-resistant (Clearfield™) rice was commercially released in 2002 [10].

At its height of acceptance (2011), 64% of planted rice in Arkansas and Mississippi was an imidazolinone-resistant variety [4]; however, the share of imidazolinone-resistant rice in Arkansas has declined in recent years to 44% in 2015 [2]. The reduction in usage can partially be attributed to imidazolinone resistance in red rice [10] and barnyardgrass [11].

With the increased pressure herbicide-resistant weeds place on current rice production systems, a new technology is needed to achieve effective control of these weeds. The development of new herbicides quickly diminished with the launch of glyphosate-resistant crops in the 1990s. Although glyphosate-resistant weeds pushed the agrichemical industry to reinvest in herbicide discovery [12], no new herbicide mechanisms of action have been commercialized in recent years, leaving growers to work with a suite of herbicides that are less effective today because of widespread resistance [13]. To help combat herbicide-resistant rice weeds, a new herbicide-resistant rice technology (Provisia™ rice) is being developed. Provisia rice is resistant to quizalofop, an acetyl coenzyme A carboxylase- (ACCase-) inhibiting herbicide [14]. Quizalofop-resistant rice is a non-GMO crop and was developed using traditional plant breeding techniques to isolate the G2096S gene, which makes the acetyl coenzyme A carboxylase enzyme resistant to ACCase-inhibiting herbicides [15].

Quizalofop is a member of the aryloxyphenoxy propionate family and commonly used for effective control of annual weedy grasses and most perennial grass weeds [16]. Quizalofop, like other ACCase-inhibiting herbicides, only has activity on grass species, with broadleaf species having natural tolerance [17]. Quizalofop is currently labeled for use in multiple broadleaf crop and noncrop areas, where 35 to 84 g ai ha^{-1} can be applied postemergence in soybean (*Glycine max* (L.) Merr.) and up to 112 g ai ha^{-1} in noncrop areas [16]. The use rate for quizalofop in quizalofop-resistant rice will range from 100 to 138 g ai ha^{-1} for a single application and 240 g ai ha^{-1} as a maximum yearly application [18]. Although quizalofop can provide moderate residual grass control [16], quizalofop will be restricted to only postemergence applications in quizalofop-resistant rice [19].

Herbicide resistance modeling has been used to predict that ACCase-inhibiting rice herbicides such as cyhalofop and fenoxaprop have a lower risk for resistance when compared to acetolactate synthase- (ALS-) inhibiting herbicides, such as those used in imidazolinone-resistant rice [20]. These findings support the hypothesis that quizalofop could be a successful selective herbicide in quizalofop-resistant rice if properly integrated with strategies to mitigate resistance. However, there are cases of grass weed species that have already evolved resistance to ACCase-inhibiting herbicides, including barnyardgrass in the Arkansas [21]. These instances of resistance can be attributed to both metabolic (non-target-site) and target-site resistance [22]. While no resistance to quizalofop has been confirmed in Arkansas, common rice weeds such as barnyardgrass (Mississippi), Amazon sprangletop (*Leptochloa panicoides* (J. Presl) A. S. Hitchc., Louisiana), and junglerice (*Echinochloa colona* (L.) Link, Arkansas) have been confirmed resistant to ACCase-inhibiting herbicides [21, 23]. Likewise, gene flow between a quizalofop-resistant rice variety and red rice could transfer herbicide resistance to red rice [10]. Although outcrossing percentage is low (0.109–0.434%), this could result in several hundred resistant plants per hectare. Hence, proper stewardship of this technology is imperative for prolonged effectiveness.

Stewardship of this technology can be accomplished through yearly crop rotation. Crop rotation restriction will be unlikely with applications of quizalofop; however, quizalofop-resistant rice cannot be planted after imidazolinone-resistant rice due to a lack of stacking of resistance traits to confer resistance to the imidazolinone herbicides [14]. With the commercialization of quizalofop-resistant rice in 2018, multiple experiments were conducted to determine the baseline response of Arkansas barnyardgrass populations to quizalofop as well as the efficacy of the quizalofop-resistant rice system compared to current grass weed control standards used in Arkansas rice production.

2. Materials and Methods

2.1. Barnyardgrass Accession Screening to Quizalofop. Barnyardgrass panicles were collected from 126 agricultural fields across the Mississippi delta region of Arkansas in the autumn of 2014. Populations were designated as B (barnyardgrass) and given a number value (1 to 126). Samples B1–B74 were personally collected with samples B75–B126 being sent in by University of Arkansas county extension agents. The number of panicles collected per accession was dependent on barnyardgrass density within a field. On average, 30 to 40 panicles were collected per accession. GPS coordinate and crop in the field at time of sampling were recorded. Populations were dried in the greenhouse (32/22 C) for 7 days, and then seeds were threshed from panicles and combined into single composite samples for each accession.

Approximately 50 seeds were sown into 8 by 14 by 5 cm pots containing a commercial potting mix (Professional Growing Mix, LC1 mix, Sun Gro Horticulture Distribution Inc., Bellevue, WA 98008). Pots were then placed in the greenhouse under conditions of 32/22 C day/night temperatures with a 16-h photoperiod. The experiment was conducted as a randomized complete block design with four replications. Quizalofop (Targa™ herbicide, Gowan Company, Yuma, AZ) was applied at the 3- to 4-leaf growth stage at 80 g ai ha^{-1} with 1% v/v crop oil concentrate (COC) (Agri-Dex, Helena Chemical Company, West Helena, AR 72390). Applications were made inside a stationary spray chamber calibrated to deliver 187 L ha^{-1} at 276 kPA with 800067 nozzles.

Visual barnyardgrass control assessments were taken at 14 and 21 days after treatment (DAT). Control was assessed on a scale of 0 to 100%, where 0% was equivalent to no response and 100% was complete plant death compared to a nontreated check of each accession. Total emerged plants were counted for each tray. Mortality (%) was calculated at 21 DAT. Alive plants were any plant with living tissue remaining after treatment. Due to collection method and greenhouse space constraints, only 45 populations were evaluated within one run. Within a run, 4 replications were included with nontreated checks for each accession. No formal analyses were conducted on individual runs to compare populations.

2.2. Efficacy of Quizalofop Compared to Currently Registered Rice Graminicides. A greenhouse and field experiment were conducted to compare the efficacy of quizalofop to currently registered rice graminicides. The greenhouse experiment was conducted in the autumn of 2014 and spring of 2015 at the University of Arkansas Research and Extension Center in Fayetteville, AR, to determine the effect of growth stage at application and choice of ACCase-inhibiting herbicide on control of common grass weeds found in Arkansas rice production systems. The experiment was conducted as a two-factor factorial, randomized complete block design (RCBD), with factor-A being growth stage of grass species at time of application and factor-B being ACCase-inhibiting herbicide treatment with four replications. Growth stages at application were 2- to 3-leaf, 5- to 6-leaf, and 12- to 16-leaf grasses. Tested herbicides were quizalofop at 80, 120, and 160 g ai ha^{-1}, fenoxaprop (Ricestar® HT herbicide, Bayer Crop Science LP, Research Triangle Park, NC) at 122 g ai ha^{-1}, and cyhalofop (Clincher® SF herbicide, Dow AgroSciences LLC, Indianapolis, IN) at 313 g ai ha^{-1}. A COC at 1% v/v was added to quizalofop and cyhalofop treatments. Tested weed species were barnyardgrass, broadleaf signalgrass, fall panicum (*Panicum dichotomiflorum* L.), and Amazon sprangletop. Approximately 20 seeds per 8 by 14 by 5 cm tray were sown into a commercial potting mix and watered daily under greenhouse conditions of 32/22 C with a 16-h photoperiod. After emergence, plants were thinned to 5 plants $tray^{-1}$. Herbicide applications were similar to the previous experiment.

Visual assessment of control and plant biomass were collected similarly to the previous experiment. For data that met the assumptions for ANOVA, means were separated using Fisher's protected LSD ($\alpha = 0.05$). For data that did not meet the assumptions for ANOVA, only treatment means are presented. Few differences were observed between 14 and 21 DAT control ratings; thus only the 21 DAT rating will be presented.

A field experiment was conducted in 2014 and 2015 to determine the efficacy of quizalofop compared to other ACCase-inhibiting rice herbicides with and without clomazone preemergence in quizalofop-resistant rice. The experiment was located at the Pine Tree Research Station near Colt, AR, on a Calloway silt-loam soil (Fine-Silty, mixed, active, thermic Aquic Fraglossudalfs). An experimental quizalofop-resistant rice variety (Provisia rice, BASF Corp., Research Triangle Park, NC) was planted on May 2, 2014, and April 30, 2015, at a seeding rate of 67 seeds m^{-1} row. Plots consisted of 9 drill seeded rows on 18 cm centers, 7.6 m long.

The experimental design was a RCB factorial with four replications and three factors: presence or absence of clomazone preemergence, sequential application of quizalofop versus fenoxaprop followed by cyhalofop, and timing of the sequential herbicide application. Plots either had clomazone (Command™ herbicide, FMC Corporation, Philadelphia, PA) applied at 336 g ai ha^{-1} preemergence or no preemergence herbicide. Herbicide regimes consisted of either sequential applications of quizalofop at 120 g ha^{-1} each or a sequential application of fenoxaprop at 122 g ha^{-1} followed by cyhalofop at 313 g ha^{-1}. The initial postemergence application was always made at the 3- to 4-leaf stage of rice, whereas the sequential application was either made before flood or 2 weeks after flood. Herbicide treatments were applied using a CO_2-backpack sprayer calibrated to deliver 143 L ha^{-1} at 276 kPA with XR110015 nozzles. All quizalofop and cyhalofop treatments included a COC at 1% v/v. Broadleaf weeds and sedges were controlled by overspraying the entire test with 2,4-D at 533 g ae ha^{-1} (Weedar™ herbicide, Nufarm Americas INC, Alsip, IL) and halosulfuron at 21 g ai ha^{-1} (Permit® herbicide, Gowan Company LLC, Yuma, AZ).

Visual weed control and crop injury were assessed as previously noted. Ratings for control of a natural population of barnyardgrass and broadleaf signalgrass were taken 14 and 21 days after each herbicide application in both 2014 and 2015. Stand counts of emerged rice seedlings per meter row were taken at 14 days after planting. The experiment was terminated before the rice reached panicle exertion because a nonregistered rice variety was planted. Before termination weed biomass of barnyardgrass and broadleaf signalgrass per m^2 was taken, biomass samples were weighed after being oven-dried at 65 C for two weeks. Data for this experiment were analyzed using the Fit Model procedure in JMP Pro 12.1. Year and replication nested within years were considered random effects. No parameters for this experiment resulted in significant interactions or main effects; hence, only treatment means are presented.

2.3. Best Rate Structure for Sequential Applications of Quizalofop in Quizalofop-Resistant Rice. An experiment was conducted to determine the best rate structure of sequential applications of quizalofop to quizalofop-resistant rice, when applied initially to either 2- or 6-leaf grasses. The field experiment was conducted in 2014 and 2015 on a Dewitt silt-loam soil (Fine, smectitic, thermic Typic Albaqualfs) at the Rice Research and Extension Center near Stuttgart, AR. The experiment was conducted as a two-factor RCBD with factor-A being growth stage of grass weeds at time of first application and factor-B being quizalofop rate structure with four replications. An experimental quizalofop-resistant rice variety was planted on 18-cm width rows, in 1.8 by 6.1 m plots, at a rate of 67 seeds m^{-1} row. Planting occurred on April 26, 2014, and April 21, 2015. Quizalofop applications were applied to a natural population of grasses, with no preemergence herbicide applied to insure a high weed density. Quizalofop was applied as previously noted with broadleaf weeds and sedges controlled in a similar manner to the previous experiment. The quizalofop rate structure was 80, 120, or 160 g ha^{-1} followed by a sequential application of 80, 120, or 160 g ha^{-1}. All rate combinations evaluated did not exceed the maximum yearly application rate of 240 g ha^{-1} to be applied over both combinations. For instance, quizalofop at 160 g ha^{-1} followed by 120 g ha^{-1} was not evaluated because this treatment would have exceeded the allowable yearly maximum [18].

Visual estimates of weed control and crop injury were rated. Barnyardgrass and broadleaf signalgrass control were

TABLE 1: Effect of ACCase-inhibiting herbicide and growth stage at time of application for control of barnyardgrass, broadleaf signalgrass, fall panicum, and Amazon sprangletop at 21 DAT.[a]

Growth stage	Herbicide	Rate	Control[b]			
			ECHCG	BRAPP	PANDI	LEFPA
		g ai ha^{-1}	%			
2- to 3-leaf	Quizalofop	80	100 a	99 a	100 a	100 a
	Quizalofop	120	100 a	100 a	100 a	100 a
	Quizalofop	160	100 a	100 a	100 a	100 a
	Fenoxaprop	122	96 ab	99 a	100 a	99 a
	Cyhalofop	313	96 ab	98 a	100 a	98 ab
5- to 6-leaf	Quizalofop	80	91 bc	100 a	99 a	96 b
	Quizalofop	120	97 a	100 a	100 a	100 a
	Quizalofop	160	99 a	100 a	100 a	100 a
	Fenoxaprop	122	89 c	85 b	76 c	85 c
	Cyhalofop	313	48 d	56 d	39 d	44 g
12- to 16-leaf	Quizalofop	80	36 e	54 d	92 b	70 e
	Quizalofop	120	40 e	64 c	90 b	73 e
	Quizalofop	160	53 d	86 b	95 ab	80 d
	Fenoxaprop	122	27 f	33 f	75 c	61 f
	Cyhalofop	313	19 g	47 e	29 e	38 h

[a]DAT: days after treatment, ECHCG: barnyardgrass, BRAPP: broadleaf signalgrass, PANDI: fall panicum, LEFPA: Amazon sprangletop. [b]Means within a column followed by the same letter are not different according to Fisher's LSD ($\alpha = 0.05$).

evaluated in 2014 and 2015 along with red rice in 2015. Rice plant heights were taken multiple times throughout the experiment for both 2014 and 2015. No grain yield were collected due to crop termination before crop maturity. Data for this experiment were analyzed using the Fit Model procedure in JMP Pro 12.1. For data that met assumptions for ANOVA, means were separated using Fisher's protected LSD ($\alpha = 0.05$) and preplanned contrasts were conducted for select treatments to compare between total yearly amounts of quizalofop applied ($\alpha = 0.05$). Years were analyzed and will be presented separately for 2014 and 2015.

3. Results and Discussion

3.1. Barnyardgrass Population Screening. Overall, quizalofop at 80 g ai ha^{-1} was effective for controlling the populations tested. At 21 DAT, barnyardgrass control across populations was 99% (data not shown). Of the 126 populations evaluated, 113 were completely controlled by quizalofop (100%), with no living tissue remaining at 21 DAT. For the 13 populations that were not completely controlled, quizalofop achieved at least 92% control (data not shown). In these 13 populations, the lowest mortality was only 80% with accession B91. Even with an 80% mortality rate, live plants only had a small portion of living tissue and most likely would not have been competitive in a field setting. Although ACCase-resistant barnyardgrass has been confirmed in the Mid-South (21), all populations evaluated were adequately controlled with quizalofop. With use rates being 100 to 138 g ha^{-1} for a single application [18], quizalofop is expected to be an effective herbicide to control barnyardgrass in rice.

3.2. Efficacy of Quizalofop Compared to Registered Rice Graminicides: Greenhouse Experiment. A significant growth stage by herbicide interaction was observed for visual control and biomass for all grass species. ACCase-inhibiting herbicides were effective for controlling all species evaluated at the 2- to 3-leaf growth stage (>96%) (Table 1), with no significant difference between treatments. Likewise, no difference was observed between herbicides for relative biomass at the 2- to 3-leaf timing of any grass species (Table 2). For applications to larger grass, there did appear to be differences in efficacy among the herbicides evaluated. At the 5- to 6-leaf and 12- to 16-leaf growth stages, quizalofop across rates consistently provided greater control compared to fenoxaprop and cyhalofop (Table 1). Only the lowest rate of quizalofop (91%) was similar to fenoxaprop (89%) for control of 5- to 6-leaf barnyardgrass. Quizalofop, regardless of rate, provided a high level of control (>90%) for all grass species at the 5- to 6-leaf growth stage. Similarly, broadleaf signalgrass, fall panicum, and Amazon sprangletop treated at the 5- to 6-leaf stage usually had less biomass following quizalofop treatments compared to fenoxaprop and cyhalofop (Table 2).

Control drastically decreased for all graminicides applied to 12- to 16-leaf grasses (Table 1). For barnyardgrass, control from the high rate of quizalofop was reduced from 99% at the 5- to 6-leaf growth stage to only 53% at the 12- to 16-leaf stage (Table 1) while relative biomass increased from 6 to 55% (Table 2). Likewise, these herbicides were not effective for controlling broadleaf signalgrass at the largest growth stage (33 to 64% control), except for the high rate of quizalofop (86%). Fall panicum was still highly susceptible to quizalofop at the 12- to 16-leaf growth stage, with all quizalofop

TABLE 2: Effect of ACCase-inhibiting herbicide and growth stage at time of application on relative biomass of barnyardgrass, broadleaf signalgrass, fall panicum, and Amazon sprangletop at 21 DAT.[a]

Grass size	Herbicide	Rate	Biomass[bc]			
			ECHCG	BRAPP	PANDI	LEFPA
		g ai ha^{-1}	% of untreated check			
2- to 3-leaf	Quizalofop	80	5 e	7 e	4 d	4 f
	Quizalofop	120	7 e	6 e	3 d	3 f
	Quizalofop	160	5 e	4 e	3 d	3 f
	Fenoxaprop	122	8 e	6 e	4 d	5 f
	Cyhalofop	313	8 e	7 e	6 d	4 f
5- to 6-leaf	Quizalofop	80	7 e	8 e	4 d	7 ef
	Quizalofop	120	8 e	7 e	5 d	5 f
	Quizalofop	160	6 e	7 e	4 d	4 f
	Fenoxaprop	122	9 e	20 d	20 c	16 de
	Cyhalofop	313	43 d	41 c	63 a	60 a
12- to 16-leaf	Quizalofop	80	61 bc	57 ab	20 c	27 c
	Quizalofop	120	58 c	51 bc	19 c	27 c
	Quizalofop	160	55 c	22 d	18 c	22 cd
	Fenoxaprop	122	67 b	65 a	39 b	44 b
	Cyhalofop	313	84 a	62 ab	69 a	67 a

[a]DAT: days after treatment, ECHCG: barnyardgrass, BRAPP: broadleaf signalgrass, PANDI: fall panicum, LEFPA: Amazon sprangletop. [b]Means within a column followed by the same letter are not different according to Fisher's protected LSD ($\alpha = 0.05$). [c]Data expressed as percent relative biomass compared with nontreated control for each grass species and growth stage. Nontreated check was harvested on the same day as treated plots.

treatments producing >90% control (Table 1) with ≤20% biomass relative to the untreated control (Table 2). Overall, at the rates tested, quizalofop appears to have greater grass activity than either fenoxaprop or cyhalofop, which is similar to previous experiments which have often shown quizalofop to outperform other ACCase-inhibiting herbicides [24–26]. Moreover, quizalofop remained more effective on the grass weeds evaluated at larger growth stages (5- to 6-leaf, 12- to 16-leaf), whereas fenoxaprop and cyhalofop efficacy quickly diminished.

3.3. Field Experiment. Overall, no parameter evaluated for this experiment produced a significant interaction or main effect for either 2014 or 2015. Grass weeds were effectively controlled by all treatments, with all control ratings for both barnyardgrass and broadleaf signalgrass being >96% (data not shown). At 21 days after the sequential application, barnyardgrass and broadleaf signalgrass control ranged from 97 to 99% for all treatments. Presence or absence of clomazone preemergence did not affect the emergence of rice, with no significant difference between rice stand counts at 14 days after planting (data not shown). The experimental quizalofop-resistant rice variety showed no symptoms of injury from any ACCase-inhibiting herbicide applied. Visual injury ratings taken 14 and 21 days after graminicide application were never higher than 5% for any treatment, with injury symptoms being small chlorotic spotting consistent with injury caused from the adjuvant. Only nontreated check plots had grass weeds present at the time of test termination; thus they were the only plots in which weed biomass were harvested. Nontreated checks resulted in 43.2 and 28.7 g m^{-2} oven-dried biomass averaged over both years for barnyardgrass and broadleaf signalgrass, respectively.

Averaged over both years, density of barnyardgrass and broadleaf signalgrass was only 4.2 and 3.6 plants m^{-2}, respectively, at time of the initial postemergence application (3- to 4-leaf rice). Previous research has shown that as weed density decreases, efficacy of herbicides can increase. This is especially true with ACCase-inhibiting herbicides, where Ndou (2009) [27] found that as large crabgrass (*Digitaria sanguinalis* (L.) Scop.) density decreased, percent mortality with clethodim increased.

3.4. Best Rate Structure for Sequential Applications of Quizalofop in Quizalofop-Resistant Rice. In 2014, there was not a significant interaction of quizalofop rate structure by growth stage; however, there were significant main effects of rate structure and growth stage for both barnyardgrass and broadleaf signalgrass control. At 21 days after the sequential application, the 120 fb 120 g ha^{-1} rate structure controlled barnyardgrass 98% but was only significantly different from the 80 fb 80 g ha^{-1} structure which produced 89% control (Table 3). Similarly, the highest control of broadleaf signalgrass was produced with the 120 fb 120 g ha^{-1} rate structure but again was only significantly different from the 80 fb 80 g ha^{-1} rate structure which resulted in 91% control. Based on an orthogonal contrast, using the full seasonal quizalofop use rate of 240 g ha^{-1} significantly increased both barnyardgrass and broadleaf signalgrass control compared to seasonal use rates of 200 and 160 g ha^{-1}. When the initial application of quizalofop was made at the 2-leaf growth stage it resulted in 98% control of barnyardgrass, averaged over

TABLE 3: Effect of quizalofop application structure on barnyardgrass, broadleaf signalgrass, and red rice control at Stuttgart, AR, in 2014 and 2015 averaged over time of first application followed by contrast between total quizalofop usage rates.

Application structure[a]	Control[b] 2014[c] ECHCG	BRAPP	2015[d] ECHCG	BRAPP	ORYSA
g ai ha^{-1}			%		
80/80	89 b	91 b	94a	97a	95a
80/120	90 ab	95 ab	96a	98a	99a
80/160	91 ab	98 a	97a	100a	98a
120/80	91 ab	96 ab	94a	99a	98a
120/120	98 a	99 a	98a	100a	99a
160/80	95 a	98 a	98a	100a	98a
Contrast[e] g ai ha^{-1}					
240 vs. 160	<0.0001	0.0032	0.0415	NS	0.0150
240 vs. 200	0.0049	0.0311	NS	NS	NS
200 vs. 160	0.0246	0.0099	NS	NS	NS

[a]First rate applied followed by (/) second rate applied 2 weeks later. [b]ECHCG: barnyardgrass, BRAPP: broadleaf signalgrass, ORYSA: red rice. [c]Means within a column followed by the same letter are not different according to Fisher's protected LSD (α = 0.05). [d]2015 resulted in no significant difference between quizalofop application structure for any weed species. [e] Total yearly amount of quizalofop applied.

TABLE 4: Effect of grass growth stage at time of first quizalofop application on barnyardgrass, broadleaf signalgrass, and red rice control in 2014 and 2015 at Stuttgart, AR.

Growth stage[b]	Control[a,c] 2014 ECHCG	BRAPP	2015 ECHCG	BRAPP[d]	ORYSA
2-leaf	98 a	98 a	98 a	98a	99 a
6-leaf	87 b	95 b	93 b	97a	97 b

[a]ECHCG: barnyardgrass, BRAPP: broadleaf signalgrass, ORYSA: red rice. [b]Growth stage of grasses at the first application of quizalofop with a subsequent applications 14 days later. [c]Means within a column followed by the same letter are not different according to Fisher's protected LSD (α = 0.05). [d]2015 resulted in no significant difference between growth stage at initial application for broadleaf signalgrass (α = 0.05).

rates; however, when the application occurred at the 6-leaf growth stage, control declined to 87% (Table 4). The same trend was apparent for broadleaf signalgrass as well, where when initially applied at the 2-leaf growth stage, control averaged 98% over rate structure, but when initiated at the 6-leaf growth stage, control declined to 95%.

In 2015, the experiment contained red rice in addition to barnyardgrass and broadleaf signalgrass. The overall weed density was less in 2015, which may have contributed to differing results between years. No significant difference was observed in quizalofop rate structure for barnyardgrass, broadleaf signalgrass, or red rice. Control of barnyardgrass, broadleaf signalgrass, and red rice ranged from 94 to 98%, 97 to 100%, and 95 to 99%, respectively, at 21 days after sequential application (Table 4). Based on a preplanned contrast, there was a difference in full seasonal quizalofop use rate (240 g ha^{-1}) compared to the low seasonal use rate (160 g ha^{-1}) for barnyardgrass and red rice control. Likewise, there was a main effect of growth stage of initial application for barnyardgrass and red rice control. When the initial application of quizalofop was made at the 2-leaf growth stage, barnyardgrass grass control was 98%; however, when it was initiated at the 6-leaf growth stage, control decreased to 93% (Table 4). Similarly, red rice control was 99% when applied at the 2-leaf growth stage but was reduced to 97% when the first application was to 6-leaf plants. Although this seems like a small difference, due the potential for gene flow from quizalofop-resistant rice to red rice, even a few escapes of red rice within a field can lead to the rapid evolution of resistance [28].

Results from this experiment support applications of quizalofop in quizalofop-resistant rice as an effective option for controlling annual grass weed species. Moreover, the results support the current labeled single application rates of quizalofop at 100 to 138 g ha^{-1} as well as the total seasonal use rate of 240 g ha^{-1} [18].

4. Conclusions

Quizalofop-resistant rice can be an effective weed control technology for annual grass control in Arkansas rice production systems. Quizalofop alone effectively controlled all 126 barnyardgrass populations from across the state of Arkansas, even at a lower than labeled rate (80 g ha^{-1}) for quizalofop-resistant rice. However, there have already been cases of ACCase-resistant grass weed species within the Mid-South.

To reduce the risk of resistance development, following a strict crop rotation plan is the best for producers. It is currently suggested to grow quizalofop-resistant rice once every 3 years in conjunction with imidazolinone-resistant rice and soybean [19]. This approach would limit the amount of quizalofop and would allow for the use of multiple herbicide modes of action for control of annual grass species.

Quizalofop also generally outperformed other currently labeled rice graminicides, especially on larger grasses. Over multiple years and locations, quizalofop-resistant rice exhibited high levels of tolerance to quizalofop (<5% injury), meaning the likelihood for injury to commercial cultivars from the herbicide should be low. For optimum efficacy, employing sequential applications is the best that can be done, where the first application targets two-leaf or smaller grasses using the full seasonal rate of $240\,g\,ha^{-1}$. Delaying applications or reducing the use rate would likely increase the risk for resistance in barnyardgrass and gene flow from red rice. Extra precautions will need to be taken in respect to control of red rice due to the risk of gene flow through cross-pollination of quizalofop-resistant rice with red rice [29]; these can include managing red rice on field borders, minimizing carryover of weed seed bank to the following crop, and using clean crop seed (certified seed) [18]. Within the full seasonal quizalofop use rate, sequential applications of $120\,g\,ha^{-1}$ followed by $120\,g\,ha^{-1}$ performed best for the grass weeds evaluated.

Acknowledgments

Funding for this research was provided by the Arkansas Rice Research and Promotion Board. This publication is part of a thesis titled "Evaluation of Provisia Rice for Arkansas Rice Production Systems" by Zachary Lancaster for fulfillment of a degree from the University of Arkansas.

References

[1] [NASS] National Agricultural Statistics Service, "Arkansas acreage report 2016," https://www.nass.usda.gov/Statistics_by_State/Arkansas/Publications/Crop_Releases/Acreage/2016/aracreage16.pdf.

[2] J. T. Hardke, *B.R. Wells Arkansas Rice Research Studies 2015*, R. J. Norman and K. A. K. Moldenhauer, Eds., University of Arkansas Agricultural Experiment Station Research Series, Fayetteville, NC, USA, 2016.

[3] J. T. Hardke, *Arkansas Rice Production Handbook. Arkansas Cooperative Extension Service Miscellaneous Publications*, vol. 129, University of Arkansas, Little Rock, AR, USA, 2012.

[4] J. K. Norsworthy, J. Bond, and R. C. Scott, "Weed management practices and needs in Arkansas and Mississippi rice," *Weed Technology*, vol. 27, no. 3, pp. 623–630, 2013.

[5] L. G. Holm, J. V. Pancho, J. P. Herberger, and D. L. Plucknett, *The Worlds Worst Weeds*, University Press of Hawaii, Honolulu, HI, USA, 1977.

[6] V. F. Carey, R. E. Hoagland, and R. E. Talbert, "Verification and distribution of propanil-resistant barnyardgrass (Echinochloa crus-galli) in Arkansas," *Weed Technology*, vol. 9, no. 2, pp. 366–372, 1995.

[7] M. L. Lovelace, R. E. Talbert, R. E. Hoagland, and E. F. Scherder, "Quinclorac absorption and translocation characteristics in quinclorac- and propanil-resistant and -susceptible barnyardgrass (Echinochloa crus-galli) biotypes," *Weed Technology*, vol. 21, no. 3, pp. 683–687, 2007.

[8] J. K. Norsworthy, R. C. Scott, and K. L. Smith, *B.R. Wells Arkansas Rice Research Studies*, R. J. Norman, K. A. K. Moldenhauer, and B. R. Wells, Eds., vol. 560 of *University of Arkansas Agriculture Experiment Station Research Series*, Fayetteville, NC, USA, 2007.

[9] F. L. Baldwin, B. A. Huey, and G. L. Morris, "Distribution of three weed species in Arkansas rice fields," in *Proceedings of the Southern Weed Science Society*, vol. 30, p. 398, 1977.

[10] N. R. Burgos, J. K. Norsworthy, R. C. Scott, and K. L. Smith, "Red rice (Oryza sativa) status after 5 years of imidazolinone-resistant rice technology in Arkansas," *Weed Technology*, vol. 22, no. 1, pp. 200–208, 2008.

[11] D. S. Riar, J. K. Norsworthy, J. A. Bond, M. T. Bararpour, M. J. Wilson, and R. C. Scott, " Resistance of ," *International Journal of Agronomy*, vol. 2012, pp. 1–8, 2012.

[12] S. O. Duke, "Why have no new herbicide modes of action appeared in recent years?" *Pest Management Science*, vol. 68, no. 4, pp. 505–512, 2012.

[13] R. E. Talbert and N. R. Burgos, "History and management of herbicide-resistant barnyardgrass (Echinochloa crus-galli) in Arkansas rice," *Weed Technology*, vol. 21, no. 2, pp. 324–331, 2007.

[14] J. Guice, C. Youman, A. Rhodes et al., "Provisia™ rice system; weed management strategies for rice," in *Proceedings of Southern Weed Science Society the annual meeting*, vol. 197, Southern Weed Science Society, Savanna, GA, USA, 2015.

[15] M. Hinga, S. Griffin, M. S. Moon, R. D. Rasmussen, and F. Cuevas, *Methods and compositions to produce rice resistant to ACCase inhibitors*, US patent 20130023416 A1, 2013.

[16] DL. Shaner, *Herbicide Handbook*, DL. Shaner, Ed., Weed Science Society of America, Lawrence, Mass, USA, 10th edition, 2014.

[17] T. Konishi and Y. Sasaki, "Compartmentalization of two forms of acetyl-CoA carboxylase in plants and the origin of their tolerance toward herbicides," *Proceedings of the National Academy of Sciences of the United States of America*, vol. 91, no. 9, pp. 3598–3601, 1994.

[18] Anonymous, "Provisia™ herbicide product label," in *NVA 2017-04-522-0004*, Research Triangle Park, NC. BASF Corporation, 2017.

[19] C. Youman, J. Guice, A. Rhodes, A. Schultz, and J. Harden, "Provisia rice production system efficacy and stewardship," in *Proceedings of the in Proceedings of the Southern Weed Science Society annual conference*, vol. 278, San Juan, PR, 2016.

[20] M. V. Bagavathiannan, J. K. Norsworthy, K. L. Smith, and P. Neve, "Modeling the simultaneous evolution of resistance to ALS-and ACCase-inhibiting herbicides in barnyardgrass (Echinochloa crus-galli) in Clearfield® rice," *Weed Technology*, vol. 28, no. 1, pp. 89–103, 2014.

[21] I. Heap, "Trends in the development of herbicide-resistant weeds," in *Integrated weed management for sustainable agriculture*, Burleigh Dodds Series in Agricultural Science, Burleigh Dodds Science Publishing, 2017.

[22] S. S. Kaundun, "Resistance to acetyl-CoA carboxylase-inhibiting herbicides," *Pest Management Science*, vol. 70, no. 9, pp. 1405–1417, 2014.

[23] C. E. Rouse, N. Burgos, A. Lawton-Rauh, and R. A. Salas, "Herbicide resistance mechanisms of multiple resistant junglerice (Echinochloa colona) from Arkansas," in *Proceedings of the in Proceedings of the Southern Weed Science Society annual conference*, San Juan, PR, USA, 2016.

[24] R. E. Blackshaw, K. N. Harker, G. W. Clayton, and J. T. O'Donovan, "Broadleaf herbicide effects on clethodim and quizalofop-P efficacy on volunteer wheat (Triticum aestivum)," *Weed Technology*, vol. 20, no. 1, pp. 221–226, 2006.

[25] W. Deen, A. Hamill, C. Shropshire, N. Soltani, and P. H. Sikkema, "Control of volunteer glyphosate-resistant corn (Lea mays) in glyphosate-resistant soybean (Glycine max)," *Weed Technology*, vol. 20, no. 1, pp. 261–266, 2006.

[26] B. W. Minton, D. R. Shaw, and M. E. Kurtz, "Postemergence Grass and Broadleaf Herbicide Interactions for Red Rice (Oryza sativa) Control in Soybeans (Glycine max)," *Weed Technology*, vol. 3, no. 2, pp. 329–334, 1989.

[27] A. M. Ndou, *Interaction of weed emergence, weed density, and herbicide rate in soybean [Ph.D. thesis]*, Kansas State University, Manhattan, 2009.

[28] D. H. Gealy, H. Agrama, and M. H. Jia, "Genetic analysis of atypical U.S. red rice phenotypes: Indications of prior gene flow in rice fields?" *Weed Science*, vol. 60, no. 3, pp. 451–461, 2012.

[29] V. K. Shivrain, N. R. Burgos, M. A. Sales et al., "Factors affecting the outcrossing rate between clearfield™ rice and red rice (oryza sativa)," *Weed Science*, vol. 57, no. 4, pp. 394–403, 2009.

8

Effect of Seed Inoculation with Actinomycetes and Rhizobium Isolated from Indigenous Soybean and Rhizosphere on Nitrogen Fixation, Growth, and Yield of Soybean

Asmiaty Sahur,[1] Ambo Ala,[1] Baharuddin Patandjengi,[2] and Elkawakib Syam'un[1]

[1]*Department of Agronomy, Faculty of Agriculture, Hasanuddin University, Makassar, Indonesia*
[2]*Research Center (PKP), Hasanuddin University, Makassar, South Sulawesi 90245, Indonesia*

Correspondence should be addressed to Asmiaty Sahur; asmiatyasmiaty@gmail.com

Academic Editor: Yuanhu Xuan

The present study was initiated to determine whether isolates from soil and roots of soybean plants can express nitrogenase activity when grown in the absence of plant host. The study was conducted to answer the question "can benefit gained by" the interaction between Actinomycetes and Rhizobium symbiosis with legume. Thirty-five isolates identified as Rhizobium and twenty-one Actinomycetes were isolated from the rhizosphere of soybean plants and identified by morphological character, biochemical content identified. Fifty-six isolates were tested for their capabilities of N_2 fixation and siderophore production. The isolated rhizobacteria were grown in N-free media, and twelve of them showed a good growth on the Burk's N-free media. Almost all strains produced siderophores; however, the production level was very low, and only the strain AK 10 released considerable amounts of this metabolite. One strain of Actinomycetes was selected to test their interactions with Rhizobium. Coinoculation of Actinomycetes and Rhizobium produced synergic benefits on plant growth and get protection from the production of siderophore.

1. Introduction

Beneficial interaction effect between Rhizobium symbioses, especially in legumes, has been developed since the late 1800s. The use of microbial inoculants more involved Rhizobium and legume crops. Compounds produced by Rhizobium have been well researched and the number of Rhizobium strains that effectively form nodules has been identified. According to Garbaye and Frey-Klett [1], it is well established that some strains of Rhizobium can exhibit nitrogenase activity when grown in vitro on different media, but such activity has not been reported for Actinomycetes strains isolated from root nodules of N2-fixing nonlegumes [1].

Soybean growth and production can be promoted by improving soil fertility in terms of physical-chemical and biological characterization. In order to achieve such condition, endophytic microorganisms such as Rhizobium and Actinomycetes should be utilized to promote the growth in the production of soybean plants. In the last few years, Actinomycetes have been well known as commercial microorganisms that can produce antibiotics and other useful metabolites.

A previous study conducted reported that Actinomycetes of *Streptomyces lydicus* species WYEC108 colonizes the roots of peas *Pisum sativum*. Such root colonizations are potentially important for the improvement and growth of *Pisum sativum* plant by improving the process of the nodule formulation of Leguminosae plants because it promotes nodule formation. Actinomycetes *Streptomyces lydicus* WYEC108 infected root by affecting root nodulation of pea *Pisum sativum* by increasing the frequency of root nodulation; it is thought to occur at the rate of *Rhizobium* spp infection. *Streptomyces lydicus* also colonizes and then sporulates the surface of nodule cell [2]. The colonization may cause an increase in the average size of the nodule formation and increase the strength of the nodule by increasing the assimilation of nodular iron and also improve soil nutrition [2].

Endophytic actinomycetes also play an important role in agriculture and forestry as they produce biopesticides instead of chemical pesticides. This helps to preserve the life and environment from toxic chemicals. Many people are interested in knowing about the existence of the Actinomycetes. There are two species of Actinomycetes which have been isolated and known to have a good bioactive activity, namely, *Frankia* sp and *Streptomyces scabies*. Genus *Frankia* is a nitrogen-fixing bacteria nonleguminous plant isolated in 1886 and *S. scabies* is found in 1890 as a phytopathogen [3]. As discussed by Sardi et al. [4], Actinomycetes endophyte isolated from the root surface sterilized from 28 plant species from northwest Italy and the most of the result of isolation process is *Streptoverticillium*, *Nocardia*, *Micromonospora*, and *Streptosporangium*.

Coomb et al. [5] isolated the endophytic Actinomycetes from the surface of the root tissue of healthy wheat plants (*Triticum aestivum* L.) collected from South Australia. They found that most isolates obtained are of the genus *Streptomyces* and of other species such as *Microbispora*, *Micromonospora*, and *Nocardiodes*. In addition, there are several studies that had isolated the endophytic Actinomycetes from other parts of the plant, such as De Araujo et al. [6] that isolated the endophyte from roots and leaves of corn. *Microbispora* sp is the most widely genus found, followed by *Streptomyces* sp. and *Streptosporangium* sp. Furthermore, Taechowisan et al. [7] has done the same process of isolation of the endophytic Actinomycetes from roots, stems, and leaves of herbaceous and woody plants in the sterile environment of Chiang Mai, Thailand, where the largest group of isolates they found were *Microbispora* sp, *Streptomyces*, *Nocardia*, and *Micromonospora*. Okazaki et al. [8] also found the genus *Microbispora* spp which isolated from leaves of the plant and the results were higher compared with isolates from the ground.

In 2010, several researchers tested thirty isolates for their capabilities of solubilizing/mineralizing sparingly phosphate sources, N2 fixation, and/or siderophore production, typical traits of the so-called plant growth-promoting rhizobacteria (PGPR). Phosphate-solubilizing ability was widely exhibited by the isolated. All of them produced acid phosphatase and thirteen of them produced alkaline phosphatase. Ten strains grew in N-free media. Almost all strains produced siderophores; however, the production level was in general very low and only the strain Thermobifida MCR24 released considerable amounts of this metabolite ([9], p. 210).

The objective of the present study was to isolate and identify Actinomycete strains with the intent of testing their abilities as a plant growth promoter. Therefore, the isolated Actinomycetes and Rhizobium were screened for their ability to produce siderophore typical secondary metabolites involved in rhizosphere colonization competence and tested for their capabilities to produce nitrogen in the absence of host plants. Then, the effects of selected Actinomycete strains on interaction with Rhizobium on plant growth and yield of soybean were also investigated.

2. Materials and Methods

Isolation process was carried out by taking root and soil samples of top soil at 0–20 cm in the three areas of soybeans rhizosphere in South Sulawesi, Indonesia. In each area, one healthy soybean was taken and placed in the plastic bag for isolation process. The soil samples around the rhizosphere area were collected in one bag and labeled based on the location criteria.

The method of serial dilution was employed to isolate the bacteria from rhizosphere of indegeneous soybeans. In this process, 10 grams of soil was dissolved in ninety mL of sterile water, and the solution was mixed for thirty minutes. One ml of suspension was inserted to the reaction tube containing nine mL of sterile water to get suspension with solution concentration at 10^{-2}. The same process was done for several levels of concentration, such as 10^{-3}, 10^{-4}, and then 10^{-6}, respectively.

Identification of bacterial isolates was conducted using methods developed by Schaaf et al. [10] as follows: characterization of morphology, determination of the morphological characteristics based on the shape and color of colonies on culture media nutrient agar (NA), and observations on a microscope. Identification of physiological and biochemical characteristics was performed through the following steps:

(a) Detect the criteria of gram reaction of tested bacteria by taking pure cultures using a needle loop and smeared on the glass objects. Then, the cultures on the object glass was added with aqueous KOH and stirred for 5 to 10 seconds. If the cultures showed slimy colonies, then they were identified as Gram-positive bacteria. On the contrary, the cultures which were not forming slimy colonies will be identified as Gram-negative bacteria.

(b) Morphology of different strains grown on tested media was analyzed for the mycelial organization and sporulation, under light microscopy. Cultural characteristics of Actinomycetes were studied on media natrium broth (NB). Colors of aerial and substrate mycelia were determined.

Identification of the molecular basis of the 10 isolates PGPR can be performed best by going through the following stages.

2.1. Testing Nitrogen Fixation. The ability of free nitrogen-binding bacteria carried by inoculation of Burk's N-free media should be tested. The isolates were able to grow, meaning that these isolates have the ability to bind free nitrogen for growth. Bacterial isolates were grown in media that Burk N-free for 24 hours at 28 ± 2°C. The pH value of the media is set to be 7.0 ± 0.1 before being autoclaved at 121°C for 15 minutes.

To measure the amount of N_2 fixation, bacterial isolates were tested by growing the bacteria in a liquid Burk's N-free medium and were placed on an orbital shaker for 24 hours at a temperature of 28 ± 20°C [8]. The method of Kjeldahl will be employed in order to measure nitrogen fixation [11].

Sample culture of bacteria was incorporated into the liquid Burk's N-free medium in 5 ml digestion tubes and then 1 g of a mixture of selenium and 3 ml of concentrated sulfuric acid were added and destructed up to 350°C temperature (3-4 hours). Destruction is considered complete when the steam out is white and when extracts obtained are clear (about 4 hours). The extracts were removed, cooled, and then diluted with distilled water up to exactly 50 ml and then shaken until it becomes homogeneous and left overnight so that the particles settle. The extract is used for the measurement of N by distillation.

Therefore, the extract was transferred into the boiling flask to release the NH3 by using Erlenmeyer flask containing 10 ml of 1% boric acid and added three drops of Conway (red) indicator. With a measuring cup, NaOH 40% as much as 10 mL into the boiling flask containing samples should be added and immediately cover with the lid. The samples were distillated until the container volume reached 50–75 ml (green). Distillation titration examples (Vc) and blank mo (Vb) then calculate the nitrogen content (%) by using the formula:

$$\text{Levels of nitrogen}\,(\%) = (\text{Vc} - \text{Vb}) \times \text{N} \times \text{bst N} \times 100\,\text{mg}. \tag{1}$$

Example 1

$(\text{Vc} - \text{Vb}) \times \text{N} \times 14 \times 100 \cdot 500^{-1}$
$(\text{Vc} - \text{Vb}) \times \text{N} \times 2.8$

Information

Vc, b = mL sample and blank titration
N = normality of standard solution H_2SO_4
14 = equivalent weight of nitrogen
100 = conversion to percentage (%)

Siderophore production was estimated by the method described by Reeves et al. [12]. Natrium Broth sterile medium was prepared as much as 100 ml in 250 ml Erlenmeyer flask. Bacterial isolates were inoculated with a loop into the medium and incubated at 37°C for 7 days. After 7 days of incubation, the liquid culture was centrifuged at 10,000 g for 20 minutes. The supernatant was used to estimate the siderophores. 20 ml of the culture supernatant was taken, and the pH was adjusted to 2.0 with dilute HCl. To 20 ml of the supernatant, 20 ml of ethyl acetate is added, and extraction is done twice. Prepare reagent of Hathway (1 ml of 0.1 M $FeCl_2$ and 1 mL of 0.1 N HCl is added to 100 ml of distilled water and added with 5 ml of reagent Hathway and absorbance was measured at 560 nm with sodium salicylate as standard in the range of 0.5 to 2 ppm to estimate salicylate kind of siderophore).

Twenty-one Actinomycete strains were preselected to investigate their plant growth promotion effects. Those are the higher ability of these strains for nitrogen testing and to produce siderophores. To investigate the plant promotion effects of these Actinomycetes, a pot experiment was carried out. This involved soybean plants (*Glycine max.* Linn) that were grown in a sterilized soil-sand mixture (300 ml pot). One of the Actinomycete strains was inoculated on the coated seed of soybean at a concentration of 10^4 cells/pot. There were thirteen treatments (one Actinomycete strain and one Rhizobium) with three replications for each treatment. Surface-sterilized seeds of soybean were germinated on the sterilized sand-soil substrate, and four seedlings were planted in each pot, watered as needed, and grown for 30 days. Pots were placed in a growth chamber in a randomized block design. After the growth period, the plants were harvested, dried at 60°C for 1 day, and weighed.

3. Results

3.1. Isolation and Identification of Actinomycetes and Rhizobium. The isolates are traditionally tentatively identified based on morphological criteria including the characteristics of the colonies on a petri dish, the morphology of the substrate and hypa air, spore morphology, and pigment produced [13], and also by referring to the Bergey's manual, [14], fifty isolates were identified (Figure 1) based on the morphological characteristics, which consist of 35 which have similarities in terms of colonies formed, colony color, colony edge, the surface of the colony, the colony growth rate, Gram test, the ability to form hypa air, spore morphology, and pigment produced. The other twenty-one isolates were similar in terms of the form of colonies, the edge of the colony, Gram test, and slime-forming bacteria (Figure 2).

The percentage (100%) of *Streptomyces* found in soil was very high compared with that of genera of Actinomycetes. The BLAST analysis was performed for all Actinomycetes. The results showed the confidence of sequencing analysis using the F1 and R5 primers correlating with those from the restriction endonuclease digestion technique used by Cook and Meyers [14].

3.2. Testing Nitrogen Fixation. As shown in Table 1, it is clear the growth ability for the twenty-one of identified Actinomycete strains and thirty-five of identified as Rhizobium on the Burk's N-free media. Ten strains were able to grow in these N-deficient media, suggesting that they could be nitrogen-fixing bacteria.

3.3. Siderophore Production. The production of siderophores was tested after evaluation of isolates on Burk's N-free media. The vast majority of isolates had produced two kinds of siderophores such as salicylate and catechol, as shown in Table 2, and it was clear that the solution showed the brown color on reaction tube containing tested isolates [15].

3.4. Interaction between Actinomycetes and Rhizobium. One of the isolated Actinomycetes was chosen in order to investigate their plant growth promotion effects on soybean plants. Consequently, the candidate of isolate was used to investigate some interactions with *Bradyrhizobium*. Interaction between Actinomycetes and Rhizobium experiments demonstrated the production of stems, seed, and mycelial development, as summarized in Table 3. The results of other

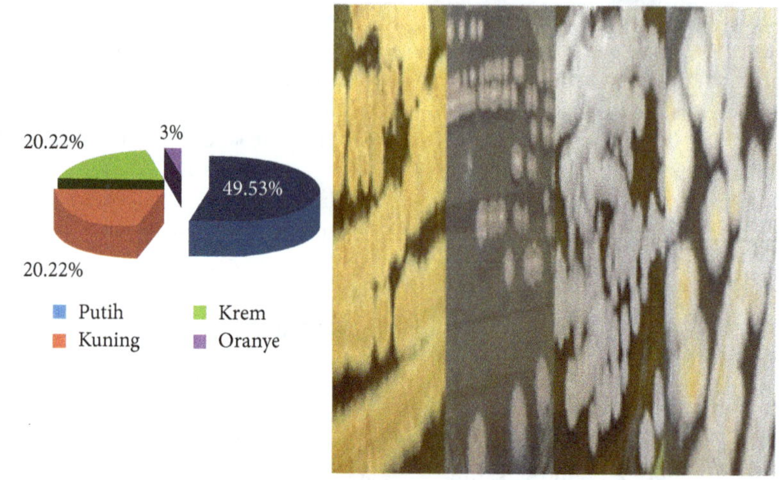

Figure 1: The percentage of bacteria identified as Gram-positive colony.

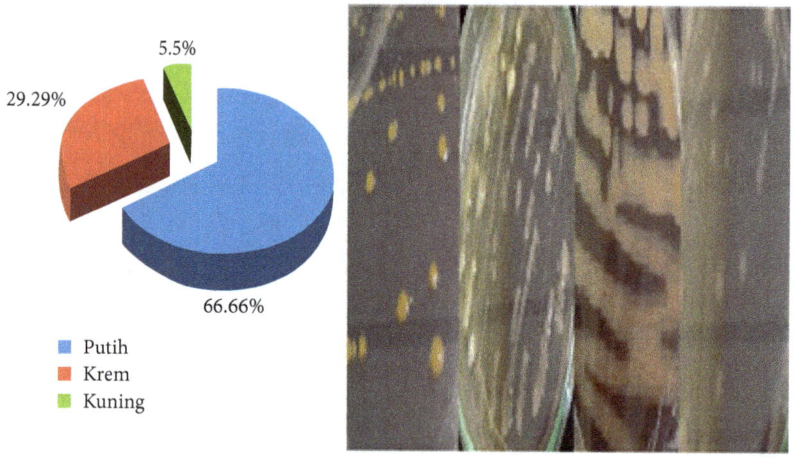

Figure 2: The percentage of bacteria identified as Gram-negative colony.

assays, which were designed to investigate any growth promotion effects using soybean plants, can be summarized as recorded in Tables 3 and 4. In general, the inoculated Actinomycetes coinoculate with Rhizobium and perform an excellent growth.

4. Discussion

Based on the results of electrophoresis by comparing the band of 16S ribosomal RNA gene amplification along with 234 base pairs which appeared on the test sample with 100 base pair marker on the left and right side, the PCR products were further sequenced to determine the basic sequence of each isolate tested and compared with the *Streptomyces* species found on GenBank (National Center for Biotechnology Information/NCBI) [16]. It is clear that the F1 and R5 primers, used for amplification of 16S rDNA gene of the Actinomycetes isolates in here reported experiments, confirmed their efficiency to amplify all sequences of 16S rDNA gene of target bacteria [14]. Results reported here maintain the use of the Cook methodology [14] because a partial 16S rDNA sequence of a new Actinomycetes isolate can be obtained quickly and at low cost to give a clear identification of the genus to which the isolate belongs, as can be seen in Table 5.

The invention of siderophores is a mechanism used by PGPR for rhizosphere colonization capability and has been revealed to be related with nitrogen fixation. Actually, symbiotic nitrogen fixers such as *Bradyrhizobium japonicum*, *Rhizobium leguminosarum*, and *Sinorhizobium meliloti* were reported as producers of siderophore [4, 16]. In the present study, the fifty-six isolates investigated showed positive result with regard to siderophore production, and ten isolates had an ability to fix nitrogen.

Several studies had proven that the differential effect can be shown by specific fungus whose the communities are related with the fungal hyphae, fungal spores, or/and with mycorrhizosphere of mycorrhizal plant [6, 18, 19, 20, 21]. It is clear that the condition of rhizosphere bacteria is under assortment to develop a fungus-specific character with the intention of giving a viable advantage during colonization of fungal surfaces. Bacteria-induced alterations to fungal development differentiation have been described in previous studies [23]. There were several effects of colonization already found by de Boer et al. [23], involving inhibition or promotion of germination; alterations to foraging

Table 1: Ability of Actinomycete isolates to grow in culture media without nitrogen source. The isolates may be diazotrophs based on their ability to grow in N-free media.

Isolate code	Nitrogen total (Kjeldahl) (%)	Growth on Burk's *N*-free media
AK 1	0.00	(−)
AK 2	0.00	(−)
AK 3	0.00	(−)
AK 4	0.00	(−)
AK 5	0.11	(+)
AK 6	0.00	(−)
AK 7	0.00	(−)
AK 8	0.00	(−)
AK 9	0.00	(−)
AK 10	0.00	(−)
AK 11	0.00	(−)
AK 12	0.00	(−)
AK 13	0.00	(−)
AK 14	0.00	(−)
AK 15	0.00	(−)
AK 16	0.20	(++)
AK 17	0.17	(+)
AK 18	0.00	(−)
AK 19	0.34	(+++)
AK 20	0.14	(+)
AK 21	0.17	(+)
RK1	0.00	(−)
RK2	0.00	(−)
RK3	0.00	(−)
RK4	0.00	(−)
RK5	0.00	(−)
RK6	0.00	(−)
RK7	0.00	(−)
RK8	0.00	(−)
RK9	0.00	(−)
RK10	0.00	(−)
RK11	0.00	(−)
RK12	0.20	(++)
RK13	0.00	(−)
RK14	0.00	(−)
RK15	0.00	(−)
RK16	0.00	(−)
RK17	0.00	(−)
RK18	0.00	(−)
RK19	0.00	(−)
RK20	0.00	(−)
RK21	0.17	(+)
RK22	0.14	(+)
RK23	0.00	(−)
RK24	0.00	(−)
RK25	0.20	(++)
RK26	0.00	(−)
RK27	0.34	(+++)
RK28	0.00	(−)
RK29	0.00	(−)
RK30	0.31	(+++)
RK31	0.00	(−)
RK32	0.00	(−)
RK33	0.00	(−)
RK34	0.22	(++)
RK35	0.00	(−)

(+++) = high nitrogen production. (++) = medium nitrogen production. (+) = low nitrogen production. (−) = no nitrogen production.

Table 2: Ability of Actinomycete isolates to produce both salicylate and catechol siderophore.

Isolate code	Siderophore production	
	Concentration of salicylate (mg/l)	Concentration of catechol (mg/l)
RK1	0.99	0.66
RK2	0.27	0.37
RK3	0.05	0.50
RK4	0.38	1.20
RK 5	0.58	0.52
RK 6	0.98	0.71
RK 7	0.58	0.60
RK 8	0.69	1.98
RK 9	0.62	1.21
RK 10	1.20	0.55
RK 11	0.64	0.41
RK 12	0.26	2.20
RK 13	0.18	0.62
RK 14	0.58	0.54
RK 15	0.17	1.63
RK 16	0.21	0.44
RK 17	0.25	0.95
RK 18	0.39	0.48
RK 19	0.41	0.56
RK 20	0.13	1.14
RK 21	0.49	0.12
RK 22	0.09	0.02
RK 23	0.41	0.26
RK 24	0.40	2.07
RK 25	0.43	0.95
RK 26	0.09	0.99
RK 27	2.82	0.92
RK 28	1.26	1.19
RK 29	0.30	1.03
RK 30	0.08	0.79
RK 31	0.03	0.76
RK 32	0.47	0.71
RK 33	0.06	0.24
RK 34	0.43	0.59
AK1	0.90	0.20
AK2	0.47	1.07
AK3	0.01	0.95
AK4	0.04	0.96
AK5	0.25	0.98
AK6	0.58	0.38
AK7	0.27	1.14
AK8	0.40	1.19
AK9	0.39	0.13
AK10	6.16	0.64
AK11	0.37	0.07
AK12	0.38	0.53
AK13	0.39	0.56
AK14	0.66	0.76
AK15	1.02	0.40
AK16	0.18	0.34
AK17	0.28	0.43
AK18	0.18	0.34
AK19	1.85	1.24
AK20	2.01	1.04
AK21	0.39	0.29
AK21	0.28	0.72
AK20	0.39	0.56
AK21	0.19	0.59

Table 3: Number of stems on treatment of Actinomycetes and Rhizobium.

Mean Actinomycetes	Mean Rhizobium			Total	NP BNJ
	*r*1	*r*2	*r*3	mean	α0.05
Number of stems					
a0	3.83	4.33	4.67	4.28a	
a1	4.83	5.67	4.67	5.06a	0.92
a2	5.67	4.67	4.83	5.06a	
a3	5.83	5.67	6.00	5.83a	
Total mean	5.04*x*	5.08*x*	5.04*x*		
NP BNJ α0.05		0.80			

Mean values followed by the same letter are not significantly different. $P < 0.005$ (Tukey test).

Table 4: Number of soybean's pods on treatment.

Mean Actinomycetes	Mean Rhizobium			Total	NP BNJ
	*r*1	*r*2	*r*3	mean	α0.05
Number of pods					
a0	30.50	38.00	34.50	34.33a	
a1	26.50	45.77	27.00	33.09a	8.31
a2	28.77	19.27	29.00	25.68a	
a3	39.50	38.00	43.27	40.26b	
Total mean	31.32*x*	35.26*x*	33.44*x*		
NP BNJ α0.05		7.19			

Mean values followed by the same letter are not significantly different. $P < 0.005$ (Tukey test).

Table 5: Number of seeds on the treatment of Actinomycetes and Rhizobium.

Mean Actinomycetes	Mean Rhizobium			Total	NP BNJ
	*r*1	*r*2	*r*3	mean	α0.05
Number of seeds per plant					
a0	91.50	76.00	103.50	90.33^{a}	
a1	53.00	91.50	61.00	68.50^{a}	25.42
a2	63.17	49.75	67.67	60.19^{a}	
a3	105.67	76.00	116.25	99.31a	
Total mean	78.33*x*	73.31*x*	87.10*x*		
NP BNJ α0.05		22.02			

Mean values followed by the same letter are not significantly different, $P < 0.005$ (Tukey test).

behavior, hyphal branching, growth, survival, reproduction, and exudates composition; and antibacterial metabolites production.

A novel plant-microbe interaction had been discovered by convincing the specific microbes which are known as *Streptomyces* sp and can induce a strong interaction between Glicyne max Merril and Rhizobium growing in soil. Numerous observations were made that show the importance of this interaction to the health of this legume when it was growing in soil. The microbial cells produced PHB in response to nitrogen limitation relative to carbon [15]. As a result of this, it induced the interaction between *Streptomyces* and Rhizobium when growing in the soil to perform a good growth of soybean and the yield of soybean. This condition supported by the significance of tricarboxylic acid cycle regulatory events like PHB synthesis and alanine synthesis also depends on the overall need to stabilize the oxidation-reduction state of pyridine nucleotides with the total carbon and nitrogen pools in the bacteroids [15].

The activities that promote plant growth could be an indirect way to influence Rhizobium and nodule symbiosis development. Actinomycete isolates improved both the number of stems and seeds. These Actinomycetes can be considered as PGPR because they may promote the colonization rate at different stages of bacterium-fungus-plant interactions. This ability was shown as the interaction between *Streptomyces* and *Pisum sativum* [23]. In the recent study in 2014, the interaction between Actinomycetes and Rhizobium on chickpea appears similar to the synergetic mutualistic interactions (Yadaf and Verma, 2014, p.70–77).

5. Conclusion

The reported experiments show that the target Actinomycete strains are able to improve plant growth and nutrition and benefit root colonization of soybean. Coinoculation with both types of microorganisms showed synergic effects at enhancing plant growth and nutrient acquisition. The results support the use of actinomycetes as plant growth-promoting and Rhizobium collaborator bacteria.

Acknowledgments

This study was partially supported by DIKTI of Indonesian Government.

References

[1] P. Frey-Klett and J. Garbaye, "Mycorrhiza helper bacteria: a promising model for the genomic analysis of fungal-bacterial interactions," *New Phytologist*, vol. 168, no. 1, pp. 4–8, 2005.

[2] S. Surange, A. G. Wollum II, N. Kumar, and C. S. Nautiyal, "Characterization of Rhizobium from root nodules of leguminous trees growing in alkaline soils," *Canadian Journal of Microbiology*, vol. 43, no. 9, pp. 891–894, 1997.

[3] L. M. Marra, S. M. Oliveira, C. R. F. S. Soares, and F. M. de Souza Moreira, "Solubilisation of inorganic phosphates by inoculant strains from tropical legumes," *Scientia Agricola*, vol. 68, no. 5, pp. 603–609, 2011.

[4] P. Sardi, M. Saracchi, S. Quaroni, B. Petrolini, G. E. Borgonovi, and S. Merli, "Isolation of the endophytic Streptomyces strains from surface-sterilized roots," *Applied and Environmental Microbiology*, vol. 58, no. 8, pp. 2681–2691, 1992.

[5] J. T. Coomb, P. P. Michelsen, and C. M. M. Franco, "Evaluation of endophytic actinobacter antagonist of Gauemannamomyces graminis var. tritici in wheat," *Biological Control*, vol. 29, no. 3, pp. 359–366, 2003.

[6] J. M. De Araujo, A. C. da Silva, and J. L. Azevedo, "Isolation of endophytic actinomycetes from roots and leaves of maize (Zea

mays)," *Brazilian Archives of Biology and Technology*, vol. 43, no. 4, pp. 447–451, 2000.

[7] T. Taechowisan, J. F. Peberdy, and S. Lumyong, "Isolation of endophytic actinomycetes from selected plants and their antifungal activity," *World Journal of Microbiology and Biotechnology*, vol. 19, no. 4, pp. 501–504, 2003.

[8] T. Okazaki, K. Takahashi, M. Kizuka, and R Enokita, "Studies on actinomycetes isolated from plant leaves," *Annual Reviews Sankyo Research Laboratory*, vol. 47, pp. 97–106, 1995.

[9] M. F. Correa, A. Quintana, C. Duque, C. Suarez, M. X. Rodriquez, and J. M. Barea, "Evaluation of actinomycete strains for key traits related with plant growth promotion and mycorrhiza helping activities," *Applied Soil Ecology*, vol. 45, no. 3, pp. 209–217, 2010.

[10] G. Schaaf, A. Honsbein, A. R. Meda, S. Kirchnert, D. Wipf, and N. vonWiren, "AtIREG2 encodes a tonoplast transport protein involved in iron-dependent Ni detoxification in *Arabidopsis thaliana* roots," *Journal of Biological Chemistry*, vol. 281, pp. 25532–25540, 2006.

[11] M. Park, C. Kim, J. Yang et al., "Isolation and characterization of diazotrophic growth promoting bacteria from rhizosphere of agricultural crops of Korea," *Microbiological Research*, vol. 160, no. 2, pp. 127–133, 2005.

[12] M. Reeves, P. L. Neilands, and A. Ballows, "Absence of siderophore activity in Leginella sp.grown in iron deficient media," *Journal of Bacteriology*, vol. 154, pp. 324–329, 1983.

[13] M. Goodfellow and T. Cross, "Classification," in *The Biology of Actinomycetes*, M. Goodfellow, M. Mordaski, and S. T. Williams, Eds., pp. 7–164, Academic Press, London, UK, 1984.

[14] N. R. Krieg, J. T. Staley, D. R. Brown et al., *Bergey's Manual of Systematic Bacteriology*, Springer, London, UK, 2nd edition, 2010.

[15] A. Cook and P. R. Meyers, "Rapid identification of filamentous actinomycetes to the genus level using genus-specific 16S rRNA gene restriction fragment patterns," *International Journal of Systematic and Evolutionary Microbiology*, vol. 53, no. 6, pp. 1907–1915, 2003.

[16] D. Gauthier, H. G. Diem, and Y. Dommergues, "In vitro nitrogenase fixation by two actinomycetes strains isolated from Casuarina nodules," *Applied and Environmental Microbiolgy*, vol. 14, pp. 306–308, 1981.

[17] C. Edwards, "Isolation properties and applications of thermophilic actinomycetes," *Applied Biochemistry and Biotechnology*, vol. 42, no. 2-3, pp. 161–179, 1993.

[18] C. Brule, P. Frey-Klett, J. C. Pierrat et al., "Survival in the soil of the ectomycorrhizal fungus Laccaria bicolor and the effects of a mycorrhiza helper Pseudomonas fluorescents," *Soil Biology and Biochemistry*, vol. 33, no. 12-13, pp. 1683–1694, 2001.

[19] K. A. El-Tarabily and K. Sivasithamparam, "Non-streptomycete actinomycetes as biocontrol agent of soil-borne fungal pathogens and as plant growth promoters," *Soil Biology and Biochemistry*, vol. 38, no. 7, pp. 1505–1520, 2006.

[20] H. Founoune, R. Duponnois, A. M. Ba et al., "Mycorrhiza helper bacteria stimulate ectomycorrhizal symbiosis of *Acacia holosericea* with Pisolithus albus," *New Phytologist*, vol. 153, no. 1, pp. 81–89, 2002.

[21] P. P. Rao, P. S. Birthal, S. Bhagavatula, and M. C. S. Bantilan, *Chick Pea and Pigeon Pea Economics in Asia: Facts, Trends and Outlook*, International Crops Research Institute for the Semi-Arid Tropics, Andhra Pradesh, India, 2010.

[22] J. M. Barea, M. J. Pozo, R. Azcon, and C. Azcon-Aguilar, "Microbial co-operation in the rhizosphere," *Journal of Experimental Botany*, vol. 56, no. 417, pp. 1761–1778, 2005.

[23] W. de Boer, L. B. Folman, R. C. Summerbell, and L. Boddy, "Living in a fungal world: impact of fungi on soil bacterial niche development," *FEMS Microbiology Reviews*, vol. 29, no. 4, pp. 795–811, 2005.

9

Amino Acids, Mineral Nutrients, and Efficacy of Vermicompost and Seafood and Municipal Solid Wastes Composts

Lord Abbey, Nana Annan, Samuel Kwaku Asiedu, Ebenezer Oluwaseun Esan, and Ekene Mark-Anthony Iheshiulo

Faculty of Agriculture, Department of Plant, Food, and Environmental Sciences, Dalhousie University, Truro, NS, Canada B2N 5E3

Correspondence should be addressed to Lord Abbey; loab07@gmail.com

Academic Editor: Manuel Tejada

Growing medium amino acids and mineral nutrients stimulate rhizosphere activities and plant growth. A greenhouse experiment was performed to compare amino acid and mineral nutrient profiles of seafood waste compost (SFWC) and municipal solid waste compost (MSWC) and vermicompost (VC). Their efficacies were also tested on onion (*Allium cepa* L. "Sweet Utah"). The control was Promix-BX™ alone. The MSWC, SFWC, and VC were composed of total of 36.4, 48.3, and 67.5 mg amino acids/100 g dry weight, respectively. Glutamic acid, aspartic acid, and glycine were the highest while methionine, histamine, and cysteine were the least in all the amendments. The VC had the highest Ca content but the least P and K contents. SFWC had the highest content of P and K and most of the determined micronutrients. The MSWC had significantly ($P < 0.05$) the highest N and leaf chlorophyll contents followed by the VC. The MSWC significantly ($P < 0.05$) increased anthocyanin content while the control recorded the least. The maximum quantum yield of photosystem II (Fv/Fm) and the potential photosynthetic capacity (Fv/Fo) were least in the VC treated plants. Dry matter was not affected by the type of amendment. Overall, plant growth was improved by the MSWC. Future research should investigate effect on secondary metabolites.

1. Introduction

Proteins are derived from amino acids and are essentially the basic component of all living cells [1, 2]. Amino acids and proteins play vital roles in the genetic and the physiological processes of organisms such as syntheses of nucleic acids, proteins, enzymes alkaloids, vitamins, terpenoids and chlorophyll, and formation of vegetative tissues [3–5]. These require specific and adequate quantities of plant tissue amino acids with each or combination of the amino acids performing specific function. Root uptake of amino acids can vary with the concentration, plant genotype, and rhizosphere microbial activities while microorganisms and plant roots compete for the available amino acids [6, 7]. For instance, onion *(Allium cepa)* is one of the few plants that use the amino acid, arginine, as a pool for nitrogen (N) during bulb formation and development [8]. This leads to an increase in chlorophyll concentration and the greening of onion leaf tissues because of the availability of free amino acids in the growing medium [9]. However, Kandil et al. [10] reported after a 2-year field trial that onion plant growth did not respond to foliar application of amino acids relative to the control treatment, but increased growth response was recorded for those onion plants that were sprayed with humic acid extracts.

Seafood such as fish, mollusk, and crustacean contains 10.3–23.3% proteins, which is composed of all the essential amino acids [11]. These proteins can be degraded during composting processes and humified into amino acids and sugars. The concept of composting seafood waste is relatively new in many parts of the world. Moreover, there are few studies that provided some information on chemical composition of seafood waste compost. On the other hand, vermicompost made by the process of earthworm (family Annelida) ingestion, digestion, and excretion, is well studied and widely used in agriculture. Many studies have confirmed the efficacy of vermicompost and its ability to enhance chemical constituents of plants such as kale (*Brassica oleracea* var. *sabellica* [12]) and aromatic and medicinal plants (*Plectranthus amboinicus* [13]) as compared to other natural amendments but excluded seafood waste compost. It was found that vermicompost increased the amino acids, flower yield, and essential oil content of *Matricaria chamomilla* [14]. Also, a study by Shaheen et al. [2] suggested that high rate of application of municipal solid waste compost increased the contents of onion bulb protein and mineral nutrients.

Onion is the second most consumed vegetable in the world with high content of antioxidants, vitamin C, folic acid, calcium (Ca), iron (Fe), and protein [15]. The content of these secondary metabolites and nutrients may be influenced by the composition of growing medium free amino acids as previously explained by Kopriva and Rennenberg [16] and Jin et al. [9]. Protein-rich seafood offal has relatively low content of connective tissue, which makes it easy to compost. During composting, proteins are broken down by microbial and enzymatic activities and humified into amino acids and sugars. The feedstock for vermicompost and municipal solid waste compost can vary, and the composition of the aggregate may not be richer in protein than seafood compost feedstock. Therefore, the objective of this study was to compare the amino acid and mineral nutrient profiles of seafood waste compost, municipal solid waste compost, and vermicompost and their agronomic impact on "Sweet Utah" onion.

2. Materials and Methods

2.1. Composting and Compost Utilization. Composting of seafood waste obtained from a local grocery store started in September 2015 and finished in January 2016. The seafood waste compost was made by adding approximately 10 kg of shredded fish and crustacean waste including the offal to a 10 L rubber-maid container with perforations on the walls for aeration. Dry rye straw was cut into pieces and added to the seafood waste as source of carbon (C) with the aim of achieving a C : N ratio of 25–30 : 1. Estimation of the C : N ratio was based on the formula by Richard and Trautmann [17]. These materials were added in layers, that is, the seafood waste on top of the straw followed by 100 g of finished seafood waste compost as a booster, and then strawed before sprinkling 1 L of water to moisten the mass. The layers were repeated in that same order until the container was one-third full before placing on a lid. The mass was turned and mixed once a month. The compost reached a peak temperature of 58°C (thermophilic stage) before cooling down at the mesophilic and maturation stages to ≤20°C after four months and allowed to cure for another month. The finished seafood waste compost was then stored at 4°C. For comparative analysis, mature municipal solid waste compost (7 months old) from Fundy Compost (Brookfield, Nova Scotia) and vermicompost (6 months old) produced by red wiggler worms *(Eisenia fetida)* from Cathy's Crawly Composters (Bradford, Ontario) were the other treatments used in the study.

2.2. Growing Medium Preparation. Based on previous Manure compost study by Abbey and Appah [18], 25% (w/w) of the seafood waste compost, municipal solid waste compost, and the vermicompost were individually incorporated into 400 g of Promix-BX soilless potting mix (Premier Horticulture Inc., Quakertown, PA, USA) purchased from Halifax Seeds, Nova Scotia, using 25.4 cm diameter plastic pots. The control treatment was a 400 g of the Promix-BX potting mix alone without the addition of any amendment. Each of the potted medium was saturated with water and allowed to stand for 24 h to attain field water-holding capacity prior to transplanting "Sweet Utah" onion seedlings.

2.3. Plant Culture. Seeds of "Sweet Utah" onion were initially sown in a 72-plastic cell-tray in the greenhouse and nurtured for five weeks after germination. Seedlings were not fertilized until after transplanting into the individual organic amendments. Three seedlings were transplanted into each pot, which was later thinned out to two seedlings per pot two weeks after transplanting (WAT). Each amendment was applied in two splits, that is, at planting and at 5 WAT. The greenhouse set temperature was 24°C in the day and 16°C at night and 72% relative humidity. Supplementary lighting was provided by a 600 W HS2000 high pressure sodium lamp with NAH600.579 ballast (P.L. Light Systems, Beamsville, ON, Canada). Approximately, 200 ml of tap water was added to each potted-plant every two days until the time of harvest.

2.4. Amino Acid and Mineral Nutrient Analyses. Approximately 200 g samples of each organic growing medium amendment (i.e., seafood waste compost (SFWC), municipal solid waste compost (MSWC), and vermicompost (VC) was sent to Covance Laboratories (Madison, Wisconsin, USA) for analyses of 17 amino acids (i.e., aspartic acid, threonine, serine, glutamic acid, proline, glycine, alanine, valine, isoleucine, leucine, tyrosine, phenylalanine, lysine, histidine, arginine, cysteine, and methionine). The Covance Laboratory reference method used for the amino acids analyses was Amino Acids TAALC_S.17 as described by Henderson and Brooks [19]. Additionally, approximately 200 g samples of the different types of compost and 20 g of harvested plants per treatment were sent to the Nova Scotia Department of Agriculture Laboratory Services (Truro, Nova Scotia, Canada) for the analyses of macro- and micronutrients, crude protein, and crude fat using the dry ash method based on Association of Analytical Communities method 968.08 as described in AOAC [20].

2.5. Plant Growth Determination. The onion plants were harvested as green onions and all the growth data were collected at final harvest at 8 WAT. The number of green leaves per plant was recorded for each treatment. Plant height was measured from the stem disc of the onion plant to the tip of the longest leaf. Total plant fresh weight was recorded for each treatment. Onion bulb development was estimated from the ratio of bulb-to-pseudostem diameter. The diameter of the swollen base of the pseudostem was measured as the bulb diameter while the pseudostem diameter was measured from the middle portion of the pseudostem. The root mass of each plant was washed under running tap water and the fresh weight was recorded. Leaf chlorophyll content was measured by leaf greenness using SPAD 502 Chlorophyll meter (Spectrum Technologies Inc., Illinois, USA). Leaf anthocyanin content was determined using portable ACM-200 Anthocyanin meter (OptiSciences Inc., New Hampshire,

USA). Both chlorophyll and anthocyanin contents were estimated from the 3rd and 4th leaves of each plant. Photosynthetic activities of the tagged leaves were determined using Os30p+ Chlorophyll Fluorometer (Opti-Science Inc., NH, USA) based on the methods of Kitajima and Butler [21] and Kalaji et al. [22] with some modifications. The middle portion of the 3rd and 4th leaves excluding the midrib was clipped with a dark adaptation clip with the window closed for 25 mins prior to recording of the photosynthesis indices on the Chlorophyll Fluorometer. The data collected included variable fluorescence (Fv = Fm − Fo), where Fo and Fm are minimum and maximum fluorescence, respectively. Data on maximum quantum efficiency as determined by the ratio of variable florescence to maximum florescence (Fv/Fm) and the ratio of efficiency of electron donation to photosystem II estimated by the ratio of variable fluorescence to minimum fluorescence (Fv/Fo) were also recorded.

2.6. Experimental Design and Data Analysis. The single-factor experiment was laid out in a completely randomized design with three replications. Each of the experimental treatment was comprised of ten pots with each containing two plants. The pots were periodically rearranged on the greenhouse bench to offset any unforeseen variations in environmental condition. Data collected were subjected to one-way analysis of variance (ANOVA) using Minitab v. 18.1. Differences in experimental treatments were identified using the Fisher's least significance difference (LSD) method when the ANOVA suggested a significant difference at $P \leq 0.05$. Graphs with standard error bars were also plotted using Microsoft Excel.

3. Results and Discussion

The concentrations of the individual 17 determined amino acids were consistently and significantly ($P < 0.05$) highest in the VC and least in the MSWC, that is, VC > SFWC > MSWC (Figure 1). Moreover, the sum total of the 17 analyzed amino acids in the MSWC, SFWC, and VC was approximately 36.4, 48.3, and 67.5 mg/g dry weight, respectively. These amino acids were part of the humic substances in the MSWC, SFWC, and the VC. The differences in the total and individual amino acid concentrations can be attributed to variations in the starting feedstock, the biodegradation process employed, the structure of the microbial community, and overall activities of hydrolyzing enzymes [23, 24]. According to Fuentes et al. [23], the composition of amino acids in compost may range between 4% and 20%. For instance, the three amino acids with the highest concentrations in the different types of organic amendments were glutamic acid (5.55–6.90 mg/g), aspartic acid (4.56–9.14 mg/g), and glycine (3.37–6.59 mg/g). On the contrary, the three amino acids with the least concentrations were cysteine (0.44–0.81 mg/g), histamine (0.62–0.98 mg/g), and methionine (0.67–0.78 mg/g). All the other remaining 11 amino acids were intermediate in amounts (Figure 1). The significance in the differences in the amounts and the types of amino acids can have an impact on the chemical properties and efficacies of the three different organic amendments.

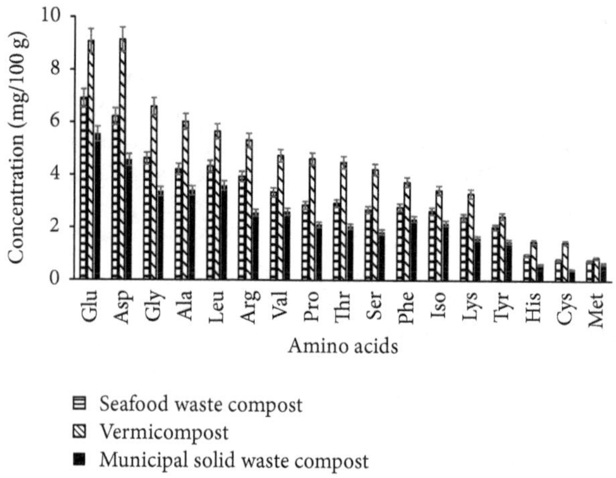

FIGURE 1: Amino acids composition in seafood waste compost, vermicompost, and municipal solid waste compost. Glu, glutamic acid; Asp, aspartic acid; Gly, glycine; Ala, alanine; Leu, leucine; Arg, arginine; Val, valine; Pro, proline; Thr, threonine; Ser, serine; Phe, phenylalanine; Iso, isoleucine; Lys, lysine; Tyr, tyrosine; His, histidine; Cys, cysteine; and Met, methionine.

The high amino acids in the VC as compared to the MSWC and SFWC can be attributed to microbial and enzymatic activities in the gut of the red wiggler worms during the process of worm ingestion, digestion, and excretion [25]. Furthermore, Vialykh et al. [24] suggested that variations in the composition of amino acids in soils and humic substances were due to the presence of wide range of amino acids in microbial biomass. This can explain the high amino acids content in the VC, that is, 40% greater than the SFWC and 85% greater than the thermophilically produced MSWC (Figure 1). The amino acid concentrations of the SFWC were also 33% greater than those of the MSWC due to the protein-rich seafood feedstock [11]. Comparatively, glutamic acid and aspartic acid were highest in all the three organic amendments. However, methionine was the least in the VC while histamine and cysteine were the least in the SFWC and the MSWC. It was reported that free amino acids improve microflora growth, which in turn enhanced soil nutrients assimilation [5, 7]. Thus, the variations in amino acid composition in the three different organic growing amendments can influence nutrient uptake by the plants growing in them.

The mineral nutrient contents of the SFWC, MSWC, and VC were significantly ($P < 0.05$) different (Figures 2 and 3). Overall, Ca and N were highest in all the organic amendments while magnesium (Mg) and phosphorus (P) were the least (Figure 2). Comparatively, the VC had the highest Ca content but the least P and potassium (K) contents while N, P, and K were highest in the SFWC (Figure 3). This was expected because of the high protein and mineral nutrient contents of seafood [11].

Onion plant tissue content of K was the highest followed by Ca and N, and the least were Mn and Zn, irrespective of the amendment (Figure 4). However, it was found that the uptake of K and Zn were highest in plants grown in the SFWC whereas plants grown in the MSWC had the highest

TABLE 1: Onion "Sweet Utah" plant growth and development parameters as affected by different growing medium organic amendments.

Treatment	% DM[2]	CP	CF	SPAD value	Anthocyanin	Fv/Fm	Fv/Fo
SFWC[1]	10.7[a]	6.04[b]	2.11[ab]	46.3[b]	6.8[b]	0.791[a]	3.805[a]
VC	10.5[a]	6.77[ab]	2.15[ab]	47.1[ab][3]	7.0[b]	0.768[b]	3.241[b]
MSWC	10.7[a]	7.95[a]	2.22[a]	51.4[a]	8.6[a]	0.783[ab]	3.626[a]
Control	11.0[a]	5.83[c]	1.91[b]	43.2[b]	5.2[c]	0.780[ab]	3.566[a]
P value	ns	0.05	0.05	0.02	0.01	0.03	0.01

[1] SFWC, seafood waste compost; VC, vermicompost; MSWC, municipal solid waste compost; [2] DM, dry matter content; CP, crude protein; CF, crude fat. [3] Means in columns followed by the same letter are not significantly different at $\alpha = 0.05$, Least Significant Difference; ns, not significantly different at $\alpha = 0.05$.

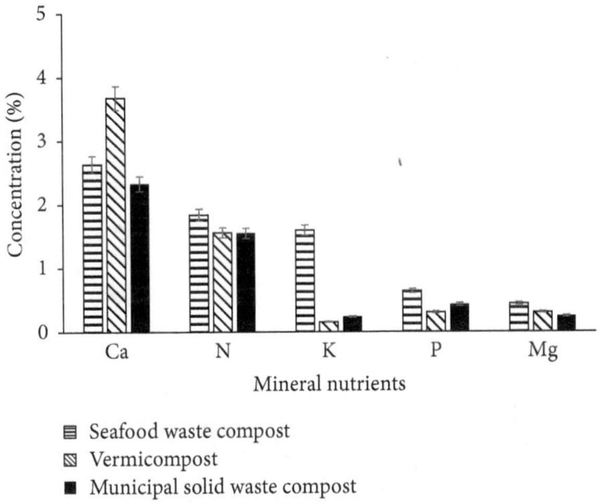

FIGURE 2: Macronutrients composition in seafood waste compost, vermicompost, and municipal solid waste compost. Calcium (Ca); nitrogen (N); potassium (K); phosphorus (P); and magnesium (Mg).

uptake of Mn, sodium (Na), N, Ca, and Mn as compared to the other treatments. Magnesium uptake did not vary among the plants irrespective of the growing medium amendment. Also, plant uptake of N, Mn, and P were not different between the SFWC and VC. Previous report suggested that cysteine and methionine were involved directly in the assimilation of N, S, and C in plants [16]. This suggested that the high amino acid content of the VC can be associated with an increase in N and crude protein contents of plants grown in the VC as compared to onion plants grown in the other amendments. It was reported that N uptake may be inhibited by amino acids accumulation in the cytoplasm of plant cells [26], which was contrary to the present result. Onion plant tissue N content was reduced by the low N content of the control treatment, that is, Promix-BX soilless potting mix alone (Figure 4). Percentage dry matter and crude fat contents of onion "Sweet Utah" were similarly affected by the three different types of organic amendments with moderate increase by the MSWC and moderate reduction by the control (Table 1). There was no significant ($P > 0.05$) association between percentage dry matter, crude fat, and crude protein contents in the present study.

Leaf greenness (chlorophyll content) as determined by SPAD value and anthocyanin content were significantly ($P < 0.05$) increased by the MSWC followed by the VC although values for the VC were not significantly ($P > 0.05$) different from those for the SFWC. Similar observation was made in previous onion bulb studies by Jin et al. [9]. The low pH of the control Promix-BX soilless potting mix (5.8) and the SFWC (5.9) may be the cause of the reduced leaf greenness and leaf tissue anthocyanin content. The lack of correlation (not presented) between growth and yield responses versus the amino acid and mineral nutrient compositions of the three different organic amendments conformed to the report by Kandil et al. [10].

The maximum quantum yield of PSII photochemistry and the potential photosynthetic capacity as determined by Fv/Fm and Fv/Fo, respectively, were significantly ($P < 0.05$) reduced in onion plants grown in the VC as compared to the other amendments including the control plants (Table 1). Lower values of these physiological indices were indications of lower chloroplasts photochemical activities and reduced photosynthesis [27]. The ratio of Fv/Fo represented the structural alterations in PSII and also indicated the electron donation efficiency [28, 29]. Although there were significant ($P < 0.05$) differences in photosynthetic activities between treatments, the percentage dry matter content (Table 1), and plant growth and yield components such as the number of leaves, plant height, pseudostem and bulb diameters, bulbing ratio, and roots and foliage, fresh weights were not significantly ($P > 0.05$) affected (data not presented) as reported by Kandil et al. [10]. The reason for the relatively low photosynthetic efficiency recorded for VC despite the high amino acids and mineral nutrients contents is not clear and will be investigated in future studies. However, it is possible that acquired chemicals were used in secondary metabolism instead of primary metabolism as there is a tradeoff between the two.

In conclusion, the findings demonstrated distinct variations in amino acid and mineral nutrient profiles of the SFWC, MSWC, and VC. Plant growth was improved by the MSWC as compared to the others. Comparatively, the VC had the highest amount of amino acids and mineral nutrients followed by the SFWC. Although the VC did not improve plant growth, previous research confirms VC enhancement of plant biosynthesis of secondary metabolites. Results from earlier studies were contradictory. Some studies showed that amino acids application increased plant growth and yield [4, 16] while others such as the study by Kandil et al. [10] did not find any significant difference. Unlike previous investigations, the current study did not apply pure amino acids as experimental treatment and as such the effects were

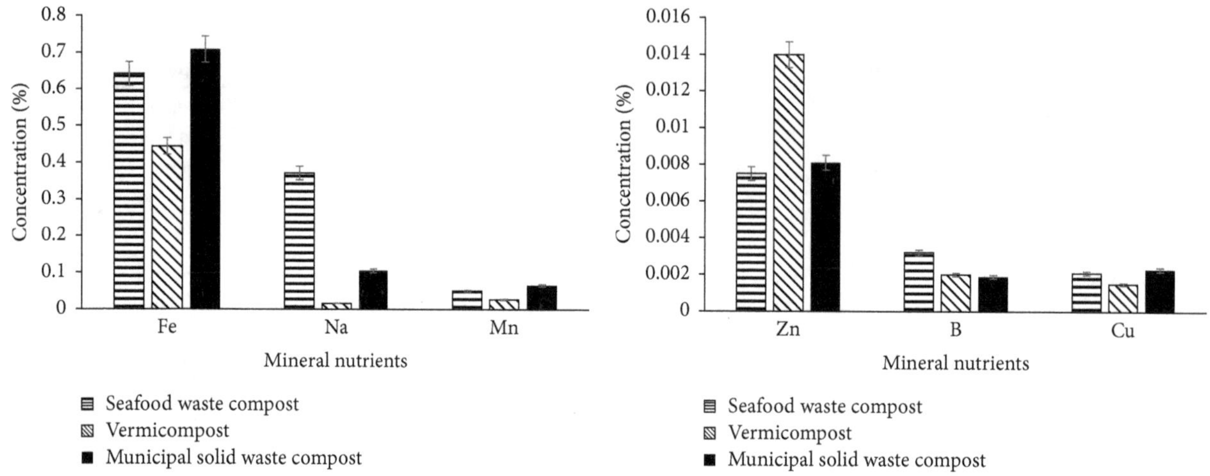

FIGURE 3: Micronutrients composition in seafood waste compost, vermicompost, and municipal solid waste compost. Iron (Fe); sodium (Na); manganese (Mn); zinc (Zn); boron (B); and copper (Cu).

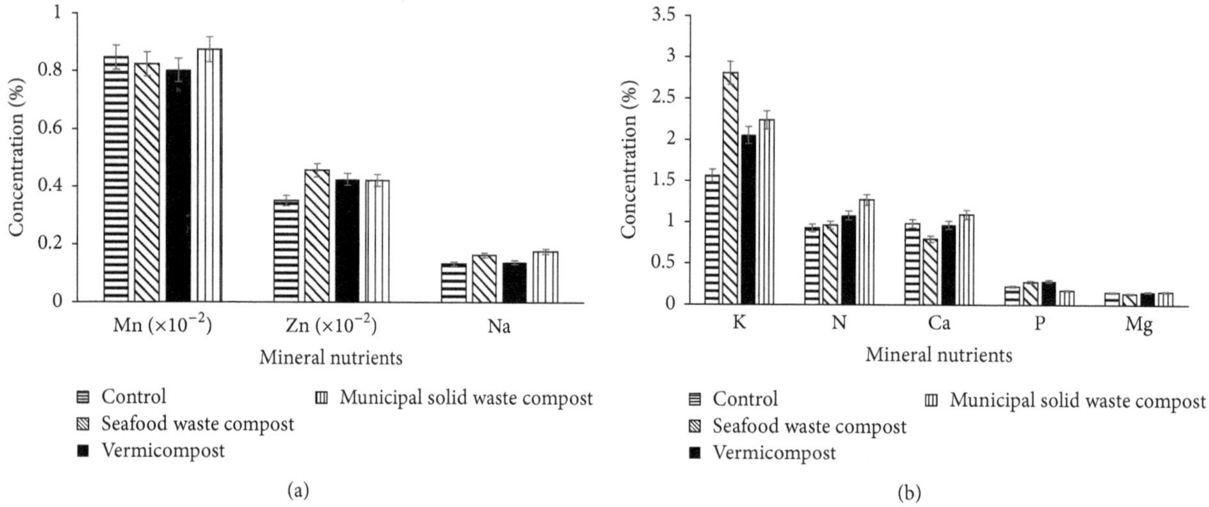

FIGURE 4: Micronutrients (a) and macronutrients (b) uptake by onion "Sweet Utah" plants grown in peat moss (control), seafood waste compost, vermicompost, and municipal solid waste compost. Manganese (Mn × 10^{-2}%); zinc (Zn × 10^{-2}%); sodium (Na); potassium (K); calcium (Ca); phosphorus (P); and magnesium (Mg).

not expected to be the same. Also, Hansen [8] reported reductions in amino acid content of onion bulbs as the plant ages. Therefore, the early harvest of the plants as immature green onions in the present study might be a contributory factor to the no-effect on plant growth and yield. Thus, future research should investigate and compare SFWC, MSWC, and VC effect on secondary metabolite biosynthesis in mature onion plants.

Acknowledgments

The authors wish to thank Ms. Lindsey Gillard for providing the compostable seafood waste for this project. This research benefitted from the start-up fund provided by Dalhousie University Faculty of Agriculture.

References

[1] A. M. Shaheen, A. R. Fatma, G. B. Awatef, M. K. Nagwa, H. H. Hassan, and H. H. Foly, "Total and exportable bulbs yield of onion as affected by msw compost and urea fertilizers," *Journal of Applied Sciences Research*, vol. 9, no. 1, pp. 156–162, 2013.

[2] A. M. Shaheen, A. R. Fatma, M. S. Omaima, and M. O. Bakry, "Sustaining the quality and quantity of onion productivity throughout complementarity treatments between compost tea and amino acids," *Middle East Journal of Agriculture Research*, vol. 2, no. 1, pp. 108–115, 2013.

[3] F. A. Abo Sedera, A. Amany, A. El-Latif, L. A. A. Bader, and S. M. Rezk, "Effect of NPK mineral fertilizer levels and foliar application with humic and amino acids on yield and quality of

strawberry," *Egypt Journal of Applied Science*, vol. 25, pp. 154–169, 2010.

[4] S. A. Faten, A. M. Shaheen, A. A. Ahmed, and R. M. Asmaa, "The effect of foliar application of urea and amino acids mixtures as antioxidants on the growth and yield and characteristics of squash," *Research Journal of Agriculture Biological Science*, vol. 6, no. 5, pp. 583–588, 2010.

[5] S. A. El-Desouky, F. H. Ismaeil, A. L. Wanas, E. S. L. Fathy, M. M. Abd El-All, and M. M. Abd, "Effect of yeast extract, amino acids and citric acid on physioanatomical aspects and productivity of tomato plants grown in late summer season," *Minufiya Journal of Agriculture Research*, vol. 36, no. 4, pp. 859–884, 2011.

[6] D. L. Jones, D. Shannon, T. Junvee-Fortune, and J. F. Farrar, "Plant capture of free amino acids is maximized under high soil amino acid concentrations," *Soil Biology & Biochemistry*, vol. 37, no. 1, pp. 179–181, 2005.

[7] V. C. Hawkes, K. M. DeAngelis, and M. K. Firestone, "Root interactions with soil microbial communities and processes," in *The Rhizosphere—An Ecological Perspective*, M. Bonkowski, Ed., pp. 1–29, Academic Press, Cambridge, Mass, USA, 2007.

[8] S. L. Hansen, "Content of free amino acids in onion (Allium cepa L.) as influenced by the stage of development at harvest and long-term storage," *Acta Agriculturae Scandinavica, Section B—Soil & Plant Science*, vol. 51, no. 2, pp. 77–83, 2001.

[9] L. E. Jin, K. S. Yoo, and B. S. Patil, "Identification of candidate amino acids involved in the formation of pink-red pigments in onion (Allium cepa L.) juice and separation by HPLC," *Journal of Food Science*, vol. 75, no. 8, pp. C684–C689, 2010.

[10] A. A. Kandil, A. E. Sharief, and H. F. Fathalla, "Onion yield as affected by foliar application with amino and humic acids under nitrogen fertilizer levels," *ESci Journal of Crop Production*, vol. 2, no. 1, pp. 62–72, 2013.

[11] FHF, *A Sea of Health—Nutritional Content and Health Benefits of Seafood*, 20i533, The Fishery and Aquaculture Industry Research Fund, Oslo, Norway, 2010.

[12] L. Abbey, T. H. Pham, N. Annan, A. Leke-Aladekoba, and R. H. Thomas, "Chemical composition of kale as influenced by dry vermicast, potassium humate and volcanic minerals," *Food Research International*, vol. 107, pp. 726–737, 2018.

[13] B. Zhang, N. M. Wijesundara, L. Abbey, and H. P. V. Rupasinghe, "Growing medium amendments effect on growth, secondary metabolites and anti-streptococcal activity of two species of Plectranthus," *Journal of Applied Research on Medicinal and Aromatic Plants*, vol. 5, pp. 53–59, 2017.

[14] M. Reza, H. S. Hadi, M. T. Darz, Z. Gh, and G. Riazi, "Effects of vermicompost and amino acids on the flower yield and essential oil production from Matricaria chamomile L," *Journal of Medicinal Plant Research*, vol. 5, no. 23, pp. 5611–5617, 2011.

[15] NAO, *Onions—Phytochemicals and Health Properties*, National Onion Association, Greely, Colo, USA, 2011.

[16] S. Kopriva and H. Rennenberg, "Control of sulphate assimilation and glutathione synthesis: Interaction with N and C metabolism," *Journal of Experimental Botany*, vol. 55, no. 404, pp. 1831–1842, 2004.

[17] T. Richard and N. Trautmann, *C/N Ratio*, Cornell Waste Management Institute, Cornell University, Ithaca, NY, USA.

[18] L. Abbey and P. Appah, "Pot-grown Swiss chard and kale responses to variable rate of manure compost in mycorrhizal fungi inoculated medium," *Acta Horticulturae*, vol. 1164, pp. 241–248, 2017.

[19] J. W. Henderson and A. Brooks, "Improved amino acid methods using Agilent Zorbax Eclipse plus C18 Columns for a variety of Agilent LC instrumentation and separation goals," Agilent Application Note 5990-45472010, 2010.

[20] AOAC, *Official Methods of Analysis Minerals Analysis Flame Atomic Absorption Spectroscopy (FAAS)—Item 58*, Method 968.08, Association of Analytical Communities, Gaithersburg, Md, USA, 17th edition, 2006.

[21] M. Kitajima and W. L. Butler, "Quenching of chlorophyll fluorescence and primary photochemistry in chloroplasts by dibromothymoquinone," *Biochimica et Biophysica Acta*, vol. 376, no. 1, pp. 105–115, 1975.

[22] H. M. Kalaji, Govindjee, K. Bosa, J. Kościelniak, and K. Zuk-Gołaszewska, "Effects of salt stress on photosystem II efficiency and CO2 assimilation of two Syrian barley landraces," *Environmental and Experimental Botany*, vol. 73, no. 1, pp. 64–72, 2011.

[23] M. Fuentes, R. Baigorri, F. J. González-Vila, G. González-Gaitano, and J. M. García-Mina, "Pyrolysis-gas chromatography/mass spectrometry identification of distinctive structures providing humic character to organic materials," *Journal of Environmental Quality*, vol. 39, no. 4, pp. 1486–1497, 2010.

[24] E. A. Vialykh, S. A. Ilarionov, H. M. Abdelrahman, and I. A. Vialykh, "Changes in amino acids content of humic acids sequentially extracted from peat and sod-Podzolic soil," *Canadian Journal of Soil Science*, vol. 94, no. 5, pp. 575–583, 2014.

[25] C. Lazcano and J. Domínguez, "The use of vermicompost in sustainable agriculture: Impact on plant growth and soil fertility," *Soil Nutrients*, pp. 211–234, 2011.

[26] A. J. Barneix and H. F. Causin, "The central role of amino acids on nitrogen utilization and plant growth," *Journal of Plant Physiology*, vol. 149, no. 3-4, pp. 358–362, 1996.

[27] F. Garousi, B. Kovacs, and S. Veres, "Investigation of photosynthesis status of sunflower plants up-taking different forms of selenium," *Advances in Plants and Agriculture Research*, vol. 3, no. 1, Article ID 00083, 2016.

[28] M. Havaux and R. Lannoye, "Drought resistance of hard wheat cultivars measured by a rapid chlorophyll fluorescence test," *The Journal of Agricultural Science*, vol. 104, no. 3, pp. 501–504, 1985.

[29] F. Babani and H. K. Lichtenthaler, "Light-induced and age-dependent development of chloroplasts in etiolated barley leaves as visualized by determination of photosynthetic pigments, CO2 assimilation rates and different kinds of chlorophyll fluorescence ratios," *Journal of Plant Physiology*, vol. 148, no. 5, pp. 555–566, 1996.

10

Impact of Ridge-Furrow Planting in Pakistan: Empirical Evidence from the Farmers Field

Imtiaz Hussain,[1] Akhter Ali,[1] Ansaar Ahmed,[2] Hafiz Nasrullah,[3] Badar ud Din Khokhar,[4] Shahid Iqbal,[5] Azhar Mahmood Aulakh,[6] Atta ullah Khan,[7] Jamil Akhter,[8] and Gulzar Ahmed[9]

[1]*CIMMYT, CSI, NARC, Park Road, Islamabad, Pakistan*
[2]*CIMMYT, Cereal Crop Research Institute, Nowshera, Pakistan*
[3]*Agronomic Research Station, Bahawalpur, Pakistan*
[4]*Wheat Research Institute Sindh, Sakrand, Pakistan*
[5]*Agriculture Research Institute, DI Khan, Pakistan*
[6]*Arid Zone Research Institute, Bhakkar, Pakistan*
[7]*Arid Zone Research Institute, Umerkot, Pakistan*
[8]*National Rural Support Program, Islamabad, Pakistan*
[9]*Cereal Crop Research Institute, Nowshera, Pakistan*

Correspondence should be addressed to Imtiaz Hussain; i.hussain@cgiar.org

Academic Editor: Manuel Tejada

Wheat is planted through the broadcast method on 7.8 million ha and irrigated through low-efficiency flood basin irrigation methods. With decreased water availability, there is a need to adopt water use efficient planting techniques. The current paper uses two-year trials data set as well as farmer survey data to demonstrate benefits of ridge-furrow planting wheat and farmer perception regarding the technology in Pakistan. During 2014–15 and 2015–16, ridge-furrow planting of wheat was compared with farmer practice of planting on flat surface through broadcasting at 162 sites located in 15 districts of Pakistan. In a survey, 134 farmers shared their experiences about ridge-furrow planting. Data collected from field trials and demonstrations showed that better tillering and crop growth contributed towards 12% higher wheat grain yield with ridge-furrow planting in comparison with the farmer practice of flat planting in all 15 districts during both years. Furrow-irrigated ridge planting of wheat helped farmers in saving of 30–35% water in comparison with farmer practice. Farmers adopting ridge-furrow wheat planting earned USD 100.63 per hectare more profit than nonadopter farmers. Adoption of this technology on one million hectare can help in saving 1 million acre foot of water and produce additional 0.36 million tons of wheat per year.

1. Introduction

Agriculture sector employs about 42.3% of the labor force and contributes about 19.5% to the GDP (gross domestic product) of Pakistan [1]. The total cultivated area is 22.12 million hectares, and out of this, 19.23 million hectares are being irrigated through various sources [1]. In Pakistan, 5.96 million hectares are irrigated through canals, whereas the rest of 8.15, 3.71, and 0.65 million hectares are irrigated through canal tube wells, tube wells, and wells, respectively [1].

Wheat, a major staple food crop, planted on more than 9.22 million hectares during 2015–16 contributed 1.9% to GDP [1]. Wheat is generally planted by the broadcast method [2–4] during *rabi* season in irrigated areas of Pakistan, and drill sowing is minimal. Irrigated wheat covers 8 million hectare area that accounted for 85% of the total wheat area with an average productivity of 3.0 t/ha.

In Pakistan, water availability in the normal year for *rabi* season was around 36.9 MAF. During the last decade (2007–08 to 2016–17), water availability for *rabi* season was

5–45% lower in comparison with the normal year [1]. In canal-irrigated areas, farmers get fixed turn time to irrigate crop once in a week [3], which is not enough to irrigate their whole cropped area, and generally rely on either tube well water or rainfall. Majority of the wheat area is irrigated through flood or the basin irrigation method [3] which has 30–50% irrigation efficiency. Because of uncertainty and limited water availability at critical growth stages, farmers tend to apply too much water to the wheat crop when available, which also affects the wheat growth negatively. Availability of water is reducing, and there is a need to adopt water-saving techniques.

Furrow-irrigated raised bed technology is one of the efficient irrigation methods in which water moves in furrow, and crop is planted on raised beds. In the Yaqui Valley of Mexico, farmers have shifted to a bed and furrow system for wheat planting due to scarcity of water [5]. The technique resulted improvement in irrigation and nutrient management, saving in water, better crop stand, lower seed rate, and reduction in lodging [5]. Bed and furrow planting systems improved water distribution [6], reduced water requirement for irrigation, improved water use efficiency in the rice-wheat cropping system of Indo-Gangetic plains [7, 8], and provided drainage in water logged fields [5]. Raised bed planting provides an opportunity for mechanical weed control, permits band application of fertilizers, improves nitrogen use efficiency, and reduces crop lodging [9–12]. Selection of the variety is also important in raised bed planting. Wheat varieties with the broad leaf area and ground cover had higher grain yield in comparison with upright and compact structure varieties lower in bed planting [10].

Raised bed planting under the ZT condition in maize-based systems had 13–28% higher system productivity and net return of USD 312 ha-1 in comparison with the conventional planting system [13]. Integration of short duration mung bean crop in the maize-wheat system on permanent beds enhanced the system productivity by 29%, net returns by 38%, and water productivity by 24% compared to the conventional system [14].

In Pakistan, wheat sown on 70 cm bed and furrow system in the rice-wheat area of Punjab produced good yields due to better spike length, number of grains per spike [15]. Mollah et al. [16] studied the raised bed planting method of wheat in Bangladesh and found higher wheat productivity by planting wheat on a 70 cm bed in the rice-wheat cropping system. Waraich et al. [17] found an increase of 18–45% higher WUE (water use efficiency) without yield increase under bed planting with furrow irrigation in comparison with conventional flat planting with flood irrigation.

Water Management Research Center at University of Agriculture, Faisalabad, introduced bed and furrow planter to plant four wheat rows on a 90 cm bed and furrow system [11], and the results showed around 49% saving in irrigation water timing and 18% higher yield with raised bed planting in comparison with flat sowing [11, 18, 19]. Adopting farmers had benefit of Rs.17000/ha and 50% saving in water with an additional cost of Rs.1350/ha [20]. PARC in collaboration with ACIAR planted 7 rows of wheat or 2 rows of maize on the 180 cm bed and furrow and observed 30–32% increase in grain yield and water saving [21].

In wheat-based systems, wheat is planted after rice/cotton/sugarcane/maize in Pakistan. Presence of previous crop residue hampers tillage and seeding operations for wheat. This had a negative effect on adoption of drill and bed planting techniques, and in resultant, majority of farmers are forced to plant wheat through the broadcast method with heavy tillage. In ridge-furrow planting of maize and cotton, ridge and furrows formed with a ridger and cotton or maize planted manually on the top or the side of the ridge increased germination and yield of cotton in comparison with flat planting [22] and is widely practiced in Pakistan.

Ridge planting was evaluated in comparison with bed planting and drill sowing, and 17–24% higher wheat yield was observed with bed panting in comparison with the conventional method of broadcasting. Raised bed and ridge sowing methods of wheat planting saved 22% irrigation water over flat sowing [23]. In the WPEP (Wheat Productivity Enhancement Program) project, ridge planting was evaluated for wheat planting and ridge planting of wheat had an improvement of 23% in yield and 40% in saving of irrigation water in comparison with the farmer's technique of broadcasting [24].

The ridge-furrow planting comprises land preparation, broadcasting of seed, and formation of ridges and furrow with the ridger, while the farmer's practice includes land preparation and broadcasting of the seed and shallow cultivation with the cultivator and planking. Bed or ridge planting of wheat is being adopted across all the four provinces in Pakistan, that is, Punjab, Sindh, Khyber Pakhtunkhwa (KP), and Balochistan. Keeping in view the advantage of technology and availability of ridgers with farmers, ridge-furrow planting of wheat was widely evaluated in farmers' fields in various districts of the wheat cropping system of Pakistan.

The main purpose of the current paper was to document the impact of the raised bed/ridge planting technology by using two types of data set, that is, field trials data set and the data set collected from the farmers. The rest of the paper is presented in two more sections that included methodology and empirical results with discussion and conclusions. The current paper has many novel aspects as the results are also verified across different data sets and are consistent and robust.

2. Materials and Methods

The paper uses integrated approaches to document the evidences of ridge-furrow planting of wheat technology, that is, one from the field evaluation or demonstration data and the other from the farmer perception through survey. From the field, two years trails were conducted and the data were collected from the sites comparing the ridge-furrow planting of wheat with farmers practice. Field evaluation of ridge-furrow planting in comparison with farmer practice was conducted on a total of 162 farmer fields and research farms that included 68 sites in 2014–15 and 94 sites in the 2015–16 wheat season. These farms were located in districts of

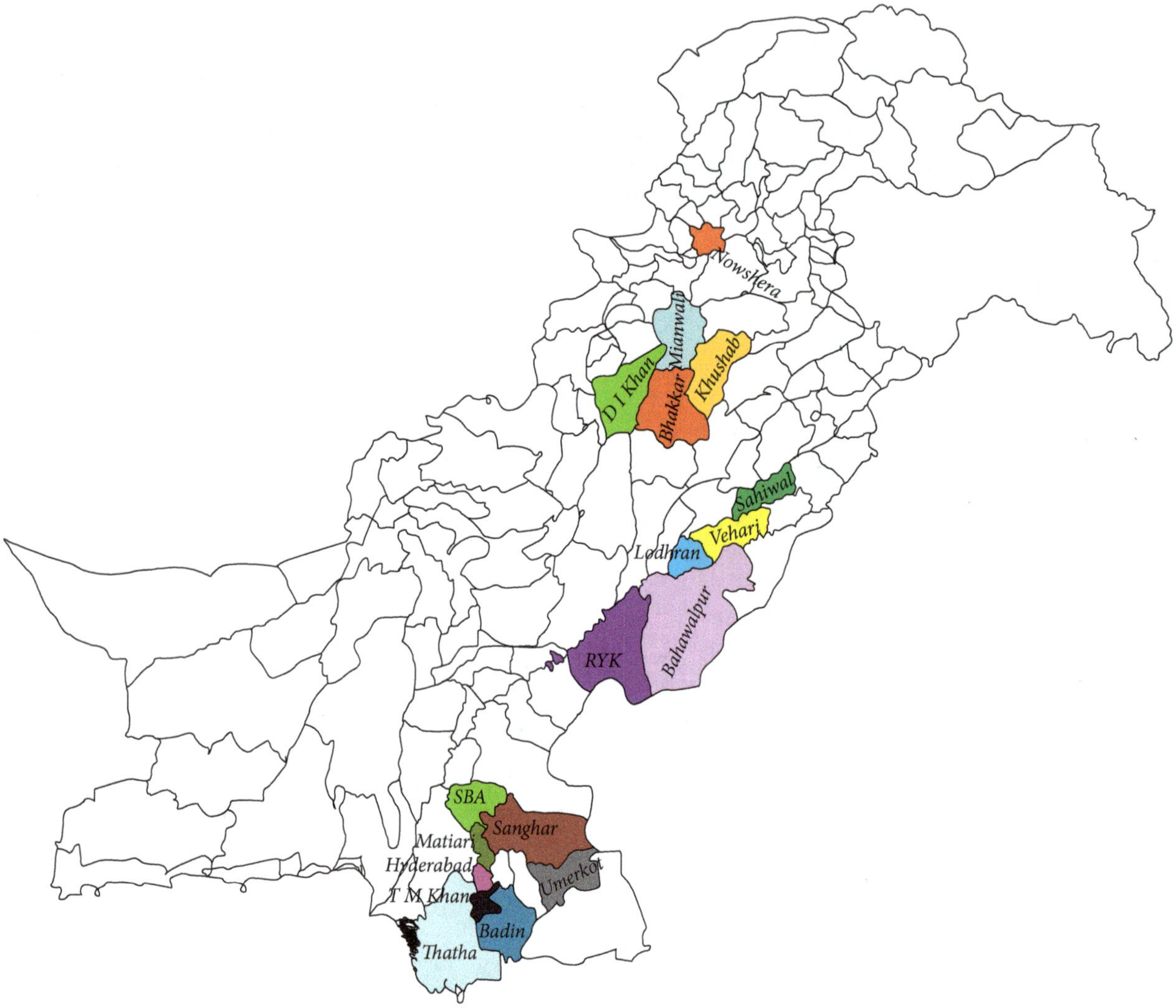

FIGURE 1: Map showing the surveyed and technology evaluation districts in Pakistan.

Khushab, Bhakkar, Mianwali, Vehari, Sahiwal, Bahawalpur, and Lodhran in Punjab Province; Shaheed Benazir Abad, Tando Muhammad Khan, Umerkot, Hyderabad, Matiari, and Sanghar districts in Sind province; and Nowshera and DI Khan districts in Khyber Pakhtunkhwa Province of Pakistan (Figure 1).

These trails were planted in various wheat-based cropping systems of Pakistan that included rice-wheat, cotton-wheat, maize-wheat, mung-wheat, and sugarcane-wheat cropping systems. Before planting of wheat, these sites were under either of autumn crops like rice, cotton, maize, sugarcane, millets, and mung. At each site, ridge-furrow planting of wheat was compared with farmers' practice of flat planting. Farmers' practice of wheat planting involved removing of the residue or burning of the residue followed by incorporation of remaining or partially burnt residue with 1–2 passes of rotavator or disc plow. Basal dose of a fertilizer (85 kg·P_2O_5/ha and 33 kg of N) in the form of DAP was broadcasted, and fields were prepared with 1–2 operations of the cultivator followed by planking. Afterwards, 125 kg of the wheat seed per hectare was broadcasted and mixed with one shallow cultivation with planking.

In ridge-furrow planting of wheat, remaining residue after burning or removal was incorporated with the help of 1–2 passes of rotavator or disk plow that was followed by broadcasting of basal dose of the fertilizer (85 kg·P_2O_5/ha and 33 kg of N) in the form of DAP and seed bed preparation with 1–2 cultivators and planking. Afterwards, 125 kg wheat seed per hectare was broadcasted, and ridges with furrows were made through the ridger. During the ridge-furrow making process, most of the seed and fertilizer gathered on ridges and the field is converted into furrow and ridges that is very similar to bed and furrow planting. These ridgers have 3–4 shovels for furrow making, and generally distance between the furrows is around 68–75 cm. Generally, ridges are narrow on the top, whereas the bed is wider than ridges on the top. In some fields, where sowing was done under dry conditions, fields were irrigated in a way that water does not go to top of the ridge and move within the furrow.

In these field trials, 120 kg·N and 85 kg·P_2O_5/ha in the form of DAP (diammonium phosphate) and urea were applied for wheat crop. All of phosphorus 85 kg·P_2O_5/ha and 33 kg of N in the form of DAP was applied before planting wheat. However, the remaining 87 kg of N/ha in the form of

TABLE 1: Effect of ridge-furrow planting and farmer's practice on wheat grain yield, tillering, and irrigation time.

Season	Outcome	Ridge planting	Farmers practice	Difference	t-values	Significance
2014–15	Yield (kg)	3939	3499	440	2.04	**
	Tillers (m^{-2})	331	295	36	2.67	***
	Irrigation time (minutes)	121	186	−65	3.11	***
2015–16	Yield (kg)	4146	3702	445	1.92	*
	Tillers (m^{-2})	337	296	41	2.53	***
	Irrigation time (minutes)	126	178	−52	−2.47	***

Note. The results are significant at ***1, **5, and *10% levels, respectively.

TABLE 2: District-wise comparative analysis of grain yield (kg/ha) from farmer practice and ridge planting of wheat in Pakistan.

District	2014–15		2015–16	
	Farmer practice	Ridge planting	Farmer practice	Ridge planting
Badin	3200	3550	—	—
Bahawalpur	4023	4617	3908	4547
Bhakkar	2950	3263	3223	3257
DI Khan	3378	3878	3756	4481
Hyderabad	4117	4567	4300	4700
Khushab	2455	2677	2725	2929
Lodhran	3906	4589	4648	5436
Matiari	4633	5553	3867	4533
Mianwali	2549	2855	3856	4007
Nowshera	2993	3077	—	—
Sanghar	3383	3677	—	—
Shaheed Benezirabad	4400	4970	3840	4560
Sahiwal	—	—	4150	4200
TM Khan	—	—	3050	3550
Thatta	—	—	3100	3800
Umerkot	—	—	3683	4194
Vehari	—	—	3715	3850
Average	3499	3939	3702	4146
RP-FP (kg/ha)	440		444	
Significance	***		**	

Note. The results are significant at ***1 and **5% levels, respectively.

urea was applied in two equal doses at the first and second irrigation that coincided with crown root initiation and the tillering stage. The data regarding the wheat yield, numbers of tillers, and time needed to irrigate one hectare were collected. Statistical analysis was done using the paired t-test between the treatment using Stata version 10.1 software [25].

To document the farmer's perception regarding ridge planting of wheat, a detailed comprehensive survey was carried out and 231 farmers were interviewed to collect information regarding ridge planting of wheat technology, as well as other socioeconomic, household, and farm level characteristics. Before carrying out the formal survey, the pretesting of the questionnaire was carried out and the questionnaire was refined in the light of pretesting results.

3. Results and Discussion

Results regarding the ridge-furrow planting effect on the wheat grain yield, numbers of tillers m^{-2} and time required to irrigate wheat crop for both growing seasons 2014–15 and 2015–16 are presented in Table 1. Average wheat grain yield from all the multilocation trials in 2014–15 and 2015–16 was 12% higher with ridge-furrow planting as compared to conventional planting (Table 1).

The detailed analysis of the wheat grain yield in all districts during 2014–15 and 2015–16 showed significant increase in the wheat grain yield with ridge-furrow planting in comparison with farmer's practice (Table 2). Higher wheat grain yield with ridge-furrow planting in all districts under various wheat cropping systems showed that the ridge-furrow planting technique was better than farmer's practice of flat planting in varying soil and climatic conditions (Table 2). The wheat grain yield with ridge-furrow planting was higher than 0.1 t/ha in all districts except Nowshera in 2014–15 and Bhakkar and Sahiwal in 2015–16. The wheat grain yield was more than 0.5 t/ha in comparison with farmer's practice in districts of Bahawalpur, DI Khan, Lodhran, Matiari, Shaheed Benazir Abad, Umerkot, and Thatta (Table 2).

During 2014–15, the wheat yield was significantly higher with ridge-furrow planting as compared to farmers practice. In ridge-furrow planting, the average difference of 440 kg/ha was noted in comparison with farmer practice and the results are significant at the 5% level of significance (Table 1). This higher yield with ridge-furrow planting was attributed to higher numbers of wheat tillers per meter square in ridge planting as compared to farmers practice (Table 1).

During 2015–16, average wheat grain yields were also significantly higher with ridge-furrow planting as compared to farmers practice and the difference in yields was 444 kg per hectare (Table 1). The grain yield with ridge-furrow planting was higher at 10% level of significance. Similarly, the number of tillers per unit area was also significantly higher with ridge-furrow planting in comparison with farmer practice in 2015–16 (Table 1). Better tillering and crop growth on ridges during the season contributed to 12% higher grain yield. Previous studies observed more than 20% higher grain yield with ridge planting and bed planting in comparison with flat planting [15, 23, 24].

Furrow-irrigated ridge planting technology helps farmers in saving irrigation water. Average irrigation time for one hectare field under ridge-furrow planting in comparison with farmer practice was 62 and 52 minutes lower in 2014–15 and 2015–16, respectively (Table 1). This reduction in time to irrigate one hectare field of ridge that planted wheat in comparison with farmer practice was highly significant at 1% level of significance. This showed that ridge-furrow planting helped farmers in saving 29–35% of water in comparison with farmer practice of flat planting. Farmers applied 4–5 irrigations during wheat growing season, the technology helps farmers in saving 208–260 minutes of irrigation time in a season, and this could also help in reducing energy cost for water pumping. Irrigation water application was also measured with the help of cut throat flume at three farmer fields, and 32–43% of water saving with ridge-furrow planting in comparison with farmer practice was noted. Raised bed and ridge sowing techniques saved 22% [23] and 40% of irrigation water [24] in comparison with flat sowing. The results are also in line with research work [11, 18, 19, 21] who reported saving in irrigation time and water with ridge bed and ridge-furrow methods in comparison with farmer practice.

Because of climate change, the availability of the irrigation water is decreasing, and hence, ridge-furrow planting might help to improve irrigation water use efficiently. In Pakistan, wheat is sown on about 7.8 million hectares under irrigated conditions with average numbers of irrigations in the wheat crop cycle are in the range of 4–5. The ridge-furrow planting technique helps the farmer in saving 30% of water and improving 12% of grain yield in comparison with the prevailing planting technique. The adoption of ridge-furrow or bed furrow on one million hectare area could help to save one million acre foot of water and increase 0.36 million tone in wheat production. This water-saving aspect coupled with the yield increase of 9–12% is an important driving force for farmers to adopt the technology. If ridge planted wheat yields are equal with farmer practice, even then water saving is also an advantage for the promotion of the ridge planting technology.

The variety-wise comparison of the ridge planting and the conventional planting indicated that the grain yield of different wheat varieties was significantly higher under ridge-furrow planting as compared to farmer practice (Table 3).

Table 3: Effect of ridge planting and farmer practice on grain yield (kg/ha) of wheat varieties in 2014–15 and 2015–16.

Wheat variety	Ridge planting	Farmer practice	Difference	Significance
AARI-2011	2733	2503	230	**
Lasani-2008	2797	2542	255	**
Punjab-2011	3112	2877	235	**
Imdad-2005	3900	3300	600	***
NARC-2011	4400	3600	800	***
Pakistan-2013	4200	3650	550	***
TD-1	4000	3475	525	***
Benazir-2013	4463	3944	519	***
Millat-2011	3337	3117	220	**

Note. The results are significant at ***1 and **5% levels, respectively.

Wheat g1rain yields was highest with Benazir-2013 and lowest with AARI-2011. Wheat varieties Imdad-2005, NARC-2011, Pakistan-2013, TD-1, and Benazir-2013 had 525–800 kg/ha higher grain yield with ridge-furrow planting in comparison with farmer practice. However, wheat Millat-2011, Punjab-2011, AARI-2011, and Lasani-2008 had 220–255 kg/ha higher wheat grain yield with ridge-furrow planting in comparison with farmer practice. For all these wheat varieties, the grain yields were higher for the ridge planting technology followed by the farmers practice. Higher wheat grain yield of wheat varieties Imdad-2005, NARC-2011, Pakistan-2013, TD-1, and Benazir-2013 on ridges might be due to better tillering and better growth under raised bed condition in comparison with flat planting. These wheat varieties could have better ground cover and broad leaf area that might have contributed to better grain yield in comparison with other varieties with erect and compact structure [10].

3.1. Farmer Survey on Ridge-Furrow Planting. Field level data were also supplemented by the farmer's survey data regarding farmer perception of the ridge planting technology, technology feasibility (cost benefit ratios), and constraints in the adoption of the technology.

The empirical results regarding adoption of the ridge planting technology are presented in Table 4. Analysis of the results showed that educated, young farmers with large land holding and tube well adopted ridge planting of wheat technology (Table 4). In this model, the dependent variable is technology adoption, that is, Score 1, if the farmers adopt the ridge planting technology and score 0, otherwise. A set of independent variables are included in the model.

The age was included in the numbers of years with a negative coefficient indicating that mostly the young farmers adopted ridge planting technology. The education was also included in the numbers of years with a positive coefficient with significance at 1% level which indicated that mostly the educated farmers have adopted the ridge planting technology. The family size coefficient is positive and nonsignificant. The land holding coefficient was positive and highly significant at 1% level of significance indicating that mostly the larger farmers have adopted the ridge planting

TABLE 4: Determinants of the ridge planting technology adoption (profit estimates).

Variable	Coefficient	t-values
Age	−0.07**	−2.15
Education	0.03***	2.78
Family size	0.02	1.44
Land holding	0.01***	2.59
Agriculture extension	0.02	1.36
Credit	0.04	1.28
Car	0.06	1.44
Tractor	0.01**	2.10
Tube well	0.02*	1.45
Constant	0.03	1.57
Number of observations	138	
Pseudo R-square	0.37	
LR-chi-square	283.57	
Probability > chi-square	0.000	

Note. The results are significant at ***1, **5, and *10% levels, respectively.

TABLE 5: Details regarding cost of production and cost benefit analysis of ridge planting and farmer practice in USD per hectare.

Input (USD ha^{-1})	Ridge planting	Farmer practice	Ridge-FP
Land preparation cost	105.61	103.74	1.87
Seed cost	61.68	56.07	5.61
Planting cost	23.36	12.85	10.51
Fertilizer cost	239.25	240.11	−0.86
Weedicide cost	16.79	18.97	−2.18
Irrigation cost	21.03	26.28	−5.25
Harvesting cost	83.71	72.54	11.17
Production cost—total	551.44	530.57	20.87
Wheat yield (kg/ha)	34.11	30.37	3.74
Wheat price (USD/kg)	0.30	0.30	0.00
Gross revenue wheat	1108.64	987.15	121.50
Net revenue (profit)	557.21	456.58	100.63
Cost benefit ratio	2.01	1.86	

Note. One USD (United States dollar) is equal to 107 PKR (Pakistani rupees).

technology. Agriculture extension, credit, and ownership of car are positive and nonsignificant. The tractor and tube well coefficients were significantly positive indicating that farmers having tractor and tube well ownership were mostly adopting the ridge planting technology. Farmer with tube well for irrigation not only got benefit of water saving but also reduced energy cost per hectare. The R-square value l was 0.37 indicating that 37% variation in the dependent variable is due to independent variables included in the model. The LR chi square is highly significant at 1% level of significance indicating the robustness of the variables included in the model.

3.2. Benefit Cost Analysis of Ridge-Furrow Planting. Benefit cost analysis of wheat crop sown by the conventional method and ridge planting showed that ridge planting of wheat was a profitable technique for farmers in comparison with the conventional planting technique (Table 5).

$$\text{Benefits cost ratio (BCR)} = \frac{PV_{\text{Benefits}}}{PV_{\text{Costs}}}, \quad (1)$$

where PV_{Benefits} = present value of benefits and PV_{Costs} = present value of costs.

The results showed that the average land preparation cost for ridge planting adopter and conventional farmers was USD/ha 105.61 and 103.74, respectively. Seed cost for ridge growers and conventional farmers was determined as USD/ha 61.68 and 56.07, respectively. Ridge planting adopters use 12.5 kg/ha more seed in comparison with nonadopter to improve crop stand. Planting cost for ridge planting farmers and conventional farmers was determined as USD/ha 23.36 and 12.85, respectively. In ridge planting, farmer used the ridger for making furrow and ridges, so the planting cost for the ridge was higher (USD 10.51/ha) in comparison with farmer practice (Table 5). Wheat growers applied DAP and urea fertilizer to meet phosphorus and nitrogen demand of the wheat crop, and the cost of fertilizer application was same for both methods. Weedicide application and irrigation cost for the beneficiary farmer was USD 16.79 and 21.03 per hectare that was lower than the non-beneficiary farmers. Harvesting cost in the beneficiary case was recorded USD 83.71 that was higher than the conventional technique farmers. Harvesting charges through combine harvestor or manual labor were higher for ridge-furrow planted wheat in comparison with farmer practice. This is the perception of service providers that there could be more wear and tear of machinery on ridges or raised beds. Total wheat production for ridge farmers was USD 551.44/ha that was higher as compared to nonadopter cost of USD 530.57/ha. However, ridge planting adopters had a wheat grain yield of 3650 kg/ha that was better than the nonadopter yield of 3250 kg/ha. This was also in line with the trial data of both years that showed that wheat yields were higher on ridges in comparison with wheat planted on flat surface. Overall profit was USD 557.21 and 456.58 per hectare in beneficiary and nonbeneficiary case, respectively. In these analysis, farmer management cost and rent of land were not included in the cost of production, and farmers were able to get USD 100.63 per hectare more net profit with ridge-furrow planting of wheat in comparison with farmer practice. Raised bed planting under the ZT condition had 13–28% higher system productivity and net return of USD 312 ha-1 in comparison with the conventional planting system [13, 14, 20].

Ridge-furrow planting adopter farmers paid additional cost of USD 17.29 ha for extraseed, ridge making, and harvesting charges in comparison with nonadopters. However, adopters saved USD 5.25/ha from irrigation cost when they used tube well for irrigation once in the season. In addition, they also saved time for the irrigating field. The cost benefit analysis indicates that ridge planting is more profitable as compared to the conventional method of planting. The cost benefit ratio for the ridge planting is well above 2, while for the conventional planting method, its only 1.86.

There are a number of constraints enlisted by farmers in adopting ridge planting of wheat. The most important is the lack of awareness, the efficient extension services, as well as the low availability of ridgers with service providers at village level.

Some of the adopting farmers think that combine harvesting on ridge planted wheat is an issue, and service providers avoid those field because of fear to damage to combine harvest. In this process, serviced providers delay the harvesting of these fields and usually agree for harvesting at the end of season with higher charges.

4. Conclusions

Ridge planting of wheat in an irrigated area of Pakistan helped farmers to obtain 12% higher wheat grain yield and 30–35% water saving and ease in irrigation management in comparison with farmer practice. Educated farmers with sizable land holding and tube well are the initial adopter of technology for its simplicity and availability of ridgers at farm level, water saving, and better yield. Ridge-furrow planting not only saves water but also improves drainage in the event of heavy rainfall during wheat season. The ridge-furrow planting is being demonstrated to the farming community through active involvement of Department of Agriculture Extension Punjab in the cotton-wheat and rice-wheat system. The adoption of ridge planting on one million hectare area could help to save one million acre foot of water that can be used to irrigate more area. In addition, it could also help in producing 0.36 million tons of additional wheat grain per year. Farmer community moving from flat planting to ridge planting also provides an opportunity for transforming ridge planting farmers into users of mechanized bed planting with planting, bed making, and fertilizer application in one operation and could also help in reducing cost and improving fertilizer use efficiency at farmer fields in future. Precision planting can make inroad with the availability of suitable bed and furrow planter that can plant wheat in residue of preceding crop like cotton, maize, and rice. Agricultural extension needs to create awareness and build capacity regarding this important technology in Pakistan.

Acknowledgments

The research work was carried out under Agricultural Innovation Program (AIP) for Pakistan funded by USAID.

References

[1] Government of Pakistan, *Pakistan Economic Survey 2016-17*, Economic Adviser's Wing, Finance Division, Government of Pakistan, Islamabad, Pakistan, 2017.

[2] D. Byerlee, P. R. Hobbs, B. R. Khan, A. Majid, R. Akhtar, and N. I. Hashmi, *Increasing Wheat Productivity in the Context of Pakistan's Irrigated Cropping Systems*, PARC/CIMMYT, Islamabad, Pakistan, 1986.

[3] M. A. Khan, *Wheat Crop Management for Yield Maximization, Vol. 44*, Wheat Research Institute, Faisalabad, Pakistan, 2003.

[4] I. Hussain, H. Shah, M. A. Khan, W. Akhtar, A. Majid, and M. Y. Mujahid, "Productivity in rice-wheat crop rotation of Punjab: an application of typical farm methodology," *Pakistan Journal of Agricultural Research*, vol. 25, pp. 1–11, 2012.

[5] K. D. Sayre and O. H. Mreno-Ramos, *Application of Raised Bed Planting System to Wheat*, Wheat Program Special Report No. 31, CIMMYT, Mexico, DF, Mexico, 1997.

[6] P. R. Hobbs, K. D. Sayre, and J. I. Ortiz-Monasterio, *Increasing Wheat Yields Sustainably through Agronomic Means*, CIMMYT NRG Paper, Mexico, DF, Mexico, 1998.

[7] R. J. Connor, J. Timsina, and E. Humphreys, "Prospects for permanent beds in the rice-wheat system," in *Improving the Productivity and Sustainability Of Rice–Wheat System: Issues and Impacts*, pp. 197–210, ASA Inc., CSSA Inc., SSSA Inc., Madison, WI, USA, 2003.

[8] B. U. Choudhury, B. M. Bouman, and A. K. Sigh, "Yield and water productivity of rice-wheat on raised beds at New Delhi, India," *Field Crops Research*, vol. 100, no. 2-3, pp. 229–239, 2007.

[9] B. Govaerts, K. D. Sayre, J. M. Ceballos-Ramirez et al., "Conventionally tilled and permanent raised beds with different crop residue management: effects on soil C and N dynamics," *Plant and Soil*, vol. 280, no. 1-2, pp. 143–155, 2006.

[10] H. Kiliç, "The effect of planting methods on yield and yield components of irrigated spring durum wheat varieties," *Scientific Research and Essays*, vol. 520, pp. 3063–3069, 2010.

[11] A. Majid, A. Muhmood, A. Niaz et al., "Bed planting of wheat (*Triticum aestivum* L.) improves nitrogen use efficiency and grain yield compared to flat planting," *Crop Journal*, vol. 3, pp. 118–124, 2015.

[12] F. Wang, W. Xuqing, and K. Sayre, "Comparison of conventional, flood irrigated, flat planting with furrow irrigated, raised bed planting for winter wheat in China," *Field Crops Research*, vol. 87, pp. 35–42, 2004.

[13] C. M. Parihar, S. L. Jat, A. K. Singh et al., "Conservation agriculture in irrigated intensive maize-based systems of north-western India: effects on crop yields, water productivity and economic profitability," *Field Crops Research*, vol. 193, pp. 104–116, 2016.

[14] R. D. Jat, H. S. Jat, R. K. Nanwal et al., "Conservation agriculture and precision nutrient management practices in maize-wheat system: effects on crop and water productivity and economic profitability," *Field Crops Research*, vol. 222, pp. 111–120, 2018.

[15] R. A. Mann and C. A. Meisner, *Proceedings of the National Workshop on Rice-Wheat Systems in Pakistan*, A Rice Wheat Consortium Paper Series 15, p. 118, Islamabad, Pakistan, 2003.

[16] M. I. U. Mollah, M. S. U. Bhuiya, M. S. Hossain, and S. M. A. Hossain, "Growth of wheat (*Triticum Aestivum* L.) under raised bed planting method in rice-wheat cropping system," *Journal of Agriculture & Rural Development*, vol. 7, pp. 23–31, 2015.

[17] E. A. Waraich, R. Ahmad, and S. Ahmad, "Raised bed planting–a new technique for enhancing water use efficiency in wheat (*Triticum aestivum* L.) in semi-arid zone," *Iranian Journal of Plant Physiology*, vol. 1, pp. 73–84, 2010.

[18] N. Ahmad and N. Mahmood, "Impact of raised bed technology on water productivity and logging of wheat," *Pakistan Journal of Water Resources*, vol. 90, pp. 7–15, 2005.

[19] M. A. Shahid, N. Ahmad, M. Saleem, and B. Akhtar, "Investigating optimum number of irrigations for wheat under raised bed Technology in a semi-arid climate," *International Journal of Applied Agricultural Sciences*, vol. 3, pp. 89–93, 2011.

[20] M. A. Shahid, N. Ahmad, M. Arshad, and M. Usman, "Enhancing land and water productivity through furrow irrigated raised bed planting–a case study of underutilized lands in Pakistan," *Pakistan Journal of Agricultural Sciences*, vol. 52, pp. 981–988, 2015.
[21] I. Hassan, Z. Hussain, and G. Akber, "Effect of permanent raised beds on water productivity for irrigated maize-wheat cropping system. Evaluation and performance of permanent raised bed cropping system in Asia, Australia and Mexico" in ACIAR Proceedings, C. H. Roth, R. A. Fischer, and C. A. Meisner, Eds., vol. 121, pp. 59–65, January 2005.
[22] S. U. Khan and K. Ullah, "Effect of various planting methods on seed cotton yield," *Pakistan Cottons*, vol. 35, no. 1, pp. 43–47, 1991.
[23] M. Ali, L. Ali, M. Q. Waqar, and M. A. Ali, "Bed planting: a new crop establishment method for wheat (*Titicum aestivum* L.) in cotton-wheat cropping system of southern Punjab," *International Journal of Applied Agricultural Sciences*, vol. 4, pp. 8–14, 2012.
[24] M. Akhtar, M. Aslam, H. M. Nasrullah, B. Ali, and M. A. Khan, "Effect of ridge sowing methods on wheat yield in cotton-wheat cropping system of southern Punjab," *Journal of Rural Development*, vol. 45, pp. 29–39, 2014.
[25] C. A. Cameron and P. K. Trivedi, *Micro Econometrics Using Stata*, Stata Press, College Station, TX, USA, 2009.

11

Effects of Light and Nutrition Manipulations on Thermal Respiratory Acclimation and Nocturnal Dynamics of Leaf Dark Respiration

Dylan N. Dillaway,[1,2] Michael C. Tyree ,[1,3] and John K. Jackson[1]

[1]*Work Performed: School of Forestry, Louisiana Tech University, Ruston, LA 71272, USA*
[2]*Center for Natural Resources Mgmt. & Protection, Unity College, Unity, ME 04988, USA*
[3]*Department of Biology, Indiana University of Pennsylvania, Indiana, PA 15705-1001, USA*

Correspondence should be addressed to Michael C. Tyree; mtyree@iup.edu

Academic Editor: Yong In Kuk

The ability of a plant to acclimate metabolically to thermal changes is necessary to maintain a positive carbon balance. It is likely that a plant's acclimatory potential is a function of leaf nitrogen and/or leaf carbohydrate status. Two important issues assessed concerning leaf dark respiration (R_D) were the effects of growth temperature, light, and fertilization on thermal respiratory acclimation and changes in respiratory parameters (indicative of acclimation) throughout the dark period. Soybean (*Glycine max* (L.) Merr.) plants were grown in greenhouses under a full factorial treatment arrangement of temperature, light, and nutrition. R_D was measured at three temperatures to estimate respiratory parameters (cool respiration R_{13}, warm respiration R_{25}, and the temperature response of respiration E_O) three times throughout the night (6 pm, 11 pm, and 4 am). Respiratory parameters did not differ throughout the night. Thermal acclimation was observed in warm grown plants under optimal growing conditions (i.e., high light and high fertilization); however, acclimation did not occur when limitations were imposed (i.e., shade or no fertilization). These findings suggest thermal acclimation will occur so long as plants do not undergo limitations. This may have major implications for natural ecosystems and may play a role in assessing an ecosystems resiliency to climate change.

1. Introduction

Tissue dark respiration (R_D) not only provides energy and carbon (C) skeletons for most plant metabolic processes [1] but also accounts for up to 50% of gross productivity [2]. Climate clearly exerts a strong influence on plant C balance as a whole, and this impact on tissue R_D is an important determinant of that balance being positive or negative [2]. In the short term, R_D has been shown to be quite sensitive to variations in temperature [3]; however, respiratory metabolism is often adjusted to minimize acceleration of C loss at higher temperatures [4]. The opposite effect has been observed when plants are exposed to low or declining temperatures [5], and the absence of thermal acclimation has been reported in other studies [6]. Clearly, the nature and extent of thermal R_D acclimation appear to differ significantly across species and growth environments [7–12]. When thermal acclimation is detected, it may be a result of changes in R_D at a predetermined base or cool temperature [10, 13], changes in the temperature response [7], or both [14].

Exponential functions are commonly used to describe the relationship between temperature and R_D. However when measured over a range of short-term temperature changes, Q_{10} (quantifying the relative rise in R_D with a 10°C increase in temperature) has been shown to introduce bias into these estimates [15], which can be avoided by using a modified Arrhenius equation to calculate an energy of activation (E_O) [16]. While both Q_{10} and E_O values can range widely across tissue types, species, and environments, they are frequently large enough to result in at least a doubling of R_D with a 10°C increase in temperature [17]. Some studies have indicated correlations between this observed temperature response and leaf chemical traits [18]; however, there are a number of studies which have suggested that the mechanisms underlying R_D may differ at warmer versus cooler growth temperatures, and as a result, E_O (shape of temperature response curve) will be a function of these rates [4, 17]. For example, Dillaway and Kruger [19] reported a strong correlation between warm respiration (R_{30}; respiration at 30°C) and mass-based photosynthesis, while cool respiration (i.e., 20°C) appeared to be a function of leaf nitrogen concentration ([N]). Thus, when assessing the relationships between leaf traits and respiratory temperature response, it is imperative that both cool and warm respiration (minimum of 10°C temperature spread) are assessed equally but separately.

Studies which assess the respiratory response of plant tissues typically do not account for changes in respiration rates and or the temperature response of respiration throughout the nocturnal period [7, 14, 19–25]. Cumulative night-time respiration is often calculated using the temperature response of respiration measured at one point-in-time (usually just after sunset or just prior to sunrise) [19]. However, it is unknown whether this temperature response of respiration at the onset of darkness is in fact identical to the temperature response of respiration found at the conclusion of the nocturnal period [26, 27]. One theory proposed to explain the observed differences in the temperature response of respiration across climate gradation is changes in [N] [2, 28, 29]. However, none of the previously correlated variables can explain why the temperature response of respiration might change either throughout the night or as a result of longer-term adjustments to growth temperature (i.e., thermal acclimation). Given that leaf [N] should be a relatively stable leaf chemical property (at least in the short-term), if a differential decline in respiration rates at contrasting temperature regimes is observed throughout the night, it should result mainly from changes in warm respiration indicating that this decline may be driven by carbohydrate concentration in the leaf.

This research addresses three important and related issues concerning leaf dark respiration:

(i) We assessed the effects of growth temperature, light, and fertilization on the temperature response of respiration and thus respiratory thermal acclimation. Thermal acclimation of respiration depends in large part on two important respiratory parameters including base respiration (cool respiration) and the temperature response of respiration (E_O, analogous to Q_{10}) [4, 17]. For the purpose of this study, we adopted the theoretical construct of Atkin and Tjoelker [4] in which two types of acclimations are possible (Type I and Type II). Type I is theorized to be a result of a shift in the temperature response of R_D (i.e., Q_{10} or E_O), and Type II is theorized to be a result of a shift in R_D at low temperatures (i.e., cool respiration (R_{13} in our study)). Based on current research, additional insight regarding the mechanistic underpinnings driving shifts in these variables is needed. We hypothesized that E_O would decrease with increased growth temperature due to a decline in warm respiration (R_{25}). Cool respiration (R_{13}) will not change as a result of growth temperature.

(ii) We assessed the temporal temperature response of respiration throughout the dark period. Cumulative night-time respiration and the nature of thermal acclimatory adjustments cannot be predicted with a high level of confidence, to date. It is unknown whether these variables change throughout the dark period of a 24-hour span. We hypothesize that E_O and R_{25} will decline throughout the night regardless of treatment but the decline will be more pronounced in the "warm" grown plants owing to faster depletion of carbohydrate stores. R_{13} will remain stable throughout the night due to its strong correlation with leaf [N].

(iii) Another principal goal of this research is to understand the fundamental mechanistic underpinnings of leaf dark respiration (warm respiration, cool respiration, and the temperature response of respiration) through manipulations of leaf carbohydrate concentration and leaf [N] status of plants growing at contrasting growth temperatures. In this regard, we hypothesized that E_O and R_{25} will be lower in the low-light treatments due to their positive correlation with total leaf nonstructural carbohydrate (TNC), while R_{13} will be nonresponsive to light treatments due to its lack of correlation with leaf TNC. Because of the strong relationship between leaf [N] and cool respiration (R_{13}), the no fertilization treatment will exhibit a lower R_{13} relative to the fertilization treatment; however, fertilization will show no apparent effect on R_{25} or E_O.

2. Materials and Methods

2.1. Study Design and Treatments. This work was conducted using university greenhouses located at Louisiana Tech University in Ruston, LA, from February to March 2010. This study employed a randomized complete block design replicated three times with the treatments arranged as a full 2 × 2 × 2 factorial. Soybean (*Glycine max* (L.) Merr.) seeds were sown on 23 February 2010, in 4.4 L pots filled with a uniform soil mixture (Metro mix-300, Sun-gro horticulture, Bellevue, Washington). Seeds were germinated in a common greenhouse under ambient CO_2 and a constant temperature of 23°C for approximately two weeks prior to treatment initiation.

Fertilizer and growth temperature treatments were initiated on 8 Mar. 2010. Seedlings were randomly moved to one of four greenhouses. Two houses were designated "warm" houses and maintained at a day/night temperature of 30/23°C, while two houses were designated "cool" houses and maintained at a day/night temperature of 20/13°C, resulting in a ~10°C spread in temperature both during the day and night periods. Duplicate houses were used for each growth temperature treatment to ensure any responses were an effect of our treatments and not random differences among greenhouses. Half of all seedlings within each house were fertilized with a controlled release fertilizer (Osmocote, 15-9-12, Scott's, Marysville, OH). In each house, fertilized

(F) and nonfertilized (NF) seedlings were grown under high light conditions. Soybean plants were grown for an additional two weeks after transitioning seedlings to treatment rooms in order to allow plants to fully acclimate to their new growth conditions and develop new foliage under their respective temperature regime. Shade treatments were initiated the morning prior to respiration measurements. Plants were shaded for approximately 14 hours prior to respiration measurement. During each measurement period, a subset of high light plants was moved into a shade house. Shade houses provided approximately 30% of full sunlight and were designed to reduce daily C fixation of "low-light" plants relative to "high-light" plants (LL and HL, respectively).

2.2. Respiration Measurements. Respiration measurements were conducted for three consecutive days (23rd, 24th, and 25th Mar.). During each measurement period, nocturnal measurements were taken three times (beginning at 1800, 2300, and 0400 h) in order to examine the variability in key respiratory parameters throughout the night. Leaf dark respiration and its response to temperature were measured, for each treatment, during three "campaigns" throughout the night, using a Li-Cor 6400 portable photosynthesis machine (Li-Cor Biosciences, Lincoln, NE) with a 6 cm^2 cuvette. Dark respiration (R_D) was determined on the youngest, fully expanded leaf from six plants per treatment, at three cuvette temperatures (13°, 18°, and 23°C). Partial pressure of CO_2 (pCO_2) inside the cuvette reference chamber was maintained at 400 ppm (~40 Pa) using the CO_2 mixer on the Li-Cor 6400. As in recent studies of leaf gas exchange [19, 30], we included a correction factor recommended by Pons and Welschen [31] in order to account for the "gasket effect" of commercially available photosynthesis systems.

The response of R_D to temperature was calculated using a modified Arrhenius equation [16, 18, 28]. The Arrhenius equation was chosen (as opposed to the Q_{10} approach) so that estimates were not dependent on measurement temperature. The use of the Arrhenius equation allows us to calculate R_D for a particular leaf (with its unique E_O and temperature-adjusted R_D) at any temperature of interest. Respiratory responses to growth environments were examined using several calculated parameters, including temperature-adjusted rates (Respiration at 25°C (R_{25}) and respiration at 13°C (R_{13})), along with the temperature response of respiration (E_O). Respiration rates at temperatures as low as or lower than 15°C are commonly used to assess "cool" respiratory metabolism in studies of R_D acclimation [4, 10, 18].

2.3. Leaf Morphology and Chemistry. At the conclusion of each respiration campaign, all measured leaves were harvested, their projected leaf areas determined, and placed in a 70°C oven. After 72 hours, the leaves were removed from the oven and weighed. Leaf area and dry mass were used to calculate specific leaf area (SLA), which was used to convert leaf R_D to a mass basis (nmol $g^{-1}\cdot s^{-1}$). Leaves were then ground to a fine-powder and sent for C and N analysis (Brookside Laboratories, Ohio).

An additional set of leaves, proximal to (and serving as a proxy for) those measured for R_D, were harvested simultaneously with individual R_D measurement campaigns. These samples were immediately frozen in a −20°C freezer. Leaves were then freeze-dried using a benchtop lyophilizer (VirTis Bench Top, SP Industries Inc., NY), ground to a fine-powder with a mortar and pestle, and analyzed for concentrations of total nonstructural carbohydrate (TNC) and its components (starch and soluble sugars), employing enzyme digestion and colorimetric assays similar to those detailed in Kaelke et al. [32]. Leaf starch and leaf [N] were then regressed against respiratory parameters (i.e., R_{25}, E_O, and R_{13}) in order to determine the variables driving changes in the temperature response of respiration and temperature-adjusted rates of respiration.

2.4. Data Analyses. Determination of respiratory thermal acclimation was analyzed using analysis of variance with repeated measures where block, growth temperature, light, and fertilization main effects and interactions were treated as fixed effects. Time of night and all associated interaction terms were treated as random effects. All subsamples were averaged and assumptions checked using plots of normality curves and residuals prior to statistical analysis. For regression analysis, outlier observations were identified using Cook's distance procedure and removed when appropriate. All analyses were conducted using PROC MIXED and PROC REG in SAS 9.3 [33].

3. Results

3.1. Comparisons of Leaf Respiration at Common Temperatures. Estimates of R_D at a common temperature were calculated for cool metabolism and warm metabolism (R_{13} and R_{25}). R_{25} was higher in the cool treatment under high light regardless of the associated fertilization treatment (Figure 1(c)).

Under low light, R_{25} was not significantly different from the "warm" grown plants under either fertilization treatment (Figure 1(c)). R_{13} in "cool" grown plants was higher than "warm" grown plants only in the high light, fertilized (HL,F) treatment but did not appear to vary under other combinations of light and fertilization (either low light or low nutrition) (Figure 1(b)). Within temperature regimes, R_{13} in "warm" grown plants was unaffected by light and nutrition manipulations ($P = 0.21$ and $P = 0.66$, respectively). R_{25} in "warm" grown plants did not differ within HL or LL, but both HL,F and HL,NF were higher than their LL counterparts ($P = 0.02$ and $P = 0.07$, respectively) (Figure 1(c)). Within "cool" grown plants, R_{13} was significantly higher in the HL,F combination than all other light and fertilization combinations ($P \leq 0.05$). R_{25} in "cool" grown plants was significantly higher in the HL,F combination than the HL,NF plants ($P = 0.01$). Both the F and NF plants under HL were significantly greater ($P < 0.0001$ and $P = 0.002$, respectively) than their LL counterparts. R_{25} showed no difference between the LL,F and LL,NF plants ($P = 0.51$; Figure 1(c)).

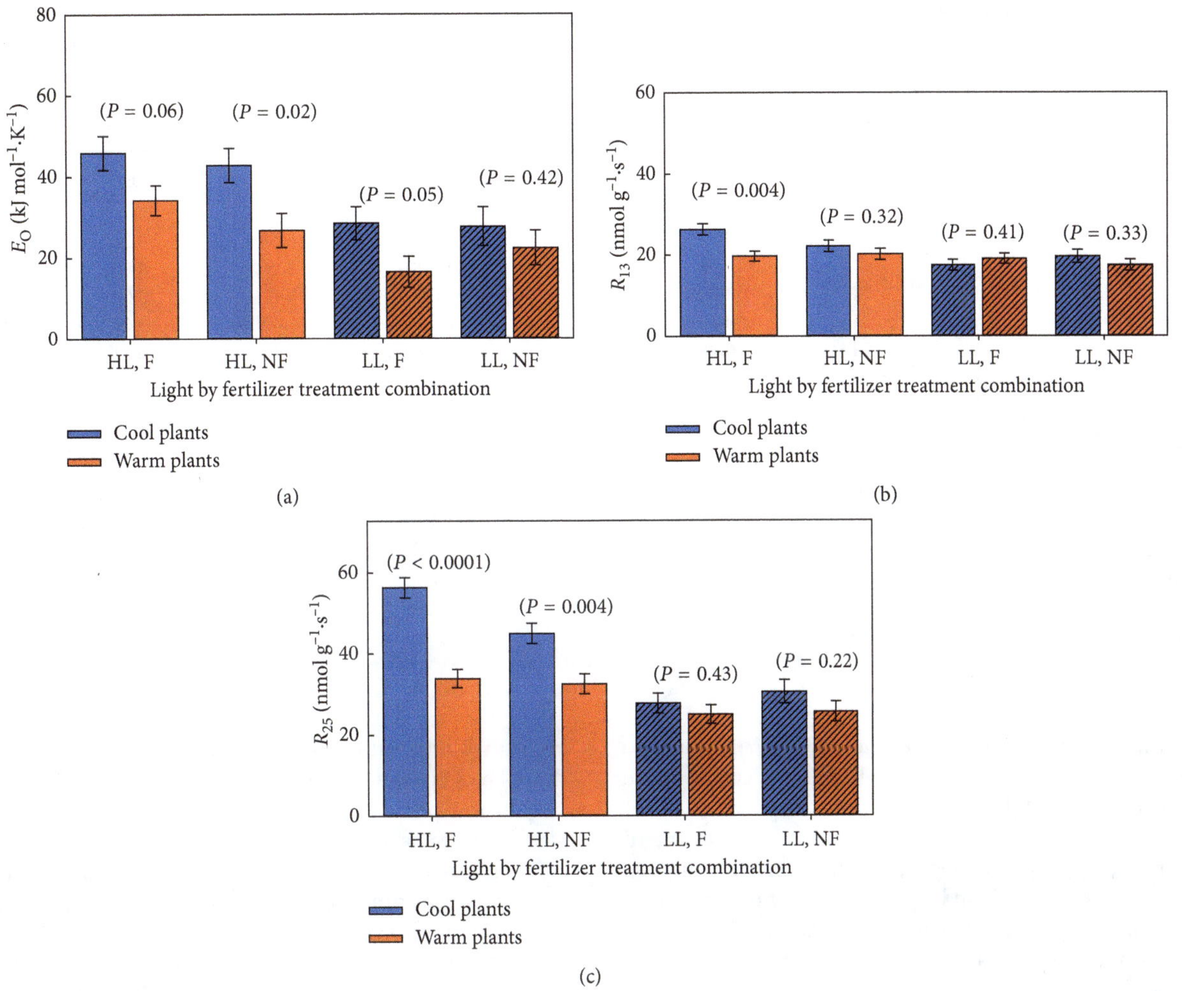

FIGURE 1: Differences among resource manipulation treatments for "warm" and "cool" grown plants (red and blue bars, respectively) for temperature response of respiration (E_O) (a), cool respiration (R_{13}) (b), and warm respiration (R_{25}) (c). Bars represent least square means ($n = 18$) throughout the night (error bars represent ± one standard error from the mean). P values show comparisons among "warm" and "cool" grown plants for each light (high light = HL and low light = LL) by fertilization (F = fertilized and NF = no fertilizer added) treatment combinations.

3.2. Temperature Sensitivity of Leaf Respiration. The temperature sensitivity of respiration (E_O) was significantly increased in "cool" grown plants relative to "warm" grown plants in all treatments with the exception of the LL,NF treatment (Figure 1(a)). Within the "cool" grown plants, E_O was significantly greater in the HL treatments than the LL treatments, regardless of the associated fertilization treatment ($P = 0.002$). Within the "warm" grown plants, there was no significant difference among any light-fertilization combinations with the exception of HL,F plants being greater than LL,F ($P = 0.01$).

3.3. Temporal Dynamics of Leaf Respiration and Leaf Carbohydrate. Throughout the night, we observed no significant change in R_{13}, R_{25}, or E_O in any treatment combination or temperature environment (Table 1). In addition, leaf starch, leaf soluble carbohydrate (TSC), and leaf total nonstructural carbohydrate (TNC) did not differ throughout the night (Table 1).

3.4. Leaf Carbohydrate Concentration. Leaf carbohydrate concentration (both starch and soluble carbohydrates) was affected by our treatments (Figure 2). Fertilization had the greatest effect on both starch ($P = 0.01$) and soluble carbohydrate ($P < 0.0001$) (Table 1). Temperature and light also had a significant effect on starch ($P = 0.03$ and $P = 0.06$, respectively) and soluble carbohydrates ($P = 0.02$ and $P = 0.01$, respectively). We did not, however, find any interactions between light, temperature, and fertilization with the exception of starch which showed an interaction between temperature and fertilization ($P = 0.01$; Table 1).

3.5. Leaf Nitrogen Concentration. Leaf [N] did not differ among fertilized plants between "cool" and "warm" grown plants in either HL or LL (Table 2). However, we did find significant differences in leaf [N] between "cool" and "warm" grown plants under HL ($P = 0.04$) and LL ($P = 0.01$). Lowest leaf [N] concentrations occurred in the HL,NF in the "warm" grown plants.

TABLE 1: Partial *P* table for leaf respiratory parameters (temperature response of respiration (E_O), basal respiration (R_{13}), and warm respiration (R_{25})) and leaf nonstructural C for "warm" and "cool" grown *Glycine max* (soybean) plants subjected to a factorial combination of light (L) and fertilization (F). Data were analyzed as a RCB design with treatments arranged as a full 2 × 2 × 2 factorial measured repeatedly throughout the night using the repeated statement in PROC MIXED (SAS 9.3).

	Num.	Den.	Resp. parameters			Nonstructural C		
	df	*df*	E_O	R_{13}	R_{25}	Sol.	Starch	Total
Between subject effects								
Block	2	13	0.89	0.05	0.12	0.32	0.26	0.18
Temp. (TC)	1	13	**0.002**	**0.03**	**<0.0001**	**0.02**	**0.03**	**0.01**
Light (L)	1	13	**0.0004**	**0.002**	**<0.0001**	**0.01**	0.06	0.23
Fert. (F)	1	13	0.65	0.42	0.21	**<0.0001**	**0.01**	**<0.0001**
TC × L	1	13	0.38	0.06	**0.002**	0.63	0.21	0.77
TC × F	1	13	0.85	0.83	0.28	0.11	**0.01**	0.93
L × F	1	13	0.20	0.31	**0.04**	0.84	0.09	0.51
TC × L × F	1	13	0.36	**0.05**	0.11	0.45	0.71	0.45
Within subject effects								
Time	2	30	0.61	0.40	0.37	0.61	0.53	0.96
TC × time	2	30	0.49	0.21	0.94	0.96	0.53	0.90
L × time	2	30	0.75	0.19	0.12	0.59	0.12	0.41
F × time	2	30	0.85	0.87	0.60	0.99	0.68	0.79
TC × L × time	2	30	0.23	0.76	0.20	0.59	0.97	0.71
TC × F × time	2	30	0.97	0.77	0.94	0.89	0.26	0.49
L × F × time	2	30	0.83	0.31	0.32	0.92	0.37	0.93

3.6. Relationships among Respiration Parameters and Associated Leaf Traits. R_{13} was not correlated with leaf [*N*] under any treatment combination with the exception of "cool" grown plants under HL which was only marginally significant ($r^2 = 0.29$; $P = 0.07$; Figure 3(a)). No relationship was evident between leaf [*N*] and R_{25} or E_O (data not shown). R_{25} was significantly correlated with leaf starch across treatments in both "warm" and "cool" grown plants ($r^2 = 0.46$; $P = 0.008$ and $r^2 = 0.29$; $P = 0.01$, respectively; Figure 4). However, within the cold grown plants the low light, fertilization treatment was anomalous and was left out of the regression analysis (Figure 4(a)). Analysis of covariance (ANACOVA) showed a slightly significant ($P = 0.10$) difference in the relationship (slopes) among our treatments.

4. Discussion

Accurately accounting for thermal acclimation in carbon cycle models can have a profound effect on annual respiration estimates [34, 35, 36]. Given that studies which report on the presence or absence of thermal respiratory acclimation are often contradictory, for example, [6, 7, 10, 11, 14, 18, 19, 24, 37–44], our study is unique in that it was designed to manipulate precursors for warm, i.e., carbohydrates, [19] and cool respiration, i.e., leaf [*N*] [11,19, 43, 45–49], in order to analyse the mechanistic underpinnings of the temperature response of respiration (E_O), warm respiration (R_{25} in our case), and cool respiration (R_{13} in our case). This was done in an attempt to help explain why these contradictory results in published findings of acclimation, or lack thereof, exist.

Our results, however, only partially supported our hypotheses in that we did observe respiratory acclimation, and this acclimation did result from a decrease in E_O with increasing growth temperature, as expected. Interestingly, these acclimatory responses were only evident in plants grown with no light or nutrition manipulations, indicating that plants growing in less than ideal situations (LL and NF) may be incapable of acclimating to increased temperatures. It was also apparent that this lack of acclimation in our LL, NF combinations were much stronger when two manipulations were combined as opposed to just one (i.e., either light or nutrition). It appears that light controlled these acclimatory responses much more strongly than nutrition evidenced by the fact that the HL,NF plants were still capable of thermal respiratory acclimation while the LL,F plants showed no evidence of acclimation. We observed no overall change in R_{13} with changes in growth temperature with very few exceptions across treatments. This was not expected due to the strong previously published relationships between cool respiration and leaf [*N*] [2, 19, 28], but we did not observe a change in R_{13} which is indicative of Type I acclimation as defined by Atkin and Tjoelker [4].

Because of the suggested link between leaf respiration rates and mitochondrial substrate supply, i.e., carbohydrates; [20, 50], we also hypothesized that we would observe a decline in both E_O and R_{25} throughout the night due to a depletion of leaf carbohydrate and that this decline would be much more pronounced in the warm grown plants. Previous reports have indicated that as much as a 50% decline in leaf starch can occur throughout a 12-hour dark period [51]. Surprisingly, we observed no discernible change in any of our respiration parameters or carbohydrate concentrations throughout the night. This stability of both the temperature response and warm respiration may be in part due to the stability of leaf carbohydrate concentrations.

We did observe a decline in E_O and R_{25} in the "low-light" (LL) treatments, regardless of fertilization, but this was not a function of leaf carbohydrate status. We found a slight increase in leaf nonstructural C (starch and soluble sugars)

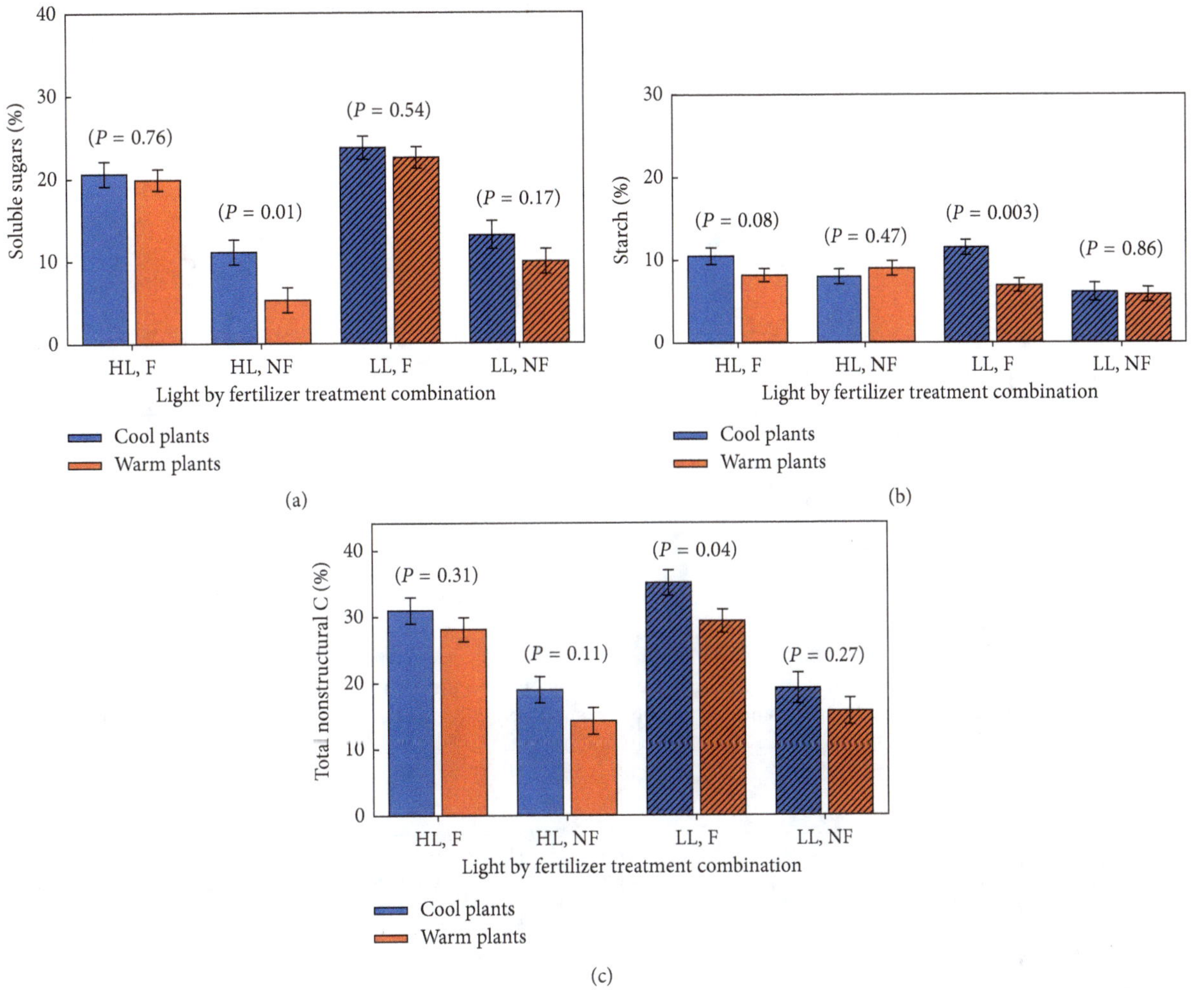

FIGURE 2: Differences among resource manipulation treatments for "warm" and "cool" grown plants (red and blue bars, respectively) for percent soluble sugars (a), starch (b), and total nonstructural C (c). Bars represent least square means ($n = 18$) throughout the night (error bars represent ± one standard error from the mean). P values show comparisons among "warm" and "cool" grown plants for each light (high light = HL and low light = LL) by fertilization (F = fertilized and NF = no fertilizer added) treatment combinations.

TABLE 2: Leaf attributes of *Glysine max* (soybean) plants exposed to different combinations of light and fertilization grown under "warm" (30°/23°C day/night) and "cool" (20°/13°C day/night) temperature treatments.

Leaf attributes	Light,Fert	Warm	Cool	P value
Mass-based leaf C ($mg{\cdot}g^{-1}$)	HL,NF	418 ± 6	417 ± 2	0.96
	HL,F	425 ± 8	422 ± 10	0.87
	LL,NF	394 ± 42	423 ± 7	0.20
	LL,F	413 ± 4	416 ± 4	0.87
Mass-based leaf N ($mg{\cdot}g^{-1}$)	HL,NF	38 ± 3	47 ± 2	**0.04**
	HL,F	59 ± 2	59 ± 3	1.00
	LL,NF	45 ± 5	56 ± 2	**0.01**
	LL,F	63 ± 2	65 ± 2	0.63
Area-based leaf N ($g{\cdot}cm^{-2}$)	HL,NF	1.48 ± 0.11	1.57 ± 1.10	0.74
	HL,F	2.18 ± 0.16	2.03 ± 0.19	0.66
	LL,NF	2.11 ± 0.22	2.25 ± 0.19	0.61
	LL,F	2.94 ± 0.28	2.63 ± 0.14	0.25
SLA ($g{\cdot}cm^{-2}$)	HL,NF	39.5 ± 2.7	33.8 ± 1.9	0.11
	HL,F	36.9 ± 2.4	34.2 ± 1.9	0.48
	LL,NF	46.9 ± 2.2	40.0 ± 1.6	0.06
	LL,F	46.3 ± 3.9	40.3 ± 2.4	0.10

in our low-light treatments, which is likely a function of reduced C use throughout the day as a result of the shade treatment. Both "warm" and "cool" grown plants across treatments showed a strong positive correlation between leaf starch and R_{25}. Within the "cool" grown plants, however, the LL,F treatment was considered an outlier and was not included in the regression analysis (Figure 4; circled observations). We hypothesize that the reasoning behind this anomalous treatment is that even though starch is likely a source of energy for the enzymatic reactions associated with leaf R_D, the starch concentration in a leaf is not affected by simply one day of reduced photosynthesis as designed in this experiment, and due to the decrease in photosynthetic rate, there was a drastic reduction in respiration rate indicating a direct link between photosynthesis and respiration, possibly through a decline in ATP-consumption rates [20]. This link between these two associated metabolic processes has been shown in numerous studies [11, 14, 19, 37, 52–57]. It is equally as likely that this same phenomenon was not observed in the warm grown plants due to the increased demand for substrate because of

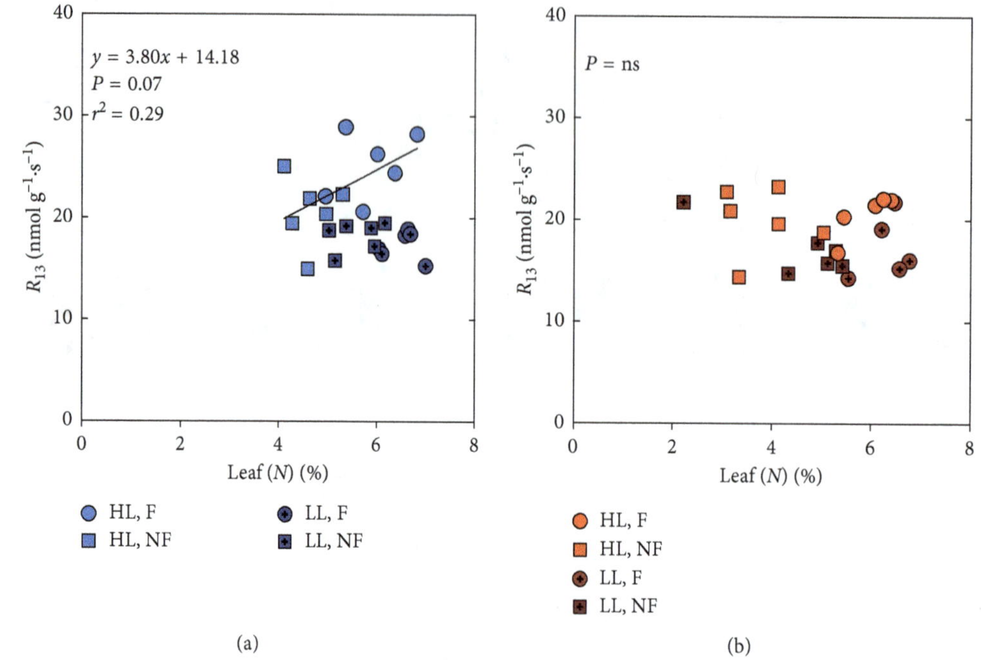

FIGURE 3: Relationship between cool respiration (R_{13}) and percent leaf N concentration ([N]) for "cool" and "warm" grown plants (represented in (a) and (b) and shown using blue and red symbols, respectively). Regressions were performed across fertilization (F,NF) treatments within a light (HL,LL) treatment.

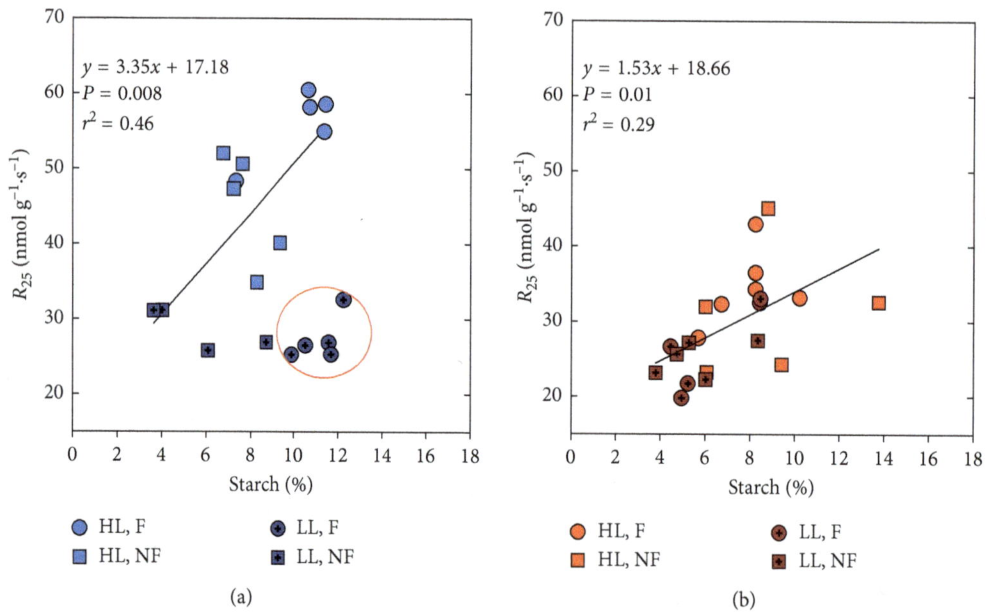

FIGURE 4: Relationship between warm respiration (R_{25}) and percent leaf starch for "cool" and "warm" grown plants (represented in (a) and (b) and shown using blue and red symbols, respectively). Regressions were performed across all treatments combined. Circled observation in (a) indicates omission of the low light, fertilized (LL,F) treatment from regression analyses.

the 10°C increase in night temperature associated with the "warm" treatment. Utilizing light response curves for soybean from Lauer et al. [57] and measuring A_{Sat} rates on HL plants (data not shown), we estimate that our LL plants experienced an approximately 60% reduction in photosynthetic rates as a result of our shade treatments (i.e., 30% of full sunlight). This estimate reasonably assumes that the short duration of the LL treatment (1 day) would not be enough time to impact leaf photochemical efficiency (slope) or the light saturation. Even with this substantial decrease in photosynthesis, starch and especially soluble carbohydrate concentrations in fertilized plants remained drastically higher than their NF counterparts in the "cool" treatment.

While the plants in our study were clearly not "starved" for nitrogen (even within the nonfertilized treatment based upon leaf N values of 3.8–5%), we did increase leaf [N] significantly with our fertilization treatment. We did observe shifts in allocation which are indicative of a nitrogen limitation. Ruffy et al. [58] suggested that nitrogen limitations can cause a general decline in the plants carbohydrate utilization efficiency which ultimately would result in a shift of allocation from roots to shoots. As reported in Ruffy et al. [58], we also observed a 22% increase in the shoot to root ratio although this was only marginally significant between fertilized and unfertilized plants ($P = 0.09$). It is important that more studies be conducted with more drastic nitrogen limitations imposed to tease out the mechanistic underpinnings of nutrient limitations on thermal acclimation of leaf dark respiration.

5. Conclusions

If the results of this study translate into similar findings in field situations and similar trends were to be found in photosynthetic acclimatory responses to stressors, this study has far-reaching implications for both managed and unmanaged (natural) landscapes. If plants are unable to acclimate metabolically to increased temperatures when grown in lower nutrition or low light (or a combination of these situations), agricultural (managed) and natural ecosystems which have these characteristics may be much more susceptible to current and future temperature increases associated with climate change. This limitation hypothesis which would potentially allow natural and managed ecosystems to be classified as to their potential resiliency to climate change needs to be thoroughly addressed in future studies.

Acknowledgments

The authors would like to thank Jessica LaGrone for her assistance with data collection and Roxie A. Jordan, Greenhouse Lab Supervisor, Louisiana Tech Department of Agriculture, for providing greenhouse space and technical support. This work was supported by National Science Foundation (NSF), EPSCoR Pilot funding program (grant number NSF(2010)-PFUND-200), and McIntire-Stennis formula funding (grant no. LAZ00074).

References

[1] J. S. Amthor, "The Mcree-de Wit-Penning Vries-Thornley respiration paradigms: 30 years later," *Annnals of Botany*, vol. 86, no. 1, pp. 1–20, 2000.

[2] M. G. Ryan, "Effects of climate change on plant respiration," *Ecological Applications*, vol. 1, no. 2, pp. 157–167, 1991.

[3] R. C. Dewar, B. E. Medlyn, and R. E. McMurtrie, "Acclimation of the respiration/photosynthesis ratio to temperature: insights from a model," *Global Change Biology*, vol. 5, no. 5, pp. 615–622, 1999.

[4] O. K. Atkin and M. G. Tjoelker, "Thermal acclimation and the dynamic response of plant respiration to temperature," *Trends in Plant Science*, vol. 8, no. 7, pp. 343–351, 2003.

[5] A. F. Armstrong, D. C. Logan, and O. K. Atkin, "On the developmental dependence of leaf respiration: responses to short- and long-term changes in growth temperature," *Annals of Botany*, vol. 93, no. 11, pp. 1633–1639, 2006.

[6] J. M. Frantz, N. N. Cornetti, and B. G. Bugbee, "Night temperature has a minimal effect on respiration and growth in rapidly growing plants," *Annals of Botany*, vol. 94, no. 1, pp. 155–166, 2004.

[7] O. K. Atkin, I. Scheurwater, and T. L. Pons, "High thermal acclimation of both photosynthesis and respiration in two lowland Plantago species in contrast to an alpine congeneric," *Global Change Biology*, vol. 12, no. 3, pp. 500–515, 2006.

[8] J. A. Arnone and C. Korner, "Temperature adaptation and acclimation potential of leaf dark respiration in two species of ranunculus in warm and cold habitats," *Arctic and Alpine Research*, vol. 29, no. 1, pp. 122–125, 1997.

[9] A. Larigauderie and C. Korner, "Acclimation of leaf dark respiration to temperature in alpine and lowland plant species," *Annals of Botany*, vol. 76, no. 3, pp. 245–252, 1995.

[10] M. G. Tjoelker, J. Oleksyn, and P. B. Reich, "Acclimation of respiration to temperature and CO_2 in seedlings of boreal tree species in relation to plant size and relative growth rate," *Global Change Biology*, vol. 5, no. 6, pp. 679–691, 1999.

[11] I. J. Wright, P. B. Reich, O. K. Atkin, C. H. Lusk, M. G. Tjoelker, and M. Westoby, "Irradiance, temperature and rainfall influence leaf dark respiration in woody plants: evidence from comparisons across 20 sites," *New Phytologist*, vol. 169, no. 2, pp. 309–319, 2005.

[12] L. H. Ziska and J. A. Bunce, "Direct and indirect inhibition of single leaf respiration by elevated CO2 concentrations: interaction with temperature," *Physiologia Plantarum*, vol. 90, no. 1, pp. 130–138, 1994.

[13] D. Bruhn, M. Schortemeyer, E. J. Edwards et al., "The apparent temperature response of leaf respiration depends on the timescale of measurements: a study of two cold climate species," *Plant Biology*, vol. 10, no. 2, pp. 185–193, 2008.

[14] B. R. Loveys, L. J. Atkinson, D. J. Sherlock, R. L. Roberts, A. H. Fitter, and O. K. Atkin, "Thermal acclimation of leaf and root respiration: and investigation comparing inherently fast- and slow-growing plant species," *Global Change Biology*, vol. 9, no. 6, pp. 895–910, 2003.

[15] M. G. Tjoelker, J. Oleksyn, and P. B. Reich, "Modelling respiration of vegetation: evidence for a general temperature-dependence Q_{10}," *Global Change Biology*, vol. 7, no. 2, pp. 223–230, 2001.

[16] J. Lloyd and J. A. Taylor, "On the temperature-dependence of soil respiration," *Functional Ecology*, vol. 8, no. 3, pp. 315–323, 1994.

[17] O. K. Atkin, D. Bruhn, V. M. Hurry, and M. G. Tjoelker, "The hot and the cold: unraveling the variable response of plant respiration to temperature," *Functional Plant Biology*, vol. 32, no. 2, pp. 87–105, 2005.

[18] K. L. Griffin, M. Turnbull, R. Murthy et al., "Leaf respiration is differentially affected by leaf vs stand level night-time warming," *Global Change Biology*, vol. 8, no. 5, pp. 479–485, 2002.

[19] D. N. Dillaway and E. L. Kruger, "Leaf respiratory acclimation to climate: comparisons among boreal and temperate tree species along a latitudinal transect," *Tree Physiology*, vol. 31, no. 10, pp. 1114–1127, 2011.

[20] O. K. Atkin, J. R. Evans, M. C. Ball, H. Lambers, and T. L. Pons, "Leaf respiration of snow gum in the light and dark. Interactions between irradiance and temperature," *Plant Physiology*, vol. 122, no. 3, pp. 915–922, 2000.

[21] O. K. Atkin, I. Scheurwater, and T. L. Pons, "Respiration as a percentage of daily photosynthesis in whole plants is homeostatic at moderate, but not high, growth temperatures," *New Phytologist*, vol. 174, no. 2, pp. 367–380, 2007.

[22] K. A. Mitchell, P. V. Bolstad, and J. M. Vose, "Interspecific and environmentally induced variation in foliar dark respiration among eighteen southeastern deciduous tree species," *Tree Physiology*, vol. 19, no. 13, pp. 861–870, 1999.

[23] P. B. Reich, M. B. Walters, M. G. Tjoelker, D. Vanderklein, and C. Buschena, "Photosynthesis and respiration rates depend on leaf and root morphology and nitrogen concentration in nine boreal tree species differing in relative growth rate," *Functional Ecology*, vol. 12, no. 3, pp. 395–405, 1998.

[24] M. G. Tjoelker, J. Oleksyn, P. B. Reich, and R. Żytkowiak, "Coupling of respiration, nitrogen, and sugars underlies convergent temperature acclimation in Pinus banksiana across wide-ranging sites and populations," *Global Change Biology*, vol. 14, no. 4, pp. 1–16, 2008.

[25] M. G. Tjoelker, J. Oleksyn, G. Lorenc-Plucinska, and P. B. Reich, "Acclimation of respiratory temperature responses in northern and southern populations of Pinus Banksiana," *New Phytologist*, vol. 181, no. 1, pp. 218–229, 2009.

[26] I. Florez-Sarasa, W. L. Araújo, S. V. Wallström, A. G. Rasmusson, A. R. Fernie, and M. Ribas-Carbo, "Light-responsive metabolite and transcript levels are maintained following a dark-adaptation period in leaves of *Arabidopsis thaliana*," *New Phytologist*, vol. 195, no. 1, pp. 136–148, 2012.

[27] J. Azcon-Bieto and C. B. Osmond, "Relationship between photosynthesis and respiration: the effect of carbohydrate status on the rate of CO2 production by respiration in darkened and illuminated wheat leaves," *Plant Physiology*, vol. 71, no. 3, pp. 574–581, 1983.

[28] M. Turnbull, D. Whitehead, D. T. Tissue, W. S. F. Schuster, K. J. Brown, and K. L. Griffin, "Responses of leaf respiration to temperature and leaf characteristics in three deciduous tree species vary with site water availability," *Tree Physiology*, vol. 21, no. 9, pp. 571–578, 2001.

[29] K. Y. Crous, G. Wallin, O. K. Atkin, J. Uddling, and A. af Ekenstam, "Acclimation of light and dark respiration to experimental and seasonal warming are mediated by changes in leaf nitrogen in Eucalyptus globulus," *Tree Physiology*, vol. 37, no. 8, pp. 1069–1083, 2017.

[30] L. F. Ow, K. L. Griffin, D. Whitehead, A. S. Walcroft, and M. H. Turnbull, "Thermal acclimation of leaf respiration but not photosynthesis in Populus deltoides x nigra," *New Phytologist*, vol. 178, no. 1, pp. 123–134, 2008.

[31] T. L. Pons and R. M. Welschen, "Over estimation of respiration rates in commercially available clamp-on leaf chambers. Complications with measurements of net photosynthesis," *Plant Cell and Environment*, vol. 25, no. 10, pp. 1367–1372, 2002.

[32] C. M. Kaelke, E. L. Kruger, and P. B. Reich, "Trade-offs in seedling survival, growth and physiology among hardwood species of contrasting successional status along a light availability gradient," *Canadian Journal of Forest Research*, vol. 31, no. 9, pp. 1602–1616, 2001.

[33] SAS, *SAS/STAT® 12.1 User's Guide*, SAS Institute Inc., Cary, NC, USA, 2012.

[34] N. G. Smith and J. S. Dukes, "Plant respiration and photosynthesis in global-scale models: incorporating acclimation to temperature and CO2," *Global Change Biology*, vol. 19, no. 1, pp. 45–63, 2013.

[35] M. C. Vanderwel, M. Slot, J. W. Lichstein et al., "Global convergence in leaf respiration from estimates of thermal acclimation across time and space," *New Phytologist*, vol. 207, no. 4, pp. 1026–1037, 2015.

[36] O. K. Atkin, K. J. Bloomfield, P. B. Reich et al., "Global variability in leaf respiration in relation to climate, plant functional types and leaf traits," *New Phytologist*, vol. 206, no. 2, pp. 614–636, 2015.

[37] J. Zaragoza-Castells, D. Sánchez-Gómez, I. P. Hartley et al., "Climate-dependent variations in leaf respiration in a dryland, low productivity Mediterranean forest: the importance of acclimation in both high-light and shaded habitats," *Functional Ecology*, vol. 22, pp. 172–184, 2008.

[38] W. Yamori, K. Noguchi, K. Hikosaka, and I. Terashima, "Cold-tolerant crop species have greater temperature homeostasis of leaf respiration and photosynthesis than cold-sensitive species," *Plant and Cell Physiology*, vol. 50, no. 2, pp. 203–215, 2008.

[39] L. F. Ow, D. Whitehead, A. S. Walcroft, and M. H. Turnbul, "Seasonal variation in foliar carbon exhchange in Pinus radiata and Populus deltoides: respiration acclimates fully to changes in temperature but photosynthesis does not," *Global Change Biology*, vol. 16, no. 1, pp. 288–302, 2010.

[40] D. A. Way and R. Oren, "Differential responses to changes in growth temperature between trees from different functional groups and biomes: a review and synthesis of data," *Tree Physiology*, vol. 30, no. 6, pp. 669–688, 2010.

[41] J. Rodriguz-Calceradda, O. K. Atkin, T. M. Robson, J. Zaragoza-Castells, L. Gil, and I. Aranda, "Thermal acclimation of leaf dark respiration of beech seedlings experiencing summer drought in high and low light environments," *Tree Physiology*, vol. 30, no. 2, pp. 214–224, 2010.

[42] A. Laragauderie and C. Korner, "Acclimation of leaf dark respiration to temperature in alpine and lowland plant species," *Annals of Botany*, vol. 76, no. 3, pp. 245–252, 1995.

[43] T. D. Lee, P. B. Reich, and P. V. Bolstad, "Acclimation of leaf respiration to temperature is rapid and related to specific leaf area, soluble sugars and leaf nitrogen across three temperature deciduous tree species," *Functional Ecology*, vol. 19, no. 4, pp. 640–647, 2005.

[44] P. V. Bolstad, A. K. Mitchell, and J. M. Vose, "Foliar temperature-respiration response functions for broad-leaved tree species in the southern Appalachians," *Tree Physiology*, vol. 19, no. 13, pp. 871–878, 1999.

[45] M. G. Ryan, "Foliar maintenance respiration of subalpine and boreal trees and shrubs in relation to nitrogen content," *Plant Cell and Environment*, vol. 18, no. 7, pp. 765–772, 1995.

[46] M. G. Tjoelker, P. B. Reich, and J. Oleksyn, "Changes in leaf nitrogen and carbohydrates underlie temperature and CO_2 acclimation of dark respiration in five boreal tree species," *Plant Cell and Environment*, vol. 22, no. 7, pp. 767–778, 1999.

[47] M. Turnbull, D. Whitehead, D. T. Tissue, W. S. F. Schuster, K. J. Brown, and K. L. Griffin, "Scaling foliar respiration in two contrasting forest canopies," *Functional Ecology*, vol. 17, no. 1, pp. 101–114, 2003.

[48] P. B. Reich, J. Oleksyn, and M. G. Tjoelker, "Needle respiration and nitrogen concentration in Scots pine populations from a broad latitudinal range: a common garden test with field grown trees," *Functional Ecology*, vol. 10, no. 6, pp. 768–776, 1996.

[49] P. B. Reich, M. G. Tjoelker, K. S. Pregitzer, I. J. Wright, J. Oleksyn, and J.-L. Machado, "Scaling of respiration to nitrogen in leaves, stems and roots in higher land plants," *Ecology Letters*, vol. 11, no. 8, pp. 793–801, 2008.

[50] O. K. Atkin and H. lambers, "Slow-growing alpine and fast growing lowland species: a case study of factors associated with variation in growth rate among herbaceous higher plants under natural and controlled conditions," in *Inherent Variation in Plant Growth: Physiological Mechanisms and Ecological Consequences*, H. lambers, H. Poorter, and M. M. I. Van Vuuren, Eds., pp. 259–288, Backhuys Publishers, Leiden, Netherlands, 1998.
[51] V. L. Lechtenberg, D. A. Holt, and H. W. Youngberg, "Diurnal variation in nonstructural carbohydrates, in vitro digestibility, and leaf to stem ratio of alfalfa," *Agronomy Journal*, vol. 63, no. 5, pp. 719–724, 1971.
[52] J. A. Bunce, "Direct and acclimatory responses of dark respiration and translocation to temperature," *Annals of Botany*, vol. 100, no. 1, pp. 67–73, 2007.
[53] I. P. Hartley, A. F. Armstrong, R. Murthy, G. Barron-Gafford, P. Ineson, and O. K. Atkin, "The dependence of respiration on photosynthetic substrate supply and temperature: integrating leaf, soil and ecosystem measurements," *Global Change Biology*, vol. 12, no. 10, pp. 1954–1968, 2006.
[54] K. Maseyk, J. M. Grünzweig, E. Rotenberg, and D. Yakir, "Respiration acclimation contributes to high carbon-use efficiency in a seasonally dry pine forest," *Global Change Biology*, vol. 14, no. 7, pp. 1553–1567, 2008.
[55] T. M. Wertin and R. O. Teskey, "Close coupling of whole-plant respiration to net photosynthesis and carbohydrates," *Tree Physiology*, vol. 28, pp. 1831–1840, 2008.
[56] D. Whitehead, K. L. Griffin, M. H. Turnbull et al., "Response of total night-time respiration to differences in total daily photosynthesis for leaves in a Quercus rubra L. canopy: implications for modelling canopy CO_2 exchange," *Global Change Biology*, vol. 10, pp. 925–938, 2004.
[57] M. J. Lauer, S. G. Pallardy, D. G. Blevins, and D. D. Randall, "Whole leaf carbon exchange characteristics of phosphate deficient soybeans (Glycine max L.)," *Plant Physiology*, vol. 91, pp. 848–854, 1989.
[58] T. W. Ruffy, S. C. Huber, and R. J. Volk, "Alterations in leaf carbohydrate metabolism in response to nitrogen stress," *Plant Physiology*, vol. 88, pp. 725–730, 1988.

12

Biochar and/or Compost Applications Improve Soil Properties, Growth, and Yield of Maize Grown in Acidic Rainforest and Coastal Savannah Soils in Ghana

Albert Kobina Mensah[1] **and Kwame Agyei Frimpong**[2]

[1]*Department of Soil Science and Soil Ecology, Institute of Geography, Ruhr-Universitaet Bochum, Universitätsstraße 150, 44801 Bochum, Germany*
[2]*Department of Soil Science, School of Agriculture, University of Cape Coast, Cape Coast, Ghana*

Correspondence should be addressed to Albert Kobina Mensah; albert.mensah@rub.de

Academic Editor: David Clay

Use of biochar for soil fertility improvement is gaining popularity due to its potential to improve soil quality, increase crop yield, and sequester carbon from the atmosphere-biosphere pool into the soil. A 40-day pot experiment was carried out to investigate the effects of corncob biochar and compost applied alone (at a rate of 2%, w/w) or in combination (1% of each, thus 1% compost + 1% biochar) on soil physicochemical properties, growth, and yield of maize on two soils of contrasting pH and texture collected from the Rainforest and Coastal Savannah agroecological zones of Ghana. Biochar and compost applied alone or in combination significantly increased soil pH, total organic carbon, available phosphorus, mineral nitrogen, reduced exchangeable acidity, and increased effective cation exchange capacity in both soils. Additionally, combined application and single application biochar or compost additions increased the plant height, stem girth, and dry matter yields of two maize (local ("*ewifompe*") and hybrid (*Obaatanpa*)) varieties used in the study. The study showed that biochar applied alone or in combination with compost offers the potential to enhance soil quality and improve maize yield.

1. Introduction

In sub-Saharan African (SSA), soil fertility decline, mainly through continuous cultivation and rapid organic matter mineralization, causes food insecurity and poverty [1, 2]. Smallholder agricultural production systems in most SSA countries including Ghana are characterized by low productivity due to low and erratic rainfall patterns, outdated agricultural practices, and low application of nutrient inputs. Agricultural intensification in Ghana highly depends on chemical fertilizer application. However, inorganic fertilizer use in Ghana is limited due to high cost, unreliable supply, and hence poor accessibility to the smallholder farmers. According to Zheng et al. [3], the use of biochar as a soil amendment is an innovative and promising soil fertility replenishment option to promote sustainable agriculture. Biochar is a form of charcoal produced through thermochemical decomposition of biomass under low or no oxygen conditions, a process referred to as pyrolysis [2, 3]. Biochar application has been reported to improve soil quality by raising soil pH [4], increasing moisture holding capacity [2], stimulating activity of more beneficial fungi and microbes [3], improving cation exchange capacity [5, 6], and retaining nutrients [7]. Incorporated biochar can also sequester carbon from the atmosphere-biosphere pool and transfer it to soil [8, 9]. Unlike compost which decomposes rapidly in humid tropical soils, biochar decomposes slowly in the soil as it contains high recalcitrant C content. The residence time for wood biochar, for instance, is in the range of 100 to 1000s of years, which is approximately 10–1000 times longer than residence times of most soil organic matter [2]. Therefore, biochar addition to soils could provide a potential sink for carbon and increase soil nutrient availability for crop use due to its higher nutrient retention and sorption capacity.

Furthermore, application of biochar to soils has been identified as a low-cost technology that can stabilize organic carbon, reduce greenhouse gas emissions [10], improve soil physical and chemical properties, and boost crop yield and productivity [11, 12] and farm incomes. Biochar application can lead to a reduction in inorganic fertilizer use by farmers [2]. Biochar research is a recent development in Ghana and so there is a paucity of information regarding its effects on soil properties, crop growth, and crop yield in Ghanaian soils.

Glaser and Birk [13] and Glaser et al. [14] reported that sole biochar application, in most cases, did not provide any substantial amount of nutrients. Other studies showed that combined application of compost and biochar enhanced soil quality, increased plant growth, provided positive synergistic effect on soil nutrient contents under field conditions, resulted in reduction of fertilizer input application, improved nutrient use efficiency, stabilized soil structure, and improved water retention capacity [13, 15–18]. Owing to the stability of biochar carbon, applying biochar with compost can enhance compost properties, leading to higher added value and a much better carbon sequestration potential compared to individual mixing of biochar or compost with soil [13, 15–18]. This study investigated the effects of corncob biochar and compost applied alone or in combination on some soil physicochemical properties, growth, and yield of maize grown on two Ghanaian soils collected from the Rainforest (Aiyinase in Western Region) and Coastal Savannah (Cape Coast in Central Region) agroecological zones.

2. Materials and Methods

2.1. Soils and Amendments. Soil samples (0–20 cm) from arable fields in Cape Coast and Aiyinase belonging to the coastal savannah and the tropical rainforest agroecological zones, respectively, were used for the study. The Cape Coast soil is classified as Alfisols [19], and the Aiyinase soil is classified as Haplic Ferralsol [20]. The physicochemical properties of the two soils are presented in Table 1.

The corncobs and composts were collected from the University of Cape Coast Teaching and Research Farm. The biochar was produced from corncobs pyrolyzed at 350°C in Lucia Biomass Pyrolytic Stoves (World Stove, Italy). The properties of the corncob biochar and compost amendments are summarized in Table 2.

2.2. Experimental Design, Treatments, and Planting. The compost and/or biochar was thoroughly mixed with 10.34 kg of air-dry soil (<2 mm) and kept in plastic pots of 9.4 m^3 volume. Biochar and compost were applied sole or in combination as 2% weight by weight compost: soil basis. The treatments involved in the experiment, compost only, biochar only, compost and biochar, and control, are presented in Table 3.

Each treatment was replicated three times, giving a total of 48 pots. The control and biochar-compost mixture-treated soils were repacked in the pots to a bulk density of 1.1 $g{\cdot}cm^{-3}$, and water was added to bring the soil water content to 20%.

TABLE 1: Physicochemical properties of the Cape Coast and Aiyinase soils used in the study.

Soil property	Aiyinase	Cape Coast
Available P ($mg{\cdot}kg^{-1}$)	7.9	16.4
TOC (%)	0.2	1.4
pH (H_2O)	4.7	5.0
Mg^{2+} ($cmol{\cdot}kg^{-1}$)	0.2	1.6
Ca^{2+} ($cmol{\cdot}kg^{-1}$)	0.9	2.8
K^+ ($cmol{\cdot}kg^{-1}$)	0.1	0.6
Na^+ ($cmol{\cdot}kg^{-1}$)	0.2	0.6
$H^+ + Al^{3+}$ ($cmol{\cdot}kg^{-1}$)	0.9	0.1
CEC ($cmol{\cdot}kg^{-1}$)	1.4	5.6
ECEC ($cmol{\cdot}kg^{-1}$)	2.3	5.8
TN (%)	0.0	0.1
Clay (%)	14.2	12.6
Silt (%)	2	23.2
Sand (%)	83.7	64.2

TABLE 2: Chemical properties of biochar and compost used in the experiment.

Property	Biochar	Compost
Available P ($mg{\cdot}kg^{-1}$)	88.2	353.7
TOC (%)	70.1	23.6
pH (H_2O)	7.0	5.3
Mg^{2+} ($cmol{\cdot}kg^{-1}$)	12.0	22.7
Ca^{2+} ($cmol{\cdot}kg^{-1}$)	44.3	162.7
K^+ ($cmol{\cdot}kg^{-1}$)	39.4	10.8
Na^+ ($cmol{\cdot}kg^{-1}$)	32.6	9.7
$H^+ + Al^{3+}$ ($cmol{\cdot}kg^{-1}$)	11.3	12.7
ECEC ($cmol{\cdot}kg^{-1}$)	139.6	218.5
Total nitrogen (%)	0.9	4.2

TABLE 3: Application rates (%) of biochar and compost used in the pot experiment.

Treatment	Application rate (%, w/w)
Compost only	2
Biochar only	2
Compost + biochar	1 + 1
Control	0

The water filled pore space (WFPS) was calculated using the following formula:

$$\text{water filled pore space} = \frac{\text{volumetric water content}}{\text{total porosity}} \times 100. \quad (1)$$

But, total porosity = 1 – (bulk density/particle density). Bulk density and particle density were 1.1 $g{\cdot}cm^{-3}$ and 2.65 $g{\cdot}cm^{-3}$, respectively.

The control and biochar- and/or compost-amended soils were incubated for two weeks at 20% WHC and 25°C temperature prior to sowing the maize test crop. During the incubation period, the pots were kept in a dark room for 14 days in a completely randomized design. At the end of the two weeks, soil samples from each treatment were collected and analyzed in the laboratory for pH, available P, total carbon, exchangeable acidity, exchangeable cations, effective cation exchange capacity, and mineral/available nitrogen contents. At the end of the two weeks' incubation period, three maize seeds were sown to a depth of 4 cm in the pots and thinned to one plant per pot 5 days after seedling emergence.

2.3. Laboratory Analyses and Plant Data Collection. Samples of the two soil types, prior to treatment application and at the end of the 14-day incubation period, the biochar and compost, were analyzed. The pH of the soils was measured in a 1 : 2.5 (w/v) soil : water ratio, with a pH meter [21]. Available P was analyzed using the Bray 1 acid method [22]. Soil extraction for exchangeable cations (Ca, Mg, K, and Na) analyses was done using the NH_4OAc method at pH 7 [23]. Soil exchangeable acidity was determined by the titration method described by Robertson et al. [24]. Soil organic carbon concentration was determined by the Walkley–Black method [25]. The NH_4^+ and NO_3^- in the soil sample were determined using the method described by Rowell [26], and soil mineral nitrogen was calculated as the sum of the NH_4^+ and NO_3^- concentrations.

Plant height, stem girth, and number of leaves were collected at weekly intervals up to the end of the experiment. The number of fully developed leaves was counted on weekly basis until the aboveground biomass was harvested. Plant height was measured with a graduated straight edge (ruler) from base of the shoot at the soil surface to the tip of the highest leaf in the foliage of the crop. Stem girth was measured using a pair of vernier calipers. The maize plant was harvested by cutting the plant at the soil level 37 days after emergence. The dry matter of the shoot biomass was determined by drying the aboveground maize plant in an oven at 60°C until constant dry weight was attained.

2.4. Statistical Analyses. One-way analysis of variance (ANOVA) was performed to assess the significant differences in soil and plant parameters between different treatments using the GenSTAT Version 15.1. Mean comparison was also made using Fisher's protected test for least significant difference after treatments were found significant at $P < 0.05$.

3. Results and Discussion

3.1. Effects of Biochar and/or Compost Applications on the Soil pH. The effect of biochar and/or compost applications on the soil pH of the Aiyinase and Cape Coast soils after the 14-day incubation period is shown in Figure 1. The results revealed a significant ($P < 0.01$) increase in the soil pH following sole and combined applications of compost and biochar in both soils. In the Aiyinase soil, the highest soil pH was recorded in the 2% sole biochar treatment, while the lowest value was recorded in the control. The soil pH increase was in the order B20 > BC > C20 > B0. Likewise, in the Cape Coast soil, the highest pH (6.1) was observed in soils treated with sole biochar, followed by the combined biochar and compost treatments (5.7). Soils with no amendments

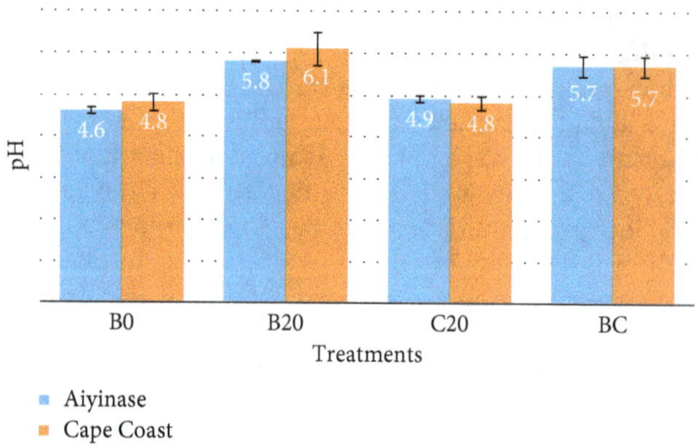

Figure 1: Changes in soil pH of the Aiyinase and Cape Coast soils following application of corncob biochar and/or compost 14 days after incubation. B0 = control with no amendments; B20 = biochar 2%, w/w; C20 = compost 2%, w/w; BC = combined biochar and compost rate 1% compost + 1% biochar. Error bars represent standard deviation of the means.

applied and soils amended only with compost had the least soil pH of 4.8.

The increases in soil pH due to sole application of biochar could be attributed to the high pH of the biochar used in the experiment. The pH of the biochar was 7.0 compared to that of the compost which was 5.3. The findings agree with those of Nigussie et al. [27] who attributed increased soil pH in biochar-amended soil to ash accretion. They further explained that ash residues are highly dominated by carbonates alkali and alkali earth metals. In this study, the ash content was not quantified, but given the conditions under which the biochar was produced, it is likely that the final product will contain some amount of ash. Zwieten et al. [28] pointed out that incorporated biochar can increase soil pH through its liming effects. In the study, soil pH in all the biochar- and/or compost-amended soils was above the threshold level pH below which Al^{3+} toxicity (4.8–5.0) occurs as mentioned by Zweiten et al. [28]. Nigussie et al. [27] attributed the increase in soil pH found in biochar-amended soils to the high surface area and porous nature of biochar that increases the cation exchange capacity (CEC) of soils, which binds Al and Fe with the soil exchange sites. This study further observed reduction in exchangeable acidity ($H^+ + Al^{3+}$) in the biochar- and/or compost-amended soils (range between 0.14 and 0.25 $cmol{\cdot}kg^{-1}$) than in the unamended soils (0.75 $cmol{\cdot}kg^{-1}$), which further explains the reason for increased acidity in the untreated soils compared to treated soils. The pH values observed in this study are also consistent with those found by Nigussie et al. [27], Zhang et al. [29], Agusalim et al. [30], and Zheng et al. [3], who concluded that biochar application can improve soil quality by increasing soil pH, which suggests that biochar could be used as an alternative option to lime materials to raise pH of acidic soils.

3.2. Soil Available Phosphorus. The application of biochar and compost, alone or in combination, significantly ($P < 0.01$) increased the soil available phosphorus contents

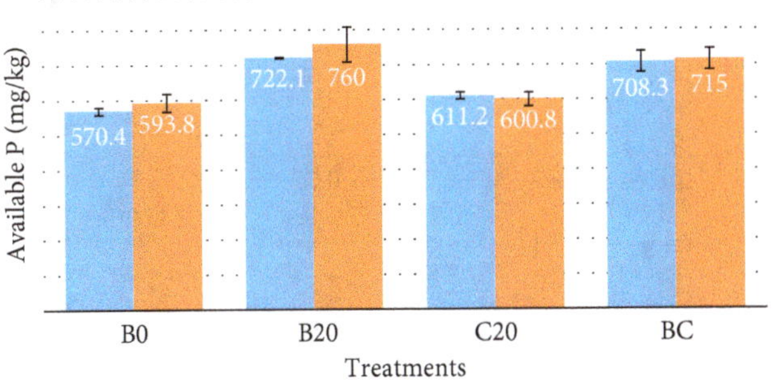

Figure 2: Changes in soil available phosphorus of Aiyinase and Cape Coast soils following application of corncob biochar and/or compost 14 days after incubation. B0 = control with no amendments; B20 = biochar 2%, w/w; C20 = compost 2%, w/w; BC = combined biochar and compost rate 1% compost + 1% biochar. Error bars represent standard deviation of the means.

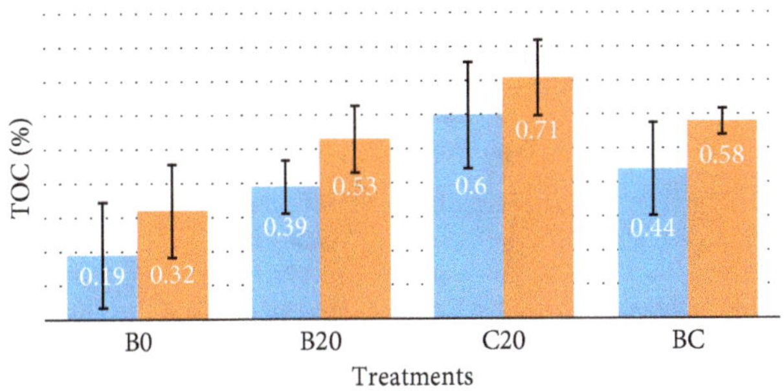

Figure 3: Changes in soil total organic carbon of Aiyinase and Cape Coast soils following application of corncob biochar and/or compost 14 days after incubation. B0 = control with no amendments; B20 = biochar 2%, w/w; C20 = compost 2%, w/w; BC = combined biochar and compost rate 1% compost + 1% biochar. Error bars represent standard deviation of the means.

in both the Aiyinase and Cape Coast soils (Figure 2). In both soils, the highest soil available P content was observed in the 2% sole biochar treatment. The increases in the soil available P observed in the soils, especially in the Aiyinase soils, could be attributed to reduced Fe and Al activity due to increase in the soil pH (4.8–5.8) and exchangeable bases/cation exchange capacity. In a similar study, Sasmita et al. [31] found that application of biochar with or without organic fertilizer linearly increased the soil available P in Indonesia acidic soil medium during a 15-day incubation period. Similar to the findings of this study, which found relatively higher concentrations in combined biochar and compost treatments for both the Aiyinase acidic soil and Cape Coast soil, Sasmita et al. [31] further reported a synergistic effect between biochar and organic fertilizer in increasing the soil available P.

Increase in soil pH would have reduced sorption of the soil available P. Agegnehu et al. [32], Nigussie et al. [27], and De Luca et al. [33] reported a greater soil available P contents in biochar-amended soils compared to unamended soils and attributed the improvement to biochar's capacity to retain and exchange phosphate ions due to its positively charged surface sites.

3.3. Soil Total Organic Carbon (TOC). The results showed that application of biochar and compost, alone or in combination, increased soil TOC contents in both the Aiyinase and Cape Coast soils (Figure 3). However, no significant differences were found among the TOC contents of the amended soils. In a similar incubation study, Frimpong et al. [34] found that TOC contents were higher in soils treated with cow dung and biochar than those of the control. Additionally, Trupiano et al. [15] found that application of compost and biochar, alone or in combination, increased soil TOC content than that in the unamended soils, which is indicative that biochar and/or compost applications to soils can enhance C accumulation and sequestration.

3.4. Soil Exchangeable Bases and Exchangeable Acidity. Results showing the effects of biochar and/or compost addition on soil exchangeable bases of both Aiyinase and Cape Coast soils are presented in Table 4. Compared to the control treatment, application of these amendments, alone or in combination, increased the exchangeable Ca, Mg, K, and Na contents in both soils. The highest Ca and Mg contents were observed in soils treated with 2% compost in both the Aiyinase and Cape Coast soils. These increments observed may be related to the exchangeable Ca and Mg, as well as the ECEC contents of the compost used in the experiment. Compared to the biochar, the compost used had greater proportions of exchangeable Ca and Mg, 162.7 and 22.7 $cmol{\cdot}kg^{-1}$, respectively. According to Fischer and Glaser [18], lone compost application has a liming effect due to its richness in alkaline cations such as Ca, Mg, and K, which are liberated from organic matter due to mineralization. In the Aiyinase soil, the highest exchangeable K content was found in the 2% biochar treatment, whereas the least was found in the control. In the Cape Coast soil, the highest exchangeable K content was recorded in the combined compost and biochar treatments. In a pot experiment, Nigussie et al. [27] found that biochar applied to chromium-polluted soils improved the soil exchangeable bases and crop yield of *Lactuca sativa* in Ethiopia. They concluded that the increase in the exchangeable bases was as a result of the presence of ash in the biochar which helps in the immediate release of mineral nutrients like Ca and K for crop use.

In terms of soil exchangeable acidity, combined biochar and compost recorded the least value and control soil with no amendments had the highest value for the acidic Aiyinase soil. The reduction was in the order BC < B20 < C20 < B0. The reduction in the soil exchangeable acidity in the amended soils may be due to the higher cation exchange capacity (CEC) of the compost and biochar, which has the capacity of binding Al and Fe with the soil exchange sites. For the Cape Coast soil, exchangeable acidity decreased in the order B20 < B0, BC < C20. Sasmita et al. [31] made similar observation in a 15-day incubation experiment using

TABLE 4: Changes in soil chemical properties of the Aiyinase and Cape Coast soils (exchangeable acidity, exchangeable cations, and ECEC) following application of biochar and/or compost 14 days after incubation.

Treatment rate	H + Al ($cmol{\cdot}kg^{-1}$)		Ca ($cmol{\cdot}kg^{-1}$)		Mg ($cmol{\cdot}kg^{-1}$)		K ($cmol{\cdot}kg^{-1}$)		Na ($cmol{\cdot}kg^{-1}$)		ECEC ($cmol{\cdot}kg^{-1}$)	
	AN	CC	AN	CC	AN	CC	AN	CC	AN	CC	AN	CC
B0	0.75	0.15	1.07	2.21	0.29	1.15	0.15	0.42	0.26	0.32	2.52	4.26
B20	0.17	0.11	1.20	2.00	0.40	0.83	1.36	1.12	0.71	1.03	3.84	5.08
C20	0.25	0.28	3.07	4.40	1.81	1.76	0.68	1.04	0.84	0.97	6.64	8.45
BC	0.14	0.15	2.93	4.29	0.72	1.55	0.81	1.35	0.64	1.22	5.25	8.56
LSD, 0.05	0.22	0.13	1.23	0.89	1.21	0.63	0.50	0.77	0.61	0.66	1.18	0.61
F. Pr.	<0.001	0.06	0.01	<0.001	0.07	0.04	0.00	0.11	0.23	0.06	<0.001	<0.001

AN = Aiyinase soil; CC = Cape Coast soil; B0 = control with no amendments; B20 = biochar 2%, w/w; C20 = compost 2%, w/w; BC = combined biochar and compost rate 1% compost + 1% biochar; LSD = least significant difference at 0.05.

rice husk biochar and attributed the reduction in the soil exchangeable Al to the liming effect from biochar. Also, the reduction of the exchangeable $H^+ + Al^{3+}$ content by biochar could be due to the formation of Al complex by the oxidized organic functional groups such as carboxylic and phenolic at the biochar surface as pointed out by Vithanage et al. [35].

3.5. Effective Cation Exchange Capacity (ECEC). The results indicated that effective cation exchange capacity (Table 4) of both the Cape Coast and Aiyinase soils was significantly increased upon application of biochar and/or compost amendments at $P < 0.01$. In the Aiyinase soil, the highest ECEC was observed in the sole compost treatment, but in the Cape Coast soil, the highest ECEC was found in the combined compost and biochar treatments. This observation could be attributed to the exchangeable acidity, pH, and exchangeable cation concentration differences of the Aiyinase and Cape Coast soils. The Aiyinase soil was found to be strongly acidic, with a higher concentration of exchangeable acidity and therefore contained relatively smaller concentrations of basic cations than the Cape Coast soil. Both the compost and biochar provide relatively higher amount of basic cations to reduce exchangeable acidity ($H^+ + Al^{3+}$) and raise the ECEC reflected in the higher pH Cape Coast soil.

Overall, biochar or compost, applied alone or in combination, significantly increased the ECEC of both the Cape Coast and Aiyinase soils. Biochar has the potential to raise the soil CEC due to its highly porous nature and higher surface area, which increases its surface sorption ability and base saturation. Liu et al. [16] and Mando and Zombré [36] indicated that compost amendment results in an increase of CEC due to input from stabilized OM being rich in functional groups such as carboxylic and phenolic acid groups being released into the soil exchange sites. The results obtained in this study agree with the findings by Liu et al. [16], who found that sole application of compost and in combination with biochar increased soil ECEC than that of the control plot with no treatment.

Combined applications of biochar and compost have numerous benefits compared to the mere mixing of biochar or compost with soil. These benefits, according to Liu et al. [16], include enhanced nutrient use efficiency, biological activation of biochar, and better material flow management; enhanced C sequestration by establishment of stable biochar-compost complexes and thus negative priming; enhanced provision of plant-available nutrients by biological N fixation; reduced nutrient leaching; and combined nutrient supply, compared to individual compost and biochar applications.

3.6. Effects of Biochar and/or Compost Applications on Soil Mineral Nitrogen. Application of biochar and/or compost led to significant ($P < 0.01$) increases in the soil mineral nitrogen concentrations measured at the end of the 14-day incubation period than in the control (Figure 4). In the Cape Coast soils, biochar and compost applied alone or in combination led to significant ($P < 0.05$) increases in the soil available nitrogen concentrations both in the first week and at the end of the 14-day incubation period than in the control soil. The increases in the soil available N concentration followed the order BC > C20 > B20 > B0.

The increased soil mineral nitrogen concentration found in the Cape Coast soils between the first and second weeks after the compost application compared to that in the sole biochar treatment is due to the fact that the compost contains relatively higher inherent N concentration and has a higher rate of nitrogen mineralization than biochar, which has a very higher C : N ratio and a lower rate of mineralization. Lehmann et al. [6] pointed out that whereas compost is easily degradable, biochar has an aromatic structure and a recalcitrant nature and is therefore very resistant to decomposition. In this regard, Duku et al. [2] reported that biochar application in soils minimizes ammonium loss through leaching and NH_3 volatilization. De Gryze et al. [10] also opined that biochar decreases the possibility of nutrient losses in soils and enhances nutrient recycling, resulting in positive impacts on crop yields in the long run through slow release to the soil.

3.7. Effects of Biochar and/or Compost Applications on Agronomic Performance of Maize. The effects of biochar and/or compost applied alone or in combination on the growth performance of maize: dry matter yield, number of leaves, plant height, and stem girth of the two maize varieties (local variety *ewifompe* and hybrid variety *Obaatanpa*), cultivated on two soil types—Aiyinase and Cape Coast soils—are presented in Tables 5 and 6, respectively. The application of biochar and/or compost resulted in significant ($P < 0.01$) increases in the number of plant leaves, plant height, and stem girth of both local and improved varieties in the Aiyinase soil.

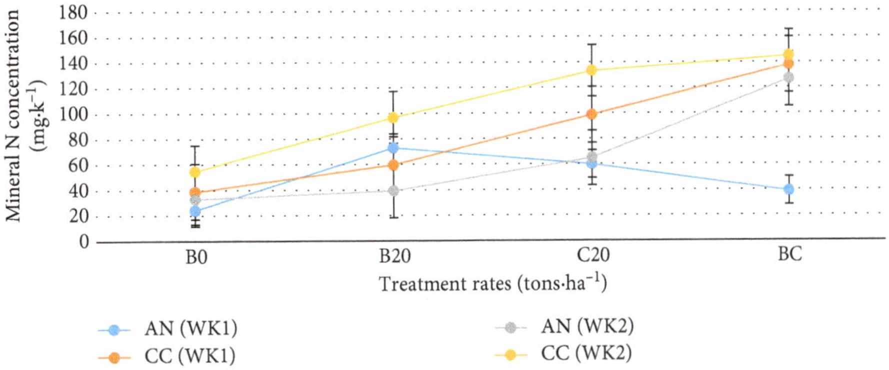

FIGURE 4: Soil mineral nitrogen concentration in Aiyinase and Cape Coast soils following application of biochar and/or compost 14 days after incubation. B0 = control with no amendments; B20 = biochar 2%, w/w; C20 = compost 2%, w/w; BC = combined biochar and compost rate 1% compost + 1% biochar. Error bars represent standard errors of the means.

TABLE 5: Effects of biochar and/or compost applications on the plant growth performance of maize crop (local and improved varieties) grown on the Aiyinase soil over 40-day period.

Treatment rate	Dry matter (g)		Number of leaves		Plant height (cm)		Stem girth (mm)	
	Improved variety	Local variety	Improved variety	Local variety	Improved variety	Local variety	Improved variety	Local variety
BC	1.3a	231.2b	3.00a	10.33b	24.7a	141.7b	3.83a	20.17b
B0	10.6a	306.8b	5.67b	12.00bc	59.3b	168.7b	10.00b	22.67bc
C20	318.4b	11.9a	10.33c	6.00a	163.3c	60.3a	23.17c	8.50a
B20	319.2b	251.2b	11.33c	12.67c	171.3c	165.3b	24.67c	24.20c
LSD, 0.05	75.9	146.1	2.31	2.24	17.55	40.21	5.16	3.21
F. Pr.	<0.001	0.01	<0.001	<0.001	<0.001	<0.001	<0.001	<0.001

B0 = control with no amendments; B20 = biochar 2%, w/w; C20 = compost 2%, w/w; BC = combined biochar and compost rate 1% compost + 1% biochar; LSD = least significant difference at 0.05. Means with different letters differ significantly at $P < 0.05$ and with same letters do not differ significantly.

TABLE 6: Effects of biochar and/or compost applications on the plant growth performance of maize crop (local and improved varieties) grown on the Cape Coast soil over 40-day period.

Treatment rate	Dry matter (g)		Number of leaves		Plant height (cm)		Stem girth (mm)	
	Improved variety	Local variety	Improved variety	Local variety	Improved variety	Local variety	Improved variety	Local variety
BC	193.5b	1.82a	10.33b	3.00a	139.3b	20.33a	18.50b	5.93a
B0	288.2bc	97.10c	11.00b	9.67c	170.3c	104.67c	21.00b	17.17c
C20	386.5c	66.73bc	10.67b	6.33b	191.0c	78.67b	25.33c	12.00b
B20	36.4a	28.94ab	5.33a	6.33b	82.7a	78.67b	12.83a	12.00b
LSD, 0.05	99.5	62.33	2.49	1.33	21.07	20.02	3.32	4.97
F. Pr.	<0.001	0.03	0.00	<0.001	<0.001	<0.001	<0.001	0.01

B0 = control with no amendments; B20 = biochar 2%, w/w; C20 = compost 2%, w/w; BC = combined biochar and compost rate 1% compost + 1% biochar; LSD = least significant difference at 0.05. Means with different letters differ significantly at $P < 0.05$ and with same letters do not differ significantly.

The increase (e.g., length, etc.) recorded in the maize grown in biochar- and/or compost-amended soils could be attributed to the increased soil pH and hence higher P availability following application of biochar or compost amendments. Oguntunde et al. [37] studied the effects of charcoal on growth and yield of maize in a Ghanaian soil and attributed the significant maize growth rate they observed in the biochar-amended soils compared to the control to biochar's ability to increase the soil pH; exchangeable cations such as Ca, Mg, K, and Na; and soil available P in these sites compared to adjacent soils with no biochar. Sohi et al. [38] reported that biochar can act as soil conditioner, which improves soil water holding capacity, enhances soil nutrient retention, and boosts plant growth. In the Cape Coast soils, plant height and stem height were greatest in the improved varieties grown in sole compost-amended soils.

The overall results showed that application of biochar had a more remarkable effect on the growth performance of maize varieties grown on the Aiyinase soil than those grown on the Cape Coast soil. This could be due to acidity differences between the two soil types. Whilst the pH of the Aiyinase soil was strongly acidic (pH 4.7), the pH of the Cape Coast soil was moderately acidic at pH 5. According to Duku et al. [2], the impact of biochar application is most seen in highly degraded acidic or nutrient-depleted soils. Results from Jeffery et al.'s [11] study also show that biochar can be a useful tool to improve crop yield in nutrient-poor and acidic soils. They concluded that biochar application is, therefore, crucial considering that approximately 30% of the world's soils are acidic, including more than 50% of potential arable land.

4. Conclusion

The study showed that biochar and compost applied alone or in combination significantly increased soil pH, total organic carbon, available phosphorus, mineral nitrogen, reduced exchangeable acidity, and increased effective cation exchange capacity in both the Rainforest and Coastal Savannah soils. Additionally, the biochar and compost additions increased the growth response (plant height and stem girth) and dry matter yield of both local and improved maize varieties. Therefore, the study demonstrated that biochar applied alone or in combination with compost offers potential to enhance soil quality and improve crop yield. In response to question about duration, we only speak to observations made in our study. This is a preliminary finding. We propose a longer term study to examine the long-term effects of the treatments on soil physicochemical properties and yield responses.

Acknowledgments

Many thanks go to Emmanuel Abban and Daniel Mensah Zorve for their assistance during laboratory analyses, field data collection, and pot setup.

References

[1] R. Lal, "Challenges and opportunities in soil organic matter research," *European Journal of Soil Science*, vol. 60, no. 2, pp. 58–169, 2009.

[2] H. M. Duku, S. Gu, and E. B. Hagan, "Biochar production potentials in Ghana—a review," *Renewable and Sustainable Energy Review*, vol. 15, pp. 3539–3551, 2011.

[3] W. Zheng, B. K. Sharma, and N. Rajagopalan, "Using biochar as a soil amendment for sustainable agriculture," Field Report, Illinois Department of Agriculture, Illinois, USA, 2010.

[4] L. Rodriguez, P. Salazar, and T. R. Preston, "Effects of biochar and digester effluent on growth of maize in acid soils," *Livestock Research for Rural Development*, vol. 21, no. 7, p. 110, 2009.

[5] J. Lehmann, "Bio-energy in the black," *Frontiers of Ecological Environment*, vol. 5, no. 7, pp. 381–387, 2007.

[6] J. Lehmann, J. Gaunt, and M. Rondon, "Biochar sequestration in terrestrial ecosystems—a review," *Mitigation Adaptation Strategies for Global Change*, vol. 11, pp. 403–427, 2006.

[7] J. Paz-Ferreiro, H. Lu, S. Fu, A. Méndez, and G. Gascó, "Use of phytoremediation and biochar to remediate heavy metal polluted soils: a review," *Solid Earth*, vol. 5, no. 1, pp. 65–75, 2014.

[8] J. Gaunt and J. Lehmann, "Energy balance and emissions associated with biochar sequestration and pyrolysis bioenergy production," *Environmental Science and Technology*, vol. 42, pp. 4152–4158, 2008.

[9] D. A. Laird, "The charcoal vision: a win–win–win scenario for simultaneously producing bioenergy, permanently sequestering carbon, while improving soil and water quality," *Journal of Agronomy*, vol. 100, no. 1, pp. 178–181, 2008.

[10] S. De Gryze, M. Cullen, L. Durschinger, J. Lehmann, D. Bluhm, and J. Six, "Evaluation of opportunities for generating carbon offsets from soil sequestration of biochar," *in An issues paper commissioned by the Climate Action Reserve*, 2010, http://www.terraglobalcapital.com/press/Soil.

[11] S. Jeffery, D. Abalos, M. Prodana et al., "Biochar boosts tropical but not temperate crop yields," *Environmental Research Letters*, vol. 12, no. 5, p. 053001, 2017.

[12] S. Jeffery, F. G. A. Verheijen, M. van der Velde, and A. C. Bastos, "A quantitative review of the effects of biochar application to soils on crop productivity using meta-analysis," *Agriculture, Ecosystems and Environment*, vol. 144, no. 1, pp. 175–187, 2011.

[13] B. Glaser and J. Birk, "State of the scientific knowledge on properties and genesis of Anthropogenic Dark Earths in Central Amazonia (terra preta de ′ındio)," *Geochimica et Cosmochimica Acta*, vol. 82, pp. 39–51, 2012.

[14] B. Glaser, J. Lehmann, and W. Zech, "Ameliorating physical and chemical properties of highly weathered soils in the tropics with charcoal—a review," *Biology and Fertility of Soils*, vol. 35, no. 4, pp. 219–230, 2002.

[15] D. Trupiano, C. Cocozza, S. Baronti et al., "The effects of biochar and its combination with compost on lettuce (*Lactuca sativa* L.) growth, soil properties, and soil microbial activity and abundance," *Hindawi International Journal of Agronomy*, vol. 2017, Article ID 3158207, 12 pages, 2017.

[16] J. Liu, H. Schulz, S. Brandl, H. Miehtke, B. Huwe, and B. Glaser, "Short-term effect of biochar and compost on soil fertility and water status of a Dystric Cambisol in NE Germany under field conditions," *Journal of Plant Nutrition, Soil Science*, vol. 175, no. 5, pp. 1–10, 2012.

[17] H. Schulz and B. Glaser, "Effects of biochar compared to organic and inorganic fertilizers on soil quality and plant growth in a greenhouse experiment," *Journal of Plant Nutrition and Soil Science*, vol. 175, no. 3, pp. 410–422, 2012.

[18] D. Fischer and B. Glaser, "Synergisms between compost and biochar for sustainable soil amelioration," in *Management of Organic Waste*, S. Kumar and A. Bharti, Eds., pp. 168–198, 2012, In press.

[19] FAO, ISRIC, and ISSS, "World reference base for soil resources," World Soil Resources Report 84, Food and Agriculture Organization of the United Nations, Rome, Italy, 1998.

[20] R. N. Issaka, J. K. Senayah, E. Andoh-Mensah, and A. E. Stella, "Assessment of fertility status of soils supporting coconut (*Cocus nucifera*) cultivation in Western and Central Regions of Ghana," *West African Journal of Applied Ecology*, vol. 20, no. 1, pp. 48–56, 2012.

[21] J. M. Anderson and J. S. I. Ingram, *Tropical Soil Biology and Fertility. A handbook of Methods*, C.A.B. International, Wallingtonford, Oxon, UK, 1993.

[22] K. J. Maghanga, K. L John, S. K. Fred, and K. O. Peter, "Comparison of soil phosphorous extraction by Olsen and double acid methods in acid soils of Western Kenya," *East African Journal of Pure and Applied Science*, vol. 2, no. 1, pp. 1–5, 2012.

[23] J. D. Rhoades, "Cation exchange capacity," in *Methods of Soil Analysis. Part 2: Chemical and Microbiological Properties*, A. L. Page, Ed., vol. 9, pp. 149–157, 2nd edition, 1982.

[24] G. P. Robertson, B. G. Solins, Ellis, and K. Lajtha, "Exchangeable ions, pH, and cation exchange capacity," in *Standard Soil Methods for Long Term Ecological Research*, G. P. Robertson, C. S. Bledsoe, D. C. Coleman, and P. Sollins, Eds., pp. 106–144, Oxford University Press, New York, NY, USA, 1999.

[25] A. Walkley and I. A. Black, "An examination of Degtjareff method for determining soil organic matter and a proposed modification of the chromic acid titration method," *Soil Science*, vol. 37, no. 1, pp. 29–37, 1934.

[26] D. L. Rowell, *Soil Science: Methods and Applications*, Longman Group UK Limited, London, UK, 1994.

[27] A. Nigussie, E. Kissi, M. Misaganaw, and G. Ambaw, "Effects of biochar application on soil properties and nutrient uptake of lettuces (*Lactuca sativa*) grown in polluted soils," *American-Eurasian Journal of Agriculture and Environmental Science*, vol. 12, no. 3, pp. 369–376, 2012.

[28] V. L. Zwieten, S. Kimber, S. Morris et al., "Effect of biochar from slow pyrolysis of paper mill waste on agronomic performance and soil fertility," *Plant and Soil*, vol. 327, no. 1-2, pp. 235–246, 2010.

[29] A. Zhang, R. Bian, G. Pan et al., "Effects of biochar amendment on soil quality, crop yield and greenhouse gas emission in a Chinese rice paddy: a field study of 2 consecutive rice growing cycles," *Field Crops Research*, vol. 127, pp. 153–160, 2012.

[30] M. Agusalim, H. U. Wani, and M. S. Syechfani, "Rice husks biochar for rice based cropping system in acid soils: the characteristics of rice husk and its influence on the properties of acid sulphate soils and rice growth in west Kalimantan," *Indonesia Journal of Agricultural Science*, vol. 2, no. 1, pp. 39–47, 2010.

[31] K. D. Sasmita, A. Iswandi, A. Syaiful, Y. Sudirman, and D. Gunawan, "Application of biochar and organic fertilizer on acid soil as growing medium for Cacao (*Theobroma cacao* L.) seedlings," *International Journal of Sciences Basic and Applied Research*, vol. 36, no. 5, pp. 261–273, 2017.

[32] G. Agegnehua, A. M. Bass, N. Paul et al., "Biochar and biochar-compost as soil amendments: Effects on peanut yield, soil properties and greenhouse gas emissions in tropical North Queensland, Australia. Agriculture," *Ecosystems and Environment*, vol. 213, pp. 72–85, 2015.

[33] T. H. De Luca, M. D. MacKenzie, and M. J. Gundale, "Biochar effects on soil nutrient transformations," in *Biochar for Environmental Management: Science and Technology*, J. Lehmann and S. Joseph, Eds., pp. 251-270, Earthscan Publications, London, UK, 2009.

[34] K. A. Frimpong, E. Amoakwah, B. A. Osei, and E. Arthur, "Changes in soil chemical properties and lettuce yield response following incorporation of biochar and cow dung to highly weathered acidic soils," *Journal of Organic Agriculture and Environment*, vol. 4, no. 1, pp. 28–39, 2016.

[35] M. Vithanage, I. Herath, S. Joseph et al., "Interaction of arsenic with biochar in soil and water: a critical review," *Carbon*, vol. 113, pp. 219–230, 2017.

[36] O. A. Mando and N. P. Zombré, "Use of compost to improve soil properties and crop productivity under low input agricultural system in West Africa," *Agriculture, Ecosystems and Environment*, vol. 84, no. 3, pp. 259–266, 2001.

[37] P. G. Oguntunde, M. Fosu, A. E. Ajayi, and N. van de Giesen, "Effects of charcoal production on maize yield, chemical properties and texture of soil," *Biology Fertility Soils*, vol. 39, no. 4, pp. 295–299, 2004.

[38] S. Sohi, S. E. Loez-Capel, E. Krull, and R. Bol, "Biochar's roles in soil and climate change: a review of research needs," *CSIRO Land and Water Science Report*, vol. 5, no. 9, pp. 1–57, 2009.

13

An Overview of Global Wheat Market Fundamentals in an Era of Climate Concerns

Aliakbar Enghiad,[1] Danielle Ufer,[2] Amanda M. Countryman,[2] and Dawn D. Thilmany[2]

[1]*Business Administration, California State University, Northridge, 18111 Nordhoff Street, Northridge, CA 91330, USA*
[2]*Department of Agricultural and Resource Economics, Colorado State University, 1200 Center Ave. Mall, Fort Collins, CO 80523-1172, USA*

Correspondence should be addressed to Amanda M. Countryman; amanda.countryman@colostate.edu

Academic Editor: Anna I. De Luca

Wheat is a key global commodity in terms of acreage and tradeable value and as a staple in household diets. Many factors affect wheat prices including climate, yields, oil prices, lagged prices, and imports. In addition to gradually and consistently increasing global wheat demand, these market drivers are posited to impact world prices and, ultimately, food security. To investigate how these factors differentially influence wheat markets, an extensive survey of literature regarding wheat market fundamentals was conducted, as well as a trend analysis using a uniquely compiled data set specific to significant wheat-producing areas. Previous studies show that imports, climate, oil prices, and past prices, among other factors, have a significant relationship with changes in the world wheat price. This study compiles and compares these same key variables from five major wheat export countries/regions for the time frame from 1980 to 2013.

1. Introduction

The international agricultural research agenda on climate-smart agriculture (CSA) explores the intersection of climate change, the reorientation of food production systems, and food security as a core mission of its programming [1]. Such priorities are a result of tension between natural resources and food production systems due to perceptions of increasing scarcity of resources and the growing demand for food that has fueled interest in analyses that inform stakeholders on food market dynamics. Clearly, the demand for all foodstuffs grows with increases in both world population and the wealth of developing countries, whereas the supplies of natural resources are limited or, if renewable, may be maintained or increase only gradually. The tension between these markets and resource systems is heightened by concerns about how climate resiliency may impact natural resources stocks and flows. In an era of increasing climate concerns and growing focus on sustainable production, this paper summarizes how wheat is produced, consumed, and traded among partners with differential exposure to climactic shocks. The purpose of this analysis is to provide a comprehensive review of the drivers that are important to consider for food security and, more specifically for this paper, the global wheat market. Food security is a complex phenomenon and is determined by distinct dimensions driven by a complex mixture of individual forces, of which volatility plays a minor role [2]. Identifying the economic and environmental factors which exert the greatest influence on the major wheat exporters (including the factors that influence the markets and price of wheat) has important implications for any policies or interventions that are intended to address global food security. This also represents a contribution to the literature by providing a synthesis of data and trends for key wheat-exporting countries with a specific focus on the importance of climate-based factors.

Grains are currently the most important contributor to human food supplies globally. Approximately 21% of the world's food depends on annual wheat crop harvests, which often have relatively low stocks [3]. Developing countries, the majority of which are net wheat importers, consume 77% of total global wheat production. Accordingly, wheat accounts for approximately 24% of food commodities imported by developing countries [4, 5]. Many of those countries pay subsidies to stabilize and/or lower food prices so that consumers are better able to consistently meet dietary needs and, ultimately, increase the level of household food security. However, increasing wheat consumption fueled by population growth and rising incomes in developing country households puts upward pressure on wheat prices [6]. Specifically, growth of the middle class in highly populated, developing countries, particularly those in the Asian region, in tandem with changes in lifestyle and consumption patterns resulting from an increase in wealth, causes an increase in demand for food, including grains.

On the supply side, climate and oil prices are two important factors affecting wheat production. Oil prices influence the cost of inputs for wheat production, and similar patterns observed in wheat and oil price fluctuations indicate high correlation between the two. Changes in climate can influence food production in a variety of ways, as the climates of major production areas may change with respect to growing season length, as well as changes in average temperatures and rainfall. Numerous international research and policy organizations (the G20 Ag Ministers and the Consultative Group for International Agricultural Research (CGIAR) research centers) have made the impact of climate change on agriculture and food security a key priority moving forward, motivating the need to further explore how one key food staple may be influenced by efforts to make the food system more resilient [1, 7]. Specifically related to this study is the likelihood that changes in precipitation patterns will affect world wheat production.

Fluctuations and patterns in wheat prices have changed in recent years compared to earlier decades, due in large part to the previously discussed factors. With increased world wheat prices, imports have become more expensive, and, subsequently, prices of foods that are largely produced using wheat as an input have increased as well. In many developing countries, wheat-based foods are a major share of household diets, meaning that increases in wheat prices will have noticeable effects on the cost of food and food security.

Understanding how different factors influence the world wheat market is essential because of its role in food security for such a large share of the global population. So, the article begins with a broad overview of global food security issues in an era of increasing concerns about climate and then gradually narrows to explore key market drivers for one specific sector of importance to global food security, wheat.

With wheat being particularly sensitive to the consequences of changes in climate, it is essential to investigate the scope of demand and supply factors that affect the global wheat market, particularly among key production regions. To describe the differences among major stakeholders in the global wheat market, this article examines key factors affecting demand and production to determine the relative importance of each. Relevant literature on factors affecting the global wheat market, including oil prices, past prices, import trends, yields, and precipitation, is also presented. A particular focus is placed on the exploration of key market determinants that interface with climate concerns in major supply regions and trading partners. To assess these wheat market dynamics, several key sources of data were compiled and analyzed for trends.

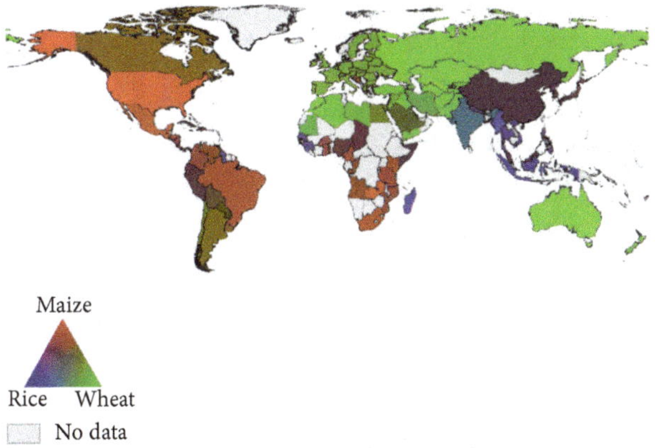

FIGURE 1: Global major grain consumption map. *Source*: New England Complex Systems Institute, 2011, and Bar-Yam et al. [12, 13]. *Note*. Colors represent the relative domestic consumption of corn, rice, and wheat in each country or indicate that no data is available.

1.1. Wheat Market Fundamentals. Wheat is one of the principal cereal grains produced and consumed globally. It is grown on more land area than any other commercial crop and continues to be the most important grain food source for human consumption [8]. World wheat production is ranked third in weight produced, after corn and rice [9]. This is likely due to the fact that wheat can be cultivated in many areas with heterogeneous types of weather, elevation, or soil. It is mostly cultivated between the latitudes of 30°N to 60°N and 27°S to 40°S [10], up to 3,000 meters above sea level, and in places with temperatures between 3° and 32° Celsius. Wheat is adapted to a broad range of moisture conditions from dry weather to seaside moisture. Although approximately three-fourths of the land area where wheat is grown receives an average of 375 to 875 mm of annual precipitation, wheat can be grown in a wider set of locations where precipitation ranges from 250 to 1750 mm [11]. Wheat production covers more than 240 million hectares (ha) globally and its gross world trade is greater than all other crops combined [8]. Wheat is a major food staple because of the wheat plant's agronomic adaptability, ease of grain storage, and ease of converting grain into flour for making staple food products. Wheat is the major source of carbohydrates in the diet of people from many countries, including Australia, most of Europe, Northern Asia, and Northern Africa (Figure 1).

Wheat production increased sharply in the 1960s and gradually afterwards, mostly as a result of higher yields per ha, in a technology shift commonly labeled the "green revolution" [14]. The green revolution resulted in the development of rust-resistant semidwarf wheat that could utilize large amounts of nitrogen fertilizer and had a higher yield. Between 1980 and 2013, the world's annual harvested area of wheat decreased by 0.24%, but yield increased by 1.41% [3].

Average annual world yields increased from 1855 kilograms per hectare (kg/ha) in 1980 to 3264 kg/ha in 2013 [3].

Most wheat is consumed within the country where it is produced, although roughly one-fifth of global annual production is exported. World wheat trade was estimated at 148 million metric tons (MMT) in 2011, most of which was imported by developing countries. Despite the increase in wheat production over the past few decades, developing countries have continued to import two-thirds of all world wheat trade flows [3].

2. Survey of the Literature and Trends

Studies which simultaneously consider the effect of global weather changes, oil prices, and imports on wheat market dynamics are somewhat rare, especially research that considers the relationship between precipitation and wheat prices on a global scale. Such an integrated analysis could provide a richer understanding of long-term wheat price fluctuations, but many studies relating to this topic consider the effect of just one of these factors at a time. Thus, a more integrated approach can help inform our discussion of cross-cutting issues for wheat markets in an era of climate change.

2.1. The Interface of Food Production, Weather Patterns, and Climate Concerns. Weather can impose a variety of changes across the climate-sensitive agricultural sector, including production challenges, global migration, economic disruption, and land use changes [15]. While positive effects may occur in some regions, drought and extreme heat [16] resulting from changes in climate in lower latitudes can reduce production yields, especially in areas with suboptimal growing conditions, subsequently increasing demand for wheat imports [5]. In recent years, Africa and the Middle East have imported approximately 45% of global wheat trade flows and are predicted to experience increases in aridity in coming decades which may drive further import needs [17]. Arid and semiarid regions account for approximately 30% of global land area and 20% of total world population, representing a large share of global wheat consumers [18]. Despite the risks of a changing climate for warmer wheat production regions, changes in climate could also have a positive effect, potentially making cold regions warmer and more suitable for wheat production [16]. While some studies forecast decreases in future production in regions such as southern Australia [19], others, such as Izaurralde et al. [20] and Zhang and Liu [21], show climate-induced increases in wheat yields in the American Northern Plains and China.

2.2. Global Food Security Policy and Programming Responses. The focus on adaptation of agriculture to climate change for food security is evident in the commitments put forward; for example, the Paris Agreement on global climate action showcased agriculture as a priority, with 94% of countries including the agricultural sector for mitigation and/or adaptation. Agriculture holds a central position in the response to climate change, as actions in this sector represent not only a political, economic, and moral imperative but also an opportunity [22]. A Brookings Institution fellow reports that, in the run up to the 21st session of the UN Conference of Parties for the Convention on Climate Change (COP 21), the CGIAR Research Program on Climate Change, Agriculture and Food Security (CCAFS) and Farming First produced a toolkit aimed at farmers' organizations, agricultural development organizations, and negotiators [23]. More broadly, the CGIAR Research Program on Climate Change, Agriculture and Food Security (CCAFS) addresses the increasing challenge of global warming and declining food security on agricultural practices, policies, and measures [1].

The G20 Ministers of Agriculture and their food security agenda are also important: these countries represent approximately 60% of all agricultural land and about 80% of world trade in agricultural products. They emphasize that there are a number of significant challenges that must be met if stable supplies of safe, nutritious, and affordable food are to be provided for the global population, which is expected to reach 8.5 billion by 2030; these challenges include climate change, urbanization, conflict, and the limited availability of energy and natural resources, such as land and water, and their increased degradation. They promote activities and innovations, such as long-term planning, investments in technologies and practices, and ecosystem-based measures that will make the agricultural sector more resilient to water-related risks such as drought, flooding, salinization, and declining water quality, which are further compounded by climate change [7]. It is critical to note that measuring food security is difficult [24], and there is a rich literature that discusses methods for evaluating the degree of food insecurity present in countries, as well as the implications of employing different metrics for food security assessments [25].

2.3. The Role of Global Economic Factors on Commodity Markets. To find a pattern and form future expectations of food market dynamics, price information from previous years and trends in market prices, both of wheat and of substitute crops, are commonly used. Like other industries, agricultural suppliers respond to price changes by increasing or decreasing production. Due to the lag in agricultural production processes, however, these responses take a longer time to appear in the market supply [26]. Haile et al. [27] found that price volatility is also an important deterrent of supply response, with producers tempering supply response in the face of greater price risk. Wheat prices from the past three decades were compiled and analyzed for trends. Table 1 includes a summary of historical world wheat prices. All prices are deflated based on the 2013 consumer price index (CPI) reported by the US Bureau of Labor Statistics [28]. The lowest deflated price for wheat was \$151.70/metric ton (MT) in 1998, for which the nominal price was \$107.96/MT. The highest deflated price of wheat was \$409.41/MT in 1980, with a nominal price of \$148/MT. Figure 2 shows the trend for annual wheat prices during the 34-year time frame. Though ultimately only one of many factors, lagged prices are a strong basis for producer decisions related to future production and are considered an important factor in wheat price forecasting.

TABLE 1: Summary of data statistics: annual wheat price in USD/MT, deflated to 2013 dollars.

	Mean	Median	Maximum	Minimum	Std. dev.	Coefficient of variation	Obs.
Price ($)	252.62	236.67	409.41	151.70	67.39	0.27	34

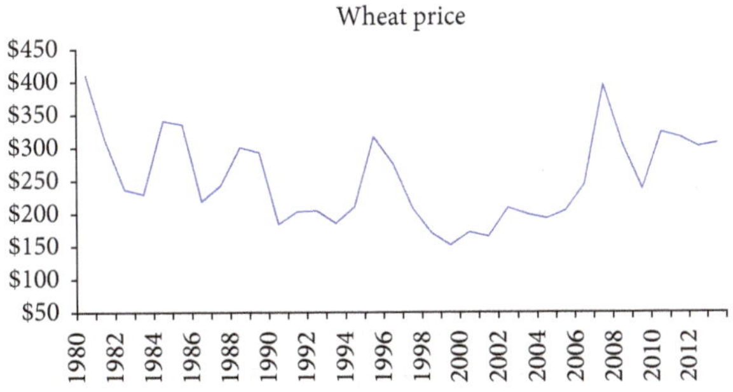

FIGURE 2: Annual world wheat price, deflated to 2013 dollars. *Source*: FAO, 2015 [3].

International commodity markets operate and adapt given fundamental ratios of supply and demand growth, but stocks can provide a buffer when historical trade patterns are disrupted. Several years of drought, including those in Europe in 2006, North America in 2006-2007, and a severe drought in Australia from 2006 to 2008, drove wheat stocks down to critical lows. This, coupled with the surge in biofuel demand, contributed to record grain prices [29] that make the scarcity of buffer stocks a concern for those who seek to stabilize world food prices.

While not directly tied to the wheat market, global events in the last three decades may have distorted markets. During the early 1980s, the economic recession in the United States, other countries' unemployment levels, energy price shocks, and corresponding monetary policies throughout the global economy may have been possible causes of changes in wheat prices that are not accounted for with the direct factors discussed previously [30–32]. Another important market distortion in the early 1980s was a US export embargo on grain trade and technology with the Soviet Union in response to the invasion of Afghanistan [33]. Though restrictions were lifted a few years later, patterns of trade changed while the embargo was in place, with the Soviet Union facing import shortages and needing to find alternative wheat suppliers. Additional domestic market interventions also followed in the US due to corresponding decreases in the world wheat price, but the long-term trade pattern implications are not clear [34].

2.4. Implications of Oil Markets for the Agricultural Sector. Oil is another factor that affects agricultural commodity prices [35–37]. The oil market affects wheat prices both directly, through production inputs, and indirectly, through demand for biofuels and the resulting substitution effects. The prices of fertilizer, farm machinery, and transportation are all affected by the crude oil price, which influences wheat production costs. Baffes and Haniotis [38] attribute the impacts of energy prices on food commodities more to these factors than to biofuels and find that the aspects of changing energy prices have exerted the greatest influence on commodity prices, including wheat, over the past few decades. Nevertheless, biofuels can also affect wheat prices as, with high oil prices, demand for biofuel increases and agricultural inputs such as land are dedicated to planting energy crops such as corn instead of wheat. This competing use of land thereby decreases wheat production, which may explain some of the volatility and upward pressure on wheat prices [39].

In addition to market dynamics, energy policy has been in a state of flux. For example, the European Union, the United States, and other countries with large agricultural sectors have been encouraging biofuels by instituting production subsidies and renewable fuel quotas, further driving up demand for biofuel input crops. Wright [40] found the impact of biofuel mandates and associated policies to have been of particular importance in driving commodity prices in recent years. Saghaian [41], Esmaeili and Shokoohi [42], Natanelov et al. [43], and Serra and Gil [44] also studied the relationship between agriculture and energy markets and described the correlation between agricultural commodities and crude oil prices.

Oil prices were collected and analyzed for trends relative to the wheat market. Annual oil price is calculated as the average of two oil prices: West Texas Intermediate oil and the Persian Gulf countries' crude oil. Oil prices are reported in US dollars per barrel and data is collected from the US Energy Information Administration [45] (Table 2). The lowest oil price during this period was $14.33/barrel in 1998, which is also the year with the lowest wheat price. The peaks in oil prices are relatively recent, occurring in 2008, 2011, and 2012, when the deflated oil price was above $100/barrel (Figure 3). When viewed in conjunction with wheat prices over time, the relationship between the two becomes apparent, though the imperfect correlation between oil and wheat prices demonstrates the complexity of the influences of energy on agricultural commodities.

2.5. Exploring the Influence of Trade on Commodity Markets. As the global population increased, wheat consumption has grown by an average of 1.6% annually over the past three decades [3]; and, as a result of increased wheat production,

Table 2: Average annual oil price statistics in USD/barrel, deflated to 2013 dollars.

	Mean	Median	Maximum	Minimum	Std. dev.	Coefficient of variation	Obs.
Oil price	$ 49.21	$ 34.60	$ 104.64	$ 14.33	$ 27.60	0.56	34

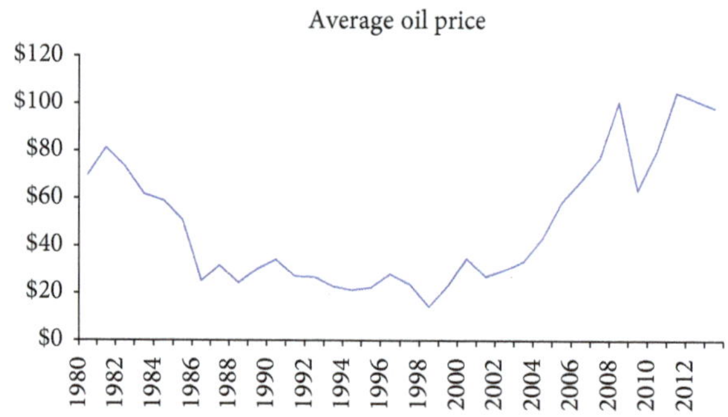

Figure 3: Average annual oil price in US dollars per barrel, deflated to 2013 dollars. *Source*: EIA, 2014 [45].

wheat trade has also grown over the past few decades. Wheat is purchased primarily for consumption and creating stocks as a buffer. The consumption share can be used both as human food or as a feed input for livestock production. Stockpiling creates an intertemporal buffer against shortages during large wheat consumption periods or low production years.

To eliminate market distortions, many trade agreements force countries to reduce tariffs and open the market to international trade. However, such agreements are not always fully successful or effective because many countries limit imports by imposing nontariff barriers, including import licenses and quality restrictions [46, 47]. In some cases, these nontariff barriers can be effective in stabilizing and insulating domestic prices of agricultural commodities against lower priced imports [48]. However, these insulating policies can also increase the volatility of global prices and, ultimately, nullify any domestic market insulation if many countries enact such policies [49].

Import tariff policies vary in different countries. In some major importing countries, such as Egypt, there are high tariffs on wheat products, like pasta and bakery products, which have resulted in increased wheat grain imports for input use to produce final consumption goods. However, in some countries, such as Kenya, the opposite is true and value added products that contain wheat have a lower average tariff than that placed on wheat grain [50–52].

Another important issue in the world wheat market is the effect of food aid or assistance programs. Most of these programs are donations from major world wheat exporters to developing countries and aim to fight famine but can indirectly affect the world wheat price if they modify demand (Figure 4). It should be noted that wheat does not have a simple geographic pattern of trade flows, since many countries are both exporters and importers of wheat. Trade patterns also change over time, as the quantity of wheat production, imports, and exports in different countries change and world prices drive secondary shocks in the market.

As previously mentioned, stocks-to-use ratios measure the relationship between available supply and demand by dividing the ending stock value by total annual use [54]. A low

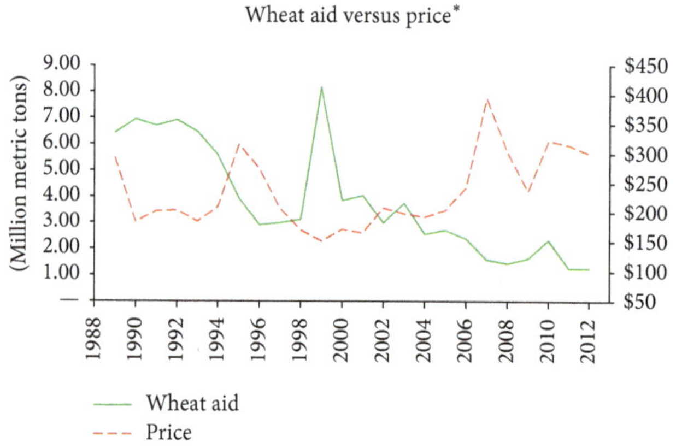

Figure 4: Wheat aid versus wheat price. *Source*: World Food Program and OECD, 2012 [53]. *Wheat aid is total of wheat and wheat flour provided and wheat prices are deflated in terms of 2013 dollars between 1989 and 2012.

stocks-to-use ratio leads to a higher market price, because low stocks reflect scarcity. While this indicator is correlated with crop prices, there is conflicting evidence as to its importance to crop price. Daugherty and Kelly [55] found the stocks-to-use ratio to be a weak factor, while Baffes and Haniotis [38] found it to be second only to energy prices in terms of explaining long-term crop price variation.

2.6. Changing Demand in Major Wheat-Importing Countries. Demand for wheat has been relatively high and stable in North Africa and the Middle East, South Asia, East and Southeast Asia, South America, and Sub-Saharan Africa. Based on FAO and USDA reports [3, 9], these five regions are defined as the world's major wheat importers, annually accounting for nearly 90% of total wheat imported in the world [9]. Specifically, Egypt was one of the top five importing countries for the 30-year time frame from 1982 to 2011, with average wheat imports of 6 MMT/year (Figure 5) [3, 4]. Other countries, such as those in Eastern Europe, had increased demand between 1980 and 2013 because of war and political

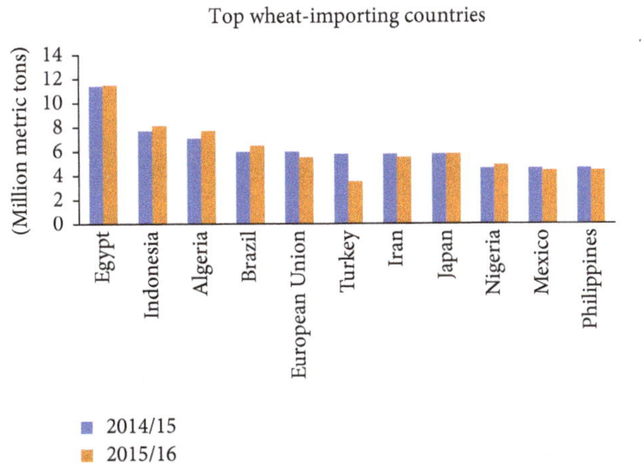

FIGURE 5: Principal importing countries of wheat, flour, and wheat products. *Source*: USDA, Foreign Agricultural Service, June 2015 [4].

FIGURE 6: Annual world wheat imports. *Source*: FAO, 2015 [3].

or financial problems but, overall, were less consistently listed among the top importers [3].

Global import data is presented in millions of metric tons (MMT) and is collected from FAO statistics for the 34-year time frame [3]. During this time frame, the maximum annual world import level is 162 MMT in 2013, and the minimum is 88.06 MMT in 1986. The average world import level during the time frame is 113 MMT, and the increasing trend in annual world imports is illustrated in Figure 6. The clear upward trend of imports demonstrates the growing global demand for wheat, having nearly doubled over the timespan evaluated. The concurrent economic expectation of such upward pressure on demand is increased pressure on supply which subsequently increases prices.

3. An Overview of Supply Drivers in Major Wheat Export Countries

Although past research shows that there are a myriad of factors that influence commodity markets and food security, a more specific focus on how wheat supplies may be influenced by climate concerns is the focus of this study. Based on USDA statistics, the five largest wheat-exporting regions between 1980 and 2013 were the United States, the European Union, Canada, Australia, and the Former Soviet Union (FSU), with 30.9, 30.2, 17.9, 13.5, and 9.8 MMT of average annual wheat exports, respectively [3]. Overall trade flows increased during this 34-year time frame, despite variations in market factors such as weather shocks, domestic and international policies, and world wheat price changes. This section describes each of these key wheat-exporting countries with particular attention to the factors that past literature highlights as important.

Precipitation data is based on National Oceanic and Atmospheric Administration records [56] and was retrieved for different stations located in the high-productivity regions of each country. The stations located in Kansas, France, South Saskatchewan, and the southwest of Western Australia represent the United States, the European Union, Canada, and Australia, respectively. Krasnodar and Stavropol represent the FSU. In general, the region with the highest yield in each country on average during the 34-year time frame from 1980 to 2013 was selected to represent the country as a whole. For more accurate results, Geographic Information System (GIS) software was used to select specific data related to each region. Data from weather stations that have been active for all, or most, of the 34-year time frame (for at least 25 years) are included.

For the United States, the European Union, Canada, and the FSU, the average of May and June monthly precipitation is considered for each year. During these months, the wheat plant is in the heading and grain development stage and has the highest water needs [58]. For the same reason, the average of November and December monthly precipitation levels is

TABLE 3: Summary of precipitation statistics (mm).

	Kansas	France	Saskatchewan	Western Australia	Krasnodar and Stavropol
Mean	939.34	651.11	640.86	166.50	721.39
Median	916.37	651.27	624.82	144.84	705.27
Maximum	1656.04	1129.09	1022.89	449.12	1082.027
Minimum	419.27	369.25	351.05	52.00	311.67
Std. dev.	269.44	205.00	163.82	88.96	195.76
Coefficient of variation	0.29	0.31	0.26	0.53	0.27

Source: National Oceanic and Atmospheric Administration, 2015 [56].

TABLE 4: Summary of precipitation statistics in +/− from average (mm).

	USA	European Union	Canada	Australia	FSU
Mean	0.0	0.0	0.0	0.0	0.0
Median	−23	0.2	−16	−22	−16
Maximum	717	478	382	283	361
Minimum	−520	−282	−290	−115	−410
Std. dev.	269.44	205.00	163.82	88.96	195.76

Source: National Oceanic and Atmospheric Administration, 2015 [56], and authors' calculations.

TABLE 5: Summary of yield data statistics (MT/ha).

	USA	Canada	European Union	Australia	FSU
Mean	2.66	2.27	4.77	1.61	1.73
Median	2.65	2.25	4.82	1.63	1.65
Maximum	3.17	3.18	5.68	2.15	2.39
Minimum	2.20	1.23	3.63	0.76	1.27
Std. dev.	0.28	0.43	0.51	0.38	0.30
Coefficient of variation	0.11	0.19	0.11	0.24	0.17

Source: OECD-FAO 2015 Agricultural Outlook and the USDA Statistical Bulletin [57].

considered for each year in Australia since it is located in the southern hemisphere and has a countercyclical agricultural season relative to the other regions [59]. Precipitation data is reported in millimeters (mm) and the average is calculated for the agricultural production year (Table 3). Among the five regions, the state of Kansas in the United States has the highest average precipitation levels and the state of Western Australia has the lowest average precipitation during the 34-year time frame. Annual precipitation in the state of Western Australia, on average, is 166 mm compared to other regions that have average annual precipitation of more than 640 mm.

To facilitate interpretation, precipitation for each year is compared to the average precipitation of that region to show how that year varied from the longer-term norm. The long-term average precipitation for every region is calculated over the 34-year time frame from 1980 to 2013 and subtracted from the raw precipitation data of that region. Therefore, the data has been converted to plus/minus the average index value (Table 4).

Annual yield data is in metric tons per hectare (MT/ha) and was retrieved from the OECD-FAO 2015 Agricultural Outlook and the USDA Statistical Bulletin [57]. After 1992, the sum of Russia, Ukraine, Kazakhstan, and Uzbekistan production and exports is used to represent the Former Soviet Union (FSU). Yield was, on average, highest in the European Union, approximately two times more than the yield in other regions, while Australia had the lowest yield during the 34-year time frame (Table 5).

The following section shows greater detail on each key wheat-exporting country to complement the discussion of factors affecting price behavior. Charts and maps accompanying the information for each country/region better demonstrate wheat production trends over time in the five major export regions and are presented in the country profiles below.

3.1. United States (Kansas). Wheat production and yield levels in the United States remained relatively stable during the 34-year time frame from 1980 to 2013. Wheat yields have increased, but, due to decreases in harvested area, total production declined by an average of 0.48% annually [3]. The US was the largest exporter for almost all of the 34-year period, with gradual narrowing of the gap from the second

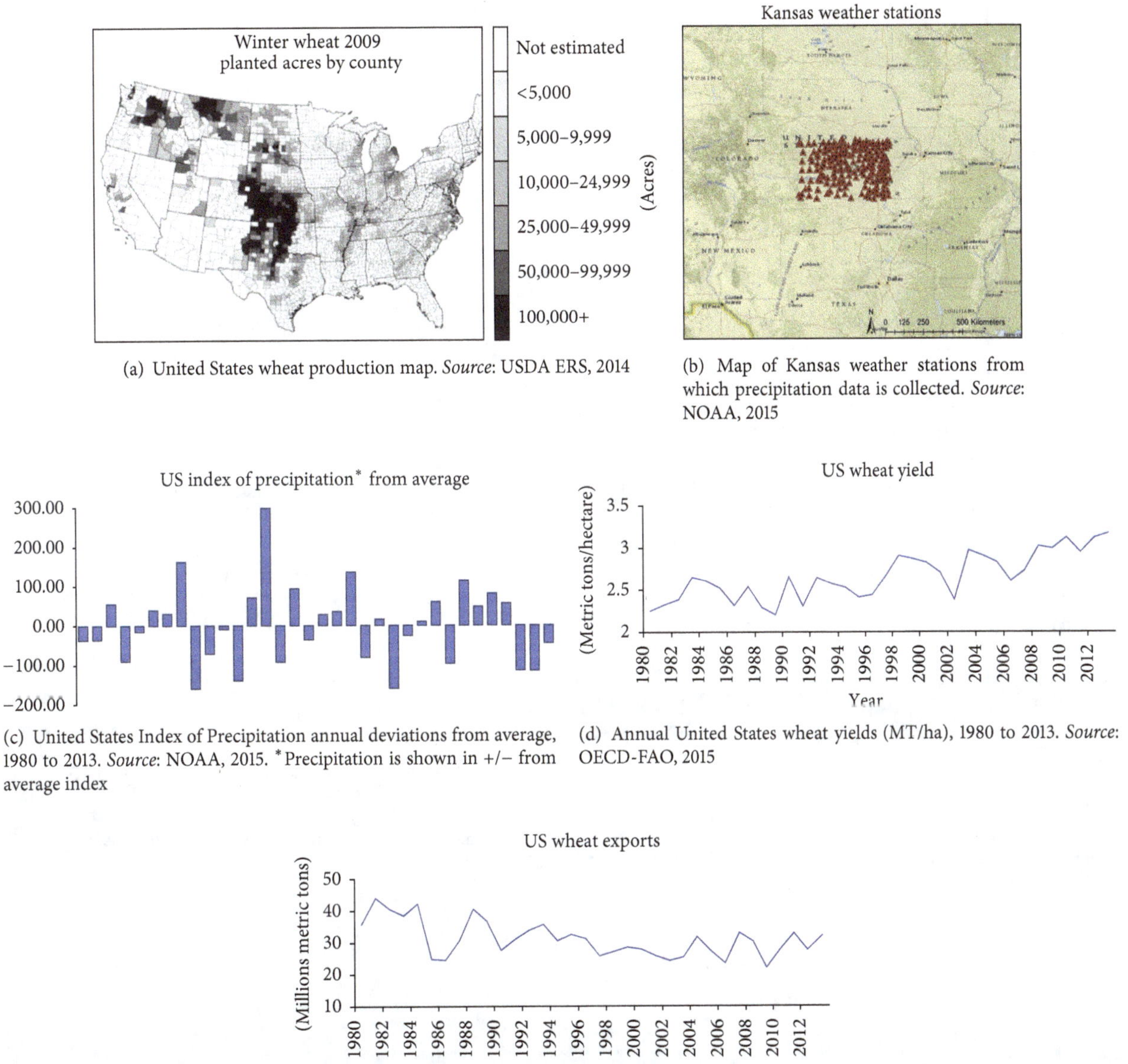

(a) United States wheat production map. *Source*: USDA ERS, 2014

(b) Map of Kansas weather stations from which precipitation data is collected. *Source*: NOAA, 2015

(c) United States Index of Precipitation annual deviations from average, 1980 to 2013. *Source*: NOAA, 2015. *Precipitation is shown in +/− from average index

(d) Annual United States wheat yields (MT/ha), 1980 to 2013. *Source*: OECD-FAO, 2015

(e) Annual United States wheat exports (MMT), 1980 to 2013. *Source*: FAOSTAT, 2015

FIGURE 7: United States.

largest exporter, the European Union [3]. In some years, such as 1994 and 2008, the US produced 10% of all wheat in the world [9]. In 2001, for instance, approximately 19.6 million ha of wheat was planted in the US and about 50% of the wheat produced was exported at a value of $9 billion [9, 60]. US wheat exports decreased over the 34-year time frame, even though yields increased gradually from 2 MT/ha to more than 3 MT/ha, and precipitation varied from the average in no discernable pattern without any long wet or dry periods. The major growing areas for hard red winter wheat are located in the Midwest, centering on Kansas. The largest wheat-producing states in the US are Kansas and North Dakota, with 10.4 MMT of wheat harvested in Kansas in 2012 alone [9]. US wheat yields, exports, and maps of production and weather stations are presented in Figure 7.

3.2. European Union (France). The European Union, which now includes 28 countries, was the second highest wheat-producing region between 1980 and 2013. The average total wheat-planted land area in this region has increased by an average of 500,000 ha every year over the time frame analyzed. In 2013, 26 million ha was planted and 139 MMT was harvested [9]. France, Germany, and Romania are the top three wheat exporters in the EU, with 66%, 15%, and 5.2% of total wheat exports, respectively. France has the highest yield,

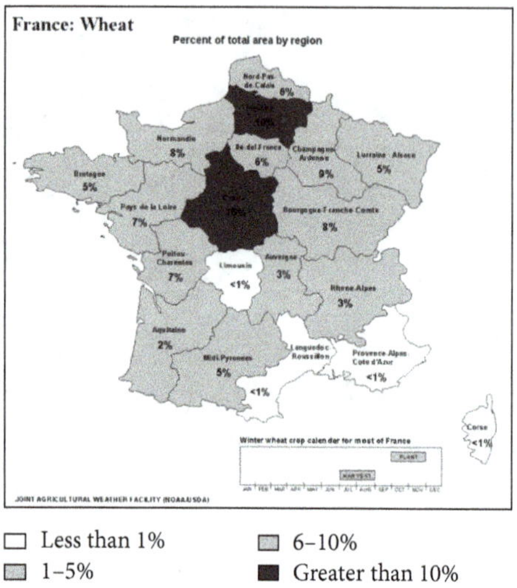

(a) France wheat production map. *Source*: USDA ERS, 2014

(b) Map of France weather stations from which precipitation data is collected. *Source*: NOAA, 2015

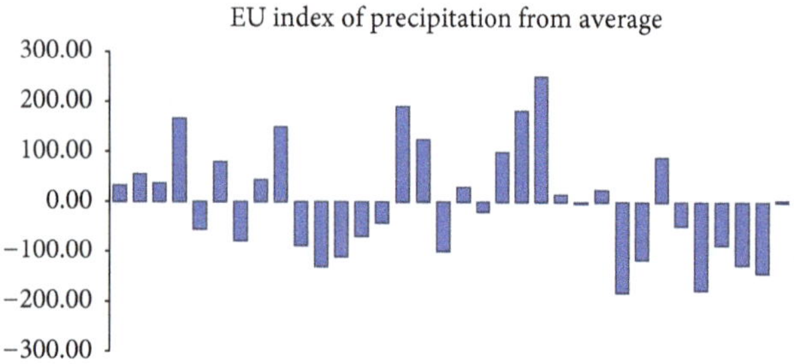

(c) European Union Index of Precipitation annual deviations from average, 1980 to 2013. *Source*: NOAA, 2015. *Precipitation is shown in +/− from average index

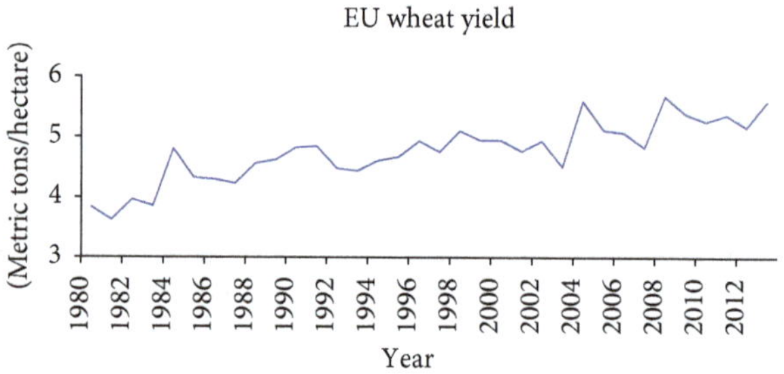

(d) Annual European Union wheat yields (MT/ha), 1980 to 2013. *Source*: OECD-FAO, 2015

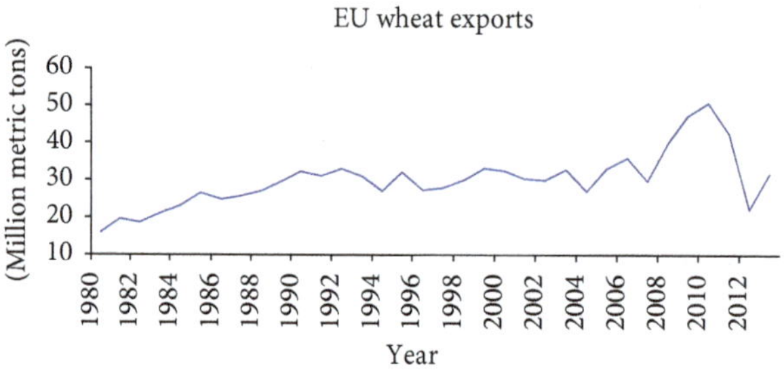

(e) Annual European Union wheat exports (MMT), 1980 to 2013. *Source*: FAOSTAT, 2015

FIGURE 8: European Union.

with an average of 7.5 MT/ha [61]. The French administrative region, Centre, leads the production of wheat in France [9]. EU wheat exports increased over time, as did wheat yields. The EU has the highest average yields in the world totaling more than 5 MT/ha. Compared to other regions, precipitation patterns in the EU appear to exhibit periodic dry and wet years during the time frame of this analysis. EU wheat yields, exports, and maps of production and weather stations are presented in Figure 8.

3.3. Canada (Saskatchewan). Canada, the third largest wheat exporter from 1980 to 2013, produces approximately 25 MMT of wheat annually, with the majority of production taking place in the provinces of Saskatchewan, Manitoba, and Alberta. Saskatchewan alone accounts for about 60% of Canada's wheat production, most of which is exported [62]. The planting and growing period is longer in Canada due to its high latitude and colder weather. The changes in exports from Canada varied during the time frame considered, while

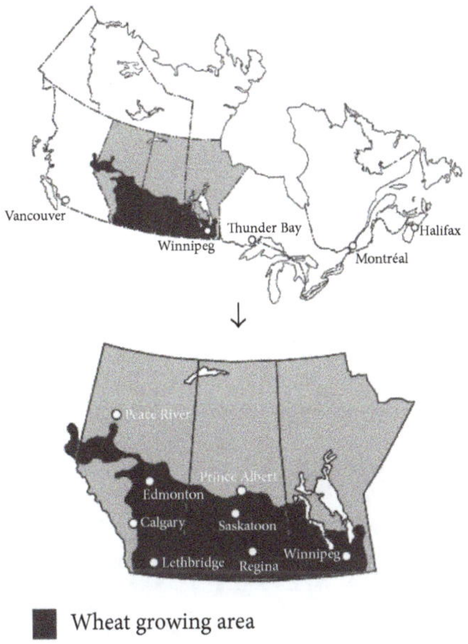

(a) Canada wheat production map. *Source*: Wheat Atlas, 2014

(b) Map of Saskatchewan weather stations from which precipitation data is collected. *Source*: NOAA, 2015

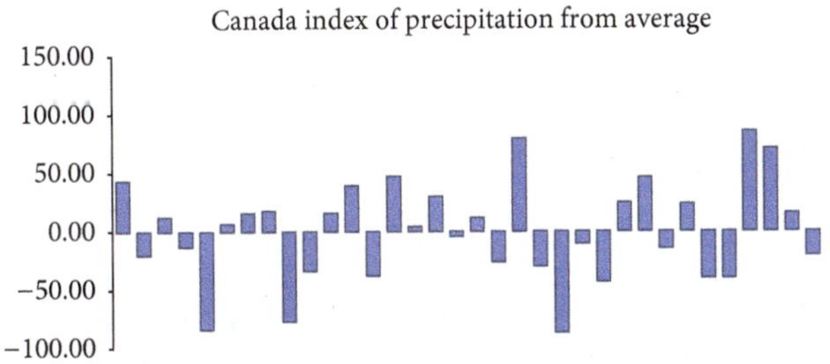

(c) Canada Index of Precipitation annual deviations from average, 1980 to 2013. *Source*: NOAA, 2015. *Precipitation is shown in +/− from average index

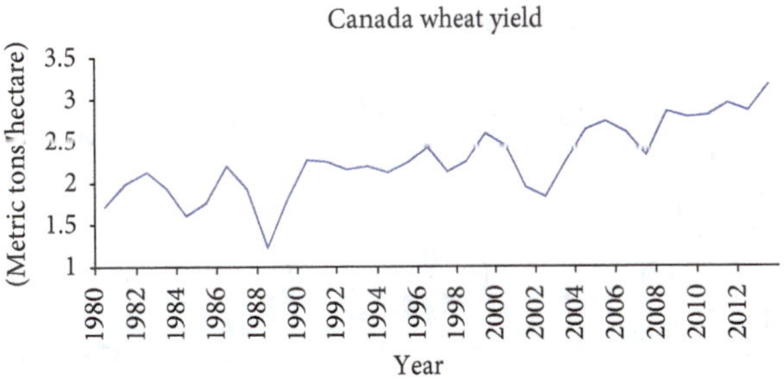

(d) Annual Canadian wheat yields (MT/ha), 1980 to 2013. *Source*: OECD-FAO, 2015

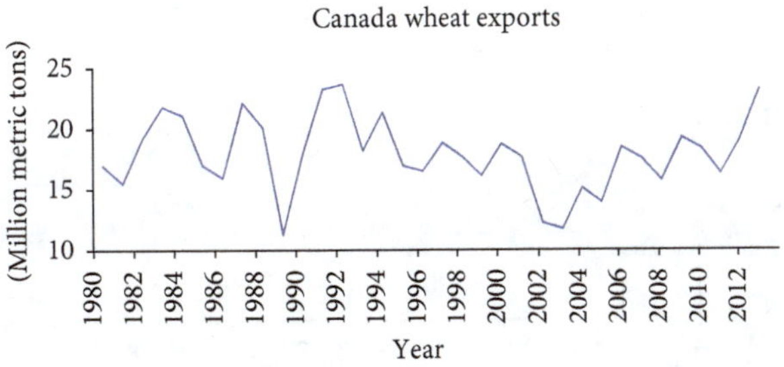

(e) Annual Canadian wheat exports (MMT), 1980 to 2013. *Source*: FAOSTAT, 2015

FIGURE 9: Canada.

precipitation levels were less variable. Like the US, yields have increased slowly over time in Canada from 1 MT/ha to 3 MT/ha. Canadian wheat yields, exports, and maps of production and weather stations are presented in Figure 9.

3.4. Australia (Western Australia). Australia's total harvested wheat-producing acreage in 2013 was 13.5 million ha, yielding 27 MMT of wheat, 18.6 MMT of which was exported. The main producing state is Western Australia; and, subsequently, the majority of Australian wheat is sold overseas from this state's supplies [63]. Australia's exports and yield have both increased gradually over time. With an average annual precipitation of 166.5 mm, Australia had the lowest precipitation levels compared to the other four key regions. Precipitation levels in recent years represent a long period of drought in the Western Australia region. Australian wheat yields, exports, and maps of production and weather stations are presented in Figure 10.

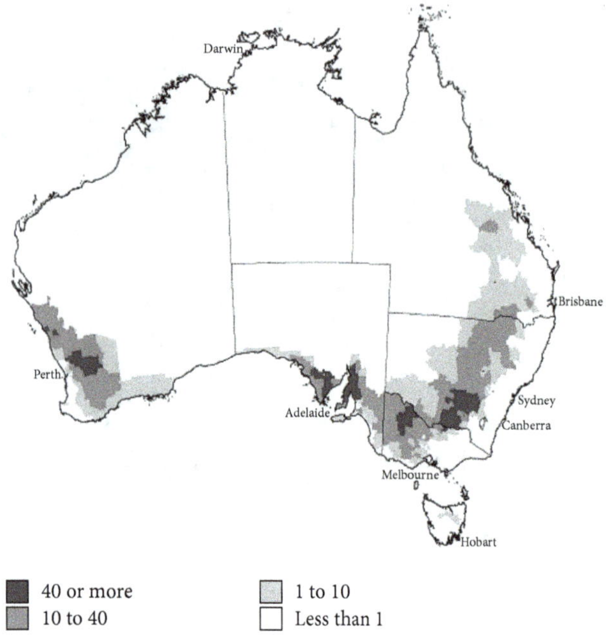

(a) Australia wheat production map. *Source*: AGDA, 2015

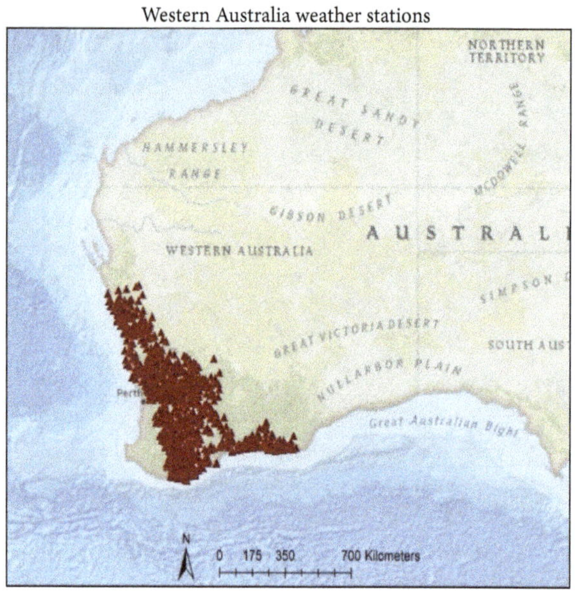

(b) Map of Western Australia weather stations from which precipitation data is collected. *Source*: NOAA, 2015

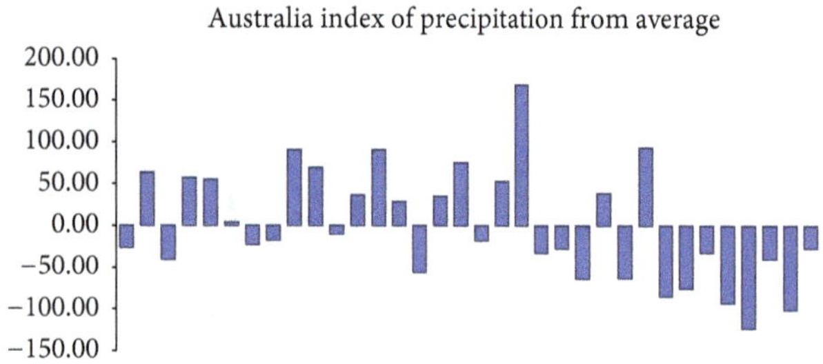

(c) Australia Index of Precipitation annual deviations from average, 1980 to 2013. *Source*: NOAA, 2015. *Precipitation is shown in +/− from average index

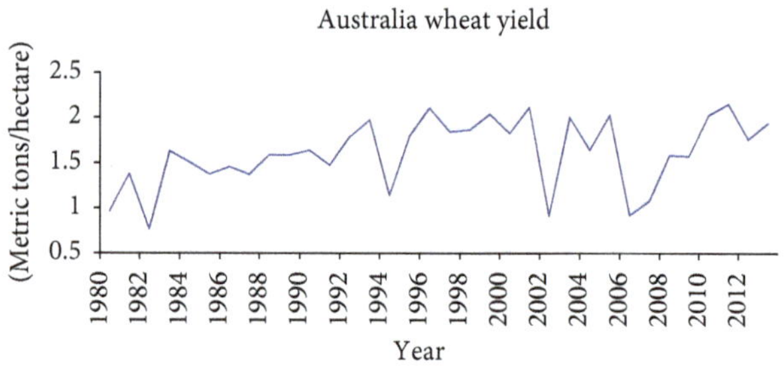

(d) Annual Australian wheat yields (MT/ha), 1980 to 2013. *Source*: OECD-FAO, 2015

(e) Annual Australian wheat exports (MMT), 1980 to 2013. *Source*: FAOSTAT, 2015

FIGURE 10: Australia.

3.5. Former Soviet Union (Southern). The Former Soviet Union (FSU) countries are the fifth largest wheat-exporting region during this 34-year time frame. Their exports have increased consistently from 1991 by an average of 1.58 MMT annually. More than 90% of wheat production in the FSU occurs in Russia, Ukraine, Kazakhstan, and Uzbekistan [9]. These four countries also account for 99% of the FSU wheat exports [64]. In 2013, 47.7 million ha of wheat was planted, with 103.9 MMT of production. Total exports that year were 37.1 MMT [9]. Almost no exports were reported for the region before the dissolution of the Soviet Union. After 1992, the region's annual exports grew to nearly 40 MMT, most of which was developed after the year 2000. Also, average yields increased over time from less than 1.5 to 2 MT/ha.

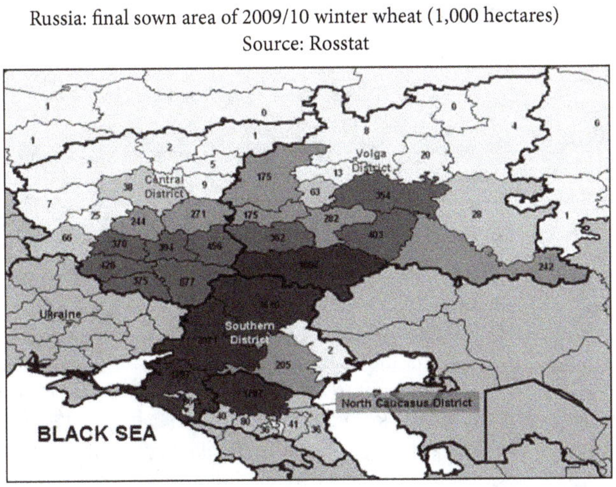

(a) Former Soviet Union wheat production map. *Source*: USDA ERS, 2014

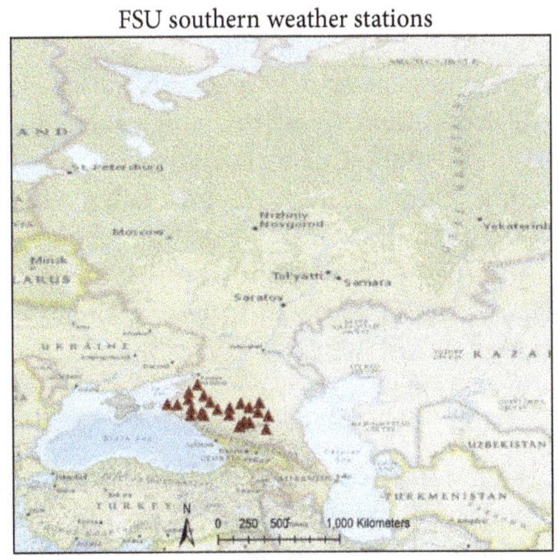

(b) Map of Former Soviet Union Southern weather stations from which precipitation data is collected. *Source*: NOAA, 2015

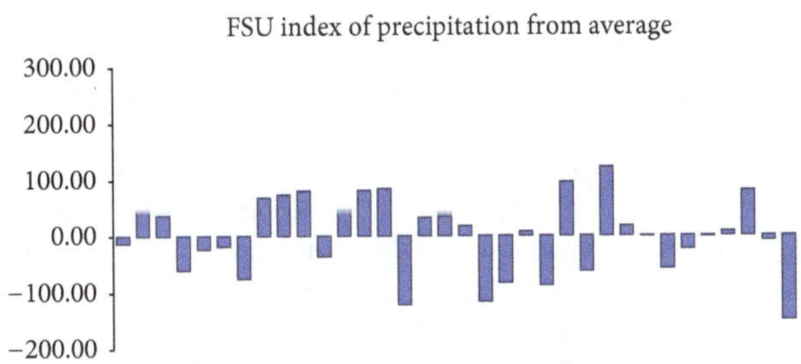

(c) Former Soviet Union Index of Precipitation annual deviations from average, 1980 to 2013. *Source*: NOAA, 2015. *Precipitation is shown in +/− from average index

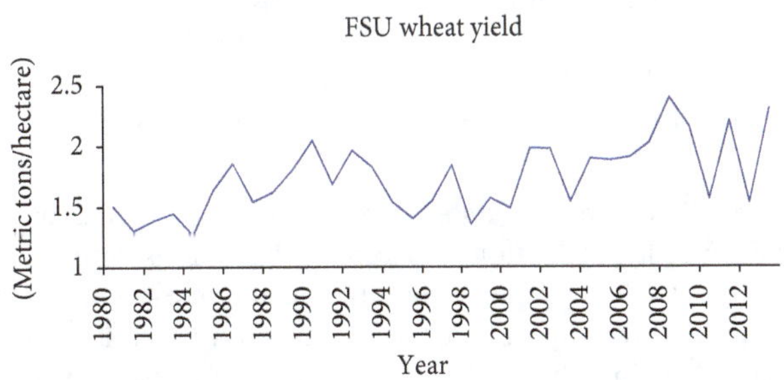

(d) Annual Former Soviet Union wheat yields (MT/ha), 1980 to 2013. *Source*: OECD-FAO, 2015

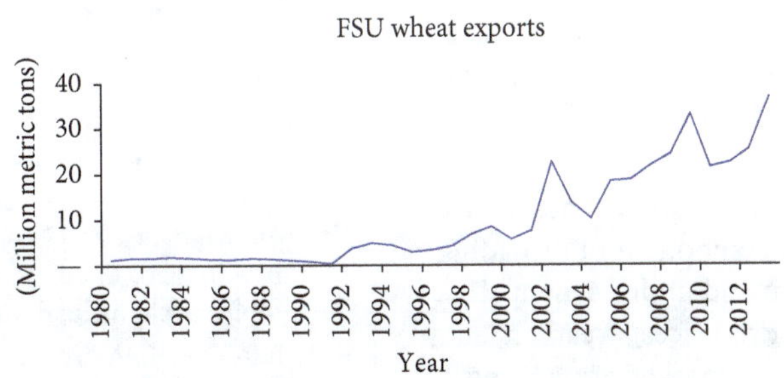

(e) Annual Former Soviet Union wheat exports (MMT), 1980 to 2013. *Source*: FAOSTAT, 2015

FIGURE 11: Former Soviet Union.

The highest levels of wheat production in the FSU were in Southern Russia. The FSU wheat yields, exports, and maps of production and weather stations are presented in Figure 11.

4. Summary and Conclusions

Global food security efforts increasingly focus on adaptation of agriculture to climate concerns, but it is clear that there are also political, economic, and cultural considerations for the effectiveness of such adaptation strategies [22, 65]. As agricultural development organizations and the countries and agribusinesses that make key management and policy decisions address the increasing challenge of declining or uncertain food security, there appear to be a few key drivers to consider if one values a systems approach.

This study explores the factors that impact wheat prices in the international market, focusing on what may have contributed most to the rapid growth in world wheat prices through 2008, as well as subsequent price fluctuations. The key contributions of this work include a comprehensive summary of global supply and demand factors that affect the wheat market. We include data and trends for wheat prices, oil prices, trade and food aid, and demand in key import countries, as well as detailed country profiles for key wheat suppliers. Increases in global wheat consumption are largely driven by population growth and rising household income in developing and emerging economies. Our work highlights that climate variability, fueled by precipitation patterns, differs among key wheat-producing regions. Past wheat prices, oil prices, and global stocks have substantial effects on wheat price determination, while policy measures including food aid and other market distortions impact wheat price variability across regions as well. Trend analyses indicate a general increase in both wheat yields and exports in the major wheat-producing regions of the world. In many cases, these increases occurred largely independent of any changes in precipitation patterns, demonstrating the influence of the other factors evaluated herein.

This paper reinforces that the G20 Ministers of Agriculture, who represent approximately 60% of all agricultural land and 80% of world trade in agricultural products, will be key players in addressing this aspect of food security [7]. Their intentions to promote activities and innovations that will make the agricultural sector more resilient are important but will only be effective if systematically framed to consider the factors and drivers that influence commodity markets in an era of globalization with complex policy dimensions.

While this study evaluates a variety of the key factors affecting wheat production and prices, limitations remain. Climate change is a complex phenomenon and merits continued study and the inclusion of factors beyond precipitation patterns when investigating climate resiliency in the wheat market. Moreover, lack of accessibility to historical data from different countries limits the scope of this analysis. For example, weather station locations have changed during the time frame, limiting access to continuous weather data. Furthermore, deeper quantitative analysis of both supply and demand factors in future work is merited.

This study explores the various factors of wheat price determination, which is an integral component of providing information for market development and improvement. This study demonstrates that past market prices, weather, and other direct influences on the wheat market are among a suite of factors policymakers must consider. Indeed, the far-reaching impacts of a changing climate, the increasing demand for fuels, both traditional and renewable, and the implications of an integrated global economy all play a part in wheat price determination. In an era of great concern about climate resiliency and food security, an understanding of global wheat market fundamentals allows policymakers and producers to plan more effectively and better prepare for potential shortages or price fluctuations. Accordingly, this work contributes to the literature and provides policy relevance by creating a synthesis of information and data regarding the aforementioned key factors in an international context. While volatility plays a minor role in the complex phenomenon of food security, recognizing how all countries interact in the global wheat market further informs the discussion regarding future dynamics in this sector.

References

[1] CGIAR Research Program on Climate Change, "About us: a strategic collaboration," Agriculture And Food Security 2017 https://ccafs.cgiar.org/about-us.

[2] C. P. Timmer, "The macro dimensions of food security: economic growth, equitable distribution, and food price stability," *Food Policy*, vol. 25, no. 3, pp. 283–295, 2000.

[3] FAO (Food And Agriculture Organization of The United Nations) Statistics (2014-15) http://faostat3.fao.org/browse/T/*/E.

[4] Foreign Agricultural Services Usda, Production, supply, and distribution statistics: world commodity imports https://apps.fas.usda.gov/psdonline/app/index.html#/app/advQuery.

[5] R. Ortiz, K. D. Sayre, B. Govaerts et al., "Climate change: can wheat beat the heat?" *Agriculture, Ecosystems & Environment*, vol. 126, no. 1, pp. 46–58, 2008.

[6] ERS (Economic Research Service), *Wheat Year Book Table 3*, USDA, Foreign Agricultural Service, Production, Supply and Distribution Database, 2015.

[7] G20 Agriculture Ministers, *G20 Agriculture Ministers' Declaration 2017: Towards food and water security: Fostering sustainability, advancing innovation*, German Federal Ministry of Food and Agriculture.

[8] B.-C. Curtis, S. Rajaram, and H. Gómez Macpherson, *Bread Wheat: Improvement And Production*, Food and Agriculture Organization of the United Nations (FAO), 2002.

[9] USDA, "(united states department of agriculture).," economic research services 2014 http://www.ers.usda.gov/topics/crops/wheat/trade.aspx#.

[10] M.-Y. Nuttonson, "Wheat-climate relationships and the use of phenology in ascertaining the thermal and photo-thermal requirements of wheat based on data of North America and of some thermally analogous areas of North America in the Soviet Union and in Finland," 1955.

[11] W. H. Leonard and J. H. Martin, "Wheat: importance, history and adaptation," in *Cereal Crops*, vol. 287, p. 227, Macmillian Co., New York, NY, USA, 1963.

[12] A. Herdağdelen, A. Gros, and Y. Bar-Yam, "Global grain consumption map," Global Grain Consumption Map, New England Complex Systems Institute, 2011.

[13] Y. Bar-Yam, L. Marco, and Y. Bar-Yam, "South African riots: repercussion of the global food crisis and US drough," in *Conflict and Complexity*, Understanding Complex Systems, pp. 261–267, Springer, New York, NY, USA, 2015.

[14] FAO (Food and Agriculture Organization of the United Nations), "Crop breeding: the green revolution and the preceding millennia," 2003 http://www.fao.org/english/newsroom/focus/2003/gmo2.htm.

[15] D. M. Liverman, "Forecasting the impact of climate on food systems: model testing and model linkage," *Climatic Change*, vol. 11, no. 1-2, pp. 267–285, 1987.

[16] J. Valizadeh, S. M. Ziaei, and S. M. Mazloumzadeh, "Assessing climate change impacts on wheat production (a case study)," *Journal of the Saudi Society of Agricultural Sciences*, vol. 13, no. 2, pp. 107–115, 2014.

[17] A. G. Dai, "Drought under global warming: a review," *Wiley Interdisciplinary Reviews: Climate Change*, vol. 2, no. 1, pp. 45–65, 2011.

[18] M. V. K. Sivakumar, H. P. Das, and O. Brunini, "Impacts of present and future climate variability and change on agriculture and forestry in the arid and semi-arid tropics," *Climatic Change*, vol. 70, no. 1-2, pp. 31–72, 2005.

[19] Q. Luo, W. Bellotti, M. Williams, and B. Bryan, "Potential impact of climate change on wheat yield in South Australia," *Agricultural and Forest Meteorology*, vol. 132, no. 3-4, pp. 273–285, 2005.

[20] R. C. Izaurralde, N. J. Rosenberg, R. A. Brown, and A. M. Thomson, "Integrated assessment of Hadley Center (HadCM2) climate-change impacts on agricultural productivity and irrigation water supply in the conterminous United States. Part II. Regional agricultural production in 2030 and 2095," *Agricultural and Forest Meteorology*, vol. 117, no. 1-2, pp. 97–122, 2003.

[21] X.-C. Zhang and W.-Z. Liu, "Simulating potential response of hydrology, soil erosion, and crop productivity to climate change in Changwu tableland region on the Loess Plateau of China," *Agricultural and Forest Meteorology*, vol. 131, no. 3-4, pp. 127–142, 2005.

[22] UNFCCC (united nations framework convention on climate change), Global Climate Action Agriculture and Food Security 2016 http://unfccc.int/files/paris_agreement/application/pdf/gca-agrfood-programme_20102016.pdf.

[23] CTA (Technical Centre for Agricultural and Rural Cooperation), "Moving agriculture further up the climate change agenda," 2015 http://www.cta.int/en/article/2015-06-01/moving-agriculture-further-up-the-climate-change-agenda.html.

[24] C. B. Barrett, "Measuring food insecurity," *Science*, vol. 327, no. 5967, pp. 825–828, 2010.

[25] F. G. Santeramo, "Food security composite indices: implications for policy and practice," *Development in Practice*, vol. 25, no. 4, pp. 594–600, 2015.

[26] R. Chand, "Wheat supply, price prospects, and food security," *Economic and Political Weekly*, vol. 42, no. 19, pp. 1659–1663, 2007.

[27] M. G. Haile, M. Kalkuhl, and J. von Braun, "Worldwide acreage and yield response to international price change and volatility: a dynamic panel data analysis for wheat, rice, corn, and soybeans," in *Food Price Volatility and Its Implications for Food Security and Policy*, vol. 98, pp. 139–165, Springer International Publishing, 2016.

[28] Bureau of Labor Statistics, "Overview of BLS statistics on inflation and prices: consumer price index," 2015 http://www.bls.gov/cpi/.

[29] A. F. McCalla, "World food prices: Causes and consequences," *Canadian Journal of Agricultural Economics*, vol. 57, no. 1, pp. 23–34, 2009.

[30] D. O. Mitchell and M. Mielke, "Wheat: The Global Market, Policies, and Priorities," in *Global Agricultural Trade and Developing Countries*, p. 11, World Bank, 2005.

[31] J. Moy, "Recent trends in unemployment and the labor force, 10 countries," *Monthly Labor Review*, vol. 108, no. 8, pp. 9–22, 1985.

[32] R. H. Rasche and J. A. Tatom, "Energy price shocks, aggregate supply and monetary policy: the theory and the international evidence," *Carnegie-Rochester Conference Series on Public Policy*, vol. 14, pp. 9–93, 1981.

[33] L. Hennemuth, "Partial grain embargo: pragmatism's centrality to argentine refusal to cooperate," *The Monitor*, pp. 44–55, 2012.

[34] C. B. Luttrell, "The russian grain embargo: dubious success," *Federal Reserve Bank of St. Louis Review*, vol. 62, no. 7, pp. 2–8, 1980.

[35] I. Jebabli, M. Arouri, and F. Teulon, "On the effects of world stock market and oil price shocks on food prices: an empirical investigation based on TVP-VAR models with stochastic volatility," *Energy Economics*, vol. 45, pp. 66–98, 2014.

[36] P. Sadorsky, "Modeling volatility and correlations between emerging market stock prices and the prices of copper, oil and wheat," *Energy Economics*, vol. 43, pp. 72–81, 2014.

[37] G. Tadasse, B. Algieri, M. Kalkuhl et al., "Drivers and triggers of international food price spikes and volatility," in *Food Price Volatility and Its Implications for Food Security and Policy*, pp. 59–82, Springer International Publishing.

[38] J. Baffes and T. Haniotis, "What explains agricultural price movements?" *Journal of Agricultural Economics*, vol. 67, no. 3, pp. 706–721, 2016.

[39] S.-T. Chen, H.-I. Kuo, and C.-C. Chen, "Modeling the relationship between the oil price and global food prices," *Applied Energy*, vol. 87, no. 8, pp. 2517–2525, 2010.

[40] B. Wright, "Global biofuels: Key to the puzzle of grain market behavior," *The Journal of Economic Perspectives*, vol. 28, no. 1, pp. 73–98, 2014.

[41] S. H. Saghaian, "The Impact of the oil sector on commodity prices: correlation or causation?" *Journal of Agricultural and Applied Economics*, vol. 42, no. 3, pp. 477–485, 2010.

[42] A. Esmaeili and Z. Shokoohi, "Assessing the effect of oil price on world food prices: application of principal component analysis," *Energy Policy*, vol. 39, no. 2, pp. 1022–1025, 2011.

[43] V. Natanelov, M. J. Alam, A. M. McKenzie, and G. Van Huylenbroeck, "Is there co-movement of agricultural commodities futures prices and crude oil?" *Energy Policy*, vol. 39, no. 9, pp. 4971–4984, 2011.

[44] T. Serra and J. M. Gil, "Price volatility in food markets: can stock building mitigate price fluctuations?" *European Review of Agricultural Economics*, vol. 40, no. 3, pp. 507–528, 2013.

[45] U.S. Energy Information Administration, Petroleum & Other Liquids 2014 http://www.eia.gov/petroleum/data.cfm#prices.

[46] J. A. Frieden and R. Rogowski, "The impact of the international economy on national policies: an analytical overview," *Internationalization and Domestic Politics*, pp. 25–47, 1996.

[47] T. W. Hertel, W. Martin, and A. M. Leister, "Potential implications of a special safeguard mechanism in the world trade organization: the case of wheat," *The World Bank Economic Review*, vol. 24, no. 2, pp. 330–359, 2010.

[48] A. Cioffi, F. G. Santeramo, and C. D. Vitale, "The price stabilization effects of the EU entry price scheme for fruit and vegetables," *Agricultural Economics*, vol. 42, no. 3, pp. 405–418, 2011.

[49] M. Ivanic and W. Martin, "Implications of domestic price insulation for global food price behavior," *Journal of International Money and Finance*, vol. 42, pp. 272–288, 2014.

[50] C. Ackello-Ogutu and P. Echessah, "Unrecorded cross-border trade between Kenya and Uganda," Tech. Rep. 59, SD Publication Series Office of Sustainable Development Bureau for Africa, USAID, 1997.

[51] M. A. Aksoy and C. B. Beghin, Eds., *Global Agricultural Trade and Developing Countries*, The International Bank for Reconstruction and Development. World Bank Publications, 2004.

[52] Market Access Map Quick Search: Tariffs, *Quick Search: Tariffs*, International Trade Centre, 2017.

[53] OECD (Organization for Economic Co-operation and Development)-FAO, Agricultural outlook 2014-2023 food

and agriculture organization of the united nation 2015 http://goo.gl/DHK14Y.

[54] E. Bobenrieth, B. Wright, and D. Zeng, "Stocks-to-use ratios and prices as indicators of vulnerability to spikes in global cereal markets," *Agricultural Economics*, vol. 44, no. 1, pp. 43–52, 2013.

[55] K. Daugherty and K. Kelly, "Stocks and the stocks-to-use ratio: are they meaningful for price determination?" *Growmark Research*, pp. 1–32, 2014, http://www.growmark.com/sites/Files/Documents/StocksToUse.pdf.

[56] NOAA (National Oceanic and Atmospheric Administration), NOAA 2015 http://www.cdc.noaa.gov/ENSO/enso.mei_index.html.

[57] J. Y. Shend, Agricultural statistics of the former USSR Republics and the Baltic States, Statistical bulletin, 1993.

[58] H. D. Rogers, "Irrigation management," in *Wheat Production Handbook*, Cooperative Extension Publication C-529, Kansas State University, Manhattan, NY, USA, 1997.

[59] Year Book Australia, *The Australian Wheat Industry. 1301.0*, Year Book Australia, 2006 http://goo.gl/Y7jKjc.

[60] VOA, *US Seeks Fast Test to Settle GM Wheat Scare*, Voice of America. Science and Technology, 2013.

[61] The World Bank, "Cereal Yield Data. World Development Indicators: Agricultural Output and Productivity," 2014 http://data.worldbank.org/indicator/AG.YLD.CREL.KG/countries.

[62] Wheat Atlas, 2014 http://wheatatlas.org/country/CAN/0.

[63] AGDA (Australian Government Department of Agriculture), "Wheat Overview," 2015 http://www.agriculture.gov.au/ag-farm-food/crops/wheat.

[64] Foreign Agricultural Services USDA, ""Grain World Markets and Trade." World Production, Markets, and Trade Reports," 2014 http://www.fas.usda.gov/data/grain-world-markets-and-trade.

[65] T. Wheeler and J. Von Braun, "Climate change impacts on global food security," *Science*, vol. 341, no. 6145, pp. 508–513, 2013.

14

Weed Control in White Bean with Pethoxamid Tank-Mixes Applied Preemergence

Nader Soltani, Lynette R. Brown, and Peter H. Sikkema

University of Guelph Ridgetown Campus, Ridgetown, ON N0P 2C0, Canada

Correspondence should be addressed to Nader Soltani; soltanin@uoguelph.ca

Academic Editor: Kassim Al-Khatib

Six field experiments were conducted during 2015 to 2017 in Ontario, Canada, to determine the efficacy of pethoxamid applied alone, and in combination with broadleaf herbicides, for the control of annual grass and broadleaved weeds in white navy bean. Visible injury was generally minimal (0 to 8%) with herbicide treatments evaluated. Weed control was variable depending on the weed species evaluated. Pethoxamid, *S*-metolachlor, halosulfuron, imazethapyr, sulfentrazone, pethoxamid + halosulfuron, pethoxamid + imazethapyr, and pethoxamid + sulfentrazone controlled redroot pigweed 82 to 98%; common ragweed 19 to 93%; common lambsquarters 49 to 84%; and green foxtail 47 to 92% in white bean. Weed biomass and weed density reductions were similar to visible control ratings for herbicides evaluated. Weed interference delayed white bean maturity and reduced yield by 50% in this study. Weed interference in plots sprayed with pethoxamid, *S*-metolachlor, and sulfentrazone reduced white bean yield 36%. White bean yield was similar to the weed-free with other herbicides evaluated. This study concludes that there is potential for the tank-mix of pethoxamid with halosulfuron, imazethapyr, or sulfentrazone for weed control in white bean production.

1. Introduction

White (navy) bean (*Phaseolus vulgaris* L.) is a valuable cash crop grown in various regions of Canada. Ontario and Manitoba farmers grow the majority of white bean produced in Canada. In 2017, Ontario growers produced white bean on 28,000 ha, with an annual production of 6,000,000 tonnes valued at $40 million [1]. Dry bean is extremely sensitive to weed competition. Effective weed management is essential to minimize white bean yield losses from weed interference [2–4]. The average yield losses from uncontrolled weeds in North America was 71%, 52%, and 52% in dry bean, corn, and soybean, respectively, in studies completed by the Weed Science Society of America (WSSA) [5]. In the same WSSA studies, the potential yield loss from uncontrolled weeds in Ontario was 56, 49 and 35% in dry bean, corn, and soybean, respectively [5]. Dry beans have greater sensitivity to herbicides compared to other legumes, such as soybean [6]. There are numerous broadleaf herbicides registered for use in soybean, but most of them cannot be used in beans because of crop injury [6]. Consequently, there are only two registered soil-applied broadleaf herbicides, imazethapyr, and halosulfuron and two postemergence broadleaf herbicides, bentazon and fomesafen for broadleaf weed control in white bean in Ontario, Canada [6]. In contrast, identity preserved (IP) soybean producers in Ontario have at least 12 herbicides for broadleaf weed control—acifluorfen, bentazon, chlorimuron, clomazone, cloransulam, flumetsulam, flumioxazin, fomesafen, imazethapyr, linuron, metribuzin, and thifensulfuron [6]. Despite the wide array of herbicides registered for use in IP soybeans, broadleaf weed control is still considered to be a challenge for soybean growers which highlights the difficulty faced by Ontario white bean growers. New herbicide options that have multiple modes of action are needed to provide effective control of troublesome weeds in white bean production.

Pethoxamid is a new Group 15 herbicide from the chloroacetamide chemical family that can control annual grasses such as foxtail species (*Setaria* spp.), large crabgrass (*Digitaria* spp.), barnyardgrass (*Echinochloa crus-galli* (L.)

P.Beauv.), and broadleaved weeds such as redroot pigweed (*Amaranthus retroflexus* L.), common lambsquarters (*Chenopodium album* L.), and ladysthumb (*Polygonum persicaria* L.) [7–9]. Pethoxamid also has activity against Groups 2, 5, and 9 herbicide-resistant weeds including palmer amaranth (*Amaranthus palmeri* S. Watson), common waterhemp (*Amaranthus Rudis* L.), and other important annual grass and broadleaf weeds [9]. Pethoxamid inhibits very long chain fatty acid (VLCFA) formation in responsive weeds and generally controls weeds prior to emergence [9]. Pethoxamid is generally applied to the soil as a preplant (PP), preplant incorporated (PPI), or pre-emergence (PRE) herbicide [9].

Halosulfuron (Group 2, ALS inhibitor), imazethapyr (Group 2, ALS inhibitor), sulfentrazone (Group 14, PPO inhibitor), and *S*-metolachlor (Group 15, VLCFA inhibitor) have potential to control problematic weeds in Ontario including redroot pigweed, common lambsquarters, common ragweed, wild mustard (*Sinapis arvensis* L.), nutsedge species (*Cyperus* spp.), ladysthumb (*Polygonum persicaria* L.), velvetleaf (*Abutilon theophrasti* Medic.), and cocklebur (*Xanthium strumarium* L.), including glyphosate and triazine-resistant biotypes [10]. Halosulfuron, imazethapyr, and sulfentrazone provide variable control of annual grasses such as *Setaria*, *Digitaria*, *Echinochloa*, and *Panicum* species [4]. The spectrum of weeds controlled with pethoxamid can be increased if tank-mixed with other herbicides such as halosulfuron, imazethapyr, or sulfentrazone.

The purpose of this study was to determine the efficacy of pethoxamid compared to *S*-metolachlor applied PRE and to determine if broadleaf weed control can be improved with the addition of halosulfuron, imazethapyr, or sulfentrazone to pethoxamid applied PRE.

2. Materials and Methods

A total of six field experiments were conducted over a three-year period (2015, 2016, and 2017) at the Huron Research Station (41°68′N, 83°11′W), Exeter, Ontario, and the University of Guelph Ridgetown Campus (42.7392° N, 81.8871° W), Ridgetown, Ontario (one site at each research station per year). The soil at Exeter was a Brookston clay loam (Orthic Humic Gleysol, mixed, mesic, and poorly drained) with 35% sand, 43% silt, 22% clay, 3.6% organic matter, and a pH of 7.6 in 2015; 41% sand, 35% silt, 24% clay, 2.9% organic matter, and a pH of 7.7 in 2016; and 32% sand, 42% silt, 26% clay, 3.2% organic matter, and a pH of 7.7 in 2017. The soil at Ridgetown was a Watford (gray-brown brunisolic, mixed, mesic, sandy, and imperfectly drained)-Brady (gleyed brunisolic gray-brown luvisol, mixed, mesic, sandy, and imperfectly drained) with 52% sand, 24% silt, 24% clay, 4.3% organic matter, and a pH of 7.3 in 2015; 59% sand, 20% silt, 21% clay, 3.4% organic matter, and a pH of 6.7 in 2016; and 49% sand, 26% silt, 25% clay, 4.3% organic matter, and a pH of 6.4 in 2017. Seedbed preparation at all sites consisted of fall moldboard plowing, followed by two passes with a field cultivator in spring.

Herbicide treatments arranged in a randomized complete block (with four replicates) included pethoxamid at 840 g ai ha^{-1}, *S*-metolachlor at 1050 g ai ha^{-1}, halosulfuron at 35 g ai ha^{-1}, imazethapyr at 75 g ai ha^{-1}, sulfentrazone at 140 g ai ha^{-1}, pethoxamid + halosulfuron at 840 + 35 g ai ha^{-1}, pethoxamid + imazethapyr at 840 + 75 g ai ha^{-1}, and pethoxamid + sulfentrazone at 840 + 140 g ai ha^{-1}. Each replicate included a weedy and a weed-free control. Each plot consisted of four rows of white bean ("T9905") spaced 0.75 m apart in rows that were 10 m long at Exeter and 8 m long at Ridgetown. Beans were planted approximately 4 cm deep at a rate of approximately 250,000 seeds ha^{-1}.

Herbicide treatments were applied to the soil surface (not incorporated) 1-2 days after planting using a CO_2-pressurized backpack sprayer calibrated to deliver 200 $L{\cdot}ha^{-1}$ at 240 kPa. The boom was 1.5 m long with four ultralow drift ULD120-02 nozzles spaced 0.5 m apart providing a spray width of 2 m.

White bean injury was estimated on a scale of 0 (no visible injury) to 100% (complete plant death) 2 and 4 weeks after crop emergence (WAE). Percent control of dominant weed species which included redroot pigweed, common ragweed, common lambsquarters, and green foxtail was estimated visually at 4 and 8 weeks after treatment application (WAT), and weed density (plants$\cdot m^{-2}$) and weed dry weight ($g{\cdot}m^{-2}$) were determined at 8 WAT from two 0.5 m^{-2} quadrats in each plot. Weeds were separated by weed species and counted before being cut at soil level. Plants were then placed in a paper bag and dried in an oven at 60°C to a constant moisture and then weighed. Yields were determined by harvesting the two middle rows of white bean in each plot using a small plot combined at crop maturity. Seed yield was adjusted to 18% seed moisture content.

Data analysis was carried out using the GLIMMIX procedure in SAS (Ver. 9.4, SAS Institute Inc., Cary, NC), and the Laplace method was used for estimation. Herbicide treatments were considered fixed effects while environment (year and location), replication within the environment, and the environment by treatment interaction were considered random effects. Random effects were chosen in order to be able to apply conclusions to a broader range of environments than those studied [11, 12]. The significance of fixed effects was tested using the *F*-test, and likelihood ratio tests were used to determine the significance of random effects. Different distributions were assessed using the AICC and Pearson chi-squared test/df (for overdispersion), as well as the examination of studentized residual plots and the Shapiro–Wilk statistic. Once the best distribution was confirmed, least square means (LSMEANS) were calculated and Tukey's adjustment was applied to pairwise comparisons to determine differences among treatment means ($p < 0.05$). Final yield and percent control of redroot pigweed, common ragweed, common lambsquarters, and green foxtail at 4 and 8 WAA were analyzed using a Gaussian distribution and identity link. Percent injury was analyzed using the poisson distribution (2 WAT) and the negative binomial distribution (4 WAT); the log link was used in both cases. Percent white bean moisture at harvest was analyzed using a gamma distribution and log link. Redroot pigweed, common ragweed, common lambsquarters, and green foxtail density and dry weight data were best described using a lognormal distribution and identity link. Injury evaluations, density, and dry weight data had a value of one added prior to analysis to accommodate for observed zero values, and the final LSMEANS were adjusted by subtracting one. Data were combined over environments. Treatment means calculated using the lognormal distribution were back-transformed for presentation due to the fact that there is no link function to display values on the mean scale. The backtransformation included a correction for log bias [13]. In all cases where a treatment was assigned a value of 0 (weedy check for percent control and weed-free check for density and dry weight) or 100 (weed-free check for percent control), it was excluded from the analysis because the data for these treatments had zero variance. However, the LSMEANS output provides information on whether each treatment least square mean differs from zero. This information was used to identify differences between the treatments included in the analysis, and the excluded check treatments, which were assigned a value of zero.

3. Results and Discussion

3.1. Crop Response. White bean injury was <9% with herbicide treatments evaluated at 2 and 4 WAE (Table 1). At 2 or 4 WAE, herbicides evaluated caused 1 to 8% white bean injury (Table 1). Weed interference delayed white bean maturity and reduced yield by 50% in this study (Table 1). Weed interference in plots sprayed with pethoxamid, *S*-metolachlor, and sulfentrazone reduced white bean yield 36%. However, other herbicide treatments evaluated provided similar seed yield as the weed-free control (Table 1). Results are similar to other studies that have shown 0 to 9% injury with pethoxamid, *S*-metolachlor, halosulfuron, imazethapyr, and sulfentrazone applied PRE alone or in combination with other herbicides in dry bean [4, 14–17]. In another study, preemergence application of pethoxamid also caused 6% injury when applied at 1200 g·ai·ha^{-1} and 10% injury when applied at 2400 g·ai·ha^{-1} in white bean [16]. In another study, *S*-metolachlor plus imazethapyr caused as much as 20% crop injury in dry bean under some environmental conditions [14].

3.2. Weed Control. Weed control was variable depending on the weed species evaluated. Herbicides evaluated controlled redroot pigweed 82 to 98% at 2 WAT and 70 to 98% at 4 WAT (Table 2). Redroot pigweed density and biomass reduction were similar to the control ratings. Herbicides evaluated reduced redroot pigweed density 79 to 98% and biomass 66 to 98%, although biomass reduction was not statistically significant than the weedy check for the imazethapyr treatment (Table 2). In other studies, *S*-metolachlor and pyroxasulfone applied PRE-controlled redroot pigweed 98% [18]. Sulfentrazone applied alone or tank-mixed with *S*-metolachlor or pyroxasulfone provided >90% control of redroot pigweed in white bean [17].

Common ragweed control was weak (except for halosulfuron treatments) with most of the herbicide treatments evaluated (Table 3). Halosulfuron and pethoxamid + halosulfuron controlled common ragweed from 89% to 93% at 2 WAT and from 88% to 92% at 4 WAT (Table 3). Halosulfuron alone and in combination with pethoxamid reduced common ragweed density 87 to 90% and biomass 91 to 95% in white bean (Table 3). However, common ragweed density and biomass were not significantly different than the weedy check with other herbicides evaluated in white bean (Table 3). In other studies, preemergence application of halosulfuron provided >90% control of common ragweed in white bean [4]. Results with other herbicides are similar to other research that has shown *S*-metolachlor and pyroxasulfone provide only 28 to 39% control of common ragweed

Table 1: Percent visible injury 2 and 4 weeks after crop emergence (WAE), moisture and yield of white bean treated with various PRE herbicides in 2015, 2016, and 2017.

Treatment	Rate (g·ai/ha)	Injury (%) 2 WAE	Injury (%) 4 WAE	Moisture (%)	Yield (t/ha)
Weedy check		0[a]	0[a]	18.6[b]	1.1[c]
Weed-free check		0[a]	0[a]	17.5[a]	2.2[a]
Pethoxamid	840	2.5[bcd]	1.0[ab]	18.2[ab]	1.4[bc]
S-metolachlor	1050	1.0[abc]	0.4[ab]	18.1[ab]	1.4[bc]
Halosulfuron	35	0.2[ab]	0.7[ab]	18.0[ab]	1.8[ab]
Imazethapyr	75	0.6[abc]	2.2[bcd]	18.0[ab]	1.7[abc]
Sulfentrazone	140	2.3[bc]	1.9[bcd]	18.1[ab]	1.4[bc]
Pethoxamid + halosulfuron	840 + 35	2.9[cd]	1.4[abc]	18.0[ab]	2.1[a]
Pethoxamid + imazethapyr	840 + 75	3.1[cd]	6.1[cd]	18.2[ab]	1.9[ab]
Pethoxamid + sulfentrazone	840 + 140	8.3[d]	7.9[d]	18.1[ab]	1.7[abc]

Data were combined over environments and years. Means followed by the same letter within a column are not significantly different according to the Tukey–Kramer multiple range test at $p < 0.05$.

Table 2: Percent control, density, and dry weight of redroot pigweed pigweed 2 and 4 weeks after herbicide treatment (WAT) in white bean treated with pethoxamid applied PRE in 2015, 2016, and 2017.

Treatment	Rate (g·ai/ha)	Control (%) 2 WAT	Control (%) 4 WAT	Density (#/m^2)	Dry weight (g/m^2)
Weedy check		0[b]	0[b]	13.6[d]	18.4[d]
Weed-free check		100[a]	100[a]	0.0[a]	0.0[a]
Pethoxamid	840	84[a]	70[a]	2.6[bc]	2.5[bc]
S-metolachlor	1050	82[a]	71[a]	2.1[bc]	2.5[bc]
Halosulfuron	35	89[a]	76[a]	1.8[bc]	2.1[abc]
Imazethapyr	75	88[a]	72[a]	2.9[c]	6.3[cd]
Sulfentrazone	140	93[a]	87[a]	0.8[abc]	1.0[abc]
Pethoxamid + halosulfuron	840 + 35	97[a]	94[a]	0.6[abc]	1.0[abc]
Pethoxamid + imazethapyr	840 + 75	91[a]	82[a]	1.4[abc]	2.2[abc]
Pethoxamid + sulfentrazone	840 + 140	98[a]	98[a]	0.3[ab]	0.3[ab]

Data were combined over environments and years. Means followed by the same letter within a column are not significantly different according to the Tukey–Kramer multiple range test at $p < 0.05$.

TABLE 3: Percent control, density, and dry weight of common ragweed 2 and 4 weeks after herbicide treatment (WAT) in white bean treated with pethoxamid applied PRE in 2015, 2016, and 2017.

Treatment	Rate (g·ai/ha)	Control (%) 2 WAT	Control (%) 4 WAT	Density (#/m^2)	Dry weight (g/m^2)
Weedy check		0^{d}	0^{e}	14.8^{b}	22.0^{cd}
Weed-free check		100	100	0.0^{a}	0.0^{a}
Pethoxamid	840	19^{cd}	25^{cd}	10.2^{b}	19.6^{cd}
S-metolachlor	1050	7^{cd}	19^{d}	18.2^{b}	35.5^{d}
Halosulfuron	35	89^{a}	88^{a}	1.5^{a}	1.9^{b}
Imazethapyr	75	57^{ab}	63^{ab}	7.7^{b}	9.6^{c}
Sulfentrazone	140	4^{cd}	25^{cd}	11.2^{b}	24.6^{cd}
Pethoxamid + halosulfuron	840 + 35	93^{a}	92^{a}	1.9^{a}	1.0^{ab}
Pethoxamid + imazethapyr	840 + 75	77^{a}	79^{ab}	6.9^{b}	9.7^{c}
Pethoxamid + sulfentrazone	840 + 140	38^{bc}	54^{bc}	10.4^{b}	21.4^{cd}

Data were combined over environments and years. Means followed by the same letter within a column are not significantly different according to the Tukey–Kramer multiple range test at $p < 0.05$.

TABLE 4: Percent control, density, and dry weight of lambsquarters 2 and 4 weeks after herbicide treatment (WAT) in white bean treated with pethoxamid applied PRE in 2015, 2016, and 2017.

Treatment	Rate (g·ai/ha)	Control (%) 2 WAT	Control (%) 4 WAT	Density (#/m^2)	Dry weight (g/m^2)
Weedy check		0^{d}	0^{c}	12.4^{d}	9.6^{c}
Weed-free check		100	100	0.0^{a}	0.0^{a}
Pethoxamid	840	53^{bc}	42^{b}	4.3^{c}	4.5^{c}
S-metolachlor	1050	49^{c}	31^{b}	5.8^{cd}	4.5^{c}
Halosulfuron	35	81^{ab}	78^{a}	1.2^{b}	0.7^{b}
Imazethapyr	75	83^{a}	83^{a}	0.7^{ab}	0.1^{ab}
Sulfentrazone	140	84^{a}	81^{a}	0.2^{ab}	0.1^{ab}
Pethoxamid + halosulfuron	840 + 35	82^{ab}	79^{a}	0.7^{b}	0.5^{ab}
Pethoxamid + imazethapyr	840 + 75	83^{a}	83^{a}	0.2^{ab}	0.1^{ab}
Pethoxamid + sulfentrazone	840 + 140	82^{a}	81^{a}	0.4^{ab}	0.2^{ab}

Data were combined over environments and years. Means followed by the same letter within a column are not significantly different according to the Tukey–Kramer multiple range test at $p < 0.05$.

[18]. Lack of common ragweed control with sulfentrazone alone or tank-mixed with imazethapyr is similar to Taziar et al. [18] who observed 35% common ragweed control with these herbicides in white bean.

Common lambsquarters was controlled the least (31 to 53%) with pethoxamid and *S*-metolachlor among the herbicide treatments evaluated (Table 4). Other herbicides evaluated controlled common lambsquarters from 81% to 84% at 2 WAT and from 78% to 83% at 4 WAT (Table 4). Pethoxamid and *S*-metolachlor provided inadequate reduction in biomass and density of common lambsquarters (Table 4). Halosulfuron, imazethapyr, sulfentrazone, pethoxamid + halosulfuron, pethoxamid + imazethapyr, and pethoxamid + sulfentrazone reduced common lambsquarters density from 90% to 98% and biomass from 93% to 99% in white bean (Table 4). Results are similar to other studies that have shown >95% control of common lambsquarters with halosulfuron or its tank-mix with trifluralin, dimethenamid-*P*, and *S*-metolachlor [4]. Sulfentrazone in combination with *S*-metolachlor or pyroxasulfone has been shown to control common lambsquarters greater than 90% in white bean [17].

Green foxtail was controlled the least with halosulfuron and sulfentrazone among the herbicide treatments evaluated (Table 5). *S*-metolachlor, imazethapyr, pethoxamid + imazethapyr, and pethoxamid + sulfentrazone provided the best control of green foxtail (88% to 92% at 2 WAT and 76% to 89% at 4 WAT) in white bean (Table 5). Halosufuron and sulfentrazone applied alone provided no significant reduction in green foxtail density and biomass except for sulfentrazone which provided 78% reduction in green foxtail biomass (Table 5). However, pethoxamid, *S*-metolachlor, imazethapyr, pethoxamid + halosulfuron, pethoxamid + imazethapyr, and pethoxamid + sulfentrazone reduced green foxtail density 81% to 93% and green foxtail biomass 86% to 97% in white bean (Table 5). In other

TABLE 5: Percent control, density, and dry weight of green foxtail 2 and 4 weeks after herbicide treatment (WAT) in white bean treated with pethoxamid applied PRE in 2015, 2016, and 2017.

Treatment	Rate (g·ai/ha)	Control (%) 2 WAT	Control (%) 4 WAT	Density (#/m^2)	Dry weight (g/m^2)
Weedy check		0^{d}	0^{d}	62.6^{d}	38.3^{e}
Weed-free check		100	100	0.0^{a}	0.0^{a}
Pethoxamid	840	78^{ab}	71^{ab}	12.2^{bc}	4.4^{bc}
s-metolachlor	1050	90^{a}	87^{a}	4.2^{b}	2.0^{bc}
Halosulfuron	35	46^{c}	47^{c}	44.1^{cd}	17.5^{de}
Imazethapyr	75	88^{ab}	85^{a}	9.2^{b}	2.6^{bc}
Sulfentrazone	140	64^{bc}	51^{bc}	14.8^{bcd}	8.3^{cd}
Pethoxamid + halosulfuron	840 + 35	81^{ab}	72^{a}	11.4^{bc}	5.3^{bcd}
Pethoxamid + imazethapyr	840 + 75	92^{a}	89^{a}	3.2^{b}	1.0^{ab}
Pethoxamid + sulfentrazone	840 + 140	89^{a}	76^{a}	7.7^{b}	3.2^{bc}

Data are combined over environments and years. Means followed by the same letter within a column are not significantly different according to the Tukey–Kramer multiple range test at $P < 0.05$.

studies, *S*-metolachlor applied PRE-controlled green foxtail 98% in dry bean [18]. Halosulfuron tank-mixed with *S*-metolachlor has been shown to provide greater than 90% control of grasses including green foxtail [4]. Additionally, sulfentrazone applied alone or tank-mixed with *S*-metolachlor or pyroxasulfone has been shown to provide >90% green foxtail control in white bean [17].

4. Conclusion

There was minimal visible injury with most of the herbicides evaluated. Pethoxamid + sulfentrazone caused the most visible injury (8%) in white bean; however, plants recovered and had no yield reduction at harvest time. Weed control was variable depending on the weed species evaluated. Pethoxamid, *S*-metolachlor, halosulfuron, imazethapyr, sulfentrazone, pethoxamid + halosulfuron, pethoxamid + imazethapyr, and pethoxamid + sulfentrazone controlled redroot pigweed 82 to 98%; common ragweed 19 to 93%; common lambsquarters 49 to 84%; and green foxtail 47 to 92%. Weed interference with pethoxamid, *S*-metolachlor, and sulfentrazone reduced white bean seed yield 36%. Generally, there was no difference in the efficacy of pethoxamid compared to *S*-metolachlor-applied PRE. The spectrum of broadleaf weeds controlled was generally increased with the addition of halosulfuron, imazethapyr, or sulfentrazone to pethoxamid-applied PRE. Results indicate that there is potential for the tank-mix of pethoxamid plus halosulfuron, imazethapyr, or sulfentrazone for the control of annual grass and broadleaf weeds in white navy bean. Pethoxamid has superior activity than *S*-metolachlor against some Group 2, 5, and 9 herbicide-resistant weeds including palmer amaranth and common waterhemp and can be used as an additional option for managing these weeds in white bean.

Acknowledgments

Funding for this project was provided in part by the Ontario Bean Growers and the GFII program of the Agricultural Adaptation Council.

References

[1] S. Mailvaganam, "Area, yield, production and farm value of specified field crops, Ontario, 2012-2017 (imperial and metric units)," 2018, http://www.omafra.gov.on.ca/english/stats/crops/estimate_new.htm#metric.

[2] V. S. Malik, C. J. Swanton, and T. E. Michaels, "Interaction of white bean (*Phaseolus vulgaris*) cultivars, row spacing, and seeding density with annual weeds," *Weed Science*, vol. 41, pp. 62–68, 1993.

[3] D. Chikoye, S. F. Weise, and C. J. Swanton, "Influence of common ragweed (*Ambrosia artemisiifolia*) time of emergence and density on white bean (*Phaseolus vulgaris*)," *Weed Science*, vol. 43, no. 3, pp. 375–380, 1995.

[4] Z. Li, R. Van Acker, D. E. Robinson, N. Soltani, and P. H. Sikkema, "Managing weeds with herbicides in white bean in Canada: a review," *Canadian Journal of Plant Science*, vol. 97, pp. 755–766, 2017.

[5] N. Soltani, J. A. Dille, R. H. Gulden et al., "Potential yield loss in dry bean crops due to weeds in the United States and Canada," *Weed Technology*, vol. 32, no. 3, pp. 342–346, 2018.

[6] Ontario Ministry of Agriculture and Food (OMAFRA), *Guide to Weed Control*, OMAFRA, Toronto, ON, Canada, 2018.

[7] G. Skrzypczak, J. A. Pudełko, and W. Waniorek, "Assessment of the tank mixture of mesotrione and pethoxamid plus terbuthylazine efficacy for weed control in maize (*Zea mays* L.)," *Journal of Plant Protection*, vol. 47, no. 4, 2007.

[8] H. G. Schlosser, H. U. N. T. Barrie, and H. B Teicher, "Combination of pethoxamid and picloram," U.S. Patent No. 9,861,103, 2018.

[9] B. Hunt, J. Barrentine, T. Hayden et al., "Pethoxamid—a new herbicide for use in agronomic and horticultural crops," in *Proceedings of 55th annual meeting of Weed Science Society of America*, vol. 55, p. 281, Lexington, KY, USA, February 2015.

[10] D. L. Shaner, *Herbicide Handbook*, Weed Science Society of America, Champaign, IL, USA, 10th edition, 2014.

[11] W. W. Stroup, *Generalized Linear Mixed Models: Modern Concepts, Methods and Applications*, CRC Press, Taylor & Francis Group, Boca Raton, FL, USA, 2012.

[12] R. C. Yang, "Towards understanding and use of mixed-model analysis of agricultural experiments," *Canadian Journal of Plant Science*, vol. 90, no. 5, pp. 605–627, 2010.

[13] P. Rothery, "A cautionary note on data transformation: bias in back-transformed means," *Bird Study*, vol. 35, no. 3, pp. 219–221, 2009.

[14] R. N. Arnold, M. Murray, E. J. Gregory, and D. Smeal, "Weed control in pinto beans (*Phaseolus vulgaris*) with imazethapyr combinations," *Weed Technology*, vol. 7, no. 2, pp. 362–364, 1992.

[15] C. P. Urwin, R. G. Wilson, and D. A. Mortensen, "Response of dry edible bean (*Phaseolus vulgaris*) cultivars to four herbicides," *Weed Technology*, vol. 10, no. 3, pp. 512–518, 2017.

[16] N. Soltani, R. E. Nurse, and P. H. Sikkema, "Sensitivity of adzuki, kidney, small red Mexican, and white beans to pethoxamid," *Canadian Journal of Plant Science*, vol. 98, pp. 141–146, 2018.

[17] A. N. Taziar, N. Soltani, C. Shropshire et al., "Response of four dry bean market classes to pre-emergence applications of pyroxasulfone, sulfentrazone and pyroxasulfone plus sulfentrazone," *American Journal of Plant Sciences*, vol. 7, no. 8, pp. 1217–1225, 2016.

[18] A. N. Taziar, N. Soltani, C. Shropshire et al., "Sulfentrazone tank mix partners for weed control in white bean (*Phaseolus vulgaris* L.)," *Canadian Journal of Plant Science*, vol. 96, pp. 1037–1044, 2016.

15

Primary Selection and Secondary Diversification: Two Key Processes in the History of Olive Domestication

Bouchaib Khadari[1,2] **and Ahmed El Bakkali**[3]

[1]*AGAP, University Montpellier, CIRAD, INRA, Montpellier SupAgro, Montpellier, France*
[2]*Conservatoire Botanique National Méditerranéen de Porquerolles (CBNMed), UMR AGAP, Montpellier, France*
[3]*INRA, UR Amélioration des Plantes et Conservation des Ressources Phytogénétiques, Meknès, Morocco*

Correspondence should be addressed to Bouchaib Khadari; khadari@supagro.fr

Academic Editor: Faruk Toklu

Knowledge on the crop domestication process is important from a cultural and agricultural standpoint since it can shed light on the origin and history of human civilizations as well as the management of genetic resources, while offering guidance for modern breeding. The olive tree (*Olea europaea* ssp. *europaea*) is the most iconic of the old crop species of the Mediterranean Basin (MB). Primary domestication from wild olive probably occurred around 6000 BP in the Middle East. However, the question remains as to whether cultivated olive derived from a single domestication event in the Levant, followed by secondary diversification, or whether it was the result of independent domestication events. Here, we analyzed a comprehensive sample collected from 35 wild populations (722 individuals) and 410 cultivars from across the MB using nuclear and plastid DNA markers. Our genetic investigations argue in favor of a single primary domestication event in the eastern MB, followed by diffusion of the first domesticated olive and diversification in the central and western MB as key processes in the olive tree history.

1. Introduction

Understanding crop domestication and diversification processes is important to infer the origin of the crop and highlight the history of human civilizations. These investigations can be useful for genetic resource management while offering guidance for modern breeding. Olive (*Olea europaea* ssp. *europaea*) is considered to be the most iconic tree in Mediterranean areas. Oliviculture is one of the oldest cropping practices developed in these areas, and olive trees have therefore accompanied the emergence of early Mediterranean civilizations [1]. According to paleobotanical, archaeological, and genetic investigations, the olive tree may have persisted around the Mediterranean Basin (MB) as part of the natural plant community since the Late Tertiary [2]. However, despite the economic, cultural, ecological, and historical importance of the species, its origin and history have yet to be clearly documented. Clarifying the olive domestication and diversification process has therefore long been a focus of active scientific research [3].

According to Carrión et al. [4], during the Middle and Late Pleniglacial (59,000–11,500 yrs BP), *Olea europaea* had persisted in three thermophilous refugia located in the southern areas of the north MB, the southern Levant, and North Africa. Due to outcrossing of olive species, wild forms (oleasters, *O. europaea* ssp. *europaea* var. *sylvestris*) probably contributed to genetic diversity at the local scale, thus facilitating secondary diversification. It is usually considered that the center of primary olive domestication, from wild progenitors, began roughly around 6000 BP in the Middle East, near the border between Turkey and Syria [1, 5]. This assumption was supported by investigations based on chloroplast DNA polymorphism [5, 6], showing that more than 90% of olive cultivars across the Mediterranean Basin share the same eastern-like haplotype, therefore indicating an east-west human-mediated diffusion of cultivars in the MB. However, the scenario of a single primary domestication center has yet to be demonstrated. Based on paleobotanical and archaeological investigations, early exploitation and use of wild olive trees from the Near East to Spain have been documented since the Neolithic period [7, 8]. The findings of several studies support multiple origins of cultivars across the Mediterranean region [9–13], but it remains unclear whether this reflects secondary diversification or multiple independent primary domestication events [13–15]. Indeed, genetic patterns observed by Diez et al. [13] suggest the occurrence of a second separate olive domestication event in the central MB. The hypothesis of a second independent domestication in the central MB remains to be explored because cultivated olive diversification may also have occurred in this area and in the western MB as the result of local independent domestication [14]. Following the response letter of Diez and Gaut [15], the hypothesis of an independent domestication event in the central MB may ultimately be confirmed or disproven. Therefore, the question as to whether there was a single olive domestication center or multiple ones has yet to be answered.

In the present study, we investigated the history of olive trees through a comprehensive sampling of genuinely wild populations and domesticated forms from across the Mediterranean using nuclear and plastid marker analysis. Our results are discussed based on the question as to whether the current cultivated olive derived from a single domestication event in the Levant followed by secondary diversification or whether it is the result of independent domestication events.

2. Materials and Methods

2.1. Plant Material. A total of 1,132 distinct genotypes were analyzed in this study, including 410 cultivars from 15 countries and 722 wild olive samples from 35 populations throughout the Mediterranean area (Figure 1; Tables S1 and S2). All cultivars are maintained in the *ex situ* Worldwide Olive Germplasm Bank at the experimental station of Tassaout, INRA, Marrakech, Morocco [6, 16]. Wild populations were sampled in natural areas far from olive agroecosystems in order to minimize admixture with cultivated olives, while taking morphological traits that differ from those of cultivated olives into consideration, such as smaller fruits with less fleshy mesocarp [17]. Thirty of the 35 wild populations were previously described and analyzed [5, 18].

2.2. Molecular Analysis. Total DNA was extracted from 100 mg of fresh leaf tissue, as described by Khadari et al. [19]. DNA quality was checked on 1% agarose gel, and the concentration was estimated using spectrofluorometry (GENios Plus, TECAN, Grödig, Austria).

A set of 16 nuclear microsatellite loci was chosen and used for sample genotyping [20–23] (Table 1). PCR amplification and product separation using a 3130XL capillary sequencer (Applied Biosystems, Foster City, CA, USA) were conducted as described by El Bakkali et al. [16]. Plastid DNA was characterized using 37 polymorphic simple sequence repeat (SSR) loci and two cleaved amplified polymorphism sites (CAPS-XapI and CAPS-EcoRI), as described by Besnard et al. [24].

2.3. Data Analysis. We computed the following genetic diversity parameters for wild and cultivated olives separately and for each genetic group: the number of alleles (Na), expected heterozygosity (He), and observed heterozygosity (Ho) using the Excel Microsatellite Toolkit v3.1 [25]. The inbreeding coefficient (Fis) was calculated using the FSTAT program v2.9.3.2b [26], whereas the allelic richness (Ar; [27]) was estimated using the ADZE program [28]. The Mann–Whitney comparison test was used to evaluate the significance of the allelic richness differences.

To investigate the genetic structure pattern within olive samples (wild and cultivated), discriminant analysis of principal components (DAPC; [29]) with the ADEGENET 1.3.1 package [30] in the R environment was applied with *a priori* grouping assumptions based on previous studies [6, 13, 18]. Unlike the STRUCTURE program, the absence of any assumption about the underlying population genetics model, in particular concerning Hardy–Weinberg equilibrium or linkage equilibrium, is one of the main assets of DAPC [29]. Based on the model-based Bayesian clustering approach implemented in the STRUCTURE program [31] as described in previous studies on olive species [6, 13, 18], wild olive was found to be structured in two groups (named western-central and eastern Mediterranean wild), whereas cultivated olive was in three groups (called western, central, and eastern Mediterranean cultivated olive), with one group shared between wild and cultivated olives (eastern Mediterranean). Hence, we set an *a priori* group number of four in the DAPC method for the whole dataset.

Moreover, once the wild and cultivated olive genotypes were assigned to their a posteriori genetic groups, relationships among genotypes and genetic groups were analyzed by principal coordinate analysis (PCoA) based on the simple matching coefficient [32], as implemented in the DARWIN v. 6.0.11 program [33]. Pairwise genetic differentiation and significance (F_{ST}, [34]) between genetic groups, as revealed by membership assignation using the DAPC method, was estimated using 100,000 permutations with the GENEPOP program [35], and the unrooted F_{ST} was plotted using the POPTREE2 program [36], with 999 bootstrap replicates with the Neighbor-joining method.

3. Results

3.1. Genetic Diversity in Wild and Cultivated Olive. Based on the analysis of 1,132 genotypes using 16 SSR markers, we identified a total of 427 alleles with an average of 26.69 alleles per locus. The number of alleles observed in wild olive (420) was higher than that in cultivated olive (276). Similarly, the expected heterozygosity (He, diversity index) was greater in wild than in cultivated olive (Table 1).

A total of 33 plastid haplotypes were identified in both wild and cultivated olives. More plastid haplotypes belonging to E1 were identified compared to E2 and E3, i.e., 18, 10, and 4 haplotypes, respectively (Table 2). Otherwise, we revealed more maternal lineages in wild (32) than in cultivated olive (12). In fact, the highest proportion of plastid haplotypes in cultivated olive was observed for E1.1 (79.9%) followed by E1.2 (8.1%), whereas those belonging to E2 and

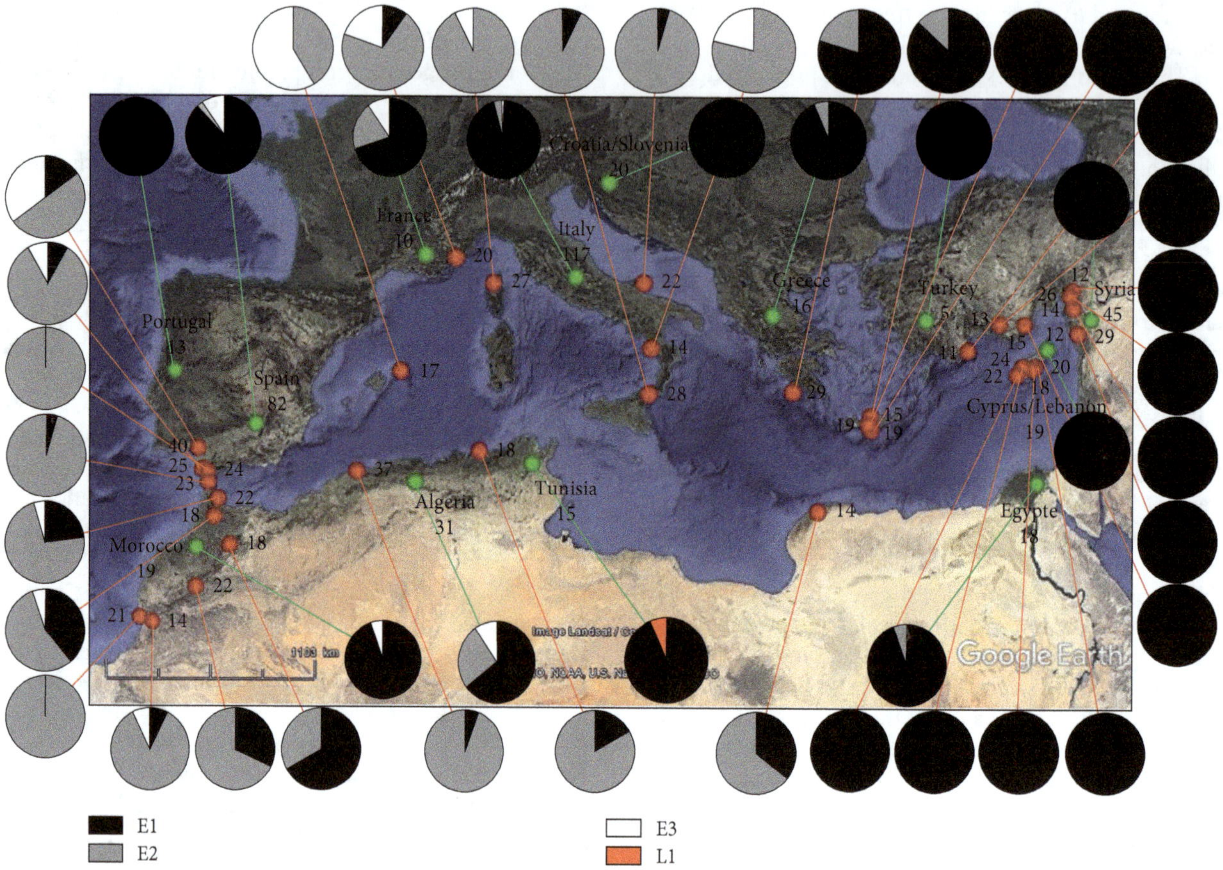

FIGURE 1: Map indicating the location, number of olive samples, and proportion of maternal lineages according to the geographic origin for wild and cultivated olives. Cultivars are shown in green and wild olives in red.

TABLE 1: Genetic parameters of the 16 SSR markers used in this study for wild and cultivated olives.

	Locus	Wild (total of 722)						Cultivated olive (total of 410)					
		Size (bp)	Na	Ho	He	Fis	Ar	Size (bp)	Na	Ho	He	Fis	Ar
1	DCA01[1]	205–275	35	0.47	0.72	0.34^{***}	31.9	205–275	21	0.71	0.62	−0.15ns	20.9
2	DCA03[1]	222–277	26	0.72	0.81	0.11^{***}	25.1	228–265	14	0.91	0.85	−0.06ns	14
3	DCA04[1]	117–196	35	0.72	0.92	0.22^{***}	33.6	117–200	33	0.65	0.85	0.23^{***}	33
4	DCA05[1]	187–245	23	0.82	0.88	0.08^{***}	20.4	192–214	12	0.5	0.5	0.01ns	12
5	DCA08[1]	118–185	32	0.91	0.93	0.03^{***}	29.3	124–167	21	0.92	0.83	-0.11^{*}	21
6	DCA09[1]	163–225	29	0.87	0.93	0.06^{***}	27.5	163–219	25	0.95	0.89	−0.07ns	25
7	DCA11[1]	124–209	40	0.74	0.82	0.11^{***}	37.2	126–201	23	0.88	0.83	-0.06^{*}	23
8	DCA14[1]	170–194	25	0.67	0.9	0.25^{***}	24.2	170–194	15	0.74	0.71	-0.04^{*}	15
9	DCA15[1]	243–266	8	0.64	0.74	0.13^{***}	7.81	243–266	6	0.7	0.64	−0.09ns	5.99
10	DCA16[1]	122–235	48	0.57	0.95	0.40^{***}	45.3	124–229	34	0.95	0.87	−0.08ns	33.9
11	DCA18[1]	155–205	26	0.88	0.91	0.03^{***}	24.7	157–191	17	0.91	0.83	−0.09ns	17
12	EMO90[2]	183–216	16	0.4	0.63	0.37^{***}	15.4	185–212	9	0.71	0.67	−0.05ns	9
13	GAPU59[3]	206–251	22	0.77	0.84	0.08^{*}	19.4	208–240	11	0.59	0.62	0.04^{***}	11
14	GAPU71A[3]	207–270	25	0.66	0.72	0.09^{***}	23.2	207–252	14	0.54	0.46	-0.17^{**}	14
15	GAPU 71B[3]	132–168	13	0.65	0.69	0.07^{***}	12.7	136–166	9	0.9	0.81	-0.12^{**}	8.99
16	UDO99-36[4]	134–170	17	0.76	0.85	0.11^{***}	15.7	140–168	12	0.69	0.71	0.023ns	12
	Mean		26.3	0.7	0.83	0.15^{***}	24.6		17.3	0.77	0.73	-0.05^{***}	17.2
	Total		420						276				

[1]Sefc et al. [20], [2]De La Rosa et al. [21], [3]Carriero et al. [22], [4]Cipriani et al. [23]. Significance level at $p \leq 0.05^{*}$, 0.01^{**}, and 0.001^{***}. ns: nonsignificant.

TABLE 2: Number and proportion of wild and cultivated genotypes assigned to each a posteriori genetic group resulting from DAPC and proportion of major maternal lineages for each group.

Genetic groups	Assigned genotypes* (%)	Maternal lineage (%)			
		E1[1]	E2[2]	E3[3]	L1.1
$Cultivated_{west}$	50 (12.2)	50 (100.0)			
$Cultivated_{center}$	138 (33.7)	129 (93.5)	3 (2.2)	6 (4.3)	
$Cultivated_{east}$	59 (14.4)	59 (100.0)			
$Cultivated_{admixed}$	150 (36.6)	129 (86.0)	14 (9.3)	6 (4.0)	1 (0.7)
Total (cultivated olive)	410	377 (91.95)	18 (4.40)	14 (3.40)	1 (0.24)
$Wild_{west/center}$	398 (55.1)	55 (13.8)	307 (77.1)	36 (9.0)	
$Wild_{east}$	265 (36.7)	265 (100.0)			
$Wild_{admixed}$	49 (6.8)	26 (53.1)	22 (44.9)	1 (2.0)	
Total (wild olive)	722	346 (47.92)	338 (46.82)	38 (5.26)	
Total	1132	723 (63.87)	356 (31.45)	52 (4.59)	1 (0.09)

*Under the assignation probability $p > 80\%$, [1]E1: E1.1, E1.2, E1.3, E1.4, E1.5, E1.6, E1.7, E1.8, E1.9, E1.10, E1.11, E1.12, E1.13, E1.14, E1.15, E1.20, E1.21, E1.27, [2]E2: E2.1, E2.2, E2.3, E2.4, E2.5, E2.6, E2.7, E2.8, E2.12, E2.14, [3]E3: E3.1, E3.2, E3.3, E3.5.

E3 (total of 6 haplotypes) were detected only in 32 cultivars (7.8%; Table S2).

3.2. Genetic Clustering. The genetic structure of Mediterranean olive was investigated using discriminant analysis of principal components (DAPC). The "find.clusters" function was used to determine the number of clusters maximizing the variation between clusters [30]. To avoid the loss of information, the function was performed with 200 principal components, accounting for more than 98% of the variance (Figure S1(a)). The Bayesian information criterion (BIC) was used to identify the optimal number of clusters, i.e., 11 clusters (Figure S1(b)). Based on these 11 clusters as a first analysis, DAPC clustering was represented according to the origin of olives classified in four *a priori* groups: western-central Mediterranean wild olive, eastern Mediterranean wild and cultivated olives, western cultivated olive, and central cultivated olive (Figure S1(c)). Western cultivated olive showed narrow genetic diversity (cluster 7), whereas those from the central Mediterranean Basin displayed high diversity included in 3 clusters (clusters 1, 3, and 6). Similarly, eastern and western-central wild olive displayed high diversity, i.e., 3 and 4 clusters, respectively (Figures S1(c) and S1(d)). Pairwise F_{ST} values among the 11 predefined clusters resulting from DAPC ranged from 0.017 (cluster 5-cluster 8) to 0.129 (cluster 4-cluster 7) (Table S3).

Based on the assignation membership probability resulting from DAPC at $p \geq 0.8$, five groups could be identified: (i) eastern Mediterranean wild (referred as $Wild_{east}$), (ii) eastern cultivated olive ($Cultivated_{east}$), (iii) western and central Mediterranean wild olive ($Wild_{west\text{-}center}$), (iv) western Mediterranean cultivated olive ($Cultivated_{west}$), and (v) central Mediterranean cultivated olive ($Cultivated_{center}$). Although they are belonging to the same pool (Figures 2 and 3), $Wild_{east}$ and $Cultivated_{east}$ were considered as two distinct groups to describe the relationships between wild and cultivated olives in the eastern MB. Otherwise, at the $p < 80\%$ assignation level, wild and cultivated olive showed admixture: $Wild_{admixed}$ and $Cultivated_{admixed}$, respectively (Figure 2; Tables 2 and S4). For admixed wild and cultivated forms, we noted the occurrence of the three plastid lineages with a high proportion of E1 for cultivated olive (86.0%), whereas for admixed wild olive, close proportions of E1 and E2 were observed (53.1% and 44.9%, respectively, Table 2). When considering the *a priori* groups, more wild admixed genotypes were noted in the western-central part of MB. Similarly, more admixed cultivars were observed in the western-central MB area compared to the east (Table S5).

To investigate genetic relationships between the five groups defined above, principal coordinate analysis (PCoA) was performed (Figure S2). Most of the variation (13.06%) was explained by the first two axes. For both wild and cultivated olives, the first axis corresponded to the east-west spatial distribution at the MB scale, where both wild olive groups ($Wild_{west\text{-}center}$ and $Wild_{east}$) were genetically distinct. The second axis separated $Cultivated_{west}$ and $Cultivated_{center}$ olives from $Wild_{east}$ and $Cultivated_{east}$ olives. The latter were clustered as one pool. Admixed genotypes for both $Wild_{admixed}$ and $Cultivated_{admixed}$ were plotted midway between the five genetic groups (Figure S2).

3.3. Genetic Variation and Relationships among Genetic Groups. The mean heterozygosity (Ho = 0.701 observed) noted for wild olive was less than expected (He = 0.826) based on the Hardy–Weinberg equilibrium findings (Fis = 0.151; $p < 0.001$; Table 3), indicating a deficit of heterozygotes, as noted for wild genetic groups ($Wild_{west\text{-}center}$ and $Wild_{east}$). These results may be explained by the subdivision of local populations into isolated and differentiated units (Wahlund effect), as revealed by the 11 predefined DAPC clusters (Figures S1(c) and S1(d)).

Allelic richness (Ar) was estimated and revealed a highly significant difference between wild olive and cultivars (24.58 vs 17.23; Mann–Whitney test, $p < 0.001$; Table 3). Otherwise, a highly significant difference was observed between $Wild_{east}$ and $Cultivated_{east}$ (12.95 vs 8.01), but not between $Wild_{west\text{-}center}$ and $Wild_{east}$ (10.67 and 12.95, respectively) or between $Cultivated_{center}$ and $Cultivated_{east}$. However, within cultivated olive, Ar for $Cultivated_{west}$ was significantly lower than for $Cultivated_{center}$ and $Cultivated_{east}$ (4.19, 7.52, and 8.01, respectively). When focusing on the maternal lineage, more maternal lineages belonging to E2 and E3 were revealed in

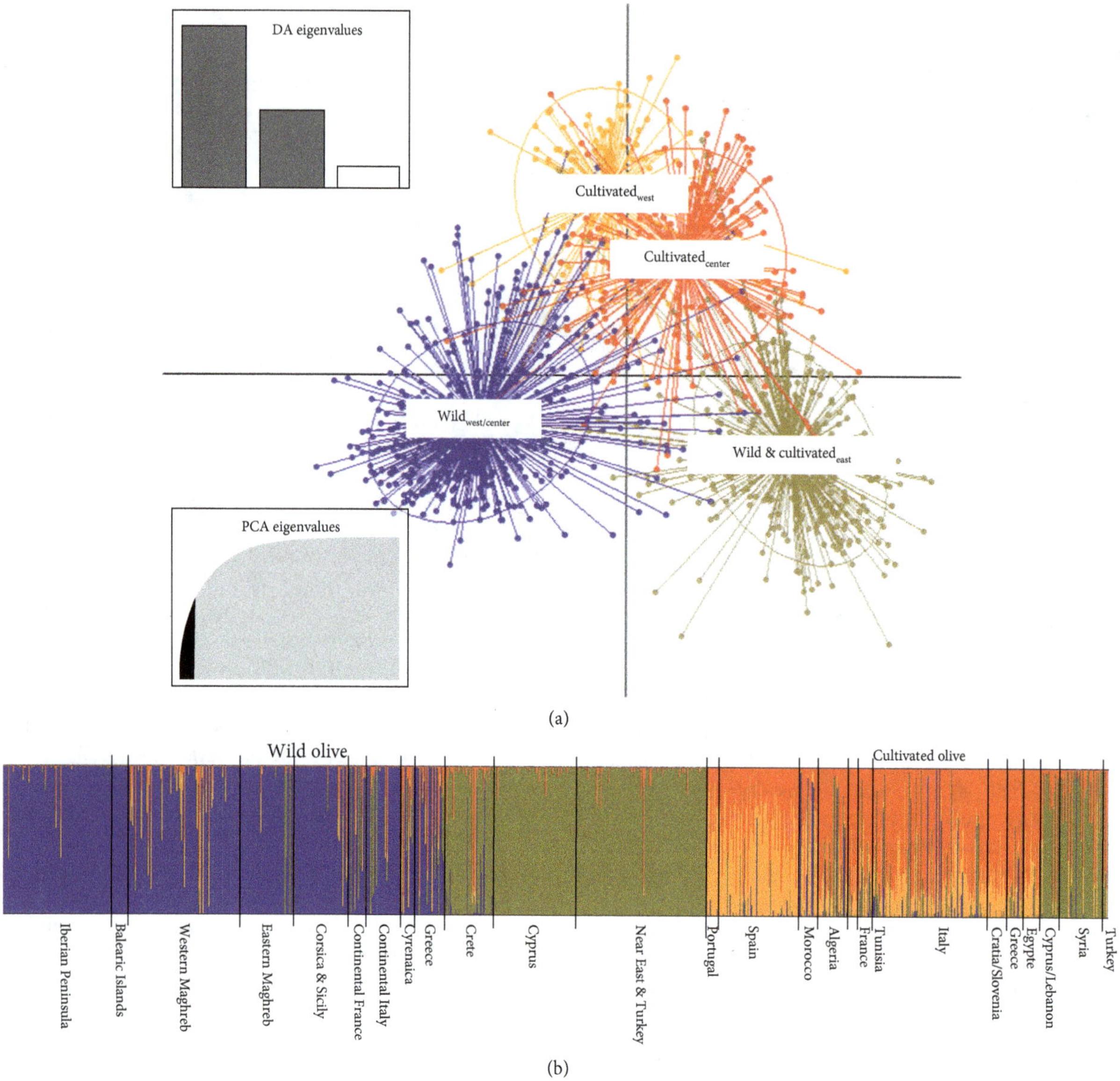

FIGURE 2: Scatterplot from a discriminant analysis of principal components (DAPC) (a) using the four *a priori* groups. The sample origins are labeled within their 95% inertia ellipses, and individuals are connected to the corresponding group centroids. The number of PCs retained and discriminant functions are shown. Bar plot from DAPC highlights the membership of individuals in different groups according to geographical zones for both wild and cultivated olives (b).

Cultivated$_{center}$ than in Cultivated$_{east}$ and Cultivated$_{west}$ olives (Table 2; Figure 3). Moreover, for cultivars identified as admixed on the basis of DAPC (membership assignation <0.8), E2 and E3 haplotypes were found in higher proportion than for other cultivated groups (Table S5).

The genetic differentiation (F_{ST}) values were significant between all pairs of the five groups (Wild$_{east}$ and Cultivated$_{east}$ treated separately, $p < 0.001$; Table 4). The mean F_{ST} was 0.105 ($p < 0.0001$) for the five groups, indicating that 10% of the total genetic variation resulted from genetic differentiation between groups. The pairwise F_{ST} ranged from 0.036 to 0.183 (Table 4). The highest values were observed between Wild$_{west\text{-}center}$ and all the other groups, whereas the lowest values were revealed between Wild$_{east}$ and Cultivated$_{east}$ groups. Relationships between the five groups showed a clear distinction between three main groups: (i) Wild$_{west\text{-}center}$, (ii) Wild$_{east}$ and Cultivated$_{east}$, and (iii) Cultivated$_{west}$ and Cultivated$_{center}$, as supported by the high bootstrap values (Figure 3). The group Wild$_{west\text{-}center}$ was highly separated from the others based on both nuclear and plastid polymorphism, with the highest proportion of maternal lineages belonging to E2 and E3 haplotypes (86.1%; Table 2; Figure 3).

4. Discussion

Over the last two decades, substantial paleobotanical, archaeological, historical, and molecular data have been accumulated on olive species and the history of its domestication [1, 4, 5, 7, 8, 11–13]. A global overview was

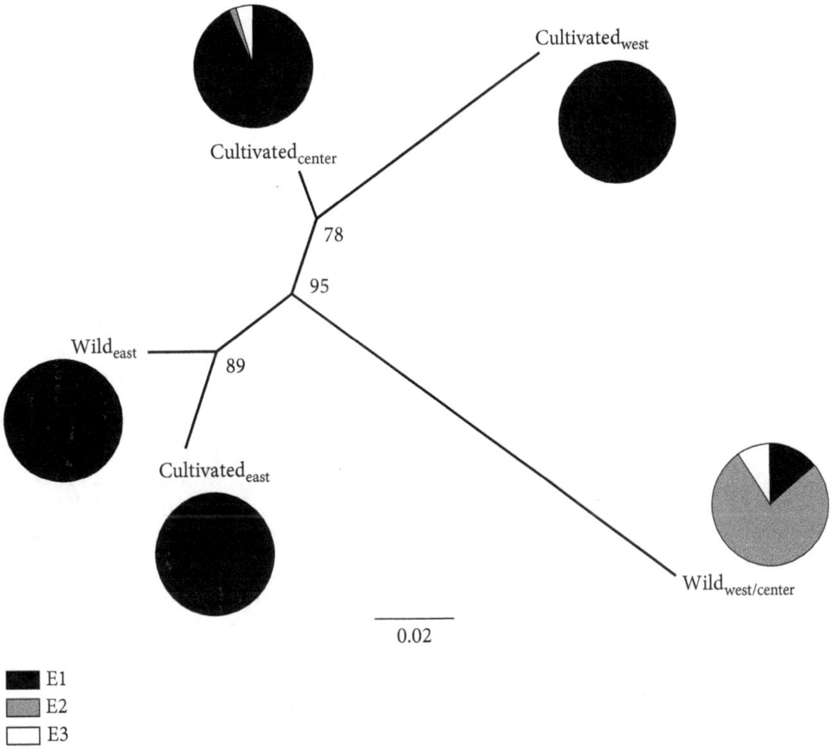

FIGURE 3: Unrooted neighbor-joining dendrogram based on the genetic differentiation index (F_{ST}), showing relationships among the five genetic groups as identified by DAPC and the proportion of the three maternal lineages (E1, E2, and E3) within each group.

TABLE 3: Summary of genetic parameters for different groups.

Groups	*n*	*A*	*A*%	Ho	He	Fis	Ar*
Wild olive	722	420	98.3	0.701	0.826	0.151***	24.58a[1]
Cultivated olive	410	276	64.6	0.765	0.731	−0.047***	17.23b[1]
Wild$_{west/center}$	398	331	77.5	0.664	0.737	0.094***	10.67ac[2]
Wild$_{east}$	265	362	84.7	0.737	0.806	0.084***	12.95a[2]
Cultivated$_{west}$	50	79	18.5	0.770	0.589	−0.310***	4.19d[2]
Cultivated$_{center}$	138	195	45.6	0.751	0.704	−0.067***	7.52b[2]
Cultivated$_{east}$	59	160	37.4	0.752	0.712	−0.057***	8.01bc[2]

n, size of the group; *A*, number of alleles; *A*%, percentage of alleles with respect to the total number of alleles observed within wild and cultivated samples (427 alleles); Ho, observed heterozygosity; He, unbiased expected heterozygosity; Fis, inbreeding coefficient; Ar, allelic richness; [1]estimated based on standardized sample size of $G = 410$; [2]estimated based on standardized sample size of $G = 50$. *Letters (a, b, c, and d) represent significance groupings after a Mann–Whitney pairwise comparison test ($p < 0.05$).

founded on the basis of several centers of primary selection across the MB [10–12]. However, the question remains unclear as to whether cultivated olives derived from a single primary domestication center followed by secondary diversification events or whether they are the result of independent primary selection events. A scenario of at least two independent primary selection centers in the eastern and central Mediterranean was proposed by Diez et al. [13]. Investigations on wild olive in the eastern Mediterranean were, however, limited to few sampled populations that were likely feral, as assumed by Diez et al. [13]. The lack of genuinely eastern wild populations has drastically limited the possibility of testing alternative complex domestication scenarios, as pointed out by Besnard and Robio de Casas [14] and Diez and Gaut [15]. Hence, investigating a comprehensive sample of wild olives throughout the MB could bring insight to help solve questions related to primary domestication and secondary diversification centers. Moreover, contrary to the findings of Diez et al. [13], our eastern wild populations were clearly distinct from the western-central wild olive populations, thus indicating their genuine status.

Olive tree history is complex, as previously highlighted by several studies (see review in Besnard et al. [3]). Instead of multiple primary domestication centers, we argue in favor of a single primary domestication in the Levant, followed by human-mediated diffusion of the first domesticated forms and admixtures with wild olives in the central and western Mediterranean Basin. However, we cannot exclude the occurrence of minor domestication centers in western and central parts of the MB, as some varieties have been found to

Table 4: Pairwise genetic differentiation F_{ST} values between the five groups as revealed by DAPC method for wild and cultivated olives.

	$\text{Cultivated}_{\text{center}}$	$\text{Cultivated}_{\text{west}}$	$\text{Cultivated}_{\text{east}}$	$\text{Wild}_{\text{east}}$
$\text{Cultivated}_{\text{west}}$	0.075***			
$\text{Cultivated}_{\text{east}}$	0.052***	0.140***		
$\text{Wild}_{\text{east}}$	0.060***	0.142***	0.036***	
$\text{Wild}_{\text{west-center}}$	**0.168*****	**0.176*****	**0.183*****	0.144***

Pairwise significance after sequential Bonferroni corrections: ***, significance at $p < 0.001$. Numbers in bold are the highest values observed and underlined are the lowest.

harbor maternal E2 or E3 lineages specific to local genetic resources, indicating their ancient local selection heritage (Figure 1, Table S4). Moreover, morphometric olive stone and charcoal analyses have revealed the use of wild olive before the Neolithic period, suggesting local domestication could have occurred in the western MB area [7]. Here, by investigating current varieties using both nuclear and plastid markers, we obtained evidence of primary selection and secondary diversification as two key processes in the history of olive domestication based on the following arguments. First, we used DAPC and identified a single group including both eastern wild olive ($\text{Wild}_{\text{east}}$) and eastern cultivars ($\text{Cultivated}_{\text{east}}$), thus indicating direct selection from wild olive populations, as suggested by Gurbuz-Veral et al. [37]. Second, most cultivated olives have an eastern-like maternal haplotype as a signature of the diffusion of the first domesticated olives from the eastern to western Mediterranean Basin [5]. Note that the above two arguments are supported by the genetic differentiation index (F_{ST}) between different groups, including wild and cultivated olive trees (Figure 3 and Table 4). Third, the allelic richness revealed highly significant differences between $\text{Wild}_{\text{east}}$ and $\text{Cultivated}_{\text{east}}$ olives. Contrary to other perennial fruit species such as apple [38], a substantial reduction in allelic diversity was observed between domesticated and wild olives across the MB (up to 30%), especially from the eastern MB (up to 38.1%; Table 3). This finding is in line with the selection pattern during the domestication process, as reviewed by Gaut et al. [39] and Besnard et al. [3] and references therein. Fourth, the genetic pattern of the $\text{Cultivated}_{\text{west}}$ and $\text{Cultivated}_{\text{center}}$ groups indicated a diversification process based on selection from crosses between the first domesticated olive forms and local olives, thus supporting the assumption of human-mediated diffusion of cultivars. Indeed, among varieties from the central MB, 35% (81 varieties) were admixed with limited gene flow from western-central wild populations (Figure 2; Table S4) and displaying the three maternal lineages (Table S5). Among varieties from the central MB, we found 11% (24) harboring maternal lineages E2 and E3. Moreover, as reported by Belaj et al. [40] and Klepo et al. [41], some central Mediterranean varieties retain wild-like phenotypic characteristics, such as low endocarp weight and a smooth endocarp surface. These findings suggest a second center of domestication, as reported by Diez et al. [13], but evidence to back this assumption has yet to be documented [3].

We argue here in favor of a diversification process occurring in the western and central MB. Fifth, Diez et al. [13] found that most first-degree relationships were from the same genetic group (i.e., western cultivated olive; 96.3%) in which two cultivars from Spain (i.e., Gordal Sevillana and Lechin de Granada) had more than 60 first-degree relationships. Varieties harboring E2 and E3 from the western MB were found to be closely related within the western cultivated group such as Lechin de Sevilla with E2.3 maternal lineage (Table S2) and five first-degree relationships [13]. Regardless of the maternal lineages, the presence of highly related varieties indicated diversification based on crosses between cultivated olives as a key olive domestication process in the central and western MB.

5. Conclusion

Beyond a single primary olive tree domestication event, our investigation underlines the importance of admixtures within cultivated olive groups from the central and western Mediterranean Basin. Clarifying the evolutionary processes responsible for these groups will help gain important insight into accurately identify the genes under selection. This will also help to design methods for sampling of Mediterranean olive germplasm, including wild olives suitable for genome-wide association studies and genomic selection under the impact of climate change and within the sustainable oliviculture setting. Moreover, identifying olive diversity hotspots in the MB could also help to develop cost-effective diversity-prioritized approaches for *in situ* olive genetic resource conservation and management.

Disclosure

First results from the present study have been presented as an oral communication within the Fourth International American Moroccan, Agriculture Science Conference "AMAS Conference IV" held in May 9–11, 2018, at Agropolis, Meknès, Morocco (see http://www.amas-conference.org/wp-content/uploads/2018/05/AMAS-IV-FinalProgramBooklet-5-3-18-RegularPrint.pdf).

Authors' Contributions

Bouchaib Khadari and Ahmed El Bakkali contributed equally to this work.

Acknowledgments

We thank A. Al Ibrahim, G. Besnard, M. Gurbuz, H.

Haouane, A. Moukhli, and J. F. Terral for providing olive samples. This work was conducted at UMR AGAP. We thank L. Essalouh, S. Santoni, and Ch. Tollon for laboratory assistance. This research was supported by the project OliveMed/Agropolis Fondation no. 1202-066 through the *Investissements d'avenir*/Labex Agro ANR-10-Labex-0001-01 managed by the French National Research Agency (ANR) and by the BeFOre project "Bioresources for Oliviculture" 2015–2019, H2020-MSCA-RISE-Marie Skłodowska-Curie Research and Innovation Staff Exchange, Grant Agreement no. 645595.

Supplementary Materials

Table S1: list of the 35 wild olive populations. The sampling locations, the number of populations, individuals per location (size), and the GPS coordinates are given. Table S2: list of the 410 Mediterranean olive cultivars analyzed in the present study, along with their origins and maternal lineages. Table S3: pairwise genetic differentiation (F_{ST}) among the 11 subclusters, as identified by DAPC using the "find.clusters" function. Table S4: number of wild and cultivated olives per region of origin and the number and proportion of individuals assigned to each group based on the DAPC findings with a membership probability of 0.8. Table S5: proportion of maternal lineages according to the *a priori* grouping clusters for both wild and cultivated olives identified as admixed genotypes by DAPC with a membership assignation of $p < 0.8$. Figure S1: discriminant analysis of principal component (DAPC) results. Cumulative variance explained by the principal component analysis (PCA) relative to the number of principal components (PCs) retained in the analysis (a). Selection of the optimal number of clusters in the DAPC using the lowest Bayesian information criterion (BIC; (b)). Comparison of clustering performed by DAPC ($K = 11$) and the *a priori* wild and cultivated olive groups (c). Squares represent the number of individuals in each pairwise comparison. Scatterplot from a DAPC of olive genotypes showing the relationships between the 11 identified clusters (d). Figure S2: principal coordinate analysis (PCoA) based on the simple matching coefficient showing the relationships among cultivated and wild olive genotypes according to groups resulting from DAPC. (*Supplementary Materials*)

References

[1] D. Kaniewski, E. Van Campo, T. Boiy, J. F. Terral, B. Khadari, and G. Besnard, "Primary domestication and early uses of the emblematic olive tree: palaeobotanical, historical and molecular evidences from the middle east," *Biological Reviews*, vol. 87, no. 4, pp. 885–899, 2012.

[2] E. Palamarev, "Paleobotanical evidences of the tertiary history and origin of the Mediterranean sclerophylldendroflora," *Plant Systematics and Evolution*, vol. 162, no. 1–4, pp. 93–107, 1989.

[3] G. Besnard, J. F. Teral, and A. Cornille, "on the origins and domestication of the olive : a review and perspectives," *Annals of Botany*, vol. 121, no. 3, pp. 385–403, 2018.

[4] Y. Carrión, M. Ntinou, and E. Badal, "*Olea europaea* L. in the North Mediterranean Basin during the Pleniglacial and the Early-Middle Holocene," *Quaternary Science Reviews*, vol. 29, no. 7-8, pp. 952–968, 2010.

[5] G. Besnard, B. Khadari, M. Navascués et al., "The complex history of the olive tree: from late quaternary diversification of Mediterranean lineages to primary domestication in the northern Levant," *Proceedings of the Royal Society B: Biological Sciences*, vol. 280, no. 1756, article 20122833, 2013.

[6] H. Haouane, A. El Bakkali, A. Moukhli et al., "Genetic structure and core collection of the World Olive Germplasm Bank of Marrakech: towards the optimised management and use of Mediterranean olive genetic resources," *Genetica*, vol. 139, no. 9, pp. 1083–1094, 2011.

[7] J. F. Terral, N. Alonso, R. B. Capdevila et al., "Historical biogeography of olive domestication (Olea europaea L.) as revealed by geometrical morphometry applied to biological and archaeological material," *Journal of Biogeography*, vol. 31, no. 1, pp. 63–77, 2004.

[8] D. Zohary, M. Hopf, and E. Weiss, *Domestication of Plants in the Old World: The Origin and Spread of Cultivated Plants in Southwest Asia, Europe, and the Mediterranean Basin*, Oxford University Press, Oxford, UK, 2012.

[9] J. F. Terral, *La domestication de l'olivier (Olea europaea L.) en Méditerranée nord-occidentale: approche morphométrique et implications paléoclimatiques*, Ph.D. thesis, Université Montpellier II, Montpellier, France, 1997.

[10] G. Besnard, P. Baradat, C. Breton, B. Khadari, and A. Berville, "Olive domestication from structure of oleasters and cultivars using nuclear RAPDs and mitochondrial RFLPs," *Genetics Selection Evolution*, vol. 33, pp. S251–S268, 2001.

[11] L. Baldoni, N. Tosti, C. Ricciolini et al., "Genetic structure of wild and cultivated olives in the central Mediterranean basin," *Annals of Botany*, vol. 98, no. 5, pp. 935–942, 2006.

[12] C. Breton, M. Tersac, and A. Berville, "Genetic diversity and gene flow between the wild olive (oleaster, Oleaeuropaea L.) and the olive: several Plio-Pleistocene refuge zones in the Mediterranean basin suggested by simple sequence repeats analysis," *Journal of Biogeography*, vol. 33, no. 11, pp. 1916–1928, 2006.

[13] C. M. Diez, I. Trujillo, N. Martinez-Urdiroz et al., "Olive domestication and diversification in the Mediterranean Basin," *New Phytologist*, vol. 206, no. 1, pp. 436–447, 2015.

[14] G. Besnard and R. Rubio de Casas, "Single vs multiple independent olive domestications: the jury is (still) out," *New Phytologist*, vol. 209, no. 2, pp. 466–470, 2016.

[15] C. M. Diez and B. S. Gaut, "The jury may be out, but it is important that it deliberates: a response to Besnard and Rubio de Casas about olive domestication," *New Phytologist*, vol. 209, no. 2, pp. 471–473, 2016.

[16] A. El Bakkali, H. Haouane, A. Moukhli, E. Costes, P. Van Damme, and B. Khadari, "Construction of core collections suitable for association mapping to optimize use of Mediterranean olive (Olea europaea L.) genetic resources," *PLoS One*, vol. 8, no. 5, Article ID e61265, 2013.

[17] H. Hannachi, C. Breton, M. Msallem, S. Ben El Hadj, M. El Gazzah, and A. Bervillé, "Differences between native and introduced olive cultivars as revealed by morphology of drupes, oil composition and SSR Polymorphisms: a case study in Tunisia," *Scientia Horticultural*, vol. 116, no. 3, pp. 280–290, 2008.

[18] G. Besnard, A. El Bakkali, H. Haouane, D. Baali-Cherif, A. Moukhli, and B. Khadari, "Population genetics of Mediterranean and Saharan olives: geographic patterns of differentiation and evidence for early generations of admixture," *Annals of Botany*, vol. 112, no. 7, pp. 1293–1302, 2013.

[19] B. Khadari, J. Charafi, A. Moukhli, and M. Ater, "Substantial genetic diversity in cultivated Moroccan olive despite a single major cultivar: a paradoxical situation evidenced by the use of SSR loci," *Tree Genetic and Genomes*, vol. 4, no. 2, pp. 213–221, 2008.

[20] K. M. Sefc, M. S. Lopes, D. Mendonca et al., "Identification of microsatellite loci in olive (Olea europaea) and their characterization in Italian and Iberian olive trees," *Molecular Ecology*, vol. 9, no. 8, pp. 1171–1173, 2000.

[21] R. De la Rosa, C. M. James, and K. R. Tobutt, "Isolation and characterization of polymorphic microsatellites in olive (Olea europaea L.) and their transferability to other genera in the Oleaceae," *Molecular Ecology Notes*, vol. 2, no. 3, pp. 265–267, 2002.

[22] F. Carriero, G. Fontanazza, F. Cellini, and G. Giorio, "Identification of simple sequence repeats (SSRs) in olive (Olea europaea L.)," *Theoretical and Applied Genetics*, vol. 104, no. 2-3, pp. 301–307, 2002.

[23] G. Cipriani, M. T. Marrazzo, R. Marconi, A. Cimato, and R. Testolin, "Microsatellite markers isolated in olive (Olea europaea L.) are suitable for individual fingerprinting and reveal polymorphism within ancient cultivars," *Theoretical and Applied Genetics*, vol. 104, no. 2-3, pp. 223–228, 2002.

[24] G. Besnard, P. Hernandez, B. Khadari, G. Dorado, and V. Savolainen, "Genomic profiling of plastid DNA variation in the Mediterranean olive tree," *BMC Plant Biology*, vol. 11, p. 80, 2011.

[25] S. D. E. Park, *Trypanotolerance in West African cattle and the population genetic effects of selection*, Ph.D. thesis, University of Dublin, Dublin, Ireland, 2001.

[26] J. Goudet, "FSTAT (Version 1.2): a computer program to calculate F-statistics," *Journal of Heredity*, vol. 86, no. 6, pp. 485-486, 1995.

[27] R. J. Petit, A. El Mousadik, and O. Pons, "Identifying populations for conservation on the basis of genetic markers," *Conservation Biology*, vol. 12, no. 4, pp. 844–855, 1998.

[28] Z. A. Szpiech, M. Jakobsson, and N. A. Rosenberg, "ADZE: a rarefaction approach for counting alleles private to combinations of populations," *Bioinformatics*, vol. 24, no. 21, pp. 2498–2504, 2008.

[29] T. Jombart, S. Devillard, and F. Balloux, "Discriminant analysis of principal components: a new method for the analysis of genetically structured populations," *BMC Genetics*, vol. 11, no. 1, p. 94, 2010.

[30] T. Jombart, "adegenet: a R package for the multivariate analysis of genetic markers," *Bioinformatics*, vol. 24, no. 11, pp. 1403–1405, 2008.

[31] J. K. Pritchard, M. Stephens, and P. Donnelly, "Inference of population structure from multilocus genotype data," *Genetics*, vol. 155, no. 2, pp. 945–959, 2000.

[32] R. R. Sokal and C. D. Michener, *A Statistical Method for Evaluating Systematic Relationships*, vol. 38 of University of Kansas Science Bulletin, University of Kansas, Lawrence, KS, USA, 1958.

[33] X. Perrier and J. P. Jacquemoud-Collet, "DARwin software," 2006, http://darwin.cirad.fr/darwin.

[34] B. S. Weir and C. C. Cockerham, "Estimating F-statistics for the analysis of population structure," *Evolution*, vol. 38, no. 6, pp. 1358–1370, 1984.

[35] M. Raymond and F. Rousset, "GENEPOP (version 1.2): population genetics software for exact tests and ecumenicism," *Journal of Heredity*, vol. 86, no. 3, pp. 248-249, 1995.

[36] N. Takezaki, M. Nei, and K. Tamura, "POPTREE2: software for constructing population trees from allele frequency data and computing other population statistics with Windows-interface," *Molecular Biology and Evolution*, vol. 27, no. 4, pp. 747–752, 2010.

[37] M. Gurbuz-Veral, A. El Bakkali, L. Essalouh et al., "New insights on olive domestication in Turkey," *Acta Horticulturae*, vol. 1199, pp. 15–20, 2018.

[38] A. Cornille, P. Gladieux, M. J. M. Smulders et al., "New insight into the history of domesticated apple: secondary contribution of the European wild apple to the genome of cultivated varieties," *PLoS Genetics*, vol. 8, no. 5, article e1002703, 2012.

[39] B. S. Gaut, C. M. Díez, and P. L. Morrell, "Genomics and the contrasting dynamics of annual and perennial domestication," *Trends in Genetics*, vol. 31, no. 12, pp. 709–719, 2015.

[40] A. Belaj, L. Leon, Z. Satovic, and R. de la Rosa, "Variability of wild olives (Olea europaea subsp. europaea var. sylvestris) analyzed by agro-morphological traits and SSR markers," *Scientia Horticulturae*, vol. 129, no. 4, pp. 561–569, 2011.

[41] T. Klepo, R. De la Rosa, Z. Satovic, L. Leon, and A. Belaj, "Utility of wild germplasm in olive breeding," *Scientia Horticulturae*, vol. 152, pp. 92–101, 2013.

16

Selected Chemical Properties of Soybean Rhizosphere Soil as Influenced by Cropping Systems, *Rhizobium* Inoculation, and the Supply of Phosphorus and Potassium after Two Consecutive Cropping Seasons

Daniel Nyoki[1,2] and Patrick A. Ndakidemi[1,2]

[1]*School of Life Science and Bio-Engineering, The Nelson Mandela African Institution of Science and Technology, P.O. Box 447, Arusha, Tanzania*
[2]*Centre for Research, Agricultural Advancement, Teaching Excellence and Sustainability (CREATES) in Food and Nutrition Security, The Nelson Mandela African Institution of Science and Technology, Arusha, Tanzania*

Correspondence should be addressed to Daniel Nyoki; dnyoki@yahoo.com

Academic Editor: Yuanhu Xuan

The field experiment was carried out in northern Tanzania to assess the effects of intercropping systems, *Rhizobium* inoculation, and fertilization with P and K on chemical properties of soybean rhizosphere soil. The experiment was laid out in split-split plot design with 2 × 4 × 7 factorial arrangement replicated thrice. The main plots had two inoculation treatments and the subplots were comprised of four cropping systems which were sole maize, sole soybean, and two intercropping at different soybean spacing (75 × 20 and 75 × 40 cm). The fertilizer levels (kg/ha) control (0 kg/ha); 20 K; 40 K; 26 P; 52 P; 26 P + 20 K; and 52 P + 40 K were assigned to sub-subplots. Statistical analysis was performed using ANOVA. Least Significant Difference was used to compare treatment means at $p = 0.05$ significance level. The results indicated that rhizosphere soil chemical properties such as pH, organic carbon (OC), and macro- and micronutrients (N, P, Ca, Mg, and Na and Fe, Cu, Mn, and Zn, resp.) were significantly increased in the *Rhizobium* inoculated soybean over the control. The supply of P and K fertilizers significantly increased the rhizosphere content of macronutrients (P, K, Ca, and Mg) and also they altered the pH and EC of the rhizosphere soil relative to control.

1. Introduction

Plants require 17 nutrients to perform different plant functions related to growth, development, and reproduction. However, most soils in sub-Saharan Africa are depleted and are deficient in mineral elements to sustain crop production [1–3]. Each of these plant nutrients is needed in deferent amount by plants and they differ in their mobility and availability in the plants and soil. In agroecosystems where most farmers prefer to grow more than one crop in the same piece of land at the same time, there are maximum interactions between plant roots and soil [4]. The concentration of plant nutrients in the zone of soil-roots interactions (rhizosphere) is reported to be different from bulk soil [5]. Compared with the bulk soil, rhizosphere soil is said to have high concentration of mineral elements and soil microorganisms (both beneficial and harmful) [6, 7]. Interactions between roots and soil during plant growth induce changes in the soil that make rhizosphere soil to differ from bulk soil [8, 9]. These changes in the rhizosphere may be caused by root uptake of nutrients, microbial activity, and/or components of root exudates [5, 10]. Plants release several low and high molecular weight organic compounds such as sugars, organic acids, amino acids, and phenolics into the rhizosphere [5, 11]. The compounds that are released can lead to dissolution of primary minerals and precipitation or crystallization of secondary compounds and/or minerals and eventually transformation of mineral components in the rhizosphere [6].

Plant Growth Promoting Rhizobacteria (PGPR) also do concentrate in the rhizosphere soil and confer the plants with beneficial effects such as solubilization of mineral nutrients, fixation of nitrogen, and disease suppression resulting from plant pathogens [12–17]. Bambara and Ndakidemi [18] reported a significant increase in soil pH, Ca, and Na following *Rhizobium* inoculation on *Phaseolus vulgaris*. Crop diversity in the field increases concentration of microbes in the rhizosphere soil compared with the single species because of specificity of microbes to plant species. Intercropping cereal with legume crops such as cowpea, mungbean, soybean, and groundnuts can fix and accumulate nitrogen ranging from 80 to 350 kg ha^{-1} per year [19], thereby improving soil fertility. Intercropping also can enhance nutrient mineralization because it improves the decomposition rates of soil organic matter [20, 21]. Biological nitrogen fixation induces changes in the soil pH resulting in limited availability of some plant nutrients in the soil [22].

Currently, there is limited information about the effect of *Rhizobium* inoculation on the chemical composition in the rhizosphere of soybean intercropped with maize, supplemented with lower and higher rates of phosphorus (P) and potassium (K) fertilizers. Therefore, the current study was carried out to determine the effects of intercropping systems, *Rhizobium* inoculation, and fertilization with different levels of P and K singly or combined application on chemical properties of soybean rhizosphere soil after two consecutive cropping seasons.

2. Material and Methods

2.1. Experimental Design and Treatments. The field experiment was conducted in the same place for two consecutive years (2015 and 2016 cropping seasons). The treatments for year one were repeated in year two in the same spots. The experimental trials were set at Tanzania Coffee Research Institute (TaCRI) farm in northern Tanzania. The experiment followed a split-split plot design with factorial arrangement and replicated thrice. The plot measured 3 × 3 m. The main plots had two Rhizobia inoculation treatments, while the subplots were comprised of the following treatments: maize (sole crop) at a spacing of 75 × 60 cm; soybean (sole crop) at a spacing of 75 × 40 cm; maize/soybean (intercropping system) at a spacing of 75 × 60 cm and 75 × 20 cm, maize, and soybean, respectively; and the last cropping system was maize/soybean (intercropping system) at a spacing of 75 × 60 cm and 75 × 40 cm, maize, and soybean, respectively. The sub-subplots were treated with the following fertilizer levels (kg ha^{-1}): control (0 kg ha^{-1}); 20 K; 40 K; 26 P; 52 P; 26 P + 20 K; 52 P + 40 K. The sources of these elements were Triple Super Phosphate (TSP for P) and Muriate of Potash (MOP) for K. The BIOFIX legume inoculants *(Bradyrhizobium japonicum)* were obtained from *MEA* Company Nairobi-Kenya, sold under the license from the University of Nairobi. The inoculants were supplied with gum Arabic for sticking as many cells as possible into the seeds. Inoculation was done following manufacturers' instructions where thirty (30) gram of gum Arabic was added to 300 ml of water and mixed to form a solution. 15 kg of soybean seeds was weighed and 300 ml of gum Arabic solution was added and mixed well. 50 gm of legume inoculants was added and mixed well so that all seeds are coated. The inoculated seeds were put under shade for 15 minutes to ensure the inoculants stick well on seeds and the seeds were then sown immediately in a wet moist soil. To avoid contamination of inoculation treatments, the uninoculated plots were sown first followed by inoculated plots.

2.2. Sample Collection. Rhizosphere soil used in this study was collected during the second year of the experimentation when the soybean was at 50% pod formation. The rhizosphere soils were sampled from five plants of middle rows for each plot excluding the border plants. This was achieved by careful excavation of soil from around each plant down to about 10–20 cm depending on root depth of the respective plant and removed with the plant and its roots intact inside the bulge of soil. The rhizosphere soil adhering to plant roots was shaken in the labeled bags, air dried in the laboratory, and sieved (2 mm) ready for determination of pH, organic carbon, and analysis of nutrients. The samples collected from each plot for the two cropping season were pooled together and well mixed to form one sample per treatment.

2.3. Determination of Plant-Available Nutrients in Rhizosphere Soil. Total N was determined by the method of micro-Kjeldahlas described by Bremner [23]. Phosphorus was determined by the molybdenum blue method [24]. Concentrations of elements such as Ca, Mg, K, and Na were determined by method described in Hesse [25]. The trace elements such as Cu, Zn, Fe, and Mn were extracted by diethylenetriamine-pentaacetic acid (DTPA) [26] and determined by an atomic absorption spectrophotometer. The rhizosphere soil pH was analysed in 1 : 2.5 (soil : water) suspension, by the electrometric method [27], and electrical conductivity (EC), measured in a 1 : 5 (soil : water) suspension, using the electrometric method [27]. The organic carbon was determined by the Walkley and Black method [28].

2.4. Statistical Analyses. The collected data was analysed using statistical software called STATISTICA. The statistical analyses were performed using analyses of variance (ANOVA) in factorial arrangement. Fisher's least significant difference (LSD) was used to compare treatment means at $p = 0.05$ level of significance [29].

3. Results

3.1. Effects of Cropping Systems on Rhizosphere Soil Chemical Properties. The results indicated that, except for sodium (Na), cropping systems had no significant effects on the chemical properties of the soybean rhizosphere soil. Only sodium was observed to be greater in the rhizosphere soil of soybean grown under maize intercropping systems compared with the rhizosphere soil of soybean pure stand (Table 1).

3.2. Effects of Rhizobium Inoculation on Rhizosphere Soil Chemical Properties. Rhizobium inoculation was observed to alter the most of the chemical properties of the rhizosphere soil of soybean compared with the rhizosphere soil collected

TABLE 1: Effects of *Rhizobium* inoculation, cropping systems, and the supply of P and K on macronutrients in the rhizosphere soil measure in 2016 season.

Treatments	N	OC	Av. P	K	Ca	Mg	Na
	%		Bray 1. mg/kg		Meq/100 g		
Crop. system							
SB	0.039 ± 0.001^{a}	2.65 ± 0.05^{a}	3.56 ± 0.56^{a}	2.20 ± 0.15^{a}	8.60 ± 0.41^{a}	1.67 ± 0.06^{a}	0.16 ± 0.01^{b}
M + B (A)	0.041 ± 0.001^{a}	2.65 ± 0.05^{a}	3.50 ± 0.59^{a}	2.25 ± 0.17^{a}	7.95 ± 0.39^{a}	1.58 ± 0.06^{a}	0.18 ± 0.01^{a}
M + B (B)	0.040 ± 0.001^{a}	2.67 ± 0.04^{a}	3.42 ± 0.66^{a}	2.27 ± 0.18^{a}	8.55 ± 0.44^{a}	1.62 ± 0.06^{a}	0.19 ± 0.01^{a}
Rhizobia							
Without	0.038 ± 0.001^{b}	2.59 ± 0.04^{b}	2.18 ± 0.40^{b}	2.32 ± 0.15^{a}	7.81 ± 0.28^{b}	1.50 ± 0.04^{b}	0.15 ± 0.01^{b}
With	0.042 ± 0.001^{a}	2.72 ± 0.03^{a}	4.81 ± 0.52^{a}	2.16 ± 0.12^{a}	8.92 ± 0.37^{a}	1.75 ± 0.05^{a}	0.20 ± 0.01^{a}
Fertilizers							
Control	0.038 ± 0.002^{a}	2.61 ± 0.06^{a}	0.60 ± 0.17^{b}	1.51 ± 0.23^{c}	6.92 ± 0.39^{c}	1.59 ± 0.06^{b}	0.14 ± 0.01^{a}
20 K	0.039 ± 0.001^{a}	2.71 ± 0.06^{a}	2.47 ± 0.95^{b}	2.52 ± 0.15^{b}	6.44 ± 0.42^{c}	1.52 ± 0.10^{b}	0.17 ± 0.01^{a}
40 K	0.039 ± 0.001^{a}	2.74 ± 0.07^{a}	0.71 ± 0.16^{b}	2.81 ± 0.19^{ab}	6.46 ± 0.40^{c}	1.42 ± 0.06^{b}	0.18 ± 0.02^{a}
26 P	0.040 ± 0.001^{a}	2.62 ± 0.07^{a}	5.60 ± 0.95^{a}	1.32 ± 0.08^{c}	10.14 ± 0.57^{ab}	1.85 ± 0.08^{a}	0.17 ± 0.02^{a}
52 P	0.041 ± 0.002^{a}	2.66 ± 0.05^{a}	5.76 ± 1.04^{a}	1.45 ± 0.12^{c}	10.46 ± 0.72^{a}	1.85 ± 0.11^{a}	0.20 ± 0.02^{a}
20 K + 26 P	0.041 ± 0.002^{a}	2.56 ± 0.08^{a}	4.69 ± 0.81^{a}	2.83 ± 0.20^{ab}	8.77 ± 0.43^{b}	1.50 ± 0.05^{b}	0.18 ± 0.02^{a}
40 K + 52 P	0.041 ± 0.001^{a}	2.69 ± 0.06^{a}	4.63 ± 0.96^{a}	3.24 ± 0.29^{a}	9.38 ± 0.57^{ab}	1.63 ± 0.10^{ab}	0.18 ± 0.01^{a}
3-way ANOVA *F*-statistics							
CroSyt	0.475^{ns}	0.06^{ns}	0.02^{ns}	0.103^{ns}	1.15^{ns}	0.62^{ns}	3.41^{*}
Rhiz	16.746^{***}	5.76^{**}	21.71^{***}	1.478^{ns}	8.40^{**}	15.97^{***}	18.79^{***}
Fert	0.810^{ns}	0.90^{ns}	8.86^{***}	19.195^{***}	11.65^{***}	4.04^{***}	1.75^{ns}
CroSyt∗Rhiz	1.091^{ns}	0.54^{ns}	0.90^{ns}	2.027^{ns}	0.48^{ns}	0.19^{ns}	0.97^{ns}
CroSyt∗Fert	0.898^{ns}	1.02^{ns}	0.66^{ns}	2.578^{**}	0.61^{ns}	0.52^{ns}	0.71^{ns}
Rhiz∗Fert	1.467^{ns}	0.58^{ns}	1.77^{ns}	1.225^{ns}	0.25^{ns}	0.19^{ns}	0.48^{ns}
CroSyt∗Rhiz∗Fert	0.830^{ns}	0.70^{ns}	1.01^{ns}	0.567^{ns}	1.41^{ns}	0.86^{ns}	0.85^{ns}

CroSyt: cropping systems; Fert: fertilizers; Rhiz: *Rhizobium*; SB: sole soybean; M + B (A): maize/soybean intercropped at a spacing of 75 × 60 cm and 75 × 20 cm, maize, and soybean, respectively; M + B (B): maize/soybean intercropped at a spacing of 75 × 60 cm and 75 × 40 cm, maize, and soybean, respectively; values presented are means ± SE; *,**,***significant at $p \leq 0.05$, $p \leq 0.01$, and $p \leq 0.001$, respectively; ns = not significant; SE = standard error. Means followed by dissimilar letter(s) in a column are significantly different from each other at $p = 0.05$ according to Fischer least significance difference (LSD).

from uninoculated soybean. Rhizosphere soil chemical properties that were significantly increased in the inoculated soybean includes macro- and micronutrients (N, P, Ca, Mg, Na, and OC and pH, Fe, Cu, Mn, Zn, resp.). In this study, *Rhizobium* inoculation treatment did not significantly alter the concentration of EC and K in the rhizosphere soil (Tables 1 and 2). Specifically, *Rhizobium* inoculation increased the concentration of chemical (N, P, Ca, Mg, Na, and OC) in the soil by 10.5, 120.6, 14.2, 16.7, 33.3, and 17.8% respectively. Similarly, the rhizosphere pH and micronutrients (Fe, Cu, Mn and Zn) were, respectively, increased by 1.3, 10.6, 31.4, 41.7, and 25% in *Rhizobium* inoculated soybean relative to uninoculated soybean.

3.3. Effects of P and K Fertilization on Rhizosphere Soil Chemical Properties. Fertilization of crop (soybean) with P and K did not significantly change the concentration nitrogen, sodium, and the organic matter content of rhizosphere soil. However, application of these fertilizers significantly increased the rhizosphere content of macronutrients such as P, K, Ca, and Mg (Table 1). P and K fertilization also altered the pH and EC of the rhizosphere soil relative to control (Table 2).

3.4. Interactions. The current study showed that there were significant interactions between cropping systems and fertilizers on rhizosphere soil K content and between cropping systems and *Rhizobium* inoculation on rhizosphere soil pH. Fertilization with two levels of K significantly increased the rhizosphere K over the plots fertilized with P and the control. The plots that were fertilized with P had statistically the same rhizosphere K with the control (Figure 1). The highest rhizosphere K level was recorded in soybean intercropped with maize at wider spacing and fertilized with doubled combined fertilizers, while the lowest level of rhizosphere K was recorded in the control plots (Figure 1). The rhizosphere soil pH was significantly higher in *Rhizobium* inoculated plots over uninoculated one throughout the cropping systems. The highest pH value was recorded in *Rhizobium* inoculated pure stand while the lowest was recorded in uninoculated pure stand soybean (Figure 2).

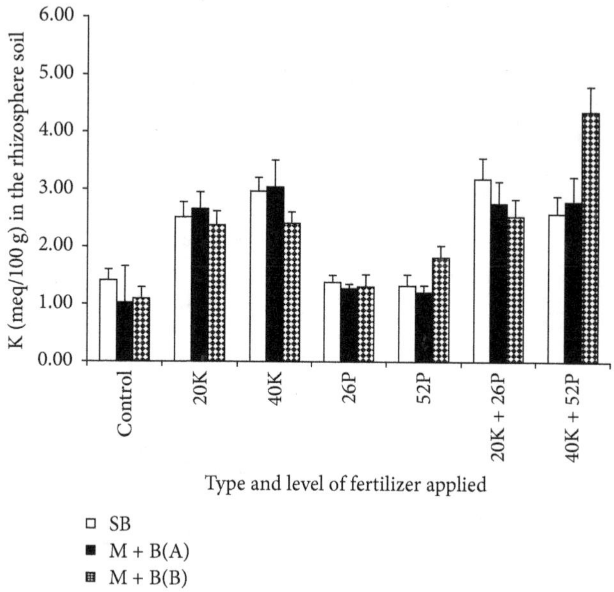

FIGURE 1: Interactive effect of cropping systems and fertilizers on rhizosphere soil K content. SB: sole soybean; M + B (A): maize/soybean intercropped at a spacing of 75 × 60 cm and 75 × 20 cm, maize, and soybean, respectively; M + B (B): maize/soybean intercropped at a spacing of 75 × 60 cm and 75 × 40 cm, maize, and soybean, respectively.

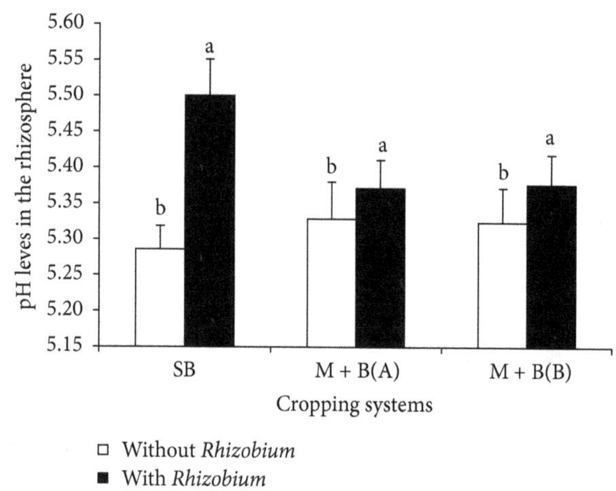

FIGURE 2: Interactive effect of cropping systems and *Rhizobium* inoculation on rhizosphere soil pH. SB: sole soybean; M + B (A): maize/soybean intercropped at a spacing of 75 × 60 cm and 75 × 20 cm, maize, and soybean, respectively; M + B (B): maize/soybean intercropped at a spacing of 75 × 60 cm and 75 × 40 cm, maize, and soybean, respectively.

4. Discussion

The findings of this study showed that *Rhizobium* inoculation altered the chemical composition of rhizospheres soil relative to uninoculated treatments. *Rhizobium* inoculation reduced the soil acidity by increasing the rhizosphere soil pH relative to uninoculated treatments. Similar findings were reported by Bambara and Ndakidemi [18] who found that *Rhizobium* inoculation significantly increased the soil pH in the rhizosphere of *P. vulgaris*. Furthermore, *Rhizobium* inoculation altered the chemical properties of rhizosphere whereby most of the mineral elements were increased in rhizosphere soils of inoculated soybean over the control. There are several possible explanations for increased concentration of macro- and micronutrients in the rhizosphere soils of inoculated soybean. Firstly, it is due to increased soil pH which favoured the availability of most plant nutrients [30, 31]. Increased availability of nutrients in the rhizosphere soil provides normal growth of plants and eventually increased yield. Normally, if there is low soil pH, the soil is acidic which results in poor plant growth and development as most of plant nutrients become unavailable for plants. Secondly, mineralization activities of rhizospheric microorganisms tend to solubilize mineral elements such as P and make it available in the soil [32]. Thirdly, *Rhizobium* produces Fe career compound called siderophores which tends to increase the Fe content in

TABLE 2: Effects of rhizobium inoculation, cropping systems, and the supply of P and K on soil pH, EC, and micronutrients in the rhizosphere soil measure in 2016 season.

Treatments	pH Water 1 : 2.5	EC mS/cm	Fe	Cu	Mn	Zn
			ppm			
Crop. system						
SB	5.39 ± 0.03^{a}	0.21 ± 0.01^{a}	40.88 ± 1.00^{a}	15.61 ± 0.72^{a}	9.51 ± 0.44^{a}	0.74 ± 0.03^{a}
M + B (A)	5.35 ± 0.03^{a}	0.24 ± 0.02^{a}	42.04 ± 1.11^{a}	15.87 ± 0.57^{a}	9.80 ± 0.48^{a}	0.69 ± 0.03^{a}
M + B (B)	5.35 ± 0.03^{a}	0.24 ± 0.02^{a}	41.34 ± 1.05^{a}	17.86 ± 1.10^{a}	10.71 ± 0.70^{a}	0.73 ± 0.03^{a}
Rhizobia						
Without	5.33 ± 0.03^{b}	0.22 ± 0.01^{a}	39.34 ± 0.67^{b}	14.22 ± 0.54^{b}	8.28 ± 0.29^{b}	0.64 ± 0.02^{b}
With	5.40 ± 0.03^{a}	0.23 ± 0.01^{a}	43.50 ± 0.89^{a}	18.68 ± 0.69^{a}	11.73 ± 0.48^{a}	0.80 ± 0.02^{a}
Fertilizers						
Control	5.21 ± 0.05^{c}	0.18 ± 0.02^{d}	41.22 ± 1.25^{a}	15.97 ± 1.28^{a}	9.16 ± 0.67^{a}	0.70 ± 0.05^{a}
20 K	5.34 ± 0.06^{b}	0.21 ± 0.01^{bcd}	42.00 ± 1.89^{a}	15.92 ± 1.36^{a}	9.90 ± 0.80^{a}	0.70 ± 0.03^{a}
40 K	5.43 ± 0.05^{ab}	0.20 ± 0.02^{cd}	42.11 ± 1.66^{a}	15.90 ± 1.13^{a}	10.08 ± 1.03^{a}	0.73 ± 0.05^{a}
26 P	5.31 ± 0.03^{bc}	0.24 ± 0.03^{abc}	43.13 ± 1.57^{a}	17.88 ± 1.13^{a}	9.09 ± 0.70^{a}	0.63 ± 0.03^{a}
52 P	5.36 ± 0.03^{b}	0.26 ± 0.02^{ab}	40.95 ± 1.26^{a}	14.37 ± 1.01^{a}	10.36 ± 0.70^{a}	0.71 ± 0.05^{a}
20 K + 26 P	5.42 ± 0.05^{ab}	0.26 ± 0.02^{ab}	39.54 ± 1.85^{a}	17.42 ± 1.10^{a}	11.93 ± 1.26^{a}	0.81 ± 0.05^{a}
40 K + 52 P	5.49 ± 0.05^{a}	0.27 ± 0.02^{a}	41.01 ± 1.36^{a}	17.67 ± 1.75^{a}	9.52 ± 0.49^{a}	0.73 ± 0.04^{a}
3-way ANOVA *F*-statistics						
CroSyt	0.7^{ns}	0.98^{ns}	0.33^{ns}	2.51^{ns}	1.57^{ns}	0.93^{ns}
Rhiz	4.9^{*}	0.33^{ns}	12.50^{*}	24.74^{***}	35.64^{***}	27.69^{***}
Fert	4.2^{***}	2.99^{*}	0.53^{ns}	1.14^{ns}	1.61^{ns}	1.77^{ns}
CroSyt∗Rhiz	5.0^{**}	2.30^{ns}	0.21^{ns}	1.05^{ns}	0.39^{ns}	0.31^{ns}
CroSyt∗Fert	1.6^{ns}	1.29^{ns}	1.36^{ns}	1.37^{ns}	0.41^{ns}	0.71^{ns}
Rhiz∗Fert	1.1^{ns}	0.90^{ns}	0.60^{ns}	0.23^{ns}	1.26^{ns}	0.84^{ns}
CroSyt∗Rhiz∗Fert	0.5^{ns}	1.55^{ns}	0.30^{ns}	0.28^{ns}	0.66^{ns}	0.79^{ns}

CroSyt: cropping systems; Fert: fertilizers; Rhiz: *Rhizobium*; SB: sole soybean; M + B (A): maize/soybean intercropped at a spacing of 75 × 60 cm and 75 × 20 cm, maize, and soybean, respectively; M + B (B): maize/soybean intercropped at a spacing of 75 × 60 cm and 75 × 40 cm, maize, and soybean respectively; values presented are means ± SE; *,**,***significant at $p \leq 0.05$, $p \leq 0.01$, and $p \leq 0.001$, respectively; ns = not significant and SE = standard error. Means followed by dissimilar letter(s) in a column are significantly different from each other at $p = 0.05$ according to Fischer least significance difference (LSD).

the rhizosphere soil [33, 34]. Fourthly, the decaying cells of microorganisms release nutrients and make them available in the rhizosphere soil [35]. Fifthly, mineral elements can be excreted in the rhizosphere soil as exudates from plant roots [36]. Sixth, the decayed plant biomasses lead to increased nutrients in the soil. The increased biomass due to inoculation means that, as it decays, it will release larger quantities of nutrients into the soil that it has accumulated [37].

All these processes are taking place in the soil and in one way or another may have attributed to the increased chemical properties of rhizosphere soil of *Rhizobium* inoculated soybean. The current study also showed that organic carbon significantly increased in the rhizosphere soil of *Rhizobium* inoculated soybean over the control. The increase in organic carbon content might be attributed to better root growth and deposition of organic materials in first cropping season since these data were taken in the second cropping season. Similar findings were reported by Sharma et al. [38] and Sharma and Verma [39]. However, our findings on organic carbon content in the rhizosphere differed with that of Yusif et al. [40] who found that *Rhizobium* inoculation decreased the soil pH and the organic carbon. They urged that decrease of organic carbon may have been attributed by increased microorganisms which hasted decomposition of organic carbon in the rhizosphere.

The results also showed that P and K fertilization increased the concentration of P, K, Ca, and Mg in the rhizosphere over the control (Table 1). Increased concentration of Ca and Mg in the rhizosphere soil may have been attributed by synergistic effect of P and K which made these nutrients to concentrate more in the rhizosphere soil. The significant increase of P and K in the rhizosphere soil has been attributed by P and K fertilization which increased the availability of these nutrients. Furthermore, root exudates may have contributed to the increased macronutrients in the plots treated with P and K. It was also noted that the rhizosphere soil pH and electrical conductivity (EC) were significantly higher in P and K fertilized plots relative to control. Soil electrical conductivity (EC) provides the measurement of the amount of salts in soil (salinity of soil) [41, 42]. EC affects crop yields, crop suitability, plant nutrient availability, and activity of soil microorganisms which influence key soil processes. However, EC does not provide a direct measurement of specific ions or salt compounds. Researchers have correlated it to concentrations of ions such as N, P, K, Ca, Mg, Na, Mn, Zn, and Cu [42–44]. Their finding are in line with our results which showed the increased EC in plots treated with P and K fertilizers compared with the control.

There were significant interactions between cropping systems and fertilizers on rhizosphere soil K content and between cropping systems and *Rhizobium* inoculation on rhizosphere soil pH. Potassium fertilization significantly interacted with cropping system and contributed to the available K in the rhizosphere soil. The highest K level was recorded in rhizosphere soil of soybean intercropped with maize at wider spacing and fertilized with doubled combined fertilizers, while the lowest level of rhizosphere K was recorded in the control plots. The rhizosphere soil pH was significantly higher in *Rhizobium* inoculated plots over uninoculated one throughout the cropping systems. The highest pH value was recorded in *Rhizobium* inoculated pure stand while the lowest was recorded in uninoculated pure stand soybean suggesting that microorganisms such as *Rhizobium* can help to reduce soil acidity.

5. Conclusion

Rhizobium inoculation altered most of the chemical properties of the rhizosphere soil of soybean in this study. The rhizosphere soil chemical properties such as pH, OC, EC, and macro- and micronutrients (N, P, Ca, Mg, and Na, and Fe, Cu, Mn, and Zn, resp.) were significantly increased in the *Rhizobium* inoculated soybean over the control. These results strongly support the use of microorganism to improve soil chemical properties for improved plant growth, development, and production. The supply of P and K fertilizers significantly increased the rhizosphere content of macronutrients such as (P, K, Ca, and Mg) and also they altered the pH and EC of the rhizosphere soil relative to control.

Acknowledgments

The authors are thankful to the Government of Tanzania through Nelson Mandela African Institution of Science and Technology for funding this study.

References

[1] P. A. Sanchez, "Soil fertility and hunger in Africa," *Science*, vol. 295, no. 5562, pp. 2019-2020, 2002.

[2] P. A. Sanchez and M. S. Swaminathan, "Hunger in Africa: The link between unhealthy people and unhealthy soils," *The Lancet*, vol. 365, no. 9457, pp. 442–444, 2005.

[3] N. Sanginga and P. L. Woomer, "Integrated Soil Fertility Management," in *Proceedings of the in Africa: Principles, Practices and Developmental Process*, p. 263, 2009.

[4] P. A. Ndakidemi, "Manipulating legume/cereal mixtures to optimize the above and below ground interactions in the traditional African cropping systems," *African Journal of Biotechnology*, vol. 5, no. 25, pp. 2526–2533, 2006.

[5] P. Hinsinger, G. R. Gobran, P. J. Gregory, and W. W. Wenzel, "Rhizosphere geometry and heterogeneity arising from root-mediated physical and chemical processes," *New Phytologist*, vol. 168, no. 2, pp. 293–303, 2005.

[6] J. Cabala, E. Teper, L. Teper, E. Małkowski, and A. Rostański, "Mineral composition in rhizosphere of plants grown in the vicinity of a Zn-Pb ore flotation tailings pond. Preliminary study," *Acta Biologica Cracoviensia Series Botanica*, vol. 46, pp. 65–73, 2004.

[7] W. Cheng and A. Gershenson, "Carbon Fluxes in the rhizosphere. The Rhizosphere," *An ecological perspective*, pp. 29–54, 2007.

[8] X. Wang and D. Zabowski, "Nutrient composition of Douglas-fir rhizosphere and bulk soil solutions," *Plant and Soil*, vol. 200, no. 1, pp. 13–20, 1998.

[9] J. H. J. R. Makoi, S. B. M. Chimphango, and F. D. Dakora, "Changes in rhizosphere concentration of mineral elements as affected by differences," *American Journal of Experimental Agriculture*, vol. 4, no. 2, pp. 193–214, 2014.

[10] X.-F. Huang, J. M. Chaparro, K. F. Reardon, R. Zhang, Q. Shen, and J. M. Vivanco, "Rhizosphere interactions: Root exudates, microbes, and microbial communitiesl," *Botany*, vol. 92, no. 4, pp. 267–275, 2014.

[11] H. Marschner and V. Römheld, "Root-induced changes in the availability of micronutrients in the rhizosphere," in *Plant Roots, The Hidden Half*, Y. Waisel, A. Eshel, and U. Kafkafi, Eds., pp. 557–579, Marcel Dekker, Inc, New York, NY, USA, 1996.

[12] A. Gupta, M. Gopal, and K. V. B. R. Tilak, "Mechanism of plant growth promotion by rhizobacteria," *Indian Journal of Experimental Biology (IJEB)*, vol. 38, no. 9, pp. 856–862, 2000.

[13] D. M. Weller, J. M. Raaijmakers, B. B. McSpadden Gardener, and L. S. Thomashow, "Microbial populations responsible for specific soil suppressiveness to plant pathogens," *Annual Review of Phytopathology*, vol. 40, pp. 309–348, 2002.

[14] J. W. Kloepper, C. Ryu, and S. Zhang, "Induced systemic resistance and promotion of plant growth by *Bacillus* spp.," *Journal of Phytopathology*, vol. 94, no. 11, pp. 1259–1266, 2004.

[15] J. Yang, J. W. Kloepper, and C.-M. Ryu, "Rhizosphere bacteria help plants tolerate abiotic stress," *Trends in Plant Science*, vol. 14, no. 1, pp. 1–4, 2009.

[16] R. Mendes, M. Kruijt, I. De Bruijn et al., "Deciphering the rhi zosphere microbiome for disease-suppressive bacteria," *Science*, vol. 332, no. 6033, pp. 1097–1100, 2011.

[17] M. I. Tahir, M. Inam-ul-Haq, M. Ashfaq, N. A. Abbasi, H. Butt, and H. Ghazal, "Screening of effective antagonists from potato rhizosphere against bacterial wilt pathogen," *International Journal of Biosciences*, vol. 8, no. 2, pp. 228–240, 2016.

[18] S. Bambara and P. A. Ndakidemi, "Changes in selected soil chemical properties in the rhizosphere of phaseolus vulgaris L. supplied with rhizobium inoculants, molybdenum and lime," *Scientific Research and Essays*, vol. 5, no. 7, pp. 679–684, 2010.

[19] M. B. Peoples and E. T. Craswell, "Biological nitrogen fixation: Investments, expectations and actual contributions to agriculture," *Plant and Soil*, vol. 141, no. 1-2, pp. 13–39, 1992.

[20] E. Blagodatskaya and Y. Kuzyakov, "Mechanisms of real and apparent priming effects and their dependence on soil microbial biomass and community structure: Critical review," *Biology and Fertility of Soils*, vol. 45, no. 2, pp. 115–131, 2008.

[21] H. R. Mobasser, M. R. Vazirimehr, and K. Rigi, "Effect of intercropping on resources use, weed management and forage quality," *International Journal of Plant, Animal and Environmental Sciences*, vol. 4, no. 2, pp. 706–713, 2014.

[22] N. S. Bolan, M. J. Hedley, and R. E. White, "Processes of soil acidification during nitrogen cycling with emphasis on legume based pastures," *Plant and Soil*, vol. 134, no. 1, pp. 53–63, 1991.

[23] J. M. Bremner, "Total Nitrogen," in *Methods of Soil Analysis*, C. A. Black, Ed., pp. 1149–1178, American Society of Agronomy, Madison, Wis, USA, 1965.

[24] J. Murphy and J. P. Riley, "A modified single solution method for the determination of phosphate in natural waters," *Analytica Chimica Acta*, vol. 27, pp. 31–36, 1962.

[25] P. R. Hesse, *A Textbook of Soil Chemical Analysis*, John Murray, London, UK, 1971.

[26] W. L. Lindsay and W. A. Norvell, "Development of a DTPA soil test for zinc, iron, manganese, and copper," *Soil Science Society of America Journal*, vol. 42, pp. 421–428, 1978.

[27] H. D. Chapman, "Cation-Exchange Capacity," in *Methods of Soil Analysis, Part 2*, C. A. Black, Ed., American Society of Agronomy, Soil Science Society of America, Madison, Wis, USA, 1965.

[28] A. Walkley and I. Black, "An examination of the Degtjareff method for determining soil organic matter, and a proposed modification of the organic acid titration method," *Soil Science*, vol. 37, no. 1, pp. 29–38, 1934.

[29] R. G. D. Steel and J. H. Torrie, *Principles and Procedures of Statistics: A Biometrical Approach*, McGraw-Hill Kogakusha, New York, NY, USA, 2nd edition, 1980.

[30] M. Bagayoko, S. Alvey, G. Neumann, and A. Buerkert, "Root-induced increases in soil pH and nutrient availability to field-grown cereals and legumes on acid sandy soils of Sudano-Sahelian West Africa," *Plant and Soil*, vol. 225, no. 1-2, pp. 117–127, 2000.

[31] L. M. Condron, H. Tiessen, C. Trasar-Cepeda, J. O. Moir, and J. W. B. Stewart, "Effects of liming on organic matter decomposition and phosphorus extractability in an acid humic Ranker soil from northwest Spain," *Biology and Fertility of Soils*, vol. 15, no. 4, pp. 279–284, 1993.

[32] F. D. Dakora and D. A. Phillips, "Root exudates as mediators of mineral acquisition in low-nutrient environments," *Plant and Soil*, vol. 245, no. 1, pp. 35–47, 2002.

[33] Y. Wang, H. N. Brown, D. E. Crowley, and P. J. Szaniszlo, "Evidence for direct utilization of a siderophore, ferrioxamine B, in axenically grown cucumber," *Plant, Cell & Environment*, vol. 16, no. 5, pp. 579–585, 1993.

[34] P. J. White and M. R. Broadley, "Biofortification of crops with seven mineral elements often lacking in human diets—iron, zinc, copper, calcium, magnesium, selenium and iodine," *New Phytologist*, vol. 182, no. 1, pp. 49–84, 2009.

[35] M. E. McCulley, "Niches for bacterial endophytes in crop plants: a plant biologists view," *Australian Journal of Plant Physiology*, vol. 28, pp. 983–990, 2001.

[36] N. Ae, J. Arihara, K. Okada, T. Yoshihara, and C. Johansen, "Phosphorus uptake by pigeon pea and its role in cropping systems of the Indian subcontinent," *Science*, vol. 248, no. 4954, pp. 477–480, 1990.

[37] D. J. Bagyaraj, C. J. Nethravathi, and K. S. Nitin, "Soil biodiversity and arthropods: Role in soil fertility," in *Economic and ecological significance of arthropods in diversified ecosystems*, pp. 17–51, Springer, Singapore, 2016.

[38] R. P. Sharma, N. Datt, and G. Chander, "Effect of vermicompost, farmyard manure and chemical fertilizers on yield, nutrient uptake and soil fertility in okra (Abelmoschus esculentus) Onion (Allium cepa) sequence in wet temperate zone of Himachal Pradesh," *Journal of the Indian Society of Soil Science*, vol. 57, no. 3, pp. 357–361, 2009.

[39] R. Sharma and M. L. Verma, "Effect of Rhizobium, farm yard manure and chemical fertilizers on sustainable production and profitability of rajmash (Phaseolus vulgaris l.) and soil fertility in dry temperate region of North-Western Himalayas," *Legume Research*, vol. 34, no. 4, pp. 251–258, 2011.

[40] S. Yusif, I. Muhammad, N. Hayatu et al., "Effects of biochar and rhizobium inoculation on selected soil chemical properties, shoot nitrogen and phosphorus of groundnut plants (arachis hypogaea l.) in sokoto state, nigeria," *Journal of Advances in Biology & Biotechnology*, vol. 8, no. 2, pp. 1–6, 2016.

[41] I. Shainberg, J. D. Rhoades, and R. J. Prather, "Effect of exchangeable sodium percentage, cation exchange capacity, and

soil solution concentration on soil electrical conductivity," *Soil Science Society of America Journal*, vol. 44, no. 3, p. 469, 1980.

[42] R. Grisso, W. G. W. Mark Alley, D. Holshouser, and W. Thomason, Precision Farming Tools: Soil Electrical Conductivity," Virginia Cooperative Extension, publication 442-508, pp. 1-6, 2009, https://www.pubs.ext.vt.edu/content/dam/pubs_ext_vt_edu/442/442-508/442-508_pdf.pdf.

[43] R. W. Heiniger, R. G. McBride, and D. E. Clay, "Using soil electrical conductivity to improve nutrient management," *Agronomy Journal*, vol. 95, no. 3, pp. 508–519, 2003.

[44] N. Hamzehpour and S. M. A. Abasiyan, "Spatial prediction of soil phosphorous using soil electrical conductivity as secondary information," *Journal of Biodiversity and Environmental Sciences*, vol. 8, no. 3, pp. 88–95, 2016.

17

Genome Editing in Plants: An Overview of Tools and Applications

Venera S. Kamburova,[1] Elena V. Nikitina,[1] Shukhrat E. Shermatov,[1] Zabardast T. Buriev,[1] Siva P. Kumpatla,[2] Chandrakanth Emani,[3] and Ibrokhim Y. Abdurakhmonov[1]

[1]*Center of Genomics and Bioinformatics, Academy of Sciences of the Republic of Uzbekistan, University Street-2, Qibray Region, 111215 Tashkent, Uzbekistan*
[2]*Dow AgroSciences LLC, Indianapolis, IN 46268, USA*
[3]*Western Kentucky University-Owensboro, Owensboro, KY 42303, USA*

Correspondence should be addressed to Ibrokhim Y. Abdurakhmonov; genomics@uzsci.net

Academic Editor: Kent Burkey

The emergence of genome manipulation methods promises a real revolution in biotechnology and genetic engineering. Targeted editing of the genomes of living organisms not only permits investigations into the understanding of the fundamental basis of biological systems but also allows addressing a wide range of goals towards improving productivity and quality of crops. This includes the creation of plants with valuable compositional properties and with traits that confer resistance to various biotic and abiotic stresses. During the past few years, several novel genome editing systems have been developed; these include zinc finger nucleases (ZFNs), transcription activator-like effector nucleases (TALENs), and clustered regularly interspaced short palindromic repeats/Cas9 (CRISPR/Cas9). These exciting new methods, briefly reviewed herein, have proved themselves as effective and reliable tools for the genetic improvement of plants.

1. Introduction

Since the advent of recombinant DNA technology in Paul Berg's laboratory [1] in 1972, genetic engineering has come a long way and achieved enormous success. Many molecular and genetic mechanisms and phenomena have been discovered and studied in detail and the knowledge accumulated now permits researchers to reproduce experiments in vitro. Several decades-long investigations in molecular genetics and biochemistry of bacteria and viruses have allowed researchers to develop new methods of manipulating DNA through creation of various vector systems and tools for their delivery into the cell. All of these developments allow successful creation of not only transgenic microorganisms but also genetically modified higher organisms including various plant and crop species. Creation of novel tools for breeding and biotechnology, an application area of genetic engineering, has received significant focus resulting in accelerated development of useful tools. However, conventional genetic engineering strategy has several issues and limitations, one of which is the complexity associated with the manipulation of large genomes of higher plants [2].

Currently, several tools that help to solve the problems of precise genome editing of plants are at scientists' disposal. In 1996, for the first time, it was shown that protein domains such as "zinc fingers" coupled with *Fok*I endonuclease domains act as site-specific nucleases (zinc finger nucleases (ZFNs)), which cleave the DNA in vitro in strictly defined regions [3]. Such a chimeric protein has a modular structure, because each of the "zinc finger" domains recognizes one triplet of nucleotides. This method became the basis for the editing of cultured cells, including model and nonmodel plants [4, 5].

Continued efforts and investigations led to the development of new genome editing tools such as TALENs (transcription activator-like effector nucleases) and CRISPR/Cas (clustered regularly interspaced short palindromic repeats). Designing TALENs requires reengineering of a new protein for each of the targets. However, the design process has been streamlined recently by making the modules of repeat combinations available that essentially reduces the cloning required for the design. On the other hand, designing and use of CRISPR are simple. Both TALEN and CRISPR systems have been shown to work in human cells, animals, and plants. Such editing systems when used for efficient manipulation of genomes could solve complex problems including the creation of mutant and transgenic plants [12, 41]. Moreover, chimeric proteins containing zinc finger domains and activation domains of other proteins and those based on the TALE DNA-binding domain and Cas9 nuclease were used in experiments for regulation of gene transcription, study of epigenomes, and the behavior of chromosome loci in cell cycle [24, 42–44].

In this review, we briefly described the mechanisms of different genome editing systems and their use for crop improvement and also highlighted the multiple advantages and applications of engineered nucleases as well as biosafety and regulatory aspects of plants generated using engineered nuclease based technologies.

2. Mechanisms of Genome Editing Systems

Novel genome editing tools, also referred to as genome editing with engineered nuclease (GEEN) technologies, allow cleavage and rejoining of DNA molecules in specified sites to successfully modify the hereditary material of cells. To this end, special enzymes such as restriction endonucleases and ligase can be used for cleaving and rejoining of DNA molecules in small genomes like bacterial and viral genomes. However, using restriction endonucleases and ligases, it is extremely difficult to manipulate large and complex genomes of higher organisms, including plant genomes. The problem is that the restriction endonucleases can only "target" relatively short DNA sequences. While such specificity is enough for short DNA viruses and bacteria, it is not sufficient to work with large plant genomes. The first efforts to create methods for the editing of complex genomes were associated with the designing of "artificial enzymes" as oligonucleotides (short nucleotide sequences) that could selectively bind to specific sequences in the structure of the target DNA and have chemical groups capable of cleaving DNA [45].

Targeted approach to address this challenge was the design of chimeric nucleases which are complex proteins containing one or two structural units, one of which catalyzes the cleavage of DNA, and the second is capable of selectively binding to specific nucleotide sequences of target molecule, providing the nuclease action to this site (Table 1) [46, 47]. These chimeric nucleases can be "produced" directly in the cell: to this end, appropriately engineered vectors encoding nucleases need to be introduced into cell. Such vectors are also supplied with nuclear localization signal which enables the nuclease to enter the cell nucleus thereby getting access to genomic DNA.

2.1. Zinc Finger Nucleases (ZFNs). ZFNs were the first generation of genome editing tools that use chimerically engineered nucleases which were developed after the discovery of the working principles of the functional Cys2-His2 zinc finger (ZF) domain [3, 4, 46, 48]. Each Cys2-His2 ZF domain consists of 30 amino acid residues, which are folded up to $\beta\beta\alpha$ configuration [48–50]. Crystallographic structure analysis showed that the Cys2-His2 ZF proteins bind to DNA by inserting an α-helix of the protein into the major groove of the DNA-double helix [51]. Each ZF protein has the ability to recognize 3 tandem nucleotides in the DNA. Generalized ZFN monomer consists of two different functional domains: artificial ZF Cys2-His2 domain at the N-terminal region and a nonspecific *Fok*I DNA cleavage domain at the C-terminal region. *Fok*I domain dimerization is critical for ZFN enzymatic activity [3]. The observation that the modular recognition of zinc finger domains presents as a series to the corresponding, consecutive three bp targets enabled the realization that each of the individual zinc finger domains could be interchangeable and that the manipulation of the order of the domains would lead to unique binding specificities to the proteins harboring them thereby enabling targeting of specific, unique sequences in the genome. For example, a ZFN dimer, consisting of two 3 or 4 ZF domains, recognizes a target sequence of 18 or 24 base pairs, which statistically form unique sites in the genomes of most organisms (Table 1).

The design and application of ZFNs involve modular design, assembly, and optimization of zinc fingers against specific target DNA sequences followed by linking of individual ZFs towards targeting larger sequences. Over the years, zinc finger domains have been generated to recognize a large number of triplet nucleotides. This enabled the selection and linking of zinc fingers in a sequence that would permit recognition of the target sequence of interest.

Since the first report on zinc fingers in 1996, they have been successfully used in several organisms including plants [4]. Examples include targeted inactivation of endogenous genes in Arabidopsis [15, 16], high frequency modification of tobacco genes [17], and precise targeted addition of a herbicide-tolerance gene as well as insertional disruption of a target locus in maize [18]. ZFNs have also been used for trait stacking in maize [52, 53].

Zinc finger nucleases have revolutionized the field of genome editing by demonstrating the ability to manipulate genomic sites of interest and opened the gates for both basic and applied research. ZFNs provide advantages over other tools with respect to efficiency, high specificity, and minimal nontarget effects and current efforts are focused on further improving design and delivery as well as expanding their applications in diverse crops of interest.

2.2. Transcription Activator-Like Effector Nucleases (TALENs). The quest for efficient and selective manipulation of target genomic DNA led to the identification of unique transcription activator-like effector (TALE) proteins that recognize and activate specific plant promoters through a set of tandem repeats that formed the basis for the creation of a new genome editing system consisting of chimeric nucleases called TALE nucleases (TALENs) [47]. TALE proteins consist of a central domain responsible for DNA binding, nuclear localization

Table 1: Comparison of plant genome editing techniques.

	ZFNs	TALENs	ODM	CRISPR/Cas9	Reference	Year
Components	Zn finger domains Nonspecific *Fok*I nuclease domain	TALE DNA-binding domains Nonspecific *Fok*I nuclease domain	Exogenous polynucleotide (chimeraplast)	crRNA, Cas9 proteins	Kumar and Jain [6] Sauer et al. [7]	2015 2016
Structural proteins	Dimeric protein	Dimeric protein	Nonprotein nature	Monomeric protein	Li et al. [8] Gaj et al. [4] Sauer et al. [7]	2013 2013 2016
Catalytic domain	Restriction endonuclease *Fok*I	Restriction endonuclease *Fok*I	There is no catalytic domain	RUVC and HNH	Jinek et al. [9] Sauer et al. [7]	2012 2016
Length of target sequence (bp)	24–36	24–59	68–88	20–22	Gaj et al. [4] Chen et al. [10] Sauer et al. [7]	2013 2016 2016
Protein engineering steps	Required	Required	Not required	Should not be complex to test gRNA	Cho et al. [11] Weeks et al. [5] Sauer et al. [7]	2013 2016 2016
Cloning	Necessary	Necessary	Not necessary	Not necessary	Cho et al. [11] Weeks et al. [5] Sauer et al. [7]	2013 2016 2016
gRNA production	Not applicable	Not applicable	Not required	Easy to produce	Cho et al. [11] Noman et al. [12] Sauer et al. [7]	2013 2016 2016

TABLE 1: Continued.

	ZFNs	TALENs	ODM	CRISPR/Cas9	Reference	Year
Mode of action	Double-strand breaks in target DNA	Double-strand breaks in target DNA	Information strand directs conversion(s) within target region	Double-strand breaks or single-strand nicks in target DNA	Li et al. [8] Mao et al. [13] Noman et al. [12] Sauer et al. [7]	2013 2013 2016 2016
Target recognition efficiency	High	High	High	High	Gaj et al. [4] Kumar and Jain [6] Sauer et al. [7]	2013 2015 2016
Mutation rate	High	Middle	Middle	Low	Li et al. [8] Gaj et al. [4] Sauer et al. [7]	2013 2013 2016
Creation of large scale libraries	Impossible	Technically difficult	Technically difficult	Possible	Hsu [14] Cho et al. [11] Sauer et al. [7]	2013 2013 2016
Multiplexing	Difficult	Difficult	Technically difficult	Possible	Li et al. [8] Mao et al. [13] Noman et al. [12] Sauer et al. [7]	2013 2013 2016 2016

signal, and a domain that serves as activator of transcription of the target gene (Table 1) [54]. For the first time, the DNA-binding ability of these proteins was described in 2007 [55], and a year later, two scientific groups have decoded the recognition code of target DNA sequence by TALE proteins [56].

It is shown that the DNA-binding domain in TALE monomers in turn consists of a central repeat domain (CRD) that confers DNA binding and host specificity. The CRD consists of tandem repeats of 34 amino acid residues and each 34-amino acid long repeat in the CRD binds to one nucleotide in the target nucleotide sequence. Two of the amino acids of the repeat, located at positions 12 and 13, are highly variable (repeat variable diresidue (RVD)) and are responsible for the recognition of specific nucleotide with degeneracy of binding several nucleotides with differential efficiency. The last tandem repeat binding to nucleotide at the 3′-end of the recognition site consists of 20 amino acid residues only and, therefore, it is named as half-repeat. While TALE proteins, in general, can be designed to bind any DNA sequence of interest, studies have demonstrated that the 5′-most nucleotide base of the DNA sequence bound by a TALE protein should always be a Thymidine and that a deviation from this requirement can affect the efficacy of TALE transcription factors (TALE-TF), TALE recombinases (TALE R), and TALENs [57].

After the DNA code recognition requirements by TALE proteins have been cracked, the very first effort undertaken was the creation of chimeric TALEN nucleases [5]. For this purpose, the sequence encoding the DNA-binding TALE domain was inserted into a plasmid vector previously used to create ZFN [58]. This resulted in the creation of a synthetic, chimeric sequence-specific nuclease genetic construct containing the DNA-binding domain of TALEs and the catalytic domain of *Fok*I restriction endonuclease. This construct helped to create artificial nucleases with DNA-binding domain and different RVDs that can target any nucleotide sequence of interest [2, 4].

In most studies, the monomers with RVDs Asn and Ile (NI), Asn and Gly (NG), two Asn (NN), and His and Asp (HD) bind to nucleotides A, T, G, and C, respectively. NN, the most common RVD that specifies G, was also found to bind to A. This suboptimal or lack of specificity is a concern for the use of engineered TALEs for targeting DNA. Another RVD NK has less functional efficiency compared to NN, although it has demonstrated guanine specificity. Several studies have also shown that the use of NH or NK RVDs for specific binding of guanine reduces the risk of nontarget effects [19, 59, 60]. It has been shown that in RVD (NI, NG, NN, or HD) the first amino acid residue, whether it is N or H, is responsible for the stabilization of spatial conformation although it does not directly bind to a nucleotide, whereas the second amino acid residue binds to a nucleotide either through hydrogen bonding with nitrogenous bases (in case of D and N amino acids) or through van der Waals forces (in case of I and G) [61].

Based on the mode of action and specificity of TALENs, it should be possible to introduce double strand breaks in any location of the genome as long as that location harbors the recognition sequence corresponding to the DNA-binding domains of TALENs. There is another condition that also needs to be met, that is, the requirement of the presence of Thymidine before the 5′ end of the intended target sequence since it has been demonstrated that the W232 residue in the N-terminal portion of the DNA-binding domain interacts with the Thymidine and influences the binding efficiency [62]. It is also possible to overcome this 5′ Thymidine constraint by developing mutant variants of TALEN N-terminal domain which can bind other nucleotides [57]. Considering the ease of site-directed manipulation using TALEN system, within a short period of time after the unraveling of the TALEN mode of action, the genes modified by this system have been used successfully in several animal and plant species and the plant examples include rice, wheat, Arabidopsis, potato, and tomato (Table 2) [63].

2.3. Oligonucleotide-Directed Mutagenesis (ODM). After first successful exploitation in mammalian systems, oligonucleotide-directed mutagenesis (ODM) has become another novel gene editing tool for plants [7, 64]. ODM, a tool for targeted mutagenesis, uses a specific 20- to 100-base long oligonucleotide, the sequence of which is identical to the target sequence in the genome except that it contains a single base pair change (intended mutation to be inserted in the genome) towards achieving site-directed editing of gene/sequence of interest (Table 1) [65]. When these synthetic oligonucleotides or repair templates with homology to a specific region of the target gene are transiently exposed to the plant cells by using a variety of specific delivery methods, they bind to the targets and activate cell's natural repair machinery which recognizes the single mismatch in the template and then copies that mismatch or mutation into the target sequence through repair process [7, 65]. This produces the desired targeted single nucleotide or base editing in the plant genome that confers novel function or trait while the plant cell degrades the repair template oligonucleotide. Using tissue culture methods, cells with edited sequences are subsequently regenerated and genome edited novel varieties with improved traits/characteristics are developed through traditional breeding (Table 2) [7, 64, 65].

2.4. Clustered Regularly Interspaced Short Palindromic Repeats (CRISPR). Another novel genome editing system that has emerged recently and has become widely popular is the clustered regularly interspaced short palindromic repeats (CRISPR)/CRISPR associated (Cas) protein system with the most prominent being the CRISPR/Cas9 (based on Cas9 protein). This is a method that utilizes adaptive bacterial and archaeal immune system, the mechanism of which relies on the presence of special sites in the bacterial genome called CRISPR loci. These loci are composed of operons encoding the Cas9 protein and a repeated array of repeat spacer sequences. The spacers in the repeat array are short fragments that are derived from foreign DNA (viral or plasmid) that have become integrated into bacterial genome following recombination [41, 66].

Table 2: Examples of application of genome editing technology in plants.

Number	Purpose or traits studied	Plant species	Method used	References	Year
1	Gene modification	*Arabidopsis thaliana*	ZFN	Osakabe et al. [15]	2010
2	Gene modification	*Arabidopsis thaliana*	ZFN	Zhang et al. [16]	2010
3	Gene editing	*Nicotiana tabacum*	ZFN	Townsend et al. [17]	2009
4	Targeted integration into endogenous loci	*Zea mays*	ZFN	Shukla [18]	2009
5	Targeted mutagenesis	*Arabidopsis thaliana*	TALEN	Christian et al. [19]	2013
6	Targeted mutagenesis	*Oryza sativa*	TALEN	Zhang et al. [20]	2016
7	Gene editing	*Oryza sativa*	TALEN	Li et al. [21]	2012
8	Targeted transcriptional gene repression	*Arabidopsis thaliana*	TALEN	Mahfouz et al. [22]	2012
9	Herbicide resistance	*Arabidopsis thaliana*	ODM	Sauer et al. [7]	2016
10	Herbicide resistance	*Nicotiana tabacum*	ODM	Sauer et al. [7]	2016
11	Herbicide resistance	*Oryza sativa*	ODM	Sauer et al. [7]	2016
12	Herbicide resistance	*Zea mays*	ODM	Sauer et al. [7]	2016
13	Herbicide resistance	*Brassica napus*	ODM	Sauer et al. [7]	2016
14	Gene knockout or editing	*Arabidopsis thaliana*	CRISPR/Cas	Li et al. [8] Mao et al. [13]	2013 2013
15	Gene knockout or editing	*Nicotiana benthamiana*	CRISPR/Cas	Li et al. [8]	2013
16	Gene knockout or editing	*Nicotiana tabacum*	CRISPR/Cas	Gao et al. [23]	2015

Table 2: Continued.

Number	Purpose or traits studied	Plant species	Method used	References	Year
17	Gene knockout or editing	*Oryza sativa*	CRISPR/Cas	Mao et al. [13]	2013
18	Multiplex genome editing	*Arabidopsis thaliana*	CRISPR/Cas	Li et al. [8] Mao et al. [13]	2013 2013
19	Multiplex genome editing	*Nicotiana tabacum*	CRISPR/Cas	Cong et al. [24] Gao et al. [23]	2013 2015
20	Gene insertion or replacement	*Arabidopsis thaliana*	CRISPR/Cas	Mao et al. [13]	2013
21	Gene insertion or replacement	*Nicotiana benthamiana*	CRISPR/Cas	Li et al. [8]	2013
22	Epigenetic gene regulation	*Arabidopsis thaliana*	CRISPR/Cas	Puchta [25]	2016
23	Targeted mutagenesis	*Glycine max*	CRISPR/Cas	Jacobs et al. [26]	2015
24	Targeted mutagenesis	*Oryza sativa*	CRISPR/Cas	Xu et al. [27]	2016
25	Improvement of nutritional value	*Camelina sativa*	CRISPR/Cas	Jiang et al. [28]	2017
26	Yield improvement	*Solanum lycopersicum*	CRISPR/Cas	Soyk et al. [29]	2017
27	Yield improvement	*Zea mays*	CRISPR/Cas	Peng et al. [30]	1999
28	Salt resistance	*Oryza sativa*	CRISPR/Cas	Osakabe et al. [31]	2016
29	Resistance to drought	*Oryza sativa*	CRISPR/Cas	Shan et al. [32]	2013
30	Herbicide resistance	*Zea mays*	CRISPR/Cas	Svitashev et al. [33]	2015
31	Biotic stress	*Triticum aestivum*	CRISPR/Cas	Wang et al. [34]	2014
32	Biotic stress	*Oryza sativa*	CRISPR/Cas	Li et al. [21]	2012
33	Biotic stress	*Arabidopsis thaliana* *Nicotiana benthamiana*	CRISPR/Cas	Ji et al. [35] Ali et al. [36]	2015 2015

TABLE 2: Continued.

Number	Purpose or traits studied	Plant species	Method used	References	Year
34	Biotic stress	*Nicotiana benthamiana*	CRISPR/Cas	Baltes et al. [37]	2015
35	Biotic stress	*Oryza sativa*	CRISPR/Cas	Liu et al. [38]	2012
36	Biotic stress	*Oryza sativa*	CRISPR/Cas	Wang et al. [39]	2016
37	Biotic stress	*Cucumis sativus*	CRISPR/Cas	Chandrasekaran et al. [40]	2016

Unlike the chimeric TALEN proteins, target site recognition by CRISPR/Cas9 system is accomplished by the complementary sequence based interaction between the guide (noncoding) RNA and DNA of the target site and the guide RNA and Cas protein complex has the nuclease activity for exact cleavage of double-stranded DNA using Cas9 endonuclease (Table 1) [9, 24, 67].

Several types of CRISPR protective systems functioning in cells of various bacteria are described in detail elsewhere [68, 69]. The most "popular" system is the CRISPR/Cas type II-A system found in the bacterium *Streptococcus pyogenes* and composed of three genes encoding CRISPR RNA (crRNA), trans-activating crRNA (tracrRNA), and Cas9 protein. Based on this system, universal genetic constructs encoding artificial elements of CRISPR/Cas "genome editor" have been created [70]. Also, a simplified version of the system, functioning as a complex of Cas9 protein and a single guide RNA, consisting of CRISPR tracrRNA and short, mature crRNA was created. The guide sequence identifies the target DNA site and binds to it based on complementarity and Cas9 cleaves the DNA in target point [71].

CRISPR system can be used for the creation of genetically modified cells grown in culture and living organisms [11]. In the first case, plasmids or viral vectors which provide high and stable synthesis of CRISPR/Cas9 system elements are introduced into cells. In the second case, cultured protoplasts and a plasmid coding CRISPR/Cas elements are used to obtain genetically modified plants [32]. Another approach, applied for plants, is the use of *Agrobacterium*, the natural "genetic engineer," that contains a special plasmid harboring CRISPR/Cas9 system [41, 44].

Thus, due to its simplicity, efficiency, and wide capabilities, in a short time CRISPR/Cas9 system has already found use in various fields of fundamental and applied biology, biotechnology, and genetic engineering.

2.5. Repair of Cleaved Genomic Sites. An important step in the genome editing process is the repair of the DNA break created by the nucleases. DNA break gets repaired by the endogenous cellular mechanisms: nonhomologous end-joining (NHEJ) or homology-dependent (or directed) repair (HDR) [14]. NHEJ is the simplest mechanism where the ends of the cleaved DNA are joined together, often resulting in the insertion or deletion of nucleotides (indels) thereby shifting the gene reading frame, resulting in a gene "knockout" [72]. If indels are not observed, the DNA is recovered, and there are no noticeable changes. On the other hand, HDR is a mechanism where a sequence containing homology to target is used as a template for repairing the break or the DNA lesion. Therefore, by providing a template that contains a desired sequence of interest flanked by sequences homologous to both sides of the break point, one can force the insertion of that desired sequence into the target site. When HDR occurs, a homologous recombination is used to enable new sequences for gene recovery or insertion [72]. This method is simple, provides the exact impact on DNA target, and can be used at almost any modern molecular biology laboratory.

3. Practical Applications of Genome Editing Systems

3.1. Application of "Genome Editors" for Functional Genomics. Several different types of genome modifications can be achieved by utilizing ZFN, TALEN, ODM, and CRISPR/Cas genome editing systems (Table 2). These include creation of point mutations, insertion of new genes in specific locations or deletion of large regions of the nucleotide sequences, and correction or substitution of individual genetic elements and gene fragments [4, 6, 10, 20, 23, 44, 73].

While introducing modifications to various genomic elements in plant cells and examining the results, scientists were able to investigate the role of individual genes in the functioning of individual cells and the organism as a whole. For example, the unique ability of CRISPR/Cas9 system to selectively bind to specific DNA sites has helped to regulate gene activity [24, 41, 44]. For this purpose, proteins activating or repressing the activity of promoters that control the gene function can be attached to the catalytically inactive mutant Cas9 protein. In one example, it was shown that complex binding to the target DNA can inhibit or stimulate the function of the target gene [44].

Furthermore, using CRISPR/Cas9 system, several genetic constructs targeted to different genome sites can simultaneously be introduced into cells [8, 24, 43]. This is a welcome feature in investigating intergenic interaction, if any, because several genes are simultaneously affected by the CRISPR/Cas9 system [44]. For example, using this approach, it was possible to identify genes involved in crop domestication process [74].

3.2. Application of Genome Editing Systems in Crop Improvement. Genome editing technologies have wide practical applications for solving one of the most important tasks of modern biotechnology—the creation of new varieties of crops, which are high-yielding and resistant to abiotic and biotic stresses and also have high nutritional value (Table 2) [31, 63, 75–80]. To this end, genome editing system has been used in plant breeding (1) to insert point mutations similar to natural SNPs [26, 27], (2) to make small modifications to gene function [13], (3) for integration of foreign genes, (4) for gene pyramiding and knockout, and (5) for the repression or activation of gene expression, as well as (6) epigenetic editing [6].

For example, the use of ZFN in *Arabidopsis thaliana* [15–17] and *Zea mays* [18] has led to the successful development of herbicide tolerant genotypes through insertion of herbicide-resistance genes into targeted sites in the genome [18]. ZFN was also used for the targeted modification of an endogenous malate dehydrogenase (MDH) gene in plants and the plants containing modified MDH have shown increased yield [81]. ODM technique has been significantly advanced through Cibus Rapid Trait Development System (RTDS) [7] and this technology has been successfully applied in several crops. Applications include but are not limited to the generation of herbicide tolerance, insect resistance, enhanced disease resistance (bacterial and viral), improved nutritional value, and enhanced yield without the introduction of foreign genes as has been used in traditional genetic engineering approach for crop development [7, 65]. A precise editing of CAC to TAC using ODM RTDS technology has been demonstrated that converts BEP to GFP by changing Histidine (H66) to Tyrosine (Y66) in GFP protein. This approach has offered a nontransgenic breeding tool for crops [7, 64].

Using the CRISPR/Cas9 technology, Jiang et al. [28] have obtained "a biotech" oil from *Camelina sativa* seeds with an improved fatty acid composition, which makes it more beneficial to human health, more resistant to oxidation, and more appropriate for the production of certain commercial chemicals including biofuels [28]. Soyk et al. [29] used targeted mutagenesis of SP5G gene of tomato to create plants with rapid flowering and more compact bush, which in turn resulted in earlier harvest. In another effort, Osakabe et al. [31], using the CRISPR-induced mutagenesis of OST2 gene in *Arabidopsis*, were able to obtain new alleles that confer salt stress resistance to plants [31].

Modulation of the gibberellin biosynthesis by genome editing methods has allowed creation of dwarf fruit trees [30], which have great potential for increasing productivity through higher density plantings and reduced labor costs. This results in a reduction of land, water, pesticide, and fertilizer use [82]. In addition, genome editing for inhibition of ethylene biosynthesis, which plays a very important role in fruit ripening process [82] or its signaling pathways, enables creation of new varieties with extended shelf life [63].

A major area of application of genome editing approaches in plant breeding is to create varieties resistant to various pathogens and/or pests. These methods have been used for the modification of the key plant immunity stages at different levels in several crops. This goal can be achieved by modifying (1) susceptibility genes (S-genes), (2) resistance genes (R-genes), (3) genes regulating the interaction between the effector and target, and (4) the genes regulating plant hormonal balance [78]. For example, wheat genotypes resistant to powdery mildew disease were obtained by TALEN- and CRISPR/Cas9-mediated genome editing on mildew-resistance locus O (MLO) [34]. Genome editing technologies have also been used to produce plants resistant to bacterial leaf blight, caused by *Xanthomonas oryzae* pv. *oryzae* [21].

The CRISPR/Cas9 system has been investigated for its efficacy in providing interference against geminiviruses by using a transient transformation system such that *N. benthamiana* degradation/suppression of curly top virus genome by single guide RNA/Cas9 (sgRNA/Cas9) has been demonstrated [35]. In other efforts, where sgRNAs specific for tomato yellow leaf curl virus (TYLCV) or bean yellow dwarf virus (BeYDV) sequences were introduced into *N. benthamiana* plants expressing Cas9 endonuclease and challenged with the corresponding viruses, it was demonstrated that the CRISPR/Cas9 system not only targeted viruses for degradation but also introduced mutations at the target sequences [36, 37] due to interference with the copy number of freely replicating viruses [78].

Metabolic pathways that regulate hormonal balance can also be modified using the genome editing technologies to enhance the immunomodulatory component of the plants immune system. This can be achieved by deactivating the ethylene-responsive factor (ERF). In particular, ethylene-dependent pathway in rice has been successfully modified by CRISPR/Cas9-mediated target OsERF922 gene mutations, resulting in increased resistance to *Magnaporthe oryzae* [38, 39].

CRISPR/Cas9 has been used to knock out eIF4E gene that encodes the eukaryotic translation initiation factor essential for translation of viruses, in *Cucumis sativus*, and that knockout confers resistance to viruses such as cucumber vein yellowing virus (CVYV), zucchini yellow mosaic virus (ZYMV), and papaya ring spot mosaic virus-W (PRSV-W) [83]. In addition, CRISPR/Cas9 was demonstrated to be an efficient system for rapid and efficient genome editing in *Phytophthora sojae*, an oomycete pathogen of Soybean, by modifying the pathogenicity gene (Avr4/6), thereby opening up an avenue for the much needed functional genomics work in *Phytophthora sojae* towards the ultimate goal of controlling this pathogen [83].

Similarly, existing genome editing methods, in particular, CRISPR/Cas9 method, have been successfully used to obtain plants resistant to herbicides [33]. For example, editing of ALS2 gene in maize (acetolactate synthase or ALS is a key enzyme in the biosynthesis of amino acids in plants and has been inhibited by sulfonylurea herbicides) allowed the creation of a mutant corn plant resistant to chlorsulfuron [33].

Another interesting area of biotechnology where CRISPR/Cas9 system has significant application is the development of plants capable of synthesizing human proteins such as insulin, necessary for patients with diabetes mellitus, or albumin, which is used in the treatment of hemorrhagic shock, burns, hypoproteinemia, and cirrhosis [84]. At present, albumin is prepared from human plasma which is in a very limited supply; however, global demand for albumin is constantly growing and currently is equal to 500 tons per year. To meet the growing needs human albumin gene is already introduced into rice genome using genomic engineering techniques [85]. Such expressed proteins can be isolated from plant and animal tissues, where it is synthesized, and after clarification, it can be used for medical purposes.

Thus, as described above and extensively referenced herein, these novel genome editing techniques are being widely used for the purpose of crop improvement including new bioenergy crop developments [86]. However, the use of tissue culture with these GEEN methods may also create complexities that could slow the process of genome editing.

4. Safety Assessment Aspects of Genome Editing Systems

4.1. Nontarget Effects. Genome editing techniques, in essence, preserve the native genomic structure and, therefore, are considered as a safe technology for crop improvement. Despite this general understanding, there are some concerns related to the biosafety of crops created using these methods. One main concern in terms of its biosafety is the possibility of nontarget effects of synthetic nucleases during genome editing.

During the biotechnological application of genome editing methods, efficiency and specificity of the engineered nucleases are the two most important functional requirements and are closely related to the choice of the target site. For each endogenous genomic locus, efficiency of DNA cleavage (both target and nontarget) depends not only on the nuclease activity (such as *Fok*I domains and *Ruv* domains of the Cas9 proteins), but also on the availability of a target site and affinity of the DNA-binding domain (e.g., TAL effector domains and guide RNA, gRNA) to the target sequence. Specificity of engineered nucleases largely depends on the binding affinity of nuclease-DNA, including the binding of zinc finger to DNA (ZFNs), TAL effector to DNA (TALENs), and hybridization of gRNA with DNA (CRISPR), although dimerization of *Fok*I domain (ZFNs and TALENs) and Cas9 interaction with the motif contiguous to protospacer adjacent motif (PAM) may also play an important role [87]. In case of ZFNs, while examples abound with respect to the binding efficiency of canonical C2H2 binding domain containing ZFNs, investigations on the utility of noncanonical ZFNs such as those containing C3H1 binding domain have demonstrated high levels of binding efficiency [88].

To minimize nontarget effects of genome editing systems, a crucial aspect is the careful selection of sites for the introduction of the double-stranded breaks by performing a prior bioinformatics analysis [89]. When choosing the desired sites, sites of repeated sequences and sites having a high homology with other regions of the genome should be avoided. In this regard, to facilitate the selection of the target sites for nucleases and experimental verification of the presence of nontarget effects, several software packages were developed that enable nuclease design and validation [79, 87, 90].

4.2. Regulation of Plants Created by Genome Editing. The novel genome editing systems help to introduce stably inherited point modifications into the plant genome, and transgenic region can be easily removed after editing a target gene. This allows creation of nontransgenic plants and improved crop varieties [22, 91–93]. These technologies are faster compared to traditional breeding methods and help to obtain the null segregant lines that have lost the transgene insertion [94–97]. Plants with targeted mutations developed by genome editing technology are nearly identical to plants obtained by classical breeding, and their safety must be assessed taking into account the resulting product rather than the process used to create them [98–100]. In this context, ODM-derived products are in many cases indistinguishable from conventionally bred or traditional mutagenesis products; therefore, such products should not be regulated in the same way as the products generated by genetic engineering methods [7, 65]. Using CRISPR-Cas9 system, it becomes possible to obtain marker-free genetically engineered crops, that is, without marker genes of antibiotic resistance [6, 100]. Thus, in the case of new varieties with targeted mutations, developed using genome editing systems, the existing operating rules for the regulation of genetically modified plants should not be applied [92, 95, 99, 100]. Currently genome editing technologies are being discussed by various advisory and regulatory authorities in the context of GMO legislation. Cultures and plants obtained using genome editing techniques are considered as nongenetically modified [95, 99, 101]. The European Commission is expected to publish a report on regulatory uncertainty of genome editing methods [100, 102, 103].

5. Multitude of Advantages and Perspectives of GEENs

Tools of genome editing have a significant impact on basic and applied research in plant biology [24, 43, 44, 73]. The simplified approach to gene/genome editing represents a valuable tool for plant researchers in functional analysis of gene(s) and for breeders in the integration of key genes in the genomes of agriculturally important crops. Genome editing systems have several attractive features including simplicity, efficiency, high specificity, minimal nontarget effects, and amenability to multiplexing and thus are very promising for use in plant breeding [6].

Site-directed mutagenesis of different genes can provide important information about their functions. Simultaneous targeting of multiple genes/loci by applying multiplex strategies can promote research to identify the role of individual genes in the intracellular signaling pathways and aid in the engineering of complex, multigenic agronomic traits in crops. The preferred use of CRISPR-Cas9 system can be exemplified in completely knockout gene function [6, 64], microRNA knockdown screening [6], and programmed editing of certain loci by genome editing systems that can provide a functional separation of cis- and trans-regulatory elements/factors with high accuracy [6]. Another prospective application of CRISPR-Cas9 system may be its use in the formation of conditional alleles, providing spatial and temporal control of gene expression to study the function of lethal genes. Use of inducible or tissue-specific promoters for expression of Cas9 and/or sgRNA can be instrumental for gene expression regulation in a specific tissue, in development stage, or in different environmental conditions [6].

CRISPR-Cas system opens up wide possibilities for labeling endogenous genes with fluorescent proteins to visualize their expression in vivo. Using fluorescent labeled dCas9, changes of genome dynamics/chromosome architectural changes during plant development and their response to environmental stimuli can be learned. These technologies can also be used for the selection of the specific cell types that greatly facilitate the study of various functional aspects [6]. Use of dCas9 can provide a new platform for the selection of activation/repression effector domains to specific genomic loci for regulating endogenous gene expression.

In addition, these technologies can be successfully used in the work on epigenome editing via the selection of proteins responsible for histone modification and DNA methylation, which has emerged as a new way of regulating cellular functions in plants [25]. For the purpose of understanding epigenetic regulation, CRISPR-Cas9 system can also be used for the enrichment of chromatin target sites for the identification of proteins attached to enriched chromatin. Likewise, CRISPR-Cas9 can be used as a tool to identify regulatory proteins binding to specific DNA sequences controlling the expression of genes.

6. Conclusion

Genome editing tools are becoming popular molecular tools of choice for functional genomics as well as crop improvement. Many examples exist currently where these editing systems are being harnessed for unprecedented understanding of plant biology and crop yield improvement through rapid and targeted mutagenesis and associated breeding [102, 104]. Because of their several attractive features such as simplicity, efficiency, high specificity, and amenability to multiplexing, genome editing technologies described here are revolutionizing the way crop breeding is done and paving the way for the next generation breeding.

Acknowledgments

The authors thank Academy of Sciences of Uzbekistan and Science and Technology Agency of Uzbekistan for Research Grants nos. FA-F5-021 and FA-F5-025.

References

[1] M. F. Singer, "Introduction and historical background," in *Genetic Engineering*, J. K. Setlow and A. Hollaender, Eds., vol. 1, pp. 1–13, Plenum, New York, NY, USA, 1979.

[2] A. A. Nemudryi, K. R. Valetdinova, S. P. Medvedev, and S. M. Zakian, "TALEN and CRISPR/Cas genome editing systems: tools of discovery," *Acta Naturae*, vol. 6, no. 22, pp. 19–40, 2014.

[3] Y.-G. Kim, J. Cha, and S. Chandrasegaran, "Hybrid restriction enzymes: zinc finger fusions to Fok I cleavage domain," *Proceedings of the National Academy of Sciences of the United States of America*, vol. 93, no. 3, pp. 1156–1160, 1996.

[4] T. Gaj, C. A. Gersbach, and C. F. Barbas III, "ZFN, TALEN, and CRISPR/Cas-based methods for genome engineering," *Trends in Biotechnology*, vol. 31, no. 7, pp. 397–405, 2013.

[5] D. P. Weeks, M. H. Spalding, and B. Yang, "Use of designer nucleases for targeted gene and genome editing in plants," *Plant Biotechnology Journal*, vol. 14, no. 2, pp. 483–495, 2016.

[6] V. Kumar and M. Jain, "The CRISPR-Cas system for plant genome editing: Advances and opportunities," *Journal of Experimental Botany*, vol. 66, no. 1, pp. 47–57, 2015.

[7] N. J. Sauer, J. Mozoruk, R. B. Miller et al., "Oligonucleotide-directed mutagenesis for precision gene editing," *Plant Biotechnology Journal*, vol. 14, no. 2, pp. 496–502, 2016.

[8] J. F. Li, J. E. Norville, and J. Aach, "Multiplex and homologous recombination-mediated genome editing in *Arabidopsis* and *Nicotiana benthamiana* using guide RNA and Cas9," *Nature Biotechnolog*, vol. 31, no. 8, pp. 688–691, 2013.

[9] M. Jinek, K. Chylinski, I. Fonfara, M. Hauer, J. A. Doudna, and E. Charpentier, "A programmable dual-RNA-guided DNA endonuclease in adaptive bacterial immunity," *Science*, vol. 337, no. 6096, pp. 816–821, 2012.

[10] B. Chen, J. Hu, R. Almeida et al., "Expanding the CRISPR imaging toolset with Staphylococcus aureus Cas9 for simultaneous imaging of multiple genomic loci," *Nucleic Acids Research*, vol. 44, no. 8, p. e75, 2016.

[11] S. W. Cho, S. Kim, J. M. Kim, and J.-S. Kim, "Targeted genome engineering in human cells with the Cas9 RNA-guided endonuclease," *Nature Biotechnology*, vol. 31, no. 3, pp. 230–232, 2013.

[12] A. Noman, M. Aqeel, and S. He, "CRISPR-Cas9: Tool for qualitative and quantitative plant genome editing," *Frontiers in Plant Science*, vol. 7, no. 2016, article no. 1740, 2016.

[13] Y. Mao, H. Zhang, N. Xu, B. Zhang, F. Gou, and J.-K. Zhu, "Application of the CRISPR-Cas system for efficient genome engineering in plants," *Molecular Plant*, vol. 6, no. 6, pp. 2008–2011, 2013.

[14] P. D. Hsu, D. A. Scott, J. A. Weinstein et al., "DNA targeting specificity of RNA-guided Cas9 nucleases," *Nature Biotechnol*, vol. 31, pp. 827–832, 2013.

[15] K. Osakabe, Y. Osakabe, and S. Toki, "Site-directed mutagenesis in Arabidopsis using custom-designed zinc finger nucleases," *Proceedings of the National Academy of Sciences of the United States of America*, vol. 107, no. 26, pp. 12034–12039, 2010.

[16] F. Zhang, M. L. Maeder, E. Unger-Wallaced et al., "High frequency targeted mutagenesis in *Arabidopsis thaliana* using zinc finger nucleases," *Proceedings of the National Academy of Sciences of the United States of America*, vol. 107, no. 26, pp. 12028–12033, 2010.

[17] J. A. Townsend, D. A. Wright, R. J. Winfrey et al., "High-frequency modification of plant genes using engineered zinc-finger nucleases," *Nature*, vol. 459, no. 7245, pp. 442–445, 2009.

[18] V. K. Shukla, Y. Doyon, J. C. Miller et al., "Precise genome modification in the crop species Zea mays using zinc-finger nucleases," *Nature*, vol. 459, pp. 437–441, 2009.

[19] M. Christian, Y. Qi, Y. Zhang, and D. F. Voytas, "Targeted Mutagenesis of Arabidopsis thaliana Using Engineered TAL Effector Nucleases," *G3: Genes, Genomes, Genetics*, vol. 3, no. 9, pp. 1697–1705, 2013.

[20] H. Zhang, F. Gou, J. Zhang et al., "TALEN-mediated targeted mutagenesis produces a large variety of heritable mutations in rice," *Plant Biotechnology Journal*, vol. 14, no. 1, pp. 186–194, 2016.

[21] T. Li, B. Liu, M. H. Spalding, D. P. Weeks, and B. Yang, "High-efficiency TALEN-based gene editing produces disease-resistant rice," *Nature Biotechnology*, vol. 30, no. 5, pp. 390–392, 2012.

[22] M. M. Mahfouz, L. Li, M. Piatek et al., "Targeted transcriptional repression using a chimeric TALE-SRDX repressor protein," *Plant Molecular Biology*, vol. 78, no. 3, pp. 311–321, 2012.

[23] J. Gao, G. Wang, S. Ma et al., "CRISPR/Cas9-mediated targeted mutagenesis in Nicotiana tabacum," *Plant Molecular Biology*, vol. 87, no. 1-2, pp. 99–110, 2015.

[24] L. Cong, F. A. Ran, D. Cox et al., "Multiplex genome engineering using CRISPR/Cas systems," *Science*, vol. 339, no. 6121, pp. 819–823, 2013.

[25] H. Puchta, "Using CRISPR/Cas in three dimensions: towards synthetic plant genomes, transcriptomes and epigenomes," *Plant Journal*, vol. 87, no. 1, pp. 5–15, 2016.

[26] T. B. Jacobs, P. R. LaFayette, R. J. Schmitz, and W. A. Parrott, "Targeted genome modifications in soybean with CRISPR/Cas9," *BMC Biotechnology*, pp. 1–10, 2015.

[27] R. Xu, R. Qin, H. Li et al., "Generation of targeted mutant rice using a CRISPR-Cpf1 system," *Plant Biotechnology Journal*, vol. 14, pp. 1–5, 2016.

[28] W. Z. Jiang, I. M. Henry, P. G. Lynagh, L. Comai, E. B. Cahoon, and D. P. Weeks, "Significant enhancement of fatty acid composition in seeds of the allohexaploid, Camelina sativa, using CRISPR/Cas9 gene editing," *Plant Biotechnology Journal*, vol. 15, no. 5, pp. 648–657, 2017.

[29] S. Soyk, N. A. Müller, S. J. Park et al., "Variation in the flowering gene SELF PRUNING 5G promotes day-neutrality and early yield in tomato," *Nature Genetics*, vol. 49, no. 1, pp. 162–168, 2017.

[30] J. Peng, D. E. Richards, and N. M. Hartley, "Green revolution genes encode mutant gibberellin response modulators," *Nature*, vol. 400, no. 6741, pp. 256–261, 1999.

[31] Y. Osakabe, T. Watanabe, S. S. Sugano et al., "Optimization of CRISPR/Cas9 genome editing to modify abiotic stress responses in plants," *Scientific Reports*, vol. 6, Article ID 26685, 2016.

[32] Q. Shan, Y. Wang, and J. Li, "Targeted genome modification of crop plants using a CRISPR-Cas system," *Nature Biotechnol*, vol. 31, pp. 686–688, 2013.

[33] S. Svitashev, J. K. Young, C. Schwartz, H. Gao, S. C. Falco, and A. M. Cigan, "Targeted mutagenesis, precise gene editing, and site-specific gene insertion in maize using Cas9 and guide RNA," *Plant Physiology*, vol. 169, no. 2, pp. 931–945, 2015.

[34] Y. Wang, X. Cheng, and Q. Shan, "Simultaneous editing of three homoeoalleles in hexaploid bread wheat confers heritable resistance to powdery mildew," *Nature Biotechnol*, vol. 32, pp. 947–952, 2014.

[35] X. Ji, H. Zhang, Y. Zhang, Y. Wang, and C. Gao, "Establishing a CRISPR-Cas-like immune system conferring DNA virus resistance in plants," *Nature Plants*, vol. 1, article 15144, Article ID 15144, 2015.

[36] Z. Ali, A. Abulfaraj, A. Idris, S. Ali, M. Tashkandi, and M. M. Mahfouz, "CRISPR/Cas9-mediated viral interference in plants," *Genome Biology*, vol. 16, no. 1, article 238, 2015.

[37] N. J. Baltes, A. W. Hummel, and E. Konecna, "Conferring resistance to geminiviruses with the CRISPR-Cas prokaryotic immune system," *Nature Plants*, vol. 1, article 15145, 2015.

[38] D. Liu, X. Chen, J. Liu, J. Ye, and Z. Guo, "The rice ERF transcription factor OsERF922 negatively regulates resistance to Magnaporthe oryzae and salt tolerance," *Journal of Experimental Botany*, vol. 63, no. 10, pp. 3899–3912, 2012.

[39] F. Wang, C. Wang, P. Liu et al., "Enhanced rice blast resistance by CRISPR/ Cas9-Targeted mutagenesis of the ERF transcription factor gene OsERF922," *PLoS ONE*, vol. 11, no. 4, Article ID e0154027, 2016.

[40] J. Chandrasekaran, M. Brumin, D. Wolf et al., "Development of broad virus resistance in non-transgenic cucumber using CRISPR/Cas9 technology," *Molecular Plant Pathology*, vol. 17, no. 7, pp. 1140–1153, 2016.

[41] F. Zhang, Y. Wen, and X. Guo, "CRISPR/Cas9 for genome editing: Progress, implications and challenges," *Human Molecular Genetics*, vol. 23, no. 1, pp. R40–R46, 2014.

[42] J. F. Petolino and J. P. Davies, "Designed transcriptional regulators for trait development," *Plant Science*, vol. 201-202, no. 1, pp. 128–136, 2013.

[43] H. Wang, H. Yang, C. S. Shivalila et al., "One-step generation of mice carrying mutations in multiple genes by CRISPR/Cas-mediated genome engineering," *Cell*, vol. 153, no. 4, pp. 910–918, 2013.

[44] L. Lowder, A. Malzahn, and Y. Qi, "Rapid evolution of manifold CRISPR systems for plant genome editing," *Frontiers in Plant Science*, vol. 7, no. 2016, article no. 1683, 2016.

[45] D. G. Knorre and V. V. Vlasov, "Reactive derivatives of nucleic acids and their components as affinity reagents," *Russian Chemical Reviews*, vol. 54, no. 9, pp. 836–851, 1985.

[46] N. J. Palpant and D. Dudzinski, "Zinc finger nucleases: Looking toward translation," *Gene Therapy*, vol. 20, no. 2, pp. 121–127, 2013.

[47] R. Jankele and P. Svoboda, "TAL effectors: Tools for DNATargeting," *Briefings in Functional Genomics*, vol. 13, no. 5, pp. 409–419, 2014.

[48] C. O. Pabo, E. Peisach, and R. A. Grant, "Design and selection of novel Cys2His2 zinc finger proteins," *Annual Review of Biochemistry*, vol. 70, pp. 313–340, 2001.

[49] T. Cathomen and J. Keith Joung, "Zinc-finger nucleases: the next generation emerges," *Molecular Therapy*, vol. 16, no. 7, pp. 1200–1207, 2008.

[50] J. F. Petolino, "Genome editing in plants via designed zinc finger nucleases," *In Vitro Cellular and Developmental Biology - Plant*, vol. 51, no. 1, 2015.

[51] N. P. Pavletich and C. O. Pabo, "Zinc finger-DNA recognition: crystal structure of a Zif268-DNA complex at 2.1 Å," *Science*, vol. 252, no. 5007, pp. 809–817, 1991.

[52] W. M. Ainley, L. Sastry-Dent, M. E. Welter et al., "Trait stacking via targeted genome editing," *Plant Biotechnology Journal*, vol. 11, no. 9, pp. 1126–1134, 2013.

[53] J. F. Petolino, A. Worden, K. Curlee et al., "Zinc finger nuclease-mediated transgene deletion," *Plant Molecular Biology*, vol. 73, no. 6, pp. 617–628, 2010.

[54] S. Schornack, A. Meyer, P. Römer, T. Jordan, and T. Lahaye, "Gene-for-gene-mediated recognition of nuclear-targeted AvrBs3-like bacterial effector proteins," *Journal of Plant Physiology*, vol. 163, no. 3, pp. 256–272, 2006.

[55] P. Römer, S. Hahn, T. Jordan, T. Strauß, U. Bonas, and T. Lahaye, "Plant pathogen recognition mediated by promoter activation of the pepper Bs3 resistance gene," *Science*, vol. 318, no. 5850, pp. 645–648, 2007.

[56] J. Boch, H. Scholze, S. Schornack et al., "Breaking the code of DNA binding specificity of TAL-type III effectors," *Science*, vol. 326, no. 5959, pp. 1509–1512, 2009.

[57] B. M. Lamb, A. C. Mercer, and C. F. Barbas III, "Directed evolution of the TALE N-terminal domain for recognition of all 50 bases," *Nucleic Acids Research*, vol. 41, no. 21, pp. 9779–9785, 2013.

[58] M. Christian, T. Cermak, E. L. Doyle et al., "Targeting DNA double-strand breaks with TAL effector nucleases," *Genetics*, vol. 186, no. 2, pp. 757–761, 2010.

[59] L. Cong, R. H. Zhou, Y.-C. Kuo, M. Cunniff, and F. Zhang, "Comprehensive interrogation of natural TALE DNA-binding modules and transcriptional repressor domains," *Nature Communications*, vol. 3, article 968, 2012.

[60] M. L. Christian, Z. L. Demorest, C. G. Starker et al., "Targeting G with TAL Effectors: A Comparison of Activities of TALENs Constructed with NN and NK Repeat Variable Di-Residues," *PLoS ONE*, vol. 7, no. 9, Article ID e45383, 2012.

[61] J. Streubel, C. Blücher, A. Landgraf, and J. Boch, "TAL effector RVD specificities and efficiencies," *Nature Biotechnology*, vol. 30, no. 7, pp. 593–595, 2012.

[62] A. N.-S. Mak, P. Bradley, R. A. Cernadas, A. J. Bogdanove, and B. L. Stoddard, "The crystal structure of TAL effector PthXo1 bound to its DNA target," *Science*, vol. 335, no. 6069, pp. 716–719, 2012.

[63] J. Xiong, J. Ding, and Y. Li, "Genome-editing technologies and their potential application in horticultural crop breeding," *Horticulture Research*, vol. 2, article 15019, 2015.

[64] I. Y. Abdurakhmonov, "Genomics Era for Plants and Crop Species—Advances Made and Needed Tasks Ahead," in *Plant Genomics*, I. Abdurakhmonov, Ed., InTech, Croatia, Balkans, 2016.

[65] CropLife International, "Oligonucleotide-Directed Mutagenesis (ODM)," LJournal, 2017.

[66] R. Barrangou, C. Fremaux, and H. Deveau, "CRISPR provides acquired resistance against viruses in prokaryotes," *Science*, vol. 315, no. 5819, pp. 1709–1712, 2007.

[67] E. Deltcheva, K. Chylinski, C. M. Sharma et al., "CRISPR RNA maturation by trans-encoded small RNA and host factor RNase III," *Nature*, vol. 471, no. 7340, pp. 602–607, 2011.

[68] D. H. Haft, J. Selengut, E. F. Mongodin, and K. E. Nelson, "A guild of 45 CRISPR-associated (Cas) protein families and multiple CRISPR/cas subtypes exist in prokaryotic genomes," *PLoS Computational Biology*, vol. 1, article e60, no. 6, pp. 0474–0483, 2005.

[69] A. F. Gilles and M. Averof, "Functional genetics for all: Engineered nucleases, CRISPR and the gene editing revolution," *EvoDevo*, vol. 5, no. 1, article no. 43, 2014.

[70] J. A. Doudna and E. Charpentier, "The new frontier of genome engineering with CRISPR-Cas9," *Science*, vol. 346, no. 6213, 2014.

[71] D. B. Graham and D. E. Root, "Resources for the design of CRISPR gene editing experiments," *Genome Biology*, vol. 16, no. 1, article no. 260, 2015.

[72] L. C. Perkin, S. L. Adrianos, and B. Oppert, "Gene disruption technologies have the potential to transform stored product insect pest control," *Insects*, vol. 7, no. 3, article no. 46, 2016.

[73] P. Perez-Pinera, D. G. Ousterout, and C. A. Gersbach, "Advances in targeted genome editing," *Current Opinion in Chemical Biology*, vol. 16, no. 3-4, pp. 268–277, 2012.

[74] L. Chen, L. Tang, H. Xiang et al., "Advances in genome editing technology and its promising application in evolutionary and ecological studies," *GigaScience*, vol. 3, no. 1, article no. 24, 2014.

[75] C. Kissoudis, C. van de Wiel, R. G. F. Visser, and G. van der Linden, "Enhancing crop resilience to combined abiotic and biotic stress through the dissection of physiological and molecular crosstalk," *Frontiers in Plant Science*, vol. 5, no. MAY, article no. 207, 2014.

[76] L. Liu and X.-D. Fan, "CRISPR-Cas system: A powerful tool for genome engineering," *Plant Molecular Biology*, vol. 85, no. 3, pp. 209–218, 2014.

[77] M. Jain, "Function genomics of abiotic stress tolerance in plants: A CRISPR approach," *Frontiers in Plant Science*, vol. 6, no. MAY, article no. 375, pp. 1–4, 2015.

[78] G. Andolfo, P. Iovieno, L. Frusciante, and M. R. Ercolano, "Genome-editing technologies for enhancing plant disease resistance," *Frontiers in Plant Science*, vol. 7, no. 2016, article no. 1813, 2016.

[79] S. Khatodia, K. Bhatotia, N. Passricha, S. M. P. Khurana, and N. Tuteja, "The CRISPR/Cas genome-editing tool: Application in improvement of crops," *Frontiers in Plant Science*, vol. 7, no. 2016, article no. 506, 2016.

[80] R. C. Nongpiur, S. L. Singla-Pareek, and A. Pareek, "Genomics Approaches for Improving Salinity Stress Tolerance in Crop Plants," *Current Genomics*, vol. 17, no. 4, pp. 343–357, 2016.

[81] V. Shukla, M. Gupta, F. Urnov, D. Guschin, M. Jan, and P. Bundock, "Targeted modifcation of malate dehydrogenase, 2013," WO Patent Publication Number: WO 2013166315 A1.

[82] C. A. Hollender and C. Dardick, "Molecular basis of angiosperm tree architecture," *New Phytologist*, vol. 206, no. 2, pp. 541–556, 2015.

[83] Y. Fang and B. M. Tyler, "Efficient disruption and replacement of an effector gene in the oomycete Phytophthora sojae using CRISPR/Cas9," *Molecular Plant Pathology*, vol. 17, no. 1, pp. 127–139, 2016.

[84] G. E. Hastings and P. G. Wolf, "The Therapeutic Use of Albumin," *Archives of Family Medicine*, vol. 1, no. 2, pp. 281–287, 1992.

[85] Y. He, T. Ning, T. Xie et al., "Large-scale production of functional human serum albumin from transgenic rice seeds," *Proceedings of the National Academy of Sciences of the United States of America*, vol. 108, no. 47, pp. 19078–19083, 2011.

[86] M. Bosch and S. P. Hazen, "Lignocellulosic feedstocks: Research progress and challenges in optimizing biomass quality and yield," *Frontiers in Plant Science*, vol. 4, article no. 474, 2013.

[87] C. M. Lee, T. J. Cradick, E. J. Fine, and G. Bao, "Nuclease target site selection for maximizing on-target activity and minimizing off-target effects in genome editing," *Molecular Therapy*, vol. 24, no. 3, pp. 475–487, 2016.

[88] Q. C. Cai, J. Miller, F. Urnov et al., "Optimized non-canonical zinc finger proteins," US Patent Number: 9,187,758. Publication date: Nov 17, 2015.

[89] A. Lombardo, D. Cesana, P. Genovese et al., "Site-specific integration and tailoring of cassette design for sustainable gene transfer," *Nature Methods*, vol. 8, no. 10, pp. 861–869, 2011.

[90] T. Koo, J. Lee, and J. Kim, "Measuring and reducing off-target activities of programmable nucleases including CRISPR-Cas9," *Molecules and Cells*, vol. 38, no. 6, pp. 475–481, 2015.

[91] Y. Gao and Y. Zhao, "Specific and heritable gene editing in Arabidopsis," *Proceedings of the National Academy of Sciences of the United States of America*, vol. 111, no. 12, pp. 4357-4358, 2014.

[92] C. Nagamangala Kanchiswamy, D. J. Sargent, R. Velasco, M. E. Maffei, and M. Malnoy, "Looking forward to genetically edited fruit crops," *Trends in Biotechnology*, vol. 33, no. 2, pp. 62–64, 2015.

[93] R.-F. Xu, H. Li, R.-Y. Qin et al., "Generation of inheritable and "transgene clean" targeted genome-modified rice in later generations using the CRISPR/Cas9 system," *Scientific Reports*, vol. 5, Article ID 11491, 2015.

[94] N. Podevin, Y. Devos, H. V. Davies, and K. M. Nielsen, "Transgenic or not? No simple answer! New biotechnology-based plant breeding techniques and the regulatory landscape," *EMBO Reports*, vol. 13, no. 12, pp. 1057–1061, 2012.

[95] M. Araki and T. Ishii, "Towards social acceptance of plant breeding by genome editing," *Trends in Plant Science*, vol. 20, no. 3, pp. 145–149, 2015.

[96] J. G. Schaart, C. C. M. van de Wiel, L. A. P. Lotz, and M. J. M. Smulders, "Opportunities for Products of New Plant Breeding Techniques," *Trends in Plant Science*, vol. 21, no. 5, pp. 438–449, 2016.

[97] J. W. Woo, J. Kim, S. I. Kwon et al., "DNA-free genome editing in plants with preassembled CRISPR-Cas9 ribonucleoproteins," *Nature Biotechnology*, vol. 33, no. 11, pp. 1162–1164, 2015.

[98] F. Hartung and J. Schiemann, "Precise plant breeding using new genome editing techniques: Opportunities, safety and

regulation in the EU," *Plant Journal*, vol. 78, no. 5, pp. 742–752, 2014.

[99] D. F. Voytas and C. Gao, "Precision genome engineering and agriculture: opportunities and regulatory challenges," *PLoS biology*, vol. 12, no. 6, p. e1001877, 2014.

[100] H. D. Jones, "Regulatory uncertainty over genome editing," *Nature Plants*, vol. 1, Article ID 14011, 2015.

[101] M. Lusser, C. Parisi, D. Plan, and E. Rodríguez-Cerezo, "Deployment of new biotechnologies in plant breeding," *Nature Biotechnology*, vol. 30, no. 3, pp. 231–239, 2012.

[102] K. Belhaj, A. Chaparro-Garcia, S. Kamoun, N. J. Patron, and V. Nekrasov, "Editing plant genomes with CRISPR/Cas9," *Current Opinion in Biotechnology*, vol. 32, pp. 76–84, 2015.

[103] J. D. Wolt, K. Wang, and B. Yang, "The regulatory status of genome-edited crops," *Plant Biotechnology Journal*, vol. 14, no. 2, pp. 510–518, 2016.

[104] S. Huang, D. Weigel, R. N. Beachy, and J. Li, "A proposed regulatory framework for genome-edited crops," *Nature Genetics*, vol. 48, no. 2, pp. 109–111, 2016.

18

Response of Sesame to Selected Herbicides Applied Early in the Growing Season

W. James Grichar,[1] **Jack J. Rose**,[2] **Peter A. Dotray**,[3] **Todd A. Baughman**,[4] **D. Ray Langham**,[5] **Kaisa Werner**,[6] **and Muthu Bagavathiannan**[6]

[1]*Texas A&M AgriLife Research and Extension Center, Corpus Christi, TX 78406, USA*
[2]*Sesaco Corporation, Austin, TX 78741, USA*
[3]*Texas A&M AgriLife Research and Extension Center, Lubbock, TX 79403, USA*
[4]*Institute for Agricultural Biosciences, Oklahoma State University, Ardmore, OK 73402, USA*
[5]*Sesame Research, LLC, San Antonio, TX 78217, USA*
[6]*Texas A&M AgriLife Research, College Station, TX 77843, USA*

Correspondence should be addressed to W. James Grichar; w-grichar@tamu.edu

Academic Editor: David Clay

Growth chamber experiments were conducted to evaluate the response of sesame to PRE and POST applications of soil residual herbicides. PRE applications of acetochlor and S-metolachlor at 1.26 and 1.43 kg ai·ha^{-1} showed little or no sesame injury (0 to 1%) 4 wks after herbicide treatments (WAT). POST treatments of acetochlor and trifluralin made 3 wks after planting (WAP) resulted in greater sesame injury (40%) compared to applications at bloom (18%). Field studies were conducted in Texas and Oklahoma during the 2014 and 2015 growing seasons to determine sesame response to clethodim, diuron, fluometuron, ethalfluralin, quizalofop-P, pendimethalin, pyroxasulfone, trifluralin, and trifloxysulfuron-sodium applied 2, 3, or 4 weeks after planting (WAP). Late-season sesame injury with the dinitroaniline herbicides consisted of a proliferation of primary branching at the upper nodes of the sesame plant (in the shape/form of a broom). Ethalfluralin and trifluralin caused more "brooming" effect than pendimethalin. Some yield reductions were noted with the dinitroaniline herbicides. Trifloxysulfuron-sodium caused the greatest injury (up to 97%) and resulted in yield reductions from the untreated check. Early-season diuron injury (leaf chlorosis and necrosis) decreased as application timing was delayed, and late-season injury was virtually nonexistent with only slight chlorosis (<4%) still apparent on the lower leaves. Sesame yield was not consistently affected by the diuron treatments. Fluometuron caused early-season injury (stunting/chlorosis), and a reduction of yield was observed at one location. Pyroxasulfone applied 2 WAP caused up to 25% sesame injury (stunting) but did not result in a yield reduction. Quizalofop-P caused slight injury (<5%) and no reduction in yield.

1. Introduction

Sesame production has typically occurred in China and India using dehiscent varieties that are prone to capsule shattering during harvest; therefore, hand harvesting is nearly compulsory to minimize seed loss [1]. Sesame was introduced to the US from Africa and was called beni/benne/benni. Betts [2] quotes letters from Thomas Jefferson that document his trials with sesame between 1808 and 1824. Jefferson stated that sesame "...is among the most valuable acquisitions our country has ever made.... I do not believe before that there existed so perfect a substitute for olive oil." He talks about the rule of thumb that still exists today that sesame will do well where cotton (*Gossypium hirsutum* L.) does well. Nondehiscent varieties were developed in the US during the 1950's to allow mechanized harvest [3], and this development has allowed hectarage to increase substantially [1].

The presence of weeds is a major obstacle in sesame production [1, 4–8] and can negatively influence sesame yield. Kropff and Spitters [9] reported that the major factor influencing sesame yield loss in a competitive situation between the crop and weed is the ratio between the relative leaf area of the weed and the crop at the time of crop canopy closure. The effects of weeds on sesame establishment and growth have been well documented. Balyan [10], Gurnah [11], and Singh et al. [12] reported weed-induced reductions of sesame yield up to 74% and a need for a critical weed-free period up to 50 days after planting. Babiker et al. [13] reported that unrestricted weed growth reduced sesame grain yield by 30% and keeping the sesame crop weed free for 2, 4, 6, and 8 weeks after planting increased the grain yield by 8, 37, 40, and 43%, respectively. They concluded that the critical period of weed control in sesame appeared to be between 2 and 6 weeks after planting. Zuhair et al. [14] found that insufficient weed control during the early growth period of sesame caused 35 to 70% yield reductions, and they added that the critical period of weed control in sesame is 2 to 3 weeks after crop emergence.

With weak seedling vigor, limited competitive ability, and a lack of inexpensive and affordable labor, the use of preemergence (PRE) and/or postemergence (POST) herbicides is essential for commercially mechanized sesame production, especially in the US [7]. Also, the long growing season for sesame requires that herbicides provide season long control [4, 6, 7]. Currently, *S*-metolachlor is the only herbicide registered for PRE use in the US but has caused sesame injury under certain conditions [15]. In Texas, *S*-metolachlor resulted in 9 to 29% sesame stand reduction at one location, while at another location, stand reduction was ≤8% [15]. At other locations, *S*-metolachlor provided 99% weed control and no injury [6]. Regardless of early-season injury issues, sesame yield from the application of *S*-metolachlor was often greatest of all herbicides tested [6].

Herbicide tolerance may be affected by many factors including application timing [1]. *S*-metolachlor-applied PRE can cause up to 47% injury to cotton on sandy soils; however, applications made after cotton emergence did not affect cotton stand [16]. Kendig et al. [17] reported that POST applications of *S*-metolachlor to cotton at the four-leaf stage caused less reduction in cotton biomass than the application at the cotyledon stage. Jefferies et al. [18] reported that, in chickpea (*Cicer arietinum* L.), the combination of imazethapyr plus imazamox caused a height reduction and decreased node development at all growth stages; however, an application at the 9 to 12 node stage caused the most severe delay.

All POST herbicides that control broadleaf weeds in sesame have caused some sesame injury or yield reduction [5, 7]. For broadleaf weed problems in sesame, the use of soil-applied herbicides still appears to be the only option [4, 6, 7]. However, since sesame hectares are increasing in areas of the US, there is a critical need to find more herbicides which can be used in sesame cultivation to extend weed control especially during the early portion of the growing season [13, 14]. Studies were undertaken to determine sesame tolerance to various PRE and POST herbicides that could be applied early in the growing stage to extend weed control and cause little sesame injury.

2. Materials and Methods

The growth chamber and field studies were conducted to compare sesame response to commonly used PRE and POST herbicides applied up to five weeks after sesame planting.

2.1. Growth Chamber Experiments. Two growth chamber experiments to evaluate the response of sesame to PRE and POST applications of soil residual herbicides were conducted in 2017 at Texas A&M University Weed Science Research Facility, College Station, TX. Each PRE and POST experiment was repeated 2x under similar growth chamber conditions and data pooled over time.

2.1.1. Sesame Variety and Soil. Five seeds of "S-39" sesame variety were planted in 15 cm diameter plastic pots, and the pots were watered as needed. In the POST-response experiment, three healthy sesame plants were maintained at the time of herbicide application. The soil used in this study was collected from Brazos Valley Research Farm near College Station, TX, with no history of residual herbicide use. The soil texture was silty clay loam with pH 8.0 and 1.2% organic matter. The growth chamber was maintained at 27/21°C day/night temperature with 14 hours of photoperiod using artificial lights.

2.1.2. Plot Design, Herbicides, and Application. The growth chamber experiments were laid out in a randomized complete-block design with four replications. The PRE herbicides were applied the day following sesame planting, while POST applications of the herbicides were made at two growth stages: 3 wks after planting (WAP) when sesame growth stage was 3-4 leaf pairs and at bloom approximately 5 WAP. The PRE applications of acetochlor (Warrant®, Monsanto Company, St. Louis, MO, 63167) at 1.26 and 2.52 kg ai·ha^{-1}, pethoxamid (not commercialized yet; FMC Corp., Philadelphia, PA, 19104) at 1.12 and 2.24 kg ai·ha^{-1}, and *S*-metolachlor (Dual II Magnum®, Syngenta Crop Protection, LLC, Greensboro, NC, 27419) at 1.43 and 2.86 kg ai·ha^{-1} were applied at 1x and 2x of the labeled US doses, whereas acetochlor at 1.26 kg ai·ha^{-1} and trifluralin (Treflan®, Helena Chemical Company, Collierville, TN, 38017) at 1.68 kg ai·ha^{-1} were applied at the 1x dose in POST-response experiment. Herbicides were applied using a single-tip spray chamber (DeVries Manufacturing Corp., Hollandale, MN, 56045) fitted with an XR110015 nozzle (TeeJet® Technologies, Spraying Systems Co., Wheaton, IL, 60139) calibrated to deliver 140 L·ha^{-1} spray volume at 207 kPa pressure at a speed of 4.8 km·h^{-1}. These studies were done twice over time.

2.1.3. Sesame Injury Evaluations and Analysis. Visual injury was evaluated at 2 and 4 wks after herbicide treatment (WAT) using a 0 to 100% scale, where 0% means no injury and 100% means complete death. Data were arcsine square root transformed before analysis; however, back-transformed real mean values are presented based on the interpretation from the transformed data. In the PRE study, ANOVA was performed using SAS 9.4 (SAS Institute. 1998. SAS user's guide. SAS Inst., Cary, NC) to evaluate the significance of herbicide treatments on sesame response. The Fishers Protected LSD at 0.05 level of probability was used for separation of mean differences. The nontreated control was included for comparison. *A priori* orthogonal contrasts (single degree of freedom) were used to compare the sesame injury following the POST herbicide applications at two growth stages (3 WAP versus bloom).

2.2. Field Research Sites. Field studies were conducted in 2014 and 2015 in south Texas near Knippa (29.396713°N, 99.609378°W), in east central Texas near College Station (30.372965°N, 96.199284°W), in the High Plains of Texas near Hale Center (34.07590°N, 102.04642°W in 2014; 34.0699°N, 101.9415°W in 2015), and in southeastern Oklahoma near Lane (34.29861°N, 95.98806°W) to evaluate sesame response to herbicides applied POST at 2, 3, and 4 WAP. Soil type near Knippa was a Winterhaven silty clay loam (fine silty, carbonatic, hyperthermic Fluventic Ustochrepts) with less than 1.0% organic matter and pH 7.8. Soil type near Hale Center was a Amarillo sandy clay loam (fine loamy, mixed, thermic Aridic Paleustalf) with 0.8% organic matter and pH 7.8, while at College Station, the soil type was a Westwood silty clay loam (fine silty, mixed, superactive, thermic Udifluventic Haplustepts) with 1.2% organic matter and pH 8.0. At Lane, the soil type was a Bosville fine sandy loam (fine, mixed, active, thermic Albaquic Paleudalfs) with 0.6% organic matter and pH 6.3.

2.2.1. Plot Design, Herbicides, and Adjuvant. A randomized complete-block experimental design was used, and treatments were replicated three to four times depending on location. Treatments consisted of a factorial arrangement of eight herbicide treatments in 2014 including clethodim (Select Max®, Valent USA Corp., Walnut Creek, CA, 94595) at 0.15 kg ai·ha^{-1}, diuron (Direx®, Alligare, LLC, Opelika, AL, 36801), fluometuron (Cotoran®, Adama USA, Raleigh, NC, 27604) at 1.12 kg·ha^{-1}, quizalofop-P (Assure II®, E. I. duPont de Nemours and Company, Wilmington, DE, 19805) at 0.08 kg·ha^{-1}, ethalfluralin (Sonalan®, Dow AgroSciences LLC, Indianapolis, IN, 46268) at 0.84 kg ai·ha^{-1}, trifluralin (Treflan EC®, Dow AgroSciences LLC, Indianapolis, IN 46268) at 0.85 kg ai·ha^{-1}, pendimethalin (Prowl H_20®, BASF Corp., Research Triangle Park, NC, 27709) at 1.06 kg ai·ha^{-1}, and pyroxasulfone (Zidua®, BASF Corp., Research Triangle Park, NC, 27709) at 0.12 kg ai·ha^{-1}. In 2015, ten herbicide treatments were evaluated including trifluralin at 0.85 and 1.7 kg ai·ha^{-1}, pendimethalin at 1.06 and 2.12 kg ai·ha^{-1}, ethalfluralin at 0.84 and 1.68 kg ai·ha^{-1}, fluometuron at 1.12 kg ai·ha^{-1}, pyroxasulfone at 0.12 kg ai·ha^{-1}, diuron at 1.12 kg ai·ha^{-1}, and trifloxysulfuron-sodium (Envoke®, Syngenta Crop Protection, LLC, Greensboro, NC, 27419) at 0.008 kg ai·ha^{-1}. The doses of the dinitroaniline herbicides (ethalfluralin, pendimethalin, and trifluralin) represent 1x and 2x of the labeled US dose, while the doses of all other herbicides represent 1x of the labeled US dose. Three POST application timings (2, 3, and 4 WAP) were used in both years. At the Lane location in 2014, herbicide applications were actually made 3, 4, and 5 WAP. A nontreated control was included for comparison at all locations.

Crop oil concentrate (Agri-Dex®, Helena, Collierville, TN, 38017) at 1.0% v/v was added to the clethodim, diuron, fluometuron, and quizalofop-P treatments. No type of adjuvant was added to the other herbicide treatments since they are considered soil-applied herbicides and have little or no foliar weed control activity.

2.2.2. Herbicide Application. Herbicides were applied in water using a CO_2-pressurized backpack sprayer with either TeeJet 11002DG, TurboTee 110015AIXR, or Turbo TeeJet 11002 flat fan spray nozzle tips (TeeJet Spraying Systems Co., P.O. Box 7900, Wheaton, IL, 60188) depending on location. Sprayers were calibrated to deliver 190 L·ha^{-1} at 207 kPa at Knippa, 140 L·ha^{-1} at 207 kPa at College Station and Hale Center, and 93.5 L·ha^{-1} at 131 kPa at the Lane location.

2.2.3. Plot Size. Each plot in south Texas was five rows (76 cm apart) by 9.1 m long and four rows (101 cm apart) by 7.3 m long in College Station, Hale Center, and Lane. Only the two middle rows were sprayed, and the other rows were untreated and served as buffers.

2.2.4. Sesame Plantings, Observations, and Harvest. The sesame variety "S-35" was planted at Knippa, Hale Center, and Lane in both years, while "S-39" was planted at College Station in 2015. Planting dates at Knippa were June 13, 2014, and July 16, 2015; at Hale Center, June 19, 2014, and July 6, 2015; at Lane, June 19, 2014, and June 20, 2015; and at College Station, June 24, 2015. The sesame cultivars were seeded approximately 1.0 to 2.0 cm deep at 9 kg·ha^{-1} at all locations. At maturity, sesame in the plots was hand harvested, dried, and threshed with a stationary harvester.

Sesame injury was evaluated 6 to 9 WAP (early season) and 11 to 19 WAP (late season) based on a scale of 0 (no sesame injury) to 100 (complete sesame death). Injury consisted of stunting, leaf chlorosis and necrosis, and brooming effect. The brooming effect is a proliferation of primary branching at the upper nodes of the sesame plant (in the shape/form of a broom). Branching typically does not occur from these nodes since they are primarily flowering nodes. When branching at these nodes does occur, flowers are not formed [19]. These branches are not typically long (<25 cm) and support 5 to 10 capsules as opposed to the primary branches found lower on the plant which can be almost a meter long and support up to 10x as many capsules [19].

2.2.5. Data Analysis. An analysis of variance was performed using the ANOVA procedure for SAS (SAS Institute. 1998. SAS user's guide. SAS Inst., Cary, NC) to evaluate the significance of herbicides and application timing on sesame response and yield. Fishers Protected LSD at the 0.05 level of

Table 1: Sesame injury with PRE and POST applications of soil residual herbicides in growth chamber experiments conducted at College Station, TX, in 2017[a].

Herbicide	Dose (kg ai·ha^{-1})	Application timing	Sesame injury[b,c] 2 WAT (%)	4 WAT
PRE application				
Nontreated control	—	—	0	0
Acetochlor	1.26	PRE	10[d]	0[c]
Acetochlor	2.52	PRE	15[c]	11[b]
Pethoxamid	1.12	PRE	10[d]	7[b]
Pethoxamid	2.24	PRE	30[a]	39[a]
S-metolachlor	1.43	PRE	10[d]	1[c]
S-metolachlor	2.86	PRE	26[b]	46[a]
POST application				
Nontreated control	—	—	0	0
Acetochlor	1.26	3 WAP	16[b]	19[b]
Acetochlor	1.26	At bloom	6[b,c]	15[b]
Trifluralin	1.68	3 WAP	40[a]	60[a]
Trifluralin	1.68	At bloom	3[c]	20[b]
Contrasts[d]				
POST: 3 WAP versus at bloom			28 versus 5**	40 versus 18**

[a]WAP, weeks after sesame planting; WAT, weeks after herbicide treatment; [b]means presented within each column with no common letter(s) are significantly different at $\alpha = 0.05$; [c]data were arcsine square root transformed before analysis; however, back-transformed real mean values are presented based on the interpretation from the transformed data; [d]a priori orthogonal contrast, **significant ($P < 0.01$).

probability was used for separation of mean differences. The untreated check was used for sesame injury ratings and yield comparison but was only included in yield data analysis.

3. Results and Discussion

3.1. Growth Chamber Results

3.1.1. Sesame Injury with PRE Herbicides. The PRE application of acetochlor, pethoxamid, and *S*-metolachlor at the 1x dose resulted in 10% sesame injury 2 WAT, whereas 15 to 30% injury was observed with these herbicides applied at the 2x dose (Table 1). Similarly, in a study conducted in Florida, Sperry et al. [1] reported that PRE application of *S*-metolachlor at 1.42 and 2.78 kg ai·ha^{-1} resulted in 41 and 65% reduction in sesame plant height, respectively. At 4 WAT, slight or no injury was observed with the PRE applications of *S*-metolachlor and acetochlor at 1.43 and 1.26 kg ai·ha^{-1}, respectively; however, pethoxamid at 1.12 kg ai·ha^{-1} resulted in 7% sesame injury (Table 1). At the 2x dose, pethoxamid and *S*-metolachlor caused the greatest injury (39–46%), while acetochlor was less injurious (11%). Injury symptoms consisted of sesame plant stunting and curling of the leaves.

3.1.2. Sesame Injury with POST Herbicides. The POST application of trifluralin (1.68 kg ai·ha^{-1}) at 3 WAP showed a higher level (≥40%) of sesame injury at 2 and 4 WAT compared to the other POST treatments (Table 1). No difference in sesame response was noted with acetochlor applied 3 WAP or at sesame bloom. Averaged across all POST herbicide treatments, when evaluated 2 and 4 WAT, sesame injury was 28 and 40%, respectively, when POST applications were made 3 WAP, compared with 5 and 18% injury, respectively, with POST applications made at sesame bloom stage. Shorter internode length of the sesame plants resulted in the stunted growth following the POST application of these herbicides.

3.2. Field Research Results

3.2.1. 2014 Results

(1) Herbicide Injury. Early-season clethodim injury at any location was no greater than 5% when applied 2 to 3 WAP; however, the 4 WAP application of clethodim caused 18% injury at the Hale Center location (Table 2). No late-season injury with clethodim was noted at Knippa with any application timings, while 5% injury was noted with the 4 WAP application at Hale Center. At Lane, when evaluated 12 WAP, injury ranged from 1 to 5% but was not different than the untreated check. Clethodim has proven equally good at controlling both annual and perennial grasses, particularly Johnsongrass (*Sorghum halepense* (L.) Pers.) in sesame [5]. There is a label in the US for clethodim (Select Max®) use in sesame which allows spraying at all growth stages except flowering [19]. Clethodim applied during flowering has resulted in lack of capsule formation along the main stem of the plant and is commonly referred to as "cap gap" [19].

Diuron caused considerable leaf necrosis and chlorosis at both Texas locations but little to no injury was observed at the Lane location when rated early season, while late-season injury was 4% or less at all locations (Table 2). Grichar et al. [6], in an earlier study, reported that diuron applied PRE at 1.12 kg·ha^{-1} reduced sesame stands and caused sesame injury in one year in the Hale Center area; however, in south Texas no adverse effects with diuron were seen in the two-year study. Typically, diuron injury with POST applications to sesame is transient, and by late-season, only slight leaf chlorosis may be noted on lower leaves [7].

TABLE 2: Early- and late-season sesame injury with selected herbicides applied 2, 3, or 4 weeks after planting at Knippa, Hale Center (Texas), and Lane (Oklahoma) in 2014.

Herbicide[a]	Dose (kg ai·ha)	Application timing WAP[b]	Early season Knippa	Early season Hale Center	Early season Lane	Late season Knippa	Late season Hale Center	Late season Lane
			WAP					
			6	6	8	19	11	12
			(%)					
Clethodim	0.15	2	0	3	1	0	0	1
		3	0	2	5	0	0	5
		4	3	18	3	0	5	3
Diuron	1.12	2	48	17	4	3	2	4
		3	50	13	4	3	3	4
		4	23	20	1	2	2	1
Fluometuron	1.12	2	—	17	3	—	2	1
		3	47	20	5	3	0	4
		4	12	27	1	2	3	1
Quizalofop-P	0.08	2	0	0	5	0	0	5
		3	0	0	5	0	0	5
		4	3	0	5	0	0	5
Ethalfluralin	0.84	2	5	5	3	2	0	1
		3	3	10	0	0	3	0
		4	3	2	1	8	0	1
Trifluralin	0.85	2	3	3	4	0	0	3
		3	5	7	1	0	0	1
		4	7	0	4	50	0	1
Pendimethalin	1.06	2	7	37	5	2	3	4
		3	2	28	0	0	2	0
		4	3	7	1	0	0	1
Pyroxasulfone	0.12	2	25	—	3	2	—	3
		3	5	—	1	0	—	2
		4	2	—	4	0	—	0
Untreated	—	—	0	0	0	0	0	0
LSD (0.05)			4	7	NS	8	5	NS

[a]Agridex at 1.0% v/v added to clethodim, diuron, fluometuron, and quizalofop-P treatments; [b]WAP, weeks after planting.

Fluometuron provided similar results to diuron with 12 to 47% early-season injury in Texas but <5% injury at the Oklahoma location when evaluated early season or late season. Grichar et al. [6] reported that fluometuron at 1.12 kg·ha^{-1} applied PRE in the High Plains region of Texas reduced sesame stand and caused injury in one of the two years, while no stand reduction or injury was noted at the south Texas location.

Sesame injury from quizalofop-P was 5% or less at all locations. Sesame injury from this herbicide has been virtually nonexistent at any rates or application timings (author's personal observation).

Ethalfluralin injury to sesame was the greatest with the 2 WAP application at Knippa (5%) and the 3 WAP application at Hale Center (10%) with little to no injury at Lane (Table 2). This injury is typically in the form of stunting and leaf malformation. Late-season injury was only noted with the 4 WAP application at Knippa. This injury manifested itself as the brooming effect which was mainly seen on the upper portion of the plant and is only seen late season.

Early-season trifluralin injury (stunting and leaf malformation) at Knippa was greatest with the 3 and 4 WAP application, while at Hale Center, the 3 WAP application caused the greatest injury. No significant injury was noted at Lane. Late-season injury at Knippa was 50% with the 4 WAP application with no injury noted at the other locations (Table 2). The injury at Knippa was in the form of the brooming effect noted earlier. This type of injury is only seen with the dinitroaniline herbicides.

Pendimethalin injury early season at Knippa was the greatest with the 2 WAP application, while at Hale Center, all application timings resulted in significant injury. This injury, in the form of stunting and leaf malformation, decreased as the sesame plant aged. No injury was noted at Lane. Late-season injury was virtually nonexistent (≤4%), and little to no brooming effect was noted.

Typically, the dinitroaniline herbicides are applied preplant incorporated (PPI) or PRE; however, POST applications of these herbicides may be used to extend residual weed control throughout the growing season [20]. Since the dinitroaniline herbicides have lower water solubility and can volatilize and photodecompose on the soil surface over time, these herbicides need to be mechanically incorporated or need rainfall or irrigation to move these herbicides into the weed seed zone [21–24]. Uptake of these herbicides is primarily through roots and emerging shoots [25]. Parker [26]

Table 3: Sesame yield at Knippa, Hale Center (Texas), and Lane (Oklahoma) as influenced by selected herbicides applied 2, 3, or 4 weeks after planting in 2014.

Herbicide[a]	Dose (kg ai·ha)	Appl timing WAP[b]	Yield (kg·ha) Knippa	Hale Center	Lane
Clethodim	0.15	2	1770	597	923
		3	1759	639	1051
		4	1773	544	760
Diuron	1.12	2	1734	636	1000
		3	1435	549	1010
		4	1787	454	902
Fluometuron	1.12	2	—	615	1085
		3	1563	602	946
		4	1559	537	983
Quizalofop-P	0.08	2	1873	611	841
		3	1869	569	926
		4	1767	560	732
Ethalfluralin	0.84	2	1805	416	923
		3	1832	630	933
		4	1893	436	749
Trifluralin	0.85	2	1713	520	655
		3	1680	608	817
		4	1730	613	732
Pendimethalin	1.06	2	1817	238	1048
		3	1802	241	936
		4	1872	560	943
Pyroxasulfone	0.12	2	1897	—	847
		3	1782	—	916
		4	1780	—	702
Untreated	—	—	1943	613	905
LSD (0.05)			270	108	214

[a]Agridex at 1.0% v/v added to clethodim, diuron, fluometuron, and quizalofop-P treatments; [b]WAP, weeks after planting.

showed that trifluralin was more inhibitory to grain sorghum (*Sorghum bicolor* (L.) Moench) when absorbed through roots than emerging shoots. It is possible that the dinitroaniline herbicides will be concentrated in the extreme upper portions of the soil profile and weed seed may be able to germinate below the zone where dinitroaniline herbicides are located [27]. In this case, emerging shoots pass through treated soil, whereas developing roots would be below the herbicide treated soil. The effectiveness of soil-applied herbicides is dependent upon several factors, including movement of the herbicide into the soil either through water provided by rainfall or irrigation or by mechanical incorporation [28, 29].

Early-season pyroxasulfone injury was greatest with the 2 and 3 WAP application at Knippa, while no injury was noted at Lane. No late-season pyroxasulfone injury was noted at any location (Table 2). Pyroxasulfone is a newly registered herbicide in the US for either preplant (PP), PPI, PRE, or early postemergence (EPOST) used in corn, cotton, soybean (*Glycine max* L.), and wheat (*Triticum aestivum* L.) and application timing is crop specific [30–32]. Pyroxasulfone controls *Amaranthus* spp., *Lolium* spp., *Urochloa* spp., goosegrass (*Eleusine indica* L.), crowfootgrass (*Dactyloctenium aegyptium* L.), and *Digitaria* spp. [33–39]. Although pyroxasulfone has a similar weed control spectrum as S-metolachlor and dimethenamid-P, it has a higher specific activity allowing for use rates approximately eight times lower than dimethenamid-P [40]. Pyroxasulfone inhibits very long chain fatty acid synthesis similar to chloroacetamide, oxyacetamide, and tetrazolinone herbicides [33].

(2) Sesame Yield. At Knippa, sesame yield was reduced from the untreated check with diuron and trifluralin applied 3 WAP and fluometuron applied 3 and 4 WAP (Table 3). At Hale Center, sesame yield was reduced when compared to the untreated check with diuron applied 4 WAP, ethalfluralin applied 2 and 4 WAP, and pendimethalin applied 2 and 3 WAP. At the Lane location, only trifluralin applied 2 WAP reduced sesame yield relative to the untreated check. The yield reductions noted with diuron at Knippa and Hale Center were probably due to the early-season injury (>13%) noted at those locations.

Grichar et al. [6] concluded that it was best if the dinitroaniline herbicides were applied PRE due to severe sesame injury when applied PPI. Of the dinitroaniline herbicides, only pendimethalin (formulated as Prowl H_20) can be applied PRE [41]; however, annual grass control following pendimethalin applied PRE is often poor [42]. Hussien et al. [43] reported that trifluralin at rates greater than 0.84 kg·ha^{-1} was harmful to sesame, while Schrodter and Rawson [44] reported that pendimethalin at 1.5 and 3.0 kg·ha^{-1} and trifluralin at 0.84 kg·ha^{-1} reduced sesame plant populations. Plant selectivity by herbicide placement is influenced greatly by the movement of the herbicide in soils [45]. If the dinitroaniline herbicides move, they may come in contact with the absorptive sites of sesame and cause sesame injury [6].

3.2.2. 2015 Results

(1) Sesame Injury. Early-season sesame injury with trifluralin was noted at Hale Center but no other locations (Table 4). At the 0.85 kg·ha^{-1} dose, sesame injury occurred with trifluralin applied 2 and 3 WAP, while at the 1.7 kg·ha^{-1} dose of trifluralin, sesame injury was noted at all application timings. When evaluated late season, sesame injury at Knippa and College Station was noted with both trifluralin doses applied 4 WAP, while at Hale Center, both doses and all application timings with the exception of trifluralin at 0.85 kg·ha^{-1} applied 3 WAP resulted in significant injury when compared to the untreated check. The late-season injury consisted of the brooming effect. No trifluralin injury was noted at Lane.

Pendimethalin injury early season at Hale Center decreased as the sesame plants aged with >40% injury when pendimethalin was applied 2 WAP and <10% injury when applied 4 WAP (Table 4). No early-season injury was noted at the other locations. When evaluated late season, sesame injury with pendimethalin at Hale Center exhibited similar trends as the early-season rating with greater injury at the early application timing. Pendimethalin applications to sesame have

TABLE 4: Early- and late-season sesame injury at Knippa, Hale Center, and College Station (Texas) and Lane (Oklahoma) with selected herbicides applied 2, 3, or 4 weeks after planting in 2015.

Herbicide[a]	Dose (kg ai·ha)	Timing WAP[b]	Early season				Late season			
			Knippa	Hale Center	Lane	College Station	Knippa	Hale Center	Lane	College Station
			WAP							
			4	7	4	6	14	18	12	18
			(%)							
Trifluralin	0.85	2	0	17	0	2	0	23	0	2
		3	0	7	0	2	0	8	0	2
		4	0	3	—	2	10	12	—	10
Trifluralin	1.70	2	5	33	0	2	0	33	0	2
		3	0	7	0	2	0	10	0	2
		4	0	12	—	2	18	12	—	25
Pendimethalin	1.06	2	0	42	0	2	4	37	0	2
		3	0	27	0	2	0	18	0	2
		4	0	8	—	2	2	2	—	2
Pendimethalin	2.12	2	0	45	0	2	0	40	0	2
		3	0	23	0	2	0	22	0	2
		4	0	5	—	2	0	10	—	2
Ethalfluralin	0.84	2	0	18	0	—	0	28	0	—
		3	0	13	0	—	0	10	0	—
		4	0	5	—	—	6	8	—	—
Ethalfluralin	1.68	2	2	10	0	—	0	32	0	—
		3	2	17	0	—	0	17	0	—
		4	0	10	—	—	12	30	—	—
Fluometuron	1.12	2	12	38	10	—	0	13	0	—
		3	0	20	0	—	0	8	0	—
		4	2	8	—	—	0	3	—	—
Pyroxasulfone	0.12	2	0	13	0	2	0	25	0	2
		3	0	17	0	—	0	10	0	—
		4	2	8	—	—	0	0	—	—
Diuron	1.12	2	12	20	10	—	0	5	0	—
		3	0	15	0	—	0	2	0	—
		4	0	18	—	—	4	3	—	—
Trifloxysulfuron	0.008	2	45	97	0	—	0	97	0	—
		3	82	67	0	—	9	5	0	—
		4	60	67	—	—	42	42	—	—
Untreated		—	0	0	0	0	0	0	0	0
LSD (0.05)			8	7	6	NS	6	9	NS	3

[a]Agridex at 1.0% v/v added to fluometuron, diuron, and trifloxysulfuron treatments. [b]WAP, weeks after planting.

resulted in less brooming effect than with other of the dinitroaniline herbicides (author's personal observation).

Early-season ethalfluralin injury to sesame was noted only at the Hale Center location (Table 4). Sesame injury with ethalfluralin at 0.84 $kg{\cdot}ha^{-1}$ decreased from 18 to 5% as herbicide application was delayed, but at 1.68 $kg{\cdot}ha^{-1}$, this trend was not seen as injury ranged from 10 to 17%. Late-season injury with ethalfluralin was noted at Knippa with both doses applied 4 WAP, and this injury was in the form of the brooming effect. At the later evaluation at the Hale Center location, similar trends as seen with the early-season evaluation were noted, with injury at the lower dose decreasing as herbicide application was delayed, but this was not true with the higher dose. The dose of a dinitroaniline herbicide can affect sesame stand establishment. The 1/2x dose of ethalfluralin, pendimethalin EC, and trifluralin or the 3/4x dose of pendimethalin H_2O resulted in higher stand counts than the 1 to 2x dose of these herbicides when incorporated with rolling cultivator mixing wheels [46]. Previous research by Grichar et al. [4, 6] reported sesame injury following the use of dinitroaniline herbicides applied PPI using various incorporation methods. Also, Grichar et al. [4] reported that ethalfluralin, pendimethalin, and trifluralin reduced sesame stand numbers when compared with the untreated check. In that study, the dinitroaniline herbicides were incorporated 2.5 cm deep with a tractor-driven power tiller. In another study, Grichar et al. [6] reported that a spring-tooth harrow, with the lack of the ability to adjust incorporation depth, caused similar problems. However, the use of rolling cultivator mixing wheels, set to a depth of less than 2.5 cm, resulted in excellent sesame stands. Therefore, only a shallow incorporation of the dinitroaniline herbicides can be done when used in sesame to ensure a good stand.

TABLE 5: Sesame yield at Knippa, Hale Center, and College Station (Texas) and Lane (Oklahoma) as influenced by selected herbicides applied 2, 3, or 4 weeks after planting in 2015.

Herbicide[a]	Dose (kg ai·ha)	Appl timing WAP[b]	Knippa	Hale Center	Lane	College Station
				(kg·ha)		
Trifluralin	0.85	2	1786	318	502	716
		3	1537	581	577	846
		4	1568	616	—	977
Trifluralin	1.70	2	1721	297	689	846
		3	1537	470	588	846
		4	1704	587	—	977
Pendimethalin	1.06	2	1772	200	563	911
		3	1620	435	568	977
		4	1601	603	—	1042
Pendimethalin	2.12	2	1512	203	717	1107
		3	1538	464	517	977
		4	1631	861	—	846
Ethalfluralin	0.84	2	1656	277	609	—
		3	1573	393	553	—
		4	1631	534	—	—
Ethalfluralin	1.68	2	1634	316	643	—
		3	1634	612	509	—
		4	1593	621	—	—
Fluometuron	1.12	2	1702	481	624	—
		3	1556	670	604	—
		4	1503	752	—	—
Pyroxasulfone	0.12	2	1679	439	621	911
		3	1469	601	609	—
		4	1560	804	—	—
Diuron	1.12	2	1630	734	619	—
		3	1576	610	572	—
		4	1541	574	—	—
Trifloxysulfuron	0.008	2	1586	0	—	—
		3	1240	387	—	—
		4	1168	251	—	—
Untreated	—	—	1616	585	637	781
LSD (0.05)			177	234	128	195

[a]Agridex at 1.0% v/v added to fluometuron, diuron, and trifloxysulfuron treatments; [b]WAP, weeks after planting.

Early-season fluometuron injury was noted with the 2 WAP application at Knippa and Lane, while at the Hale Center location, sesame injury decreased as application timing was delayed. This trend also was visible at the late-season rating at Hale Center, but no late-season injury was apparent at the other locations (Table 4).

Sesame injury with pyroxasulfone early season and late season was visible at Hale Center but no other locations. Sesame injury early season was less with the 4 WAP application then the 3 WAP application; however, at the late-season evaluation sesame injury decreased as pyroxasulfone application was delayed.

Diuron injury was noted with the 2 WAP application at Knippa, Hale Center, and Lane, but the 3 and 4 WAP applications did not cause any sesame injury at Knippa or Lane. Late-season sesame injury was 5% or less at all locations.

Trifloxysulfuron-sodium injury to early-season sesame ranged from 45 to 82% at Knippa and 67 to 97% at Hale Center, but no injury was noted at Lane (Table 4). No late-season injury at Knippa with trifloxysulfuron-sodium applied 2 WAP was noted; however, the 4 WAP application resulted in >40% injury. Sesame injury at Hale Center was inconsistent and ranged from 5% with the 3 WAP application to 97% with the 2 WAP application. In cotton, trifloxysulfuron-sodium applied POST at the one-, three-, or five-leaf stage resulted in 22, 16, and 6% injury, respectively, when evaluated 21 days after treatment; however, no effect on yield was noted [47]. Tobacco (*Nicotiana tabacum* L.) metobilized ^{14}C-trifloxysulfuron-sodium rapidly, with 60.9% of the absorbed herbicide remaining in the parent herbicide form 4 h after treatment, whereas only 12.1% remained after 72 h [48]. These results show that the limited absorption and translocation, as well as rapid metabolism, are the basis for tobacco tolerance to foliar-applied trifloxysulfuron-sodium.

(2) Sesame Yield. At Knippa, only trifloxysulfuron-sodium applied 3 and 4 WAP reduced sesame yields when compared with the untreated check (Table 5). At Hale Center, the 2

WAP application of either trifluralin at 0.85 and 1.7 kg·ha^{-1}, pendimethalin at 1.06 and 2.12 kg·ha^{-1}, or ethalfluralin at 0.84 and 1.68 kg·ha^{-1} and trifloxysulfuron-sodium applied 2 and 4 WAP reduced the yield from the untreated check (Table 5). At Lane, trifluralin at 0.85 kg·ha^{-1} applied 2 WAP and ethalfluralin at 1.68 kg·ha^{-1} applied 3 WAP reduced the sesame yield when compared with the untreated check. No yield difference from the untreated check was noted with trifluralin or pendimethalin at any application timing at the College Station location. At this location, pendimethalin at 1.06 kg·ha^{-1} applied 4 WAP or pendimethalin at 2.12 kg·ha^{-1} applied 2 WAP produced the highest yields.

4. Conclusion

As these results show, sesame has a great ability to compensate for injury. In these studies where some of the herbicides caused severe sesame injury, sesame yields were good because the plants can compensate for open space and poor growth by additional branches with capsules [7, 15, 49]. However, branching can only compensate for gaps of less than 30 cm. Wider gaps not only lead to lower yields, but also let light through the canopy to encourage late-season weed emergence and growth [49]. A limited number of herbicides are available that can help control weeds during the production of sesame. Increasing the number of effective herbicides without crop injury would be a welcomed addition in sesame production. Annual grasses can be easily controlled in sesame with clethodim and quizalofop-P without sesame injury; however, only clethodim is approved for use on sesame in the US. Also, the label states that clethodim should not be applied during sesame flowering as an inhabitation of blooming and/or capsule formation may result [50]. The use of herbicides such as diuron or fluometuron is not without issues. These herbicides can effectively control weeds but can also affect sesame growth and yield as seen in these studies [5, 7]. Also, the dinitroaniline herbicides can extend weed control when used POST but can result in sesame injury and yield reductions in some instances.

Another option that may be available to growers is the use of cultivation to control unwanted or hard-to-kill weeds. However, cultivation cannot reliably control weeds within the seed row that emerge while the sesame is emerging. Since sesame grows slowly in the first three to four weeks, many growers have waited three to four weeks to cultivate. Sesame roots follow moisture, and with rain or irrigation in the first few weeks after planting, the roots may grow laterally and stay near the surface. Cultivating too close to the plant can cut roots, and plants will wilt quickly. In times of a dry season, roots grow more vertically allowing for closer cultivation. The cultivation process can throw soil up on the base of the plant covering any small weed after sesame plants are 10 to 15 cm in height.

Acknowledgments

The authors acknowledge the improvements made by Dr. Debalin Sarangi on an earlier version of the manuscript.

References

[1] B. P. Sperry, J. A. Ferrell, R. G. Leon, D. L. Rowland, and M. J. Mulvaney, "Influence of planting depth and application timing on S-metolachlor injury in sesame (*Sesamum indicum* L.)," *Weed Technology*, vol. 30, no. 4, pp. 958–964, 2016.

[2] E. M. Betts, *Thomas Jefferson's Garden Book (1766–1824)*, Thomas Jefferson Memorial Foundation, Inc., Charlottesville, VA, USA, 1999.

[3] D. R. Langham, T. Wiemers, T. Janick, and A. Whipkey, "Progress in mechanizing sesame in the US through breeding," in *Proceedings of the Fifth National Symposium, Trends in New Crops and New Uses*, pp. 157–173, Atlanta, GA, USA, 2002.

[4] W. J. Grichar, D. C. Sestak, K. D. Brewer, B. A. Besler, C. R. Stichler, and D. T. Smith, "Sesame (*Sesamum indicum* L.) tolerance and weed control with soil-applied herbicides," *Crop Protection*, vol. 20, no. 5, pp. 389–394, 2001.

[5] W. J. Grichar, D. C. Sestak, K. D. Brewer, B. A. Besler, C. R. Stichler, and D. T. Smith, "Sesame (*Sesamum indicum* L.) tolerance with various postemergence herbicides," *Crop Protection*, vol. 20, no. 8, pp. 685–689, 2001.

[6] W. J. Grichar, P. A. Dotray, and D. R. Langham, "Sesame (*Sesamum indicum* L.) response to preemergence herbicides," *Crop Protection*, vol. 28, no. 11, pp. 928–933, 2009.

[7] W. J. Grichar, P. A. Dotray, and D. R. Langham, "Weed control and the use of herbicides in sesame production," in *Herbicides, Theory and Applications*, S. Soloneski and M. L. Larramendy, Eds., IntechOpen Limited, London, UK, 2011, http://www.intechopen.com/articles/show/title/weed-control-and-the-use-of-herbicides-in-sesame-production.

[8] Z. Ijlal, N. Akhtar, A. Ali et al., "Effects of weed crop competition period on weeds and yield and yield components of sesame (*Sesamum indicum* L.)," *Pakistan Journal of Weed Science Research*, vol. 17, pp. 51–63, 2011.

[9] M. J. Kropff and C. J. T. Spitters, "A simple model of crop loss by weed competition from early observations on relative leaf area of weeds," *Weed Research*, vol. 31, no. 2, pp. 97–105, 1991.

[10] R. S. Balyan, "Integrated weed management in oilseed crops in India," *Indian Society of Weed Science*, vol. 1, pp. 317–323, 1993.

[11] A. M. Gurnah, "Critical weed competition periods in annual crops," in *Proceedings of the Fifth East African Weed Control Conference*, pp. 89–98, Harare, Zimbabwe, 1994.

[12] D. Singh, J. C. Dagar, and B. Gangwar, "Infestation by weeds and their management in oilseed crops–a review," *Agricultural Review*, vol. 13, pp. 163–175, 1992.

[13] M. M. Babiker, O. O. Siraj, and A. E. Salah, "The critical period of weed control in sesame (*Sesamum orientale* L.)," *Journal of Forest Products and Industries*, vol. 3, pp. 66–70, 2014.

[14] I. Zuhair, T. Asif, A. Muhamed, A. Naeem, A. Farhan, and M. M. Muhamed, "Effect of weed competition on yield and yield components of sesame (*Sesamum indicum* L.)," *Pakistan Weed Science Research*, vol. 17, pp. 51–63, 2011.

[15] W. J. Grichar, P. A. Dotray, and D. R. Langham, "Sesame (*Sesamum indicum* L.) growth and yield as influenced by preemergence herbicides," *International Journal of Agronomy*, vol. 2012, Article ID 809587, 7 pages, 2012.

[16] J. W. Keeling and J. R. Abernathy, "Preemergence weed control in a conservation tillage (*Gossypium hirsutum*) cropping system on sandy soils," *Weed Technology*, vol. 3, no. 1, pp. 182–185, 1989.

[17] J. A. Kendig, R. L. Nichols, and G. A. Ohmes, "Tolerance of cotton (*Gossypium hirsutum*) seedlings to preemergence and postemergence herbicides with four modes of action," *Plant Health Progress*.

[18] M. L. Jefferies, C. J. Willenborg, and B. Tar'an, "Response of chickpea cultivars to imidazolinone herbicide applied at different growth stages," *Weed Technology*, vol. 30, no. 3, pp. 664–676, 2016.

[19] D. Langham, J. Riney, G. Smith, T. Wiemers, D. Pepper, and T. Speed, *Sesame Producers Guide*, Sesaco Corporation, Austin, TX, USA, 2010.

[20] W. C. Johnson III., E. P. Prostko, and B. G. Mullinix Jr., "Annual grass control in strip-tillage peanut production with delayed applications of pendimethalin," *Weed Technology*, vol. 24, no. 1, pp. 1–5, 2010.

[21] J. W. Wilcut, A. C York, and G. R. Wehtje, "The control and interaction of weeds in peanut (*Arachis hypogaea*)," *Review of Weed Science*, vol. 6, pp. 177–205, 1994.

[22] J. W. Wilcut, A. C. York, W. J. Grichar, and G. R. Wehtje, "The biology and management of weeds in peanut (*Arachis hypogaea*)," in *Advances in Peanut Science*, H. E. Pattee and S. T. Stalker, American Peanut Research Education Society, Stillwater, OK, USA, 1995.

[23] M. A. Ross and D. J. Childs, *Herbicide Mode of Action Summary, WS-23-W*, Purdue University, West Lafayette, IN, USA, 2017, http://www.extension.purdue.edu/extmedia/ws/ws-23-w.html.

[24] J. Weber, "Behavior of dinitroaniline herbicides in soils," *Weed Technology*, vol. 4, no. 2, pp. 394–406, 1990.

[25] A. Appleby and B. Valverde, "Behavior of dinitroaniline herbicides in plants," *Weed Technology*, vol. 3, no. 1, pp. 198–206, 1989.

[26] C. Parker, "The importance of shoot entry in the action of herbicides applied to the soil," *Weeds*, vol. 14, no. 2, pp. 117–121, 1966.

[27] W. Johnson III., E. Prostko, and B. Mullinix Jr., "Texas panicum (*Panicum texanum*) control in strip-tillage peanut (*Arachis hypogaea*)," *Peanut Science*, vol. 29, no. 2, pp. 141–145, 2002.

[28] E. Prostko, W. Johnson III, and B. Mullinix Jr., "Annual grass control with preplant incorporated and preemergence applications of ethalfluralin and pendimethalin in peanut (*Arachis hypogaea*)," *Weed Technology*, vol. 15, no. 1, pp. 36–41, 2001.

[29] M. Ross and C. Lembi, "Herbicide incorporation techniques and equipment," in *Applied Weed Science*, C. Stewart, J. Stagman, and M. Carnis, Eds., Prentice-Hall, Upper Saddle River, NJ, USA, 2nd edition, 1999.

[30] Anonymous, *Speciman Label-Zidua® Herbicide*, BASF Corporation, Research Triangle Park, NC, USA, 2015.

[31] K. E. McNaughton, C. Shropshire, D. E. Robinson, and P. H. Sikkema, "Soybean (*Glycine max*) tolerance to timing applications of pyroxasulfone, flumioxazin, and pyroxasulfone + flumioxazin," *Weed Technology*, vol. 28, no. 3, pp. 494–500, 2014.

[32] G. L. Steele, P. J. Porpiglia, and J. M. Chandler, "Efficacy of KIH-485 on Texas panicum (*Panicum texanum*) and selected broadleaf weeds in corn," *Weed Technology*, vol. 19, no. 4, pp. 866–869, 2005.

[33] Y. Tanetani, K. Kahu, K. Kawai, T. Fujioka, and T. Shiminizu, "Action mechanism of a novel herbicide, pyroxasulfone," *Pesticide Biochemistry and Physiology*, vol. 95, no. 1, pp. 47–55, 2009.

[34] S. R. King and J. O. Garcia, "Annual broadleaf control with KIH-485 in glyphosate-resistant furrow-irrigated corn," *Weed Technology*, vol. 22, no. 3, pp. 420–424, 2008.

[35] C. H. Koger, R. Bond, D. H. Poston, T. W. Eubank, J. B. Blessitt, and V. K. Nandula, "Evaluation of new herbicide chemistry: does KIH-485 have a fit in the southern cotton producing region?," in *Proceedings of the 2008 Beltwide Cotton Conference*, p. 1738, Nashville, TN, USA, 2008.

[36] S. Z. Knezevic, A. Datta, J. Scott, and P. Porpoglia, "Dose-response curves of KIH-485 for preemergence weed control in corn," *Weed Technology*, vol. 23, no. 3, pp. 340–345, 2009.

[37] R. E. Nurse, P. H. Sikkema, and D. E. Robinson, "Weed control and sweet maize (*Zea mays* L.) yield as affected by pyroxasulfone dose," *Crop Protection*, vol. 30, no. 7, pp. 789–793, 2011.

[38] A. G. Hulting, J. T. Dauer, B. Hinds-Cook, D. Curtis, R. M. Koepke-Hill, and C. Mallory-Smith, "Management of Italian ryegrass in western Oregon with preemergence applications of pyroxasulfone in winter wheat," *Weed Technology*, vol. 26, no. 2, pp. 230–235, 2012.

[39] P. W. Grier, P. W. Stahlman, and J. C. Frihauf, "KIH-485 and *S*-metolachlor efficacy comparisons in conventional and no-tillage corn," *Weed Technology*, vol. 20, no. 3, pp. 622–626, 2006.

[40] W. Curran and D. Lingenfelter, *Pyroxasulfone: The New Kid in the Neighborhood*, Penn State Extension, State College, PA, USA, 2017, http://extension.psu.edu/plants/crops/news/2013/04/pyroxasulfone-the-new-kid-in-the-neighborhood.

[41] Anonymous, *Prowl H20 Product Label*, BASF Corporation, Ludwigshafen, Germany, 2017, http://www.cdms.net/Idat/Id6CT004.pdf.

[42] J. Byrd Jr. and A. York, "Annual grass control in cotton (*Gossypium hirsutum*) with fluazifop, sethoxydim, and selected dinitroaniline herbicides," *Weed Science*, vol. 35, pp. 388–394, 1987.

[43] M. Hussien, A. El-Hatlab, M. Abd-El Raouf, S. Shaban, and M. El-Deek, "Effect of soil herbicides on weeds, yield and quality of sesame, *Sesamum indicum*," *Pflanzenbou*, vol. 152, pp. 173–185, 1983.

[44] G. Schrodter and J. Rawson, "Herbicide evaluation studies in sesame," *Australian Weeds*, vol. 3, pp. 47–49, 1984.

[45] W. Ennis Jr, "Selected toxicity in herbicides," *Weed Research*, vol. 4, no. 2, pp. 93–104, 1964.

[46] W. J. Grichar and P. A. Dotray, "Weed control and sesame (*Sesamum indicum* L.) response to preplant incorporated herbicides and method of incorporation," *Crop Protection*, vol. 26, no. 12, pp. 1826–1830, 2007.

[47] R. J. Richardson, H. P. Wilson, G. R. Armel, and T. E. Hines, "Growth stage affects cotton response to trifloxysulfuron," *Weed Technology*, vol. 21, no. 1, pp. 37–40, 2007.

[48] S. C. Troxler, L. R. Fisher, W. D. Smith, and J. W. Wilcut, "Absorption, translocation, and metabolism of foliar-applied trifloxysulfuron in tobacco," *Weed Technology*, vol. 21, no. 2, pp. 421–425, 2007.

[49] D. R. Langham, *Growth and Development of Sesame*, Sesaco Corporation, Austin, TX, USA, 2008.

[50] Anonymous, *Speciman Label-Select Max® Herbicide*, BASF Corporation, Research Triangle Park, NC, USA, 2015.

19

Variation and Relationships among Agronomic Traits in Durum Wheat [*Triticum turgidum* (L.) Thell. ssp. *turgidum* conv. *durum* (Desf.) MacKey] under South Mediterranean Growth Conditions: Stepwise and Path Analyses

Ali Mansouri,[1] Bachir Oudjehih,[1] Abdelkader Benbelkacem,[2] Zine El Abidine Fellahi,[3] and Hamenna Bouzerzour[4]

[1]*Agronomy Department, El Hadj Lakhdar University, 5000 Batna, Algeria*
[2]*Plant Breeding and Biotechnology Division, National Agronomic Research Institute of Algeria (INRAA), 25000 Constantine, Algeria*
[3]*Department of Agronomy, Faculty of Natural, Life and Earth Sciences and the Universe, University of Mohamed El Bachir El Ibrahimi, 34030 Bordj Bou Arreridj, Algeria*
[4]*Valorization of Natural Biological Resources Laboratory, University of Setif, 19000 Setif, Algeria*

Correspondence should be addressed to Ali Mansouri; aliwheat@yahoo.fr

Academic Editor: Maria Serrano

Relationships among agronomic traits and grain yield were investigated in 56 genotypes of durum wheat (*Triticum durum* Desf.). The results indicated the presence of sufficient variability nearly for all measured traits. Heritability and expected genetic gain varied among traits. Aboveground biomass, harvest index, and spike number were the most grain yield-influencing traits. Early genotypes showed above-average grain and biological yields, spike number, and lower canopy temperature. Assessed genotypes were clustered into three groups which differed mainly for biological, economical, straw, and grain yields, on the one hand, and plant height, chlorophyll content, and canopy temperature, on the other hand. Selection for direct use from clusters carrying best combinations of yield-related traits and crosses to be made between genotypes belonging to contrasted clusters were suggested to generate more variability. Selection preferentially for spike number, biological yield, harvest index, and canopy temperature to accumulate favorable alleles in the selected entries for future uses is suggested.

1. Introduction

Durum wheat (*Triticum turgidum* L. var. *durum*) is grown on around 17 million hectares worldwide, representing 8% of the world's wheat area [1]. It is among the most cultivated crops in the Mediterranean basin, where approximately 75% of the world's durum wheat area is still grown, which contributes to 50% of the worldwide production [2]. In the Mediterranean basin (characterized by highly variable environments), wheat is mainly grown under rainfed conditions, and yield is often constrained by water and heat stresses that are common during the grain-filling period, which is due to the low and unpredictable seasonal rainfalls and high temperatures during the last stages of its development cycle [3]. Terminal drought and heat stresses negatively affect wheat grain weight and yield [4, 5].

Breeders, all over the world, are traying through wheat breeding programs to develop productive and adapted cultivars to their specific environments. This was resulted in the release of several commercial varieties. However, obtaining high-yielding genotypes of durum wheat, especially for Mediterranean rainfed areas, is considered difficult due to high genotype × environment interaction [6, 7]. In Mediterranean rainfed areas, genetic improvement of durum wheat for drought and heat stresses can be achieved by direct or indirect selection for yield. Direct selection involves selecting for yield, whereas indirect selection, that is, physiological breeding, involves selecting for yield components or other associated traits.

Information on the extent of variation and character association is useful for selection purposes to improve durum wheat (*Triticum durum* Desf.) grain yield. Evaluation of breeding populations is made to determine variability, character association, heritability, and genetic advance, which are needed to measure the genetic gain that can be expected through selection for a given trait. Wheat grain yield improvement was achieved through selection of closely yield-related characters. Sign and strength of correlations of these traits, with each other and with grain yield, depend on the type of plant material assessed and the environment test. It is, therefore, useful to target traits having the greatest influence on grain yield, to maximize yield genetic gain.

Yield components are agronomic traits that provide an opportunity to improve yield under water stress, along with morphological and physiological traits. Among the phonological, morphological, and physiological traits reported in the literature with high heritability and used in breeding programs, we discuss the following ones.

Early flowering, which positively affects yield, provides the ability to escape to terminal drought stress. The reduced number of days to heading and also days to maturity are important when breeding for terminal drought stress tolerance since they allow for drought escape [2, 8–11].

Plant heights, which have permitted the important improvements in grain yield in wheat in the last years, according to Fisher [12], have been achieved through the deployment of major genes such as the *Rht* reducing plant height genes (dwarf and semidwarf, the main drivers of the Green Revolution). However, there is a minimum height below which yield limitation is evident [5]. In addition, it has been found that greater plant height has a positive effect on yield in dry Mediterranean rainfed conditions; for this reason, it is desirable that the genotypes maintain their height in arid rainfed areas [13], compared to irrigated areas [14]. However, plant cycle and size should not be too short and too small since such traits will compromise yields under both stressed and optimum conditions. This may be attributed to low capacity to accumulate sufficient stem reserves for subsequent partitioning to the grain.

Hence, high yield under stressed conditions of some earliest and shortest genotypes is due to their ability to maintain a high number of productive tillers and a relatively high thousand kernel weight. This could have resulted from a lengthy grain-filling period [14, 15].

Drought increases senescence, by accelerating chlorophyll degradation, leading to a decrease in leaf area and canopy photosynthesis. There is evidence that stay-green phenotypes with delayed leaf senescence can improve their performance under drought conditions [10].

In wheat, leaf chlorophyll content (as a proxy of leaf photosynthesis) and stomatal conductance were proposed as proper criteria for identifying drought-tolerant genotypes under field conditions [16]. These tow traits allow as well as the identification of germplasm combining improved water use efficiency and productivity under both well-watered or water-limited conditions [17].

Canopy temperature can be used as a tool to indirectly evaluate the transpiration rate of a plant [18]. Olivares-Villegas et al. [19] reported that, under drought, canopy temperature is the most drought-adaptive trait contributing to a higher performance. This trait is highly heritable and consistently associated with yield phenotypically and genetically. It can be utilized as a selection criterion to identify high-yielding wheat genotypes or as an important predictor of yield performance under drought since it epitomizes a mechanism of dehydration avoidance.

However, not all studies have demonstrated correlated benefits of canopy temperature and grain yield in wheat [20]. Further, the study of Giunta et al. [21] on durum wheat has indicated that genotypes with warmer canopy temperature, implying more closed stomata, produced the highest yields. Rebetzke et al. [17] found that preflowering canopy temperature increases were associated with reduced final plant height and increased harvest index and grain yield but were uncorrelated with aerial biomass. In contrast, postflowering, cooler canopies were associated with greater aerial biomass and increased grain number and yield.

Grain yield of cereals is the result of the development and growth of yield components during the life of the crop. The most popular approach to attempt understanding wheat yield is that dividing it into numerical components (i.e., plants per unit land area, spikes per plant, spikelets per spike, grains per spikelet, resulting in the number of grains per m^2, and average individual grain weight). This approach is simple to understand and easy to do. It only requires one sample at maturity and a minimum process of the sample to divide the components in the sample, and by multiplying these components, one can obtain the yield [22]. Grain number and grain weight are the main components affecting grain yield and are negatively correlated [23–25]. Graziani et al. [2] reported that, among yield components, number of grains m^{-2} showed the strongest correlation with grain yield.

In his study, Mohsin et al. [26] found that grain yield correlated positively with flag leaf area, plant height, biomass, number of spikes, spike length, number of spikelets per spike, number of grains per spike, 1000-grain weight, and harvest index.

Fellahi et al. [27] reported for bread wheat a positive and significant correlation of seed yield with biological yield, spike weight, and spike number. Aboveground biomass had the highest effect, explaining more than 83.0% of grain yield variation. This suggested that any increase in biological yield, particularly the spike weight fraction, affected positively grain yield. Aboveground biomass was largely determined by spike number, which explained 85.20% of total variation in biological yield. Kandel et al. [28] reported similar results for these traits. In addition, moderate to high positive and significant correlations of grain yield with number of spikes, grains per spike, and 1000-grain weight under both stressed and optimum conditions imply the direct contribution of these yield components to yield and should be considered as important target traits during selection [14, 15], and it can be confirmed that maintenance of a high number of productive tillers and kernels per spike contributes more to the grain yield when compared to the other yield components under stress because the number of grains produced per plant will compensate better for the reduction in seed weight [23]. However, under optimum conditions, all the yield components have considerable contribution to grain yield, implying that selection for any of the yield components could significantly improve the yields [18]. Straw yield was positively related to spike number while thousand kernel weight and plant height exhibited significant and negative correlations with seed number per spike [5]. This suggested that compensation effect was operating among traits, lessening the overall effect on grain yield. Fellahi et al. [29] reported that any increase in spike number improved both aboveground biomass and grain yield. As selection criteria, spike number could be easily appreciated visually, in the field. Furthermore, according to Royo et al. [30], grain yield increases have been mostly based on increases in harvest index and number of grains per unit area through an enhanced number of spikes and grain set.

Salmi et al. [31] found appreciable variability for several durum wheat traits such as days to heading, plant height, aboveground biomass, grain yield, and yield components. Broad-sense heritability was high for days to heading and plant height, intermediate for spike number and number of kernels per spike, and low for grain yield and aboveground biomass.

On the other hand, Graziani et al. [2] found high broad-sense heritability for yield, in contrast with González-Ribot et al. [25]. For the yield components (spike number, grain number, and grain number per spike) and harvest index, broad-sense heritability was low to moderate [2, 25]. Mohsin et al. [26] and Arya et al. [32] found high broad-sense heritability for harvest index. Heritability was, respectively, high to moderate for the physiological traits chlorophyll content and canopy temperature [2, 10].

Expected genetic gain, expressed as % of trait mean, varied from 4.8% for days to head to 45.9% for grain yield. Relatively to check the cultivar, the best selected entries, for the grain yield, were taller and had an above average biological yield; the harvest index and spike number were below the average thousand kernel weight [31]. In the same context, Arya et al. [32] found the lowest value for days to maturity (9.44%) while the highest gain for number of grains per spike (36.88%). For grain yield, expected genetic gain was 31.98%.

Hannachi et al. [33] reported that both additive and dominance gene actions were implicated in the expression of durum wheat traits, $H1$ component was larger than D component for days to heading, spike number, kernel per spike, and grain yield and smaller for plant height and thousand kernel weight. This indicated the complex inheritance pattern of yield and yield components, suggesting that selection to improve these traits should be conducted on nonsegregating plant material [33]. Results of several research indicated that effective progress could be made through selection of morphological characters to enhance grain yield potential [27, 34–37].

The objectives of this study were to evaluate phenotypic variability, heritability, and genetic advance, to analyze relationships existing among yield and morphophysiological traits of 56 genotypes of durum wheat (*Triticum durum* Desf.) grown under south Mediterranean conditions, and to determine the direct and indirect contributions of the important traits with the aim to select the more desirable traits which may contribute to the improvement of durum wheat and identify the characters of utmost importance which may be used as selection criteria in a wheat breeding program.

2. Materials and Methods

2.1. Site, Plant Material, and Experimental Design. Plant material under study consisted of 56 genotypes of durum wheat (*Triticum durum* Desf.) (Table 1). These genotypes were evaluated in a randomized complete block design with two replications at the Field Crop Institute Agricultural Research Station (ITGC-ARS), El-Khroub, Algeria (36°17′N, 6°38′E, and 640 m above sea level), during the 2014-2015 cropping season. Plot dimensions were 4 rows, 5 m long, and 20 cm apart. Land preparation, sowing date, fertilizer application, and weed control were as recommended for the local area. The seeds were sown in early January 2015, 100 kg/ha of 46% triple superphosphate was applied at sowing and at jointing, and 130 kg/ha of 46% urea was broadcasted. Weeds were controlled chemically with Cossack® OD (mesosulfuron-methyl(mesomaxx) + iodosulfuron-methyl-sodium + mefenpyr-diethyl) at 1 l/ha rate.

2.2. Notations and Measurements. The following morphological traits were measured. Number of days to heading (DHE, days) was counted as number of calendar days from the date of sowing to the time when 50% of the spikes were half way out from the flag leaf. Plant height (PHT, cm) was measured at maturity from the soil surface up to the tip of the spike, excluding awns. Aboveground biomass (BIO, g/m), spike number (SN, #/m), straw yield (STW, g/m), and grain yield (GY, g/m) were recorded from a vegetative sample harvested from one row, 1.0 m long per plot. Harvest index (HI, %) was derived as 100 times the ratio of grain yield to aboveground biomass:

$$\mathrm{HI}\,(\%) = 100 \times \frac{\mathrm{GY}}{\mathrm{BIO}}. \tag{1}$$

Economical yield (Y_{eco}) was deduced as grain yield plus 0.3 times straw yield, according to Annicchiarico et al. [38]: $Y_{\mathrm{eco}} = \mathrm{GY} + 0.3\mathrm{STW}$. At heading stage, chlorophyll content and canopy temperature were read with a chlorophyll meter type Yara N-Tester (Yara GmbH & Co. KG, Dülmen, Germany) and an infrared thermometer model Fluke 62 Max+ (Fluke Corporation, Everett, WA, USA), respectively.

2.3. Data Analysis. Collected data were subjected to an analysis of variance according to the additive model in a complete block design with two replicates, as per Steel and Torrie [39] (Table 2) as follows: $Y_{ij} = \mu + g_i + b_j + e_{ij}$, where

TABLE 1: List of the 56 genotypes of durum wheat evaluated in 2014-2015 at the INRAA-ITGC-ARS, El-Khroub, Algeria.

Number	Name, cross, pedigree
1	Chacan
2	Capeiti8
3	Waha
4	Fg/Palest20C/606//Mexi/3/Rabi
5	Mrb5
6	Tez/Fri//Wulp
7	Kamillaroi
8	Chen "S"
9	TP/Zb//Gdo578/3/Swan
10	Omsinina-1
11	Camadi Abou 73 N°7510
12	JoriC96
13	Omruff2
14	Hoggar (=Vitron)
15	Granizo
16	Tell76
17	Inrat 69
18	M1084
19	Daki
20	Jordan collect 8″S″N°42
21	Cocorit C71
22	Kyperounda
23	Str/AA//21563
24	Araldur
25	Belikh2
26	Sahel
27	Langdon
28	Occotillo
29	Durell
30	Ardente
31	Zb/Fg
32	Kebir
33	Gta dur
34	Simeto
35	Kucuk
36	Gidara-2
37	Sooty9/Rascon-37
38	Yavaros79
39	Ajaia-12/F3Locl (sel.ethio.135)//Plata13
40	Barba de lobo
41	Neodur
42	Bousselam
43	Ahlal
44	Ofanto
45	Acsad1103
46	Beltagy-2
47	Geruftel
48	F413/3/Arthur71/Lahn//Blk2/ Lahn/4/Quarmal/Icamor
49	Azeghar-1//1BIrn/Mrf-2
50	Mrf-2/Stj2//Gdr2/Mgnl1
51	Oss1/Sij5/Bid1/4/ Bezaiz-Shfiisd-19539/Waha/3/Stj
52	Ter-1/3/Stj3//Bcr/Lks4
53	F413/3/Arthur71/Lahn//Blk2/ Lahn/4/Quarmal/Icamor
54	Beltagy-3
55	Beltagy-2
56	Mgnl3/Ainzen-1

TABLE 2: Skeleton of the analysis of variance.

Source	df	Mean square	*F* test ratio	Expected mean square (EMS)
Block	1	MS_b	MS_b/MS_e	—
Genotype	55	MS_g	MS_g/MS_e	$\sigma_e^2 + b\sigma_g^2$
Residual	55	MS_e	—	σ_e^2

Y_{ij} is the observed value of the measured trait for ith genotype and jth block, μ is the overall mean for the measured trait, g_i is the ith genotype effect, b_j is the jth block effect, and e_{ij} is the residual associated with ith genotype and jth block.

Genotypic (σ_g^2) and phenotypic (σ_p^2) variance components were derived from the expected mean square according to Aquaah [40] as follows:

$$\sigma_g^2 = \frac{MS_g - MS_e}{b},$$
$$\sigma_p^2 = \sigma_g^2 + \sigma_e^2. \tag{2}$$

The least significance difference at 5% probability ($LSD_{5\%}$) was derived according to Steel and Torrie [39] as follows:

$$LSD_{5\%} = t_{5\%}\sqrt{\frac{2\sigma_e^2}{b}}, \tag{3}$$

where t is the tabulated t value at 5% probability with $(b-1)(g-1)$ df of the residual, b is the number of blocks, and σ_e^2 is the residual mean square. Based on the estimates of phenotypic (σ_p^2), genotypic (σ_g^2), and environmental (σ_e^2) variances, phenotypic (PCV) and genotypic coefficients of variation (GCV) were obtained according to the method outlined by Cruz et al. [41]:

$$PCV(\%) = \frac{100\sigma_p}{X_{bar}},$$
$$GCV(\%) = \frac{100\sigma_g}{X_{bar}}, \tag{4}$$

where σ_p and σ_g are the phenotypic and genotypic standard deviations and X_{bar} is the grand mean of the trait of interest. Broad-sense heritability (h_{bs}^2) was calculated using variance components according to Acquaah [40]:

$$h_{bs}^2 = \frac{\sigma_g^2}{\sigma_g^2 + \sigma_e^2} \times 100, \tag{5}$$

where σ_g^2 is the genotypic variance and σ_e^2 is the environmental variance. Selection differential (S), the expected response to selection (RS), absolute value expressed as % of the grand mean (%RS) were calculated according to Cruz et al. [41] as follows:

$$S = X_s - X_0,$$
$$RS = Sh_{bs}^2 = ih_{bs}^2\sigma_p, \tag{6}$$
$$RS(\%) = 100\left(\frac{RS}{X_0}\right),$$

Table 3: Analysis of variance mean squares for 10 characters of 56 durum wheat genotypes.

Source	Block	Genotype	Residual
DF	1	55	55
DHE	1.75	8.71**	0.64
BIO	3960.32	2865.79**	558.85
SN	405.08	207.87**	67.43
GY	987.70	375.11**	130.60
Y_{eco}	1671.07	805.83**	148.48
PHT	0.65	111.84**	1.71
STW	992.46	1507.05**	526.67
HI	24.78	64.79**	11.55
CHL	493.08	8323.08**	598.81
CT	0.10	7.39**	0.35

DHE = days to head, BIO = aboveground biomass, SN = spike number, GY = grain yield, Y_{eco} = economical yield, PHT = plant height, STW = straw yield, HI = harvest index, CHL = chlorophyll content, and CT = canopy temperature.

where X_0 is the grand mean of the trait of interest, X_s is the mean of the selected individuals ($X_0 \pm \text{LSD}_{5\%}$), i is the selection intensity ($S/\sigma_p = 2.06$) at 5% probability level, and σ_p is the phenotypic standard deviation of the trait of interest.

Trait relationship was inspected through phenotypic (r_p) and genotypic (r_g) correlation coefficients, stepwise regression, and path analyses. Phenotypic (r_p) and genotypic (r_g) correlation coefficients were calculated as ratios of covariance to root square of the product of corresponding variances, as follows:

$$r_p = \frac{\text{Cov}_p(xy)}{\sqrt{\sigma_p}(x) * \sigma_p(y)},$$
$$r_g = \frac{\text{Cov}_g(xy)}{\sqrt{\sigma_g}(x) * \sigma_g(y)}. \quad (7)$$

The genotypic covariance was derived using the property of the sum of two variables, x and y, whose expected mean square is as follows:

$$\sigma(x+y) = \sigma_g(x) + \sigma_g(y) + 2\sigma_g(xy). \quad (8)$$

This can be rewritten as follows:

$$\text{Cov}_g(xy) = \sigma_g(xy) = \frac{\sigma(x+y) - \sigma_g(x) - \sigma_g(y)}{2}. \quad (9)$$

Direct and indirect effects of independent variables (i) on dependent variable (j) were obtained by decomposing correlation coefficients between dependent (j) and independent variables (i) according to Singh and Chaudhary [42]. Statistical analyses were performed with LazStats [43], Past [44], and Cropstat [45] software.

3. Results and Discussion

3.1. Variability, Performance, Heritability, and Expected Response to Selection. Results of the analysis of variance indicated significant genotype effect, suggesting the presence of sizeable variation among genotypes for all measured traits (Table 3). Days to heading varied from a minimum of 117.5 to 125.0 days, with a range of 7.5 days, a value statistically significant compared to $\text{LSD}_{5\%}$. Phenotypic and genotypic coefficients of variation were very low in magnitude, suggesting little variability for this trait within the set of studied genotypes. Low variability was reported by Oulmi et al. [46] and Hannachi et al. [34] in durum wheat and by Fellahi et al. [47] in bread wheat, for this trait. Earliness is controlled by various genes related to vernalization, photoperiod, and earliness per se [5]. When grown under heat and drought terminal stresses, plant materials drastically reduced growth cycle, minimizing variability of this trait [47].

Based on a single-year data, broad-sense heritability of days to heading this trait was high, suggesting minor environmental effect, since genotypic variance was almost equal to phenotypic counterpart (Table 4). To minimize the effect of terminal heat and drought stresses, selection favored early heading genotypes; at the same time, the selection material must be late enough to avoid late spring frost [5, 48]. However, due to its low genotypic coefficient of variation, there is little scope to improve this trait through selection. Undoubtedly, earliness has been the predominant trait improved for Mediterranean conditions and may not be a prospective trait for future breeding [5].

Aboveground biomass, measured at maturity, varied from 81.0 to 313.5 g/m. The range of this trait was 232.4 g/m, value significantly greater than the minimum mean of 81.0 g/m, measured for the least aboveground biomass performing genotype. Both phenotypic and genotypic coefficients of variation were high, taking values of 24.11 and 19.79%, respectively.

Broad-sense heritability was just medium, being less than 70% (Table 4). Based on genotypic coefficient of variation and broad-sense heritability, aboveground biomass appeared to be amenable to improvement, within the set of studied genotypes. González-Ribot et al. [25] reported a very low heritability of 9% for aboveground biomass.

Spike number varied from 31.0 to 86.0, with a range value of 55.0 spikes/m. Both phenotypic and genotypic coefficients of variation were appreciably high, taking values of 23.44 and 16.74%, respectively. Broad-sense heritability was low to medium, (51.02%), suggesting a sizeable environmental effect on the expression of this trait. In their studies, Graziani et al. [2] and González-Ribot et al. [25] found a low heritability of 27% for spike number. Similarly, grain and straw yields expressed low broad-sense heritability, high phenotypic and genotypic coefficients of variation, and significant ranges.

Plant height, chlorophyll content, and canopy temperature had high broad-sense heritability and relatively low phenotypic and genotypic coefficients of variation. The same results have been reported by Graziani et al. [2] and Lopes et al. [10], except for canopy temperature which has a moderate heritability value. Harvest index and economical yield have medium broad-sense heritability and medium to high coefficients of variation (Table 4).

Expected response to selection was relatively low for the number of days to heading (3.17%) and high for the remaining traits. The observed values were over 30.0% for biological and economical yields, over 25.00% for grain yield, plant height, and straw yield, over 20.00% for spike number and harvest index, and over 15.0% for canopy temperature and chlorophyll content (Table 4).

TABLE 4: Arithmetic mean, minimum and maximum values, range, least significant difference, phenotypic and genotypic coefficients of variation, phenotypic and genotypic variances, broad-sense heritability, and expected response to selection for 10 morphological and agronomic traits of 56 durum wheat genotypes evaluated in 2014-2015.

Traits	Mean	Min	Max	Range	$LSD_{5\%}$	C_{Vp} (%)	C_{Vg} (%)	σ_p^2	σ_g^2	h_{bs}^2	RS (%)
DHE	121.48	117.50	125.00	7.50*	1.60	1.78	1.65	4.68	4.03	86.29	3.17
BIO	171.64	81.00	313.50	232.50*	47.37	24.11	19.79	1712.32	1153.47	67.36	33.45
SN	50.06	31.00	86.00	55.00*	16.45	23.44	16.74	137.65	70.22	51.02	24.63
GY	60.72	37.35	108.15	70.80*	22.90	26.19	18.21	252.86	122.25	48.35	26.08
Y_{eco}	94.00	59.33	169.76	110.43*	24.42	23.24	19.29	477.15	328.68	68.88	32.97
PHT	56.07	39.50	75.75	36.25*	2.62	13.44	13.24	56.77	55.07	96.99	26.85
STW	110.92	27.75	205.35	177.60*	45.99	28.75	19.96	1016.86	490.19	48.21	28.55
HI	35.94	26.57	65.74	39.18*	4.91	17.19	14.36	38.17	26.62	69.74	24.70
CHL	638.60	493.50	830.50	337.00*	49.04	10.46	9.73	4460.94	3862.14	86.58	18.65
CT	23.57	21.45	27.80	6.35*	1.19	8.35	7.96	3.87	3.52	90.84	15.62

DHE = days to head, BIO = aboveground biomass, SN = spike number, GY = grain yield, Y_{eco} = economical yield, PHT = plant height, STW = straw yield, HI = harvest index, CHL = chlorophyll content, and CT = canopy temperature.

Similar findings have been reported by Ghaed-Rahimi and Heidari [49] who observed 3.10% of expected response to selection for canopy temperature, 23.93% for harvest index, and 4.18% for days to heading. However, these authors reported low values of expected response to selection for biological yield (4.68%) and grain yield (6.95%) compared to what was found in the present study.

Expected genetic gain, expressed as % of trait mean, found by Salmi et al. [31] varied from 4.8% for days to heading to 45.9% for grain yield. High genetic advance coupled with high heritability estimates offers the most effective condition for selection. The utility of heritability therefore increases when it is used to calculate genetic advance, which indicates the degree of gain in a character obtained under a particular selection pressure [32].

3.2. Trait Relationship. Study of the relationships between traits indicated that number of days to heading exhibited negative and significant phenotypic correlations with aboveground biomass (−0.273*), spike number (−0.266*), grain yield (−0.339**), and canopy temperature (0.388**). The corresponding genetic correlations were high in magnitude, except the one relating days to head to canopy temperature (Table 5).

The weakness of the phenotypic correlations is imputable to the modifying effect of the environment which reduced the genotypic variation of the number of days to heading. Despite this, the results suggested that early genotypes tended to be more performing in terms of grain yield, aboveground biomass, and spike number and avoided terminal stresses. Low canopy temperature is associated with the ability to extract water from deeper in the soil profile [10]. This encourages selection of early plant material.

Aboveground biomass exhibited significant and positive correlation coefficients with spike number (0.709**), grain yield (0.836**), economical yield (0.965**), straw yield (0.962**), and plant height (0.272*), and negative correlation coefficients with harvest index (−0.309*). Aboveground biomass, grain yield, spike number, economical yield, and straw yield were highly interrelated, at both phenotypic and genetic levels. These results suggested aboveground biomass as an indirect selection criterion to improve grain yield, along with spike number and straw and economical yields. Phenotypic correlation indicated that plant height was significantly and positively correlated with economical yield (0.300*), but not with straw yield. The genetic correlations indicated that this trait had high negative correlations with canopy temperature (−4.281) and with harvest index (−1.569). Thus, plant height is usually negatively related to the HI [50].

This suggested that dwarf genes are linked to genes controlling high harvest index and sensitivity to high temperatures. This is suggested also by the genetic correlations of straw yield and harvest index (0.554) and straw yield and canopy temperature (2.302) (Table 5). Phenotypic correlations indicated that chlorophyll content and canopy temperature had no effect on morphological traits except that early genotypes exhibited low canopy temperature than late ones. This could be explained by the selection history of the studied genotypes, as chlorophyll content and canopy temperature were not used apart or along with grain yield and morphological traits in the selection process. But the genetic correlation indicated that genes controlling high chlorophyll content tended to be linked to those controlling high temperature tolerance. Findings about relationships between morphological and yield-related traits were in accordance with results of Leilah and Al-Khateeb [51] who reported that wheat grain yield was positively correlated with plant height, spike number, harvest index, and biological yield.

When subjecting the measured traits to regression analysis, the full model captured 94.58% of the total variation in grain yield and indicated that only aboveground biomass and harvest index had highly significant effect on the dependent variable. With the same set of traits, stepwise regression retained aboveground biomass and harvest index which explained 94.90% of the total variation observed in grain yield (Table 6). These results corroborated with those of Fellahi et al. [27] who reported that stepwise regression model retained biological yield and harvest index, which explained 99.7% of grain yield variation. In the same context, Leilah and Al-Khateeb [51] found that stepwise regression analysis identified spike yield, harvest index, spike number, and biological yield as the most important grain yield predictors.

Table 5: Phenotypic (below diagonal) and genotypic (above diagonal) correlation coefficients for ten characters of 56 genotypes of durum wheat.

	DHE	BIO	SN	GY	Y_{eco}	PHT	STW	HI	CHL	CT
DHE	—	−0.334	−0.303	−0.402	−0.359	−0.180	−0.311	−1.008	−0.919	−0.096
BIO	−0.273**	—	0.925	1.046	1.009	0.315	1.012	−0.239	−0.119	−0.134
SN	−0.266**	0.709**	—	0.955	0.927	0.143	0.942	−0.345	−0.049	−0.102
GY	−0.339**	0.836**	0.758**	—	1.015	0.401	1.105	−0.402	−0.382	−0.160
Y_{eco}	−0.316*	0.965**	0.763**	0.951**	—	0.348	1.040	−0.306	−0.230	−0.144
PHT	−0.175 ns	0.272*	0.130 ns	0.307*	0.300*	—	−0.225	−1.569	−0.230	−4.281
STW	−0.207 ns	0.962**	0.599**	0.654**	0.857**	0.222 ns	—	0.554	−0.021	2.302
HI	−0.091 ns	−0.309*	0.069 ns	0.219 ns	−0.070 ns	−0.011 ns	−0.535**	—	−0.596	−1.205
CHL	−0.029 ns	−0.067 ns	0.135 ns	−0.179 ns	−0.123 ns	0.024 ns	−0.003 ns	−0.192 ns	—	−1.080
CT	0.388**	−0.143 ns	−0.001 ns	−0.187 ns	−0.170 ns	−0.091 ns	−0.103 ns	−0.067 ns	0.107 ns	—

DHE = days to head, BIO = aboveground biomass, SN = spike number, GY = grain yield, Y_{eco} = economical yield, PHT = plant height, STW = straw yield, HI = harvest index, CHL = chlorophyll content, and CT = canopy temperature.

Table 6: Full and stepwise regression models and direct and indirect effects of independent variables on grain yield.

Full regression model							
	b	t	Prob	R^2 mult			
Constant	41.742 ± 47.85	0.872	0.392	—			
DHE	−0.630 ± 0.39	−1.613	0.121	—			
BIO	0.328 ± 0.04	9.382	0.000	—			
SN	0.126 ± 0.12	1.065	0.299	—			
PHT	0.071 ± 0.08	0.903	0.376	—			
HI	0.934 ± 0.13	7.049	0.000	—			
CHL	−0.013 ± 0.01	−1.140	0.267	—			
CT	0.173 ± 0.41	0.419	0.679	0.9458			
Stepwise regression model							
Variables available: BIO, SN, PHT, HI, CHL, and CT							
	β	$b \pm SE_b$	t	a	R^2 mult		
BIO	0.999	0.361 ± 0.012	31.22**	—	—		
HI	0.528	1.271 ± 0.077	16.49**	−46.954	0.949		
Variables available: SN, PHT, HI, CHL, and CT							
SN	0.769	1.033 ± 0.104	9.95**	—	—		
CHL	−0.287	−0.061 ± −0.016	3.75**	—	—		
PHT	0.231	0.319 ± 0.0140	2.79**	26.08	0.683		
Path analysis							
Variable	DHE	BIO	SN	PHT	HI	CHL	CT
Direct effect	−0.014	0.918	0.068	0.050	0.490	−0.033	−0.010
Indirect effects	−0.325	−0.082	0.690	0.257	−0.271	−0.146	−0.177
$r_{i/j}$	−0.339	0.836	0.758	0.307	0.219	−0.179	−0.187

DHE = days to head, BIO = aboveground biomass, SN = spike number, GY = grain yield, Y_{eco} = economical yield, PHT = plant height, STW = straw yield, HI = harvest index, CHL = chlorophyll content, and CT = canopy temperature.

Path analysis results indicated that the negative and significant correlation coefficients, relating grain yield to the number of days to heading (−0.339**), were made of sizeable negative indirect effects (−0.325) via other traits and an almost nil direct effect (−0.014). The high and significant correlation coefficients relating grain yield to aboveground biomass (0.836**) and to spike number (0.758**) were made of high direct effect for biological yield (0.918) and indirect effects for spike number (0.690). Indirect effects of biological yield (−0.082) and direct effect of spike number (0.068), on grain yield, were too small to be of interest. The significant correlation coefficient of plant with grain yield had greater indirect effects (0.27) than direct effect (0.050). Interestingly, the nonsignificant correlation coefficient relating grain yield to harvest index was made of sizeable positive direct (0.490) and negative indirect effects (−0.271). Chlorophyll and canopy temperature had no significant direct or indirect effects on grain yield (Table 6).

Khan et al. [52] reported significant and positive direct effect of spike number and large indirect effects of days to heading and plant height on grain yield. The number of spikes and biological yield appeared to be the most effective variables influencing grain yield, in the present study; therefore, it is reasonable to assume that wheat grain yield, under south Mediterranean conditions, could be increased by selecting genotypes with high spike number, biological yield, and harvest index. Classifying the 56 wheat genotypes on the basis of the measured traits yielded three clusters, showing gradation, from cluster XG2 to XG1 to XG3, for biological, economical, grain, and straw yields, spike number, and harvest index. Differences between clusters for canopy temperature and number of days to heading were not

TABLE 7: Cluster means for the 10 measured traits of the 56 durum wheat genotypes.

	BIO	SN	GY	Y_{eco}	STW	PHT	HI	CHL	CT	DHE
XG2	151.1	45.7	54.3	83.3	96.9	53.5	36.8	621.8	23.7	121.8
XG1	178.5	53.9	61.2	96.4	117.3	57.4	34.6	710.8	23.8	121.2
XG3	203.7	53.8	73.0	112.2	130.7	59.5	36.0	577.8	23.1	121.3

DHE = days to head, BIO = aboveground biomass, SN = spike number, GY = grain yield, Y_{eco} = economical yield, PHT = plant height, STW = straw yield, HI = harvest index, CHL = chlorophyll content, and CT = canopy temperature.

large. Chlorophyll content was high in cluster XG1 and low in cluster XG3 (Table 7).

Relative differences among clusters are indicated in Figure 1. Cluster XG3 carried entries having best combinations for biological, economical, grain, and straw yields, spike number, harvest index, and plant height. Therefore, selection of suitable genotypes among the 13 available within this cluster would be beneficial, if the selected genotypes show appreciable gain over currently adopted cultivars or checks. In fact, selection of the best 5 genotypes belonging to cluster XG3 brought appreciable gains, principally for biological (10.4 to 30.3%), economical (10.3 to 31.2%), grain (13.7 to 33.7%), and straw (−2.4 to 39.3%) yields, spike number (9.9 to 20.8%), and canopy temperature (−17.3 to −21.8%). No significant change in the number of days to heading (0.0 to 3.3%) was noted.

Changes in harvest index (−11.9 to 9.7%), chlorophyll content (−9.9 to 2.4%), and plant height (−3.4 to 12.6%) were variable (Figure 2). Furthermore, choosing appropriate parents, for hybridization purposes, is an important step in plant breeding program. In fact, performance of progenies depends on parental genotype. As differences between clusters could be attributed to differences in allelic forms, crosses made between genotypes belonging to different clusters are expected to generate variability within which selection to improve the targeted traits would be more efficient.

Altogether, results of the present study indicated that the existing variability varied among traits, being low for number of days to heading, plant height, chlorophyll content, and canopy temperature and high for biological, grain, economical, and straw yields and harvest index. Heritability was low for grain and straw yields, medium for biological and economical yields, spike number, and harvest index, and high for days to heading, spike number, plant height, chlorophyll content, and canopy temperature. Expected response to selection was relatively low for the number of days to heading and high for the remaining traits. These results corroborate findings reported by Salmi et al. [31] and Ghaed-Rahimi and Heidari [49].

The results of correlation coefficients indicated that early genotypes performed best in terms of grain yield, aboveground biomass, and spike number and avoided terminal heat stress, encouraging selection of early plant material. Correlation coefficient results pointed to aboveground biomass as an indirect selection criterion to improve grain yield, along with spike number, straw yield, economical yield, and harvest index. These findings were in accordance with results of Fellahi et al. [27], Salmi et al. [31], Hannachi et al. [34], and Leilah and Al-Khateeb [51]. Path analysis indicated that days to heading and spike number affected grain yield mainly through indirect effects, while aboveground biomass had high direct effect. Harvest index had both significant positive direct and negative indirect effects. Three contrasted clusters were extracted from the data set.

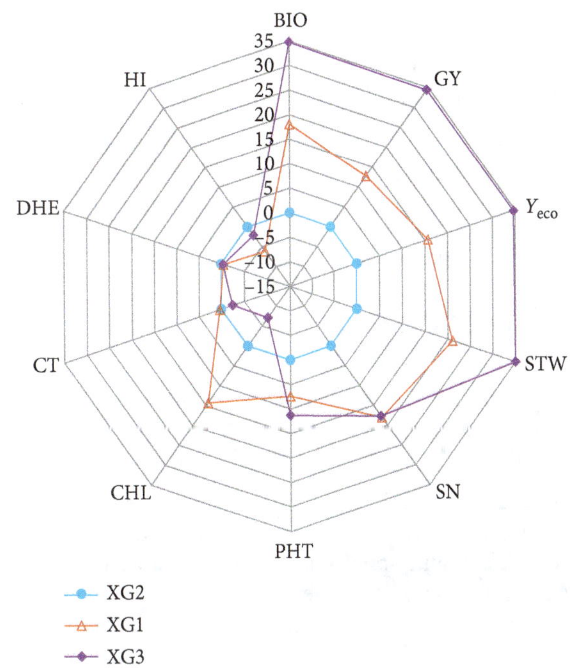

FIGURE 1: Relative deviation of the performances of clusters XG1 and XG3 as percent of mean values of cluster XG2 for the measured traits.

Crosses between genotypes belonging to different clusters to generate variability within which selection to improve the targeted traits would be more efficient are proposed. This increases possibilities to improve both yield potential and stress tolerance. Selection should be done, preferentially for spike number, biological yield, harvest index, and canopy temperature to accumulate favorable alleles in the selected entries. Selection of best genotypes, for direct uses as cultivars, from clusters carrying best combinations of yield-related traits is feasible.

4. Conclusion

Sufficient variability was noted for the measured traits in the data set obtained from the 56 assessed genotypes. Heritability

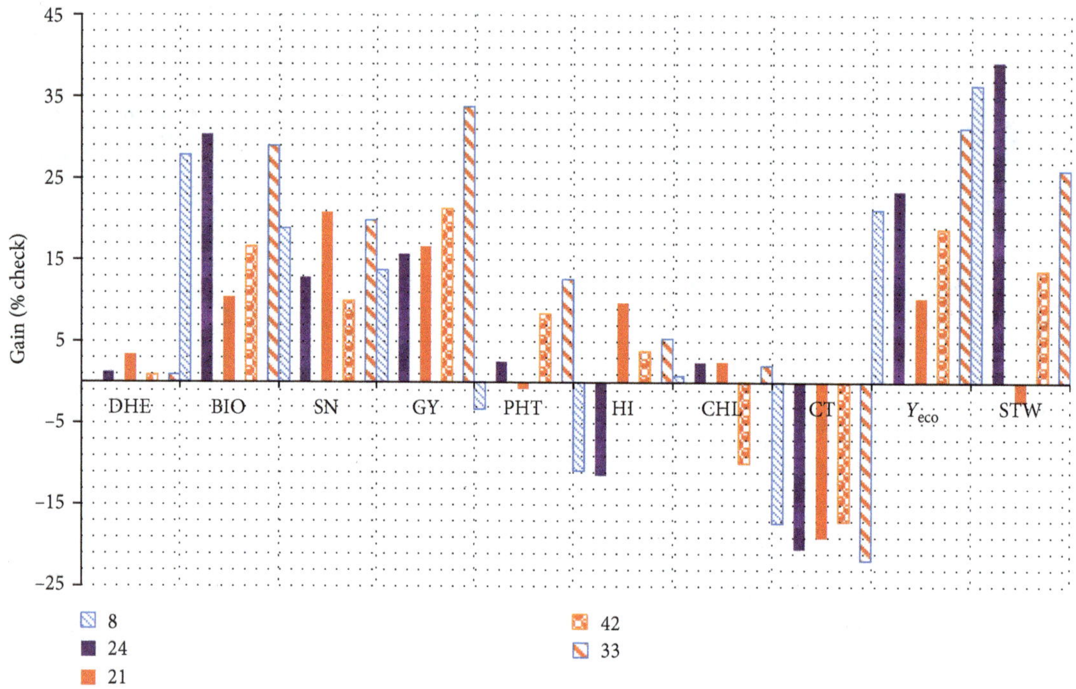

FIGURE 2: Gain (% of check) for the measured traits brought by selection of the 5 best yielding entries from cluster XG3.

and expected genetic gain varied among the measured traits. Relationship analysis designated aboveground biomass, harvest index, and spike number as the most grain yield-influencing traits. High genetic advance coupled with high heritability estimates offers the most effective condition for selection. The utility of heritability therefore increases when it is used to calculate genetic advance, which indicates the degree of gain in a character obtained under a particular selection pressure. Selection for earliness is encouraging since early genotypes exhibited above-average grain and biological yields and spike number and most importantly avoided terminal heat stress. Selection for direct use from cluster carrying best combination of yield-related traits and crosses among genotypes belonging to contrasted clusters were suggested to generate more variability. Selection should be done, preferentially for spike number, biological yield, harvest index, and canopy temperature to accumulate favorable alleles in the selected entries for future uses. The results suggest that genotypes with high expression of biomass (source), good harvest index expression, and other yield components can be used as parents in durum wheat breeding program to boost the genetic gains.

Acknowledgments

The authors would like to express special thanks to the personnel of INRAA-ITGC-ARS of El-Khroub, particularly durum wheat breeding team for kindly providing seeds and facilities and help in conducting the experiment.

References

[1] Food and Agricultural Organization of the United Nations, Statistics Division, *FAOSTAT*, 2016, http://faostat3.fao.org/home/e.

[2] M. Graziani, M. Maccaferri, C. Royo, F. Salvatorelli, and R. Tuberosa, "QTL dissection of yield components and morpho-physiological traits in a durum wheat elite population tested in contrasting thermo-pluviometric conditions," *Crop and Pasture Science*, vol. 65, no. 1, pp. 80–95, 2014.

[3] J. Subira, F. Álvaro, L. F. G. Del Moral, and C. Royo, "Breeding effects on the cultivar x environment interaction of durum wheat yield," *European Journal of Agronomy*, vol. 68, pp. 78–88, 2015.

[4] J. L. Araus, J. Bort, P. Steduto, D. Villegas, and C. Royo, "Breeding cereals for Mediterranean conditions: ecophysiological clues for biotechnology application," *Annals of Applied Biology*, vol. 142, no. 2, pp. 129–141, 2003.

[5] G.A. Slafer, J. L. Araus, C. Royo, and L. F. G. Del Moral, "Promising eco-physiological traits for genetic improvement of cereal yields in Mediterranean environments," *Annals of Applied Biology*, vol. 146, no. 1, pp. 61–70, 2005.

[6] P. Annicchiarico, *Genotype x Environment Interactions: Challenges and Opportunities for Plant Breeding and Cultivar Recommendations*, Food and Agriculture Organization, Rome, Italy, 2002.

[7] E. Acevedo and P. Silva, "Trigo candeal: calidad, mercado y zonas de cultivo," in *Serie Ciencias Agronómicas*, Universidad de Chile, Facultad de Ciencias Agronómicas, Región Metropolitana, Chile, 2007.

[8] R. Motzo, F. Giunta, and G. Pruneddu, "The response of rate and duration of grain filling to long-term selection for yield in Italian durum wheats," *Crop and Pasture Science*, vol. 61, no. 2, pp. 162–169, 2010.

[9] A. Blum, *Plant Breeding for Water-Limited Environments*, Springer Science & Business Media, New York, NY, USA, 2011.

[10] M. Lopes, M. Reynolds, M. Jalal-Kamali et al., "The yield correlations of selectable physiological traits in a population of advanced spring wheat lines grown in warm and drought environments," *Field Crops Research*, vol. 128, pp. 129–136, 2012.

[11] J. Crossa, G. de los Campos, M. Maccaferri, R. Tuberosa, J. Burgueño, and P. Pérez-Rodríguez, "Extending the marker × environment interaction model for genomic-enabled prediction and genome-wide association analysis in durum wheat," *Crop Science*, vol. 56, no. 5, pp. 2193–2209, 2016.

[12] R. A. Fisher, "Wheat physiology: a review of recent developments," *Crop and Pasture Science*, vol. 62, no. 2, pp. 95–114, 2011.

[13] C. Royo, R. Nazco, and D. Villegas, "The climate of the zone of origin of Mediterranean durum wheat (*Triticum durum* Desf.) landraces affects their agronomic performance," *Genetic Resources and Crop Evolution*, vol. 61, no. 7, pp. 1345–1358, 2014.

[14] L. Mwadzingeni, H. Shimelis, S. Tesfay, and T. J. Tsilo, "Screening of bread wheat genotypes for drought tolerance using phenotypic and proline analyses," *Frontiers in Plant Science*, vol. 7, p. 1276, 2016.

[15] D. Dodig, M. Zoriæ, V. Kandiæ, D. Peroviæ, and G. Šurlan–Momiroviæ, "Comparison of responses to drought stress of 100 wheat accessions and landraces to identify opportunities for improving wheat drought resistance," *Plant Breeding*, vol. 131, no. 3, pp. 369–379, 2012.

[16] N. N. Hamssi and A. Najaphy, "Comparison of photosynthetic components of wheat genotypes under rain-fed and irrigated conditions," *Photochemistry and Photobiology*, vol. 88, no. 1, pp. 76–80, 2012.

[17] G. J. Rebetzke, A. Rattey, G. D. Farquhar, R. A. Richards, and A. G. Condon, "Genomic regions for canopy temperature and their genetic association with stomatal conductance and grain yield in wheat," *Functional Plant Biology*, vol. 40, no. 1, pp. 14–33, 2013.

[18] M. Tattaris, M. P. Reynolds, and S. C. Chapman, "A direct comparison of remote sensing approaches for high-throughput phenotyping in plant breeding," *Frontiers in Plant Science*, vol. 7, p. 1131, 2016.

[19] J. J. Olivares-Villegas, M. P. Reynolds, and G. K. McDonald, "Drought adaptive attributes in the Seri/Babax hexaploid wheat population," *Functional Plant Biology*, vol. 34, no. 3, pp. 189–203, 2007.

[20] C. Royo, D. Villegas, L. F. G. Del Moral et al., "Comparative performance of carbon isotope discrimination and canopy temperature depression as predictors of genotype differences in durum wheat yield in Spain," *Australian Journal of Agricultural Research*, vol. 53, no. 5, pp. 561–569, 2002.

[21] F. Giunta, R. Motzo, and G. Pruneddu, "Has long-term selection for yield in durum wheat also induced changes in leaf and canopy traits?," *Field Crops Research*, vol. 106, no. 1, pp. 68–76, 2008.

[22] G. A. Slafer, V. O. Sadras, and S. Roxana, *Is Compensation between Yield Components in Wheat a Feedback Process of Responses to Availability of Resources?*, 2008, http://www.intlcss.org/files/icss/congress-proceedings/2008-papers/cs3-s1/cs3-s1-o3-gustavo-slafer.pdf.

[23] G. A. Slafer, R. Savin, and V. O. Sadras, "Coarse and fine regulation of wheat yield components in response to genotype and environment," *Field Crops Research*, vol. 157, pp. 71–83, 2014.

[24] V. O. Sadras, "Evolutionary aspects of the trade-off between seed size and number in crops," *Field Crops Research*, vol. 100, pp. 125–138, 2007.

[25] G. González-Ribot, M. Opazo, P. Silva, and E. Acevedo, "Traits explaining durum wheat (*Triticum turgidum* L. spp. *durum*) yield in dry Chilean Mediterranean environments," *Frontiers in Plant Science*, vol. 8, p. 1781, 2017.

[26] T. Mohsin, N. Khan, and F.N. Naqvi, "Heritability, phenotypic correlation and path coefficient studies for some agronomic characters in synthetic elite lines of wheat," *Journal of Food, Agriculture and Environment*, vol. 7, no. 3-4, pp. 278–282, 2009.

[27] Z. Fellahi, A. Hannachi, H. Bouzerzour, and A. Boutekrabt, "Line × tester mating design analysis for grain yield and yield related traits in bread wheat (*Triticum aestivum* L.)," *International Journal of Agronomy*, vol. 2013, Article ID 201851, 9 pages, 2013.

[28] M. Kandel, A. Bastola, P. Sapkota, O. Chaudhary, P. Dhakal, and P. Chalise, "Association and path coefficient analysis of grain yield and its attributing traits in different genotypes of wheat (*Triticum aestivum* L.)," *International Journal of Applied Sciences and Biotechnology*, vol. 5, no. 4, pp. 449–453, 2017.

[29] Z. Fellahi, A. Hannachi, H. Bouzerzour, and A. Benbelkacem, "Inheritance pattern of metric characters affecting grain yield in two bread wheat (*Triticum aestivum* L.) crosses under rainfed conditions," *Jordan Journal of Biological Sciences*, vol. 8, no. 3, pp. 175–181, 2015.

[30] C. Royo, E. M. Elias, and F. A. Manthey, "Durum wheat breeding," in *Handbook of Plant Breeding: Cereals*, M. J. Carena, Ed., pp. 199–226, Springer, New York, NY, USA, 2009.

[31] M. Salmi, L. Haddad, A. Oulmi, A. Benmahammed, and A. Benbelkacem, "Variabilité phénotypique et sélection des caractères agronomiques du blé dur (*Triticum durum* Desf.) sous conditions semi-arides," *European Scientific Journal*, vol. 11, no. 2, pp. 99–111, 2015.

[32] V. K. Arya, J. Singh, L. Kumar et al., "Character association and path coefficient analysis in wheat (*Triticum aestivum* L.)," *Indian Journal of Agricultural Research*, vol. 51, no. 3, pp. 245–251, 2017.

[33] A. Hannachi, Z. Fellahi, A. Rabti, A. Guendouz, and H. Bouzerzour, "Combining ability and gene action estimates for some yield attributes in durum wheat (*Triticum turgidum* L. var. *durum*)," *Journal of Fundamental and Applied Sciences*, vol. 9, no. 3, pp. 1519–1534, 2017.

[34] A. Hannachi, Z. Fellahi, H. Bouzerzour, and A. Boutekrabt, "Diallel-cross analysis of grain yield and stress tolerance-related traits under semi-arid conditions in durum wheat (*Triticum durum* Desf.)," *Electronic Journal of Plant Breeding*, vol. 4, pp. 1027–1033, 2013.

[35] M. Moghaddam, B. Ehdaie, and J. G. Waines, "Genetic variation for and interrelationships among agronomic traits in landraces of bread wheat from southwestern Iran," *Journal of Genetics and Breeding*, vol. 52, no. 1, pp. 73–81, 1998.

[36] V. Mollasadeghi, R. Shahryari, A. A. Imani, and M. Khayatnezhad, "Factor analysis of wheat quantitative traits on yield under terminal drought," *American-Eurasian Journal of Agriculture and Environmental Science*, vol. 10, no. 2, pp. 157–159, 2011.

[37] M. Zakizadeh, M. Esmaeilzadeh, and D. Kahrizi, "Study on genetic variation and relationship between plant characteristics and yield in long spike bread wheat (*Triticum aestivum* L.) genotypes using multivariate analysis," *Iranian Journal of Crop Sciences*, vol. 12, no. 2, pp. 18–30, 2010.

[38] P. Annicchiarico, F. Bellah, and T. Chiari, "Repeatable genotype x location interaction and its exploitation by conventional and GIS-based cultivar recommendation for durum wheat in Algeria," *European Journal of Agronomy*, vol. 24, no. 1, pp. 70–81, 2006.

[39] R. G. D. Steel and J. H. Torrie, *Principles and Procedures of Statistics*, McGraw-Hill Books, New York, NY, USA, 1982.

[40] G. Acquaah, *Principles of Plant Genetics and Breeding*, John Wiley and Sons, Hoboken, NJ, USA, 2009.

[41] C. D. Cruz, A. J. Regazi, and P. C. S. Carneiro, *Modelos Biométricos Aplicados ao Melhoramento Genético*, UFV, Viçosa, Brazil, 4th edition, 2012.

[42] R. K. Singh and B. D. Chaudhary, *Biometrical Methods in Quantitative Genetic Analysis*, Kalyani Publishers, New Delhi, India, 1985.

[43] LazStats, *Free Statistics Programs and Materials*, W. G. Miller, Iowa City, IA, USA, 2013.

[44] O. Hammer, D. A. T. Harper, and P. D. Ryan, "PAST: paleontological statistics software package for education and data analysis," *Paleontologia Electronica*, vol. 4, pp. 1–9, 2001.

[45] Cropstat, *Cropstat for Windows 2007.7.2.3*, International Rice Research Institute, Manila, Philippines, 2007.

[46] A. Oulmi, A. Benmahammed, Z. Laala, A. Adjabi, and H. Bouzerzour, "Phenotypic variability and relations between the morpho-physiological traits of three F5 populations of durum wheat (*Triticum durum* Desf.) evaluated under semi-arid conditions," *Advances in Environmental Biology*, vol. 8, pp. 436–443, 2015.

[47] Z. Fellahi, A. Hannachi, H. Bouzerzour, and A. Benbelkacem, "Genetic control of bread wheat (*Triticum aestivum* L.) traits," *Songklanakarin Journal of Science and Technology*, vol. 38, no. 1, pp. 91–97, 2016.

[48] A. Mekhlouf, F. Dehbi, H. Bouzerzour, A. Hannachi, A. Benmahammed, and A. Adjabi, "Relationships between cold tolerance, grain yield performance and stability of durum wheat (*Triticum durum* Desf.) genotypes grown at high elevation area of eastern Algeria," *Asian Journal of Plant Science*, vol. 5, pp. 700–708, 2006.

[49] L. Ghaed-Rahimi and B. Heidari, "Genetic advance prediction and multivariate analysis for antioxidants and agronomic traits in wheat," *Annual Research and Review in Biology*, vol. 4, no. 14, pp. 2427–2449, 2014.

[50] C. Royo, F. Alvaro, V. Martos et al., "Genetic changes in durum wheat yield components and associated traits in Italian and Spanish varieties during the 20th century," *Euphytica*, vol. 155, no. 1-2, pp. 259–270, 2007.

[51] A. A. Leilah and S. A. Al-Khateeb, "Statistical analysis of wheat yield under drought conditions," *Journal of Arid Environments*, vol. 61, no. 3, pp. 483–496, 2005.

[52] A. A. Khan, M. A. Alam, M. K. Alam, M. J. Alam, and Z. I. Sarker, "Correlation and path analysis of durum wheat (*Triticum turgidum* L. var. *durum*)," *Bangladesh Journal of Agriculture Research*, vol. 38, no. 3, pp. 515–521, 2013.

20

The Potential Growing Areas for *Argania spinosa* (L.) Skeels (Sapotaceae) in Argentinean Drylands

Silvia L. Falasca,[1] **Sandra Pitta-Alvarez**,[2,3] **and Ana Ulberich**[4]

[1]*Climate and Water Institute, INTA, Las Cabañas y Los Reseros, s/n, Castelar, Buenos Aires, Argentina*
[2]*Laboratorio de Cultivo Experimental de Plantas y Microalgas, Departamento de Biodiversidad y Biología Experimental, Facultad de Ciencias Exactas y Naturales, Universidad de Buenos Aires, Buenos Aires, Argentina*
[3]*Instituto de Micología y Botánica (INMIBO), Universidad de Buenos Aires-CONICET, Buenos Aires, Argentina*
[4]*Centro de Investigaciones y Estudios Ambientales (CINEA), Facultad de Ciencias Humanas (FCH), Universidad Nacional del Centro de la Provincia de Buenos Aires (UNICEN), Campus Universitario, Paraje Arroyo Seco, s/n, Tandil, Argentina*

Correspondence should be addressed to Silvia L. Falasca; slfalasca@gmail.com

Academic Editor: Othmane Merah

The argan (*Argania spinosa* L.) tree is indigenous to southwestern Morocco. Successful introduction of this species in Argentina depends on the degree of climate similarity in the natural area compared to new areas with agroclimatic suitability. Based on international bibliography, the authors outlined an agroclimatic zoning model for this tree in Argentina. To determine agroclimatic suitability, the average climatic data of all the meteorological stations (1981–2010) were analyzed. Climatological characteristics of its region of origin were considered to evaluate the requirements and biometeorological conditions. The maps were obtained using a series of previously interpolated bioclimatic variables. The agroclimatic indices, which determine different classes of suitability, were integrated in a Geographic Information System (GIS) to create thermal and moisture regions. The resulting maps were superimposed to determine the agroclimatic zoning. Eight classes of agroclimatic suitability under dry-subhumid, semiarid, and arid climatic conditions were delineated. There are no optimal zones for the argan tree in Argentina. This work can be updated by further incorporation of complementary information. Argan tree is a promising dryland crop for farmers in semiarid and arid climates of Argentina. This model may be applied in any part of the world, using the agroclimatic limits presented in this paper.

1. Introduction

The argan (*Argania spinosa* L.) belongs to the Sapotaceae family and it is the only species of the genus Argania. It is a little-known tree indigenous to intramontane areas, with very low average rainfall, in southwestern Morocco. The plant is distributed across five Moroccan provinces (the most important ones are Agadir, Essaouira, Toroudant, and Tiznit), in arid and semiarid ecosystems, with Atlantic influence, between 29° and 32° of north latitude [1, 2]. Some isolated populations can be found northeast of Morocco, 400 to 700 km away from the main ones: in Berkane in the northern slope of the mountain Beni-Snassen, close to the Mediterranean coast (800 ha), and in the Bou-Dreg plains (Kariat Arakmane), in the eastern Rif of the high valleys of Oued Grou, close to Rabat (50 ha) [2]. The argan tree is disseminated from Safi (north) to the Drâa River (south), and isolated populations extend as far as Tindouf, well inside the Western Sahara. Starting in Essaouira and extending to Goulmime, and from the High Atlas to the Atlantic Ocean, it covers more than 828,000 ha [3]. According to [4], it appears in subspontaneous form in the Sahara, between Djebel Ouarkziz and the Hamada of Tindouf (Algeria), in humid microhabitats. It can also be found in certain parts of the Spanish Levante in Alicante. These isolated populations are considered relicts of a wider distribution in the past [5].

Edible argan oil is obtained from lightly toasted kernels, and it is becoming increasingly popular in gastronomy due to its exotic aroma and fruit-like flavor, reminiscent of a mixture of nuts, hazelnuts, and sesame seeds. The color is darker and the odor stronger than argan cosmetic oil because of the previous toasting of the kernels. Edible argan oil has several sterols that are uncommon in other edible vegetable oils, such as scotenol (48%), which has antitumoral and anticancer properties, and spinasterol (42%), a compound that reduces intestinal cholesterol absorption. The white seeds contain up to 50% of light brown oil, and oil production ranges between 1000 and 2000 tons per year. The ratio of total unsaturated to total saturated fatty acids is around 4 : 5. Polyunsaturated fatty acids constitute approximately 80% of argan oil, and 30% of it is linoleic acid. This composition is particularly adept for nutritional, cosmetic, and medicinal uses. Argan oil is rich in vitamin E and it has been reported to lower cholesterol levels, stimulate blood circulation, facilitate digestion, and strengthen the body's natural defenses [6]. The dietary and nutritional properties of argan oil are superior to those of extravirgin olive oil. The oil is also used for illumination and soap.

The tree is also popular as a shade tree and it is useful for soil conservation and windbreaks. After the oil is extracted, the residue from the kernels constitutes a thick chocolate-colored paste called "amlou," with a flavor similar to that of peanut butter.

The tree is an important source of forage for sheep, goats, and cattle, but horses reject it. It contains the haemolytic saponin sapoarganine, which is slightly toxic but does not harm ruminants. However, if dairy cattle are fed with the cake, it contaminates the milk, and this may cause diarrhoea in children. Goats digest the fleshy part of the fruit, but discharge the nut because of its hard shell. The nuts are later collected by farmers to produce oil.

The wood is hard, heavy, and durable, and it results in good charcoal and firewood. The nut-shells are also burned for cooking. Argan wood cannot be destructed by insects, and it has been used for centuries in carpentry, charcoal, construction, and utensils at local level.

The international price of argan oil has stimulated the idea of cultivating *A. spinosa* in regions far from its original geographic center [7]. This could affect the argan oil business in Morocco and defeat the efforts initiated to save the argan forests. For this reason, the Moroccan government seeks to obtain a Geographical Indication (GI) on argan oil [8]. A GI would mean that only the argan oil prepared from seeds collected in Morocco, with the appropriate rules of production, could be sold as such under that name [7]. According to [7], the oil of this species was first reported in the thirteenth century by the botanist, scientist, pharmacist, and physician Ibn al-Baitar. These researchers maintain that argan tree is one of the few surviving relict-breed species of the Tertiary Period, which spanned from 65 million to 1.8 million years ago. Currently, the tree can be found in the south of Morocco, west of Algeria, north of Western Sahara, and north of Mauritania [9]. It was introduced for cultivation into Algeria, Israel, Kuwait, Libya, the semiarid Southeast Spain [10], and Mexico [6].

The argan forest is particularly sensitive to climate change and it was ignored as a specific biotope for a long time. In 1998, it was declared a Biosphere Reserve by UNESCO [11]. In some regions, it is already damaged, and its future is endangered by prolonged drought periods and desertification. The tree is in threat of extinction due to clearance, poor germination quality, and the use of the timber as fuel. According to [12], in the period 1970–2007, argan tree density decreased 44.5%. Recently, strict conservation measures have been implemented to prevent its disappearance [13].

Occasionally, new crops with attractive properties and exploitation potential capture the attention of both public and private sectors. Some of these new crops ameliorate the effects of increased atmospheric CO_2 (greenhouse effect), which turns fertile lands into deserts by increasing temperature and producing drought [15]. The resulting boom of this particular species can lead to large investments and campaigns promoting the need for its domestication and industrial scale commercialization. However, research on these plants is usually scarce because they are commonly wild nondomesticated species. Farmers tend to be cautious and avert the risks involved in developing new crops. Investors, on the other hand, frequently fail because they lack adequate information concerning the risks and uncertainties related to these ventures. For instance, they sometimes put at risk the sustainability of the ecosystem where implantation is purported, or they ignore the agroclimatic requirements of the species.

The aim of this work was to design an agroclimatic zoning model to define the potential growing areas for *Argania spinosa* in Argentina.

2. Materials and Methods

2.1. Study Area. The area studied was the Argentine Republic. The country borders to the north with Bolivia, Brazil, and Paraguay; to the south with Chile and the Atlantic Ocean; to the east with Brazil, Uruguay, and the Atlantic Ocean; and to the west with Chile (Figure 1).

Argentina's climatic diversity is remarkable due to both its vast territory and distinct geographic factors that influence the climatic characteristics of the different regions. One of these factors is the great latitudinal development: Argentina expands from 21°46′ (province of Jujuy) in the north to 55°03 in Cape San Pío (province of Tierra del Fuego), in the south. The extreme eastern limit of the country is located at 53°38′W, in the town of Bernardo de Irigoyen (province of Misiones).

The limit between the subhumid and semiarid zones is represented by the 500 mm isohyet. The area situated between 500 and 1000 mm/year presents a subhumid climate, and the climate is humid when it exceeds 1000 mm. To the west of the 500 mm isohyet, the moisture gradient provided by rainfall from the South Atlantic anticyclone decreases and the climate turns dry-subhumid, semiarid, and, finally, arid.

2.2. Ecological Requirements. According to [13], the argan tree prospers in desert or arid (Bw), steppe or semiarid (Bs) climates, between 20 and 32° latitude. The tree has several mechanisms that restrain water loss: it reaches the deepest layers of soil during dry season, it can close its stomata during the day, and it has water reservoirs in its trunk and branches [10].

The argan is a species that can be cultivated in drylands in regions where annual rainfalls are in the 200–300 mm range, with a minimum of 100 mm and a maximum of

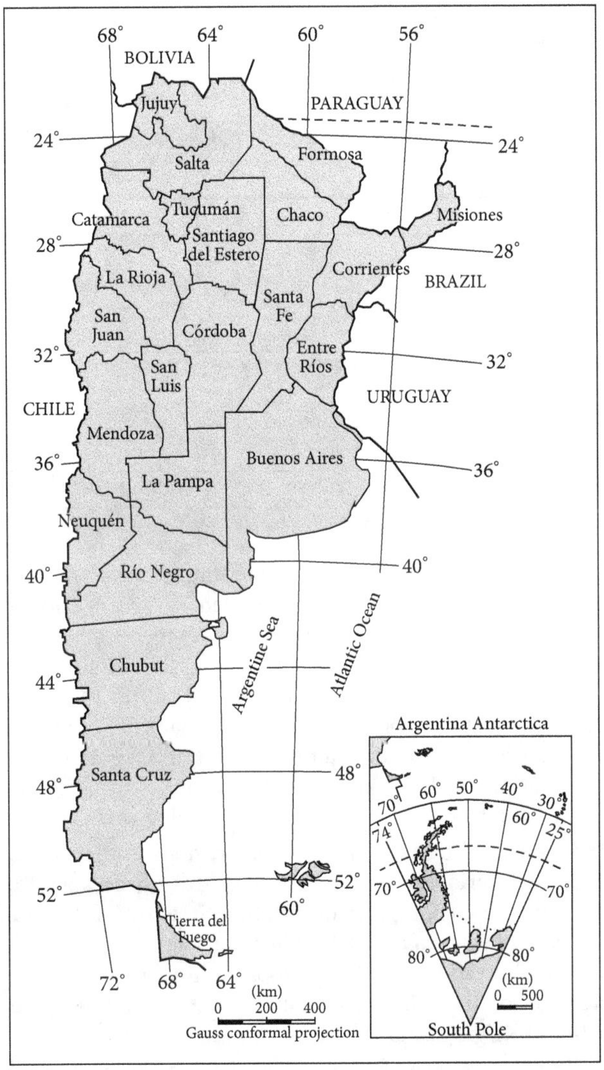

FIGURE 1: Argentina's political map.

400 mm [13]. The critical period, as far as water is concerned, is between the preflowering and ripening phases, and this usually corresponds to the period with the highest rain scarcity in the original areas.

Germination does not occur with temperatures below 10°C, and the optimum germination temperature is 25°C [16], with night temperatures above 21°C [17]. The tree flourishes in areas where annual mean temperatures fluctuate between 18 and 20°C. Best growth is observed with temperatures in the 20–30°C range, with a minimum of 10°C and a maximum of 35°C, although it can tolerate maximum temperatures of 50°C. According to [13], the tree dies with frosts of −2°C, although according to [18] it can tolerate −5°C.

The species needs neutral soils (pH of 6.5 to 7.5) and, although soil quality is not an issue, drifting sands and water-logged soils must be avoided. Argan is a saline-sensitive plant [14, 19] and adult trees do not adapt to the higher salinity associated with dry areas.

Flowering occurs mainly in spring, with great differences among individuals, and with a secondary peak in autumn. As a result, fruit production extends over a long period and fruits can be found in different ripening phases from April to September [3] in the north hemisphere.

2.3. Climate of the Original Area. Considering that the species has not yet been introduced into Argentina, the climate of its original area was studied in detail. Morocco's climate is Mediterranean, with Atlantic and Saharan influences. The climate is characterized by two gradients: decreasing precipitation with increasing temperature from north to south and from west to east [20]. Currently, the argan tree is found in an intramontane area with very low average rainfall. The land is bordered by the Sahara, the High Atlas, and the Anti-Atlas Mountains between Essaouira and Agadir Ida Outanane and open to the Atlantic on the west side. The altitudinal range of the argan tree extends from sea level up to 1,300–1,500 m.

The annual rainfall varies from 250 mm in Safi to 400 mm in Agadir. From the Souss plain to the Anti-Atlas Mountains, annual precipitation is in the 250 to 150 mm range. To the south, in desert areas, precipitations are considerably lower than 100 mm [2].

In the coastal areas where argan trees grow, there are temperate oceanic influences with moderate temperatures (annual average in Agadir is 18.5°C; minimum temperature of coldest month: 7.4°C) and low annual and daily thermal amplitude [14].

In the mountains and inside the continent, the thermal amplitude reaches higher values and the minimum temperatures of the coldest month fluctuate from 3°C to 7°C. The highest boundary for argan tree is close to the 3.8°C isotherm, which corresponds to the lower limit for snow [2]. The average maximum winter temperature ranges from 17.5°C to 22°C [21].

Most crops commonly show recognizable differences in many of their properties within their area of distribution. These are usually the result of climate variations, and even slight changes in a single climate element can wield important influences. For this reason, two extreme climatic areas were chosen to describe the "argan bioclimate."

One of these areas is Admine Forest (hereafter Coastal Browsing, 30°26′54.13″N and 9°32′49.90″W, located in the suburbs of Agadir city, 5 km inland, 107 m of altitude), which different authors [22–24] consider as the ecological optimum for the species and the center of its distribution area. It is a region traditionally used for agriculture and livestock raising. Natural vegetation is scarce, except for the dominant *A. spinosa*, and the climate is temperate due to marine influence. Maximum temperatures never surpass 30°C and the minimum of the coldest month is never below 7°C [25]. Annual rainfall is scarce (232 mm) and concentrated from October to April.

The second area is Tassademt (hereafter Mountain), a site located 70 km NE of Agadir in the first foothills of the Atlas Mountains, which is characterized by its extreme climatic conditions. This site was selected because it is situated in the proximity of the species' natural distribution limit. It is

TABLE 1: Climatic characteristics of area of origin of the argan tree (elaborated by the authors based on [14]).

City	Geographical location	North latitude	Annual minim. temper. (°C)	Annual temper. (°C)	Annual maxim. temper. (°C)	Annual rainfall (mm)
Agadir	Seaside	30°25′	13.4	18.4	23.4	291.9
Ait Melloul	Souss plain	32°00′	13.0	18.4	23.0	233.5
Argana	Southern slope of High Atlas Mountains	30°45′	9.1	18.9	28.7	421.7
Ait Baha	Northern slope of Anti-Atlas Mountains	30°30′	10.7	20.1	29.4	223.0
Essaouira	Seaside	31°30′	14.6	17.3	20.0	342.6

an open forest where argan is the predominant vegetation type and broad gaps are covered with a grass community. Scarce and scattered spiny shrubs are the main woody vegetation. The climate is more extreme than in Agadir, with a distinct continental character. Monthly average temperatures are 40°C in summer and 3°C in winter; but they can drop to −5°C. Summer is particularly harsh because temperatures are inordinately high and rainfall is scant (385 mm) and almost totally absent.

Table 1 shows some climatic characteristics of area of origin of the argan tree.

2.4. The Agroclimatic Zoning Model for Argan. The climate system of a particular place defines the crops that can be developed. To identify regions with appropriate bioclimatic conditions for cultivating *A. spinosa*, both winter and summer requirements must be considered, since resistance to frosts and drought can be the key to success. To define the suitable zones, studies must focus on whether the agroclimatic availability of an area can satisfy the biometeorological demands of this species. To achieve this, both the thermal and the moisture regime, which are fundamental climatic characteristics that affect the development of this crop, must be considered.

First, we had to identify the requirements, limits, and biometeorological tolerance and conditions to establish the potential growing areas. The climatological characteristics of the original area were analyzed, and the resulting bioclimatic indicators were extrapolated to the Argentine territory. Aspects of growth, such as development and death chances by excess or deficiency, were analyzed.

The climate of a geographic location can be depicted by long-term averages or "Normals," which are defined as 30-year averages of a climate element. In this work, a set of long-term averages, covering a range of climate elements, for the period 1981 to 2010 was assembled. Geographical location, mean annual rainfall, and temperature records available from meteorological and agrometeorological stations within the study area were also considered. Values were averaged for each month over a 30-year period to obtain the long-term average.

The values which best indicate the thermal requirements of this species and can be used as equivalents are the bioclimatic variables: annual mean temperature, annual mean maximum temperature, annual mean minimum temperature, winter mean maximum temperature, minimum temperature of the coldest month, mean annual minimum temperature, and annual rainfall. These were obtained by interpolating the data collected in the 125 meteorological stations across the country (National Meteorological Service).

To create the maps, a series of previously interpolated bioclimatic variables were used. Afterwards, these were processed by the Geographic Information System (GIS) tool of the Arc-GIS 9.3 program. Climatic interpolations were made using the "Interpolate to Raster" tool, within the "3D Analyst" extension of the GIS of the same program, following the Ordinary Kriging interpolation method.

Since the moisture factor was the first bioclimatic variable examined, the average yearly rainfall isohyets for the period 1981–2010 were observed. The limits were delineated by the 100 and 400 mm isohyets [13]. If argan tree is grown under rainfed conditions, the nonsuitable areas are those with annual precipitations below 100 mm. However, if the purpose is production of edible oil, these areas could qualify as suitable with complementary irrigation. In the 100 to 200 mm and 300 to 400 mm interval, the area is suitable and, from 200 to 300 mm, optimal.

The first index used to analyze the thermal factor was the annual mean temperature, measured over a period of 30 years. This index is calculated by averaging the 12 normal monthly temperatures over the 30-year period. If temperatures fall below 18°C the climate is deemed too cold for the argan tree, and the area is not suitable [13]. Another index is the annual mean maximum temperature, which should surpass 23.0°C to ensure an appropriate thermal regime during the growing period [14].

To the extent that the species is unable to withstand cold, the annual mean minimum temperature, the winter mean maximum temperature, and the minimum temperature of the coldest month were analyzed. The annual mean minimum temperature, an index that represents the minimum daily

Table 2: Classes of agroclimatic suitability and limits for each class.

Agroclimatic suitability	Annual temper. (°C)	Annual min. temper. (°C)	Winter max. temper. (°C)	Minim temper. coldest month (°C)	Mean annual min. temper. (°C)	Annual max. temper. (°C)	Annual rainfall (mm)
Nonsuitable	<18	<10	<17.5	<3.8	<−5	<23.0	<100
Suitable$_1$	>18	>10	17.5–22.0	3.8–7.0	>−5	>23.0	100–200
Suitable$_2$	>18	>10	17.5–22.0	3.8–7.0	>−5	>23.0	300–400
Very suitable	>18	>10	17.5–22.0	3.8–7.0	>−5	>23.0	200–300
Optimal	>18	>10	17.5–22.0	>7.0	>−5	>23.0	200–300
Marginal			>22				>500

temperatures registered over 30 years, reveals the intensity of minimum temperatures and, according to [14], it should surpass 10.0°C.

Argentina's winter lasts from June to August. According to [21], the winter mean maximum temperature should be higher than 17.5°C. This index represents the average of the maximum temperatures from June to August, over the 30-year period considered.

The minimum temperature of the coldest month (in Argentina, July) is, by definition, the minimum monthly temperature occurrence over a 30-year period (normal) and it should be above 3.8°C [2]. In the optimal zone, the minimum temperature of the coldest month must never be below 7°C [25].

A very important factor is the species' resistance to cold, since it can hinder its cultivation in a specific area. Although Argentine territory extends towards the Tropic of Capricorn (23°27′S), it lacks tropical climates. Consequently, damage by frost should be avoided when tropical species are exploited. To prevent lethal effects, the annual mean minimum temperature (that represents the long-term average daily minimum air temperature observed during a calendar month and over the year) cannot be lower than −5°C [18].

The agroclimatic zoning map was obtained by superimposing the maps related to moisture and thermal regimes. The following classes were delineated: very suitable, suitable, suitable with constraints (under dry-subhumid, semiarid, and arid regimes), suitable with constraints under irrigation, marginal, and nonsuitable.

Table 2 shows the agroclimatic indices that define each class of agroclimatic suitability. It is important to highlight that the nonsuitable areas present two or more of the constraints indicated in Table 2.

Argentina's political map, with the toponomy of the provinces, was included to aid in the identification of the classified areas with different grades of agroclimatic fitness (Figure 1).

3. Results

Figure 2 shows the moisture regions. Different classes of suitability were characterized: nonsuitable, marginal, suitable, and optimal. If rainfalls are below 100 mm, the area is deemed

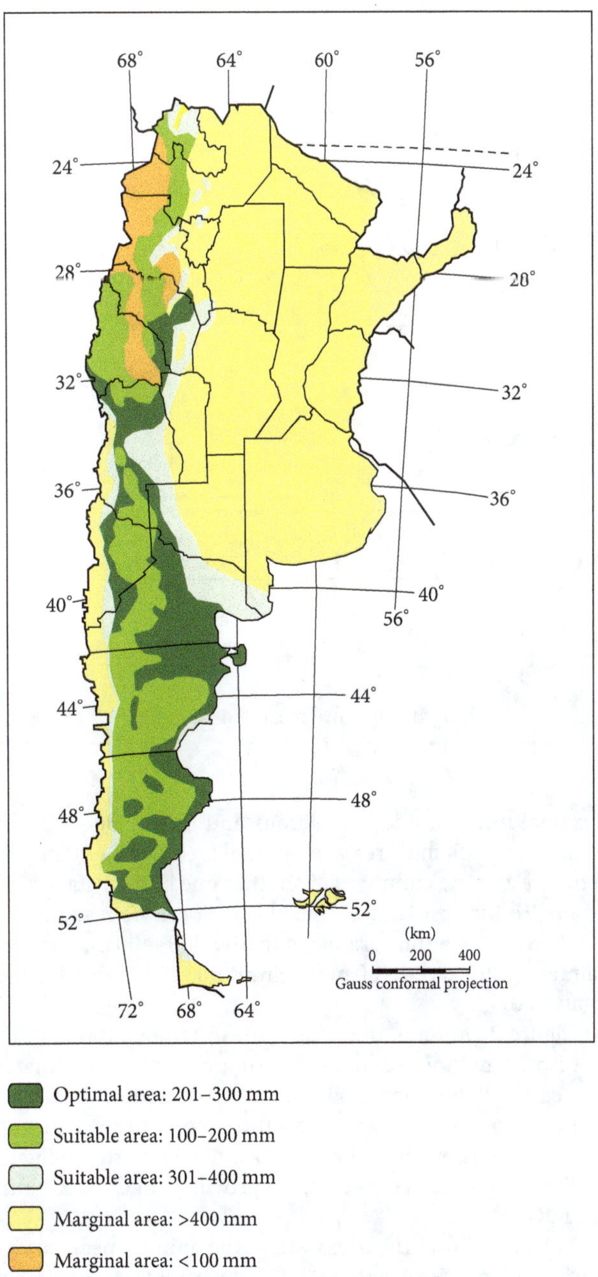

Figure 2: Annual rainfall.

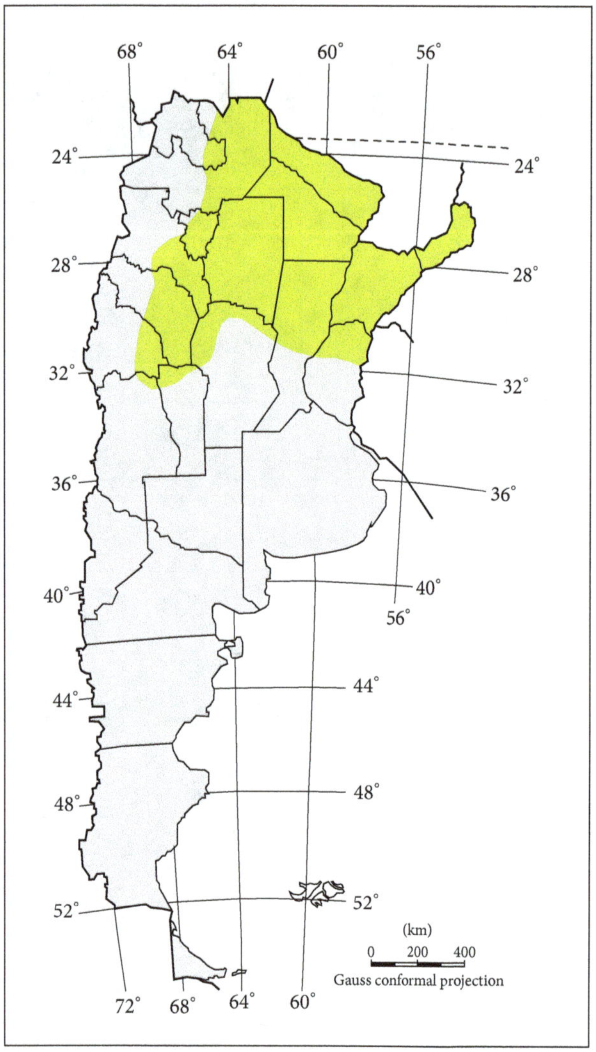

Suitable area: >18°C

Non suitable area: <18°C

FIGURE 3: Annual mean temperature.

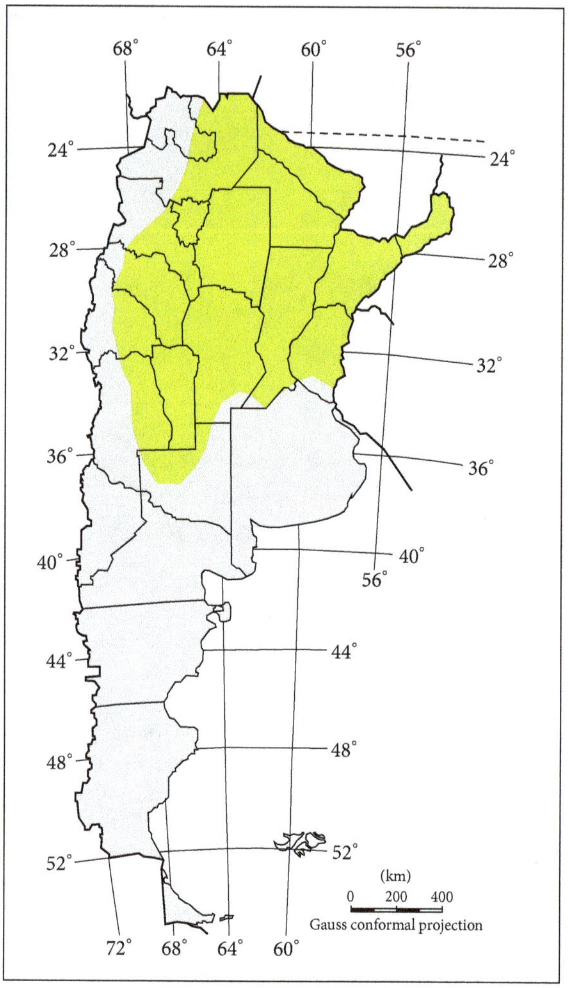

Suitable area: >23°C

Non suitable area: <23°C

FIGURE 4: Annual mean maximum temperature.

marginal under rainfed conditions and in the 100–200 mm range, it is a suitable area. An optimal area can be identified to the west of the country and another one in the Patagonian region. To the east, another suitable area can be recognized, and it receives annual rainfalls in the 300–400 mm range. Rainfalls that exceed 400 mm delimit marginal areas for the argan tree.

Figure 3 presents the annual mean temperature >18°C. This isotherm indicates the southern limit for warm climates that can still be appropriate for tropical crops. The 18°C isotherm covers north and part of the center of the country. In its path, it shows an inflection towards the north owing to the Pampeanas Saws located in the provinces of Córdoba and San Luis.

Figure 4 shows the areas where the annual mean maximum temperature surpasses 23.0°C, determining the suitable area. The 23°C isotherm covers the north and a large section of the center of the country. The great inflection towards the south in the province of La Pampa, due to the continentality factor, is noteworthy. It later diverts to the north because of the altitudinal effect imposed by the Andes Mountains. This zone, considered suitable, is influenced by the north wind year-round, which is responsible for the high temperatures over the year.

Figure 5 presents the areas where the annual mean minimum temperature is above 10°C. The 10.0°C isotherm diverts to the south influenced by the oceanity factor, which is conspicuous in the province of Buenos Aires. This area, considered suitable, reveals how the incoming cold air from the Polar Front projects to the north. This is responsible for the low minimum temperatures in all the national territory.

Figure 6 shows the winter mean maximum temperature. When it is lower than 17.5°C a nonsuitable area is delineated; in the 17.5 to 22.0°C range the area is suitable, while temperatures above 22°C define marginal areas. The suitable areas cover the eastern and center part of the country. There

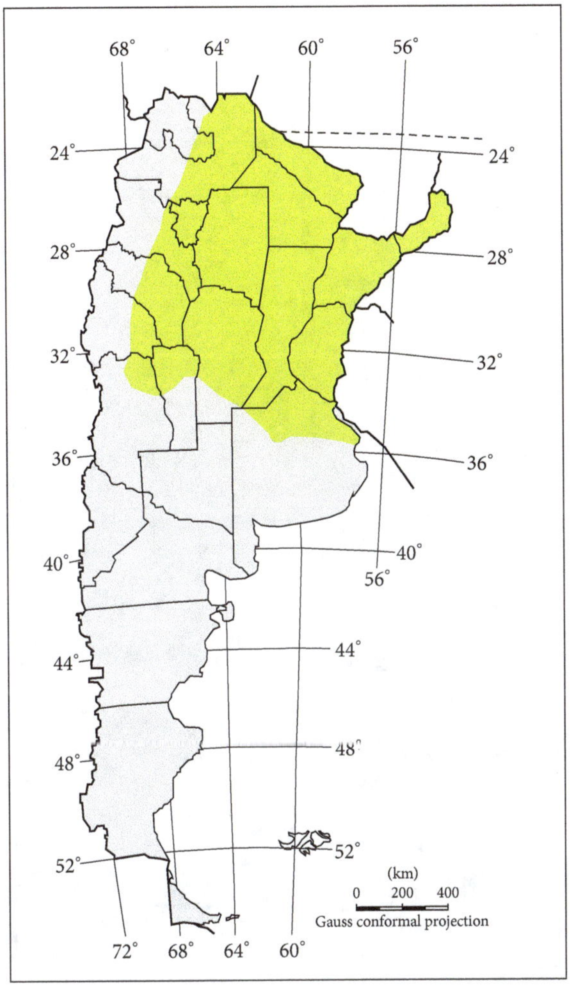

Suitable area: >10°C

Non suitable area: <10°C

FIGURE 5: Annual mean minimum temperature.

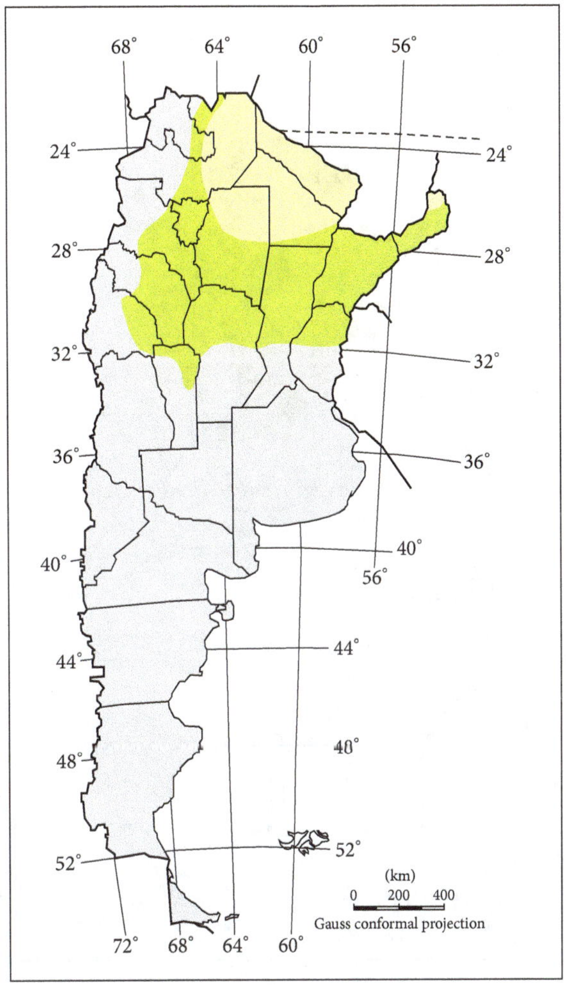

Marginal area: >22°C

Suitable area: 22–17,5°C

Non suitable area: <17,5°C

FIGURE 6: Winter mean maximum temperature.

is a marginal area in the north due to the high winter temperatures.

Figure 7 shows the minimum temperature of the coldest month (July). The north and northeast present optimum conditions with an average temperature of 7°C, while the suitable area (where minimum temperatures in July range from 3.8 to 7°C) covers the center of the country and reaches the center of the Buenos Aires province. Nonsuitable areas, whose minimum temperature of coldest month is lower than 3.8°C, cover the west, the center, and the Patagonian sector.

In Figure 8 the mean annual minimum temperature of −5°C is shown. The suitable areas are those where the average minimum temperature is above −5°C. This index is a very important constraint, because lower temperatures can be lethal to argan tree. Suitable areas cover part of the north and the center of the country, and the east of Argentina.

Finally, Figure 9 (obtained by overlaying all the above-mentioned maps) presents the agroclimatic zoning for the argan tree. Eight classes of suitability were delineated: very suitable, suitable, suitable area with constraints (under dry-subhumid, semiarid, and arid regime) and under irrigation, marginal, and nonsuitable areas.

4. Discussion

As can be observed in Figure 9, there is no optimal agroclimatic area for argan cultivation because, in the optimal zone, the minimum temperature of the coldest month is never below 7°C [25]. In the areas with agroclimatic suitability (suitable and very suitable), this index is between 3.5 and 7°C. A very suitable area was described, which covers a large portion of the province of La Rioja and a small part of Catamarca. The suitable areas cover part of the provinces of Catamarca, La Rioja, Córdoba, and San Luis and small areas in Tucumán, Santiago del Estero, and Salta. To the west of the very suitable area, there is a suitable area with constraints under semiarid regime that comprises part of the provinces of San Juan, La Rioja, and San Luis. A suitable area

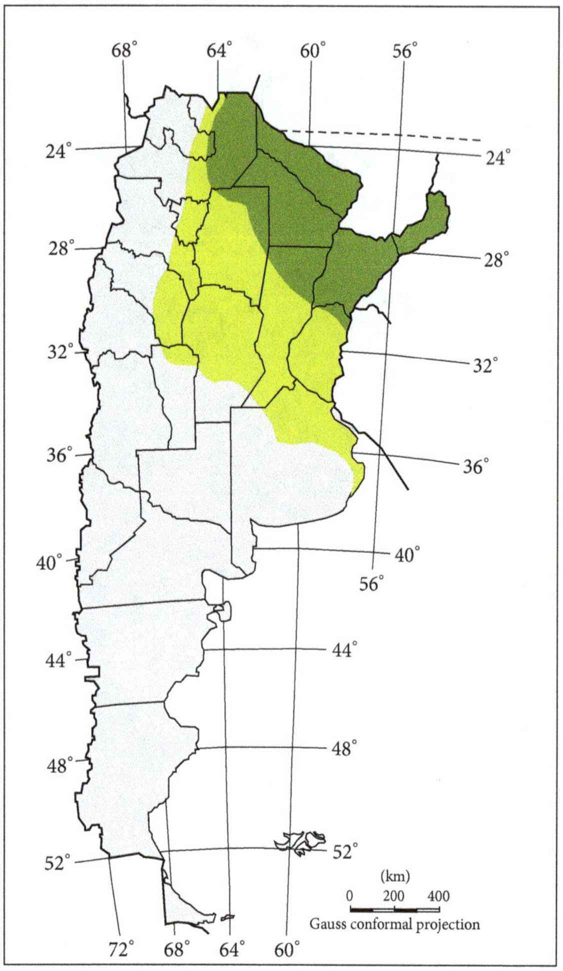

Optimal area: >7°C

Suitable area: 7–3,8°C

Non suitable area: <3,8°C

FIGURE 7: Minimum temperature of coldest month.

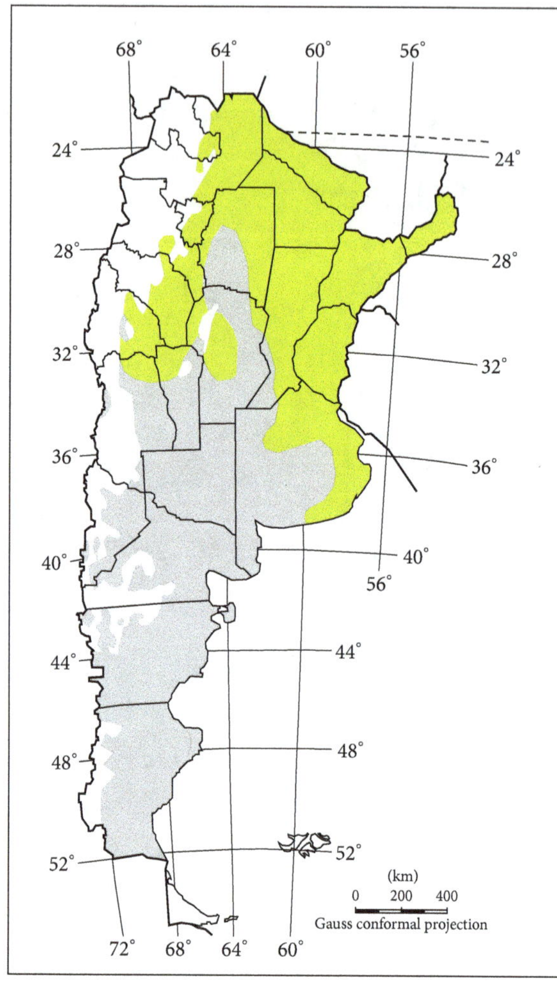

Suitable area: >−5°C

Non suitable area: <−5°C

Non data

FIGURE 8: Mean annual minimum temperature.

with constraints under dry-subhumid regime is present in the center of the provinces of Catamarca and Salta. Under arid conditions there is another area classified as suitable with constraints which covers parts of the provinces of San Mendoza, San Juan, and La Rioja. A suitable area with constraints under irrigation is in the provinces of San Juan, La Rioja, and Catamarca, and small ones can be found in Mendoza and San Luis. The suitable and very suitable zones delimited for argan in Argentina are characterized by low and variable rainfall (100 to 400 mm) and mild winter temperatures. The dry season is inordinately long, resulting in a short growing season. These are areas with mostly lithosols and regosols soils, associated with saline soils on lowlands.

The argan tree does not exist in Argentina and, consequently, there are no commercial plantations that can be used to compare the agroclimatic zones delineated in this paper.

According to [13], this species flourishes in desert or arid (Bw), steppe or semiarid (Bs) climates, between 20 and 32° latitude. This agrees with the agroclimatic zoning map presented in this paper, since the suitable area reaches the north of San Luis province (approximately 33°S).

However, Discover Life [26] cites the species in a larger latitudinal range. For example, it mentions a preserved specimen in Arizona, USA, at 34.5°N/111°W; in Spain at 37.5°N/9°W, 38.21°N/10.25°W, and 40.5°N 3°W; in United Kingdom at 52.5°N/3°W; in Kenya at 1.5°N/39°E; and in Morocco at 31.5°N/9°W and 31.5°N/3°W.

The GBIF [27] cites only 34 occurrences with georeferenced data. Some of them are in the following countries: Algeria (1.33°N/36.16°W), Netherlands (52.25°N/5.60°E), USA (34.14°N/118.05°W and 34.00°N/118.00°W). In Spain, there are specimens in Alicante, Murcia, Gran Canaria (Canary Islands), and Barcelona (38,06°N/1,17°W, 38,42°N/0,42°W, 27,10°N/15,10°W, 38,41°N/0,42°W, 38,33°N/0,54°W, 38,51°N/0,53°W, 38,41°N/0,42°W, 41,10°N/2,10°E, and 41,36N°/2,16°E). The rest of the citations correspond to Morocco

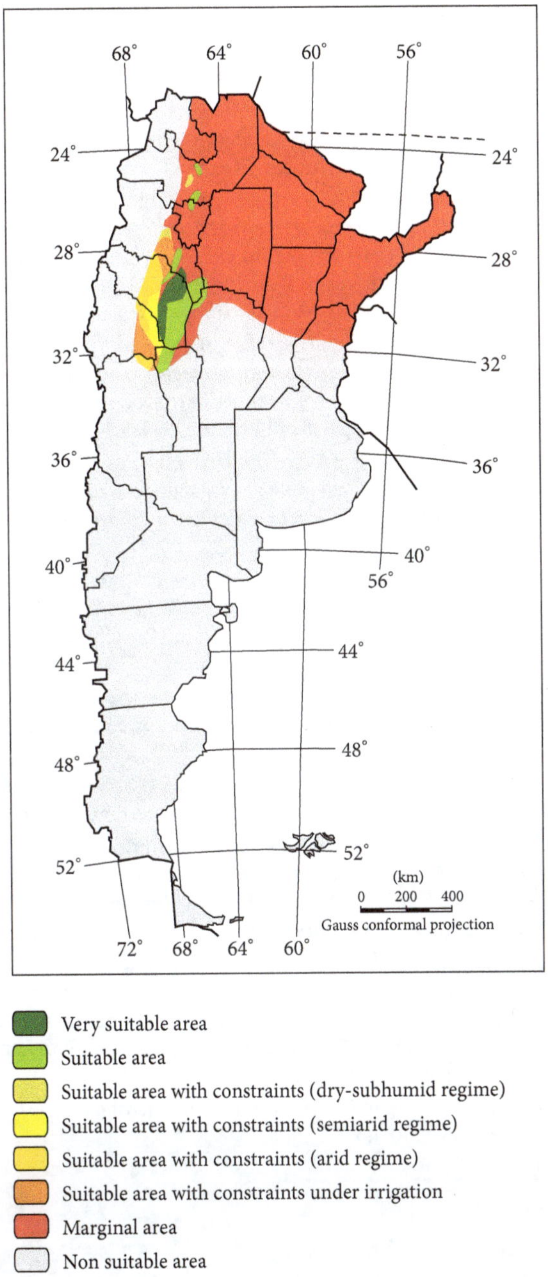

FIGURE 9: Agroclimatic zoning for the argan tree.

(30,53°N/7,89°W, 30,79°N/8,38°W, 29,70°N/9,92°W, 30,82°N/8,40°W, 30,68°N/8,17°W, 29,42°N/9,44°W, 30,37°N/9,70°W, 30,62°N/8,09°W, 29,75°N/9,28°W, etc.).

Argan is a thermoxerophile and its edaphic requirements are scarce [28]. It adapts well to hostile environments, and it can even be irrigated with salt-rich water [29]. Consequently, argan can be used as a repopulation species in semiarid and arid zones.

The suitable and very suitable areas for argan largely coexist with the Argentine olive tree cultivation region. Currently, the olive industry has been curtailed due to rising production costs (approximately 40% annually) and low foreign currency (dollar) values. As a result, many growers have ceased investing and olive farms have been abandoned. The harvested area for olives in Argentina has been decreasing permanently. Argan cultivation could reconvert the olive areas by interspersing argan and olive trees to reach a more lucrative production per hectare.

In the future, farmers could combine horticultural species cropping, some olive trees, vegetables and livestock (mainly goats) in association with the argan tree, and beekeeping. These areas have regular drought periods, which could be mitigated by argan tree since it not only provides forage for goats but also produces argan oil for local consumption and for sale.

Considering the increase in arid conditions worldwide, together with the prediction for fewer rainfalls, argan trees could be used for greenery purpose and to reduce climate change effect.

The agroclimatic suitability map presented is indicative because climate data spanning 30 years was used rather than data for a specific year. As a result, in locations where topographic and geographic traits significantly alter the microclimate, specific areas where the species may be grown could be unknowingly disregarded. Notwithstanding, this is pioneering work because the agroclimatic zoning for the argan tree in Argentina has never been done before.

5. Conclusions

The climatic requirements and tolerances of the argan tree have been outlined in relation to the environmental conditions of their natural habitat. An agroclimatic zoning model was developed. Basic data in determining this similarity were the amount of annual rainfall, the annual temperature, the annual minimum temperature, the winter maximum temperature, the minimum temperature of the coldest month, the mean annual minimum temperature, and the annual maximum temperature.

The successful introduction of this species in Argentina depends essentially on the degree of similarity of climate in the natural area to that of the new areas with agroclimatic suitability.

There are no optimal zones for the argan tree in Argentina. Very suitable, suitable, and marginal areas for this species were delimitated.

Acknowledgments

This research was supported by the National Council of Scientific and Technical Research (CONICET) of Argentina.

References

[1] T. J. Lybbert, C. B. Barrett, and H. Narjisse, "Market-based conservation and local benefits: The case of argan oil in Morocco," *Ecological Economics*, vol. 41, no. 1, pp. 125–144, 2002.

[2] F. Msanda, A. El Aboudi, and J. P. Peltier, "Biodiversité et biogéographie de l'arganeraie marocaine," *Cahiers Agricultures*, vol. 14, no. 4, pp. 357–364, 28 May 2016, http://www.jle.com/fr/revues/agr/e-docs/biodiversite_et_biogeographie_de_larganeraie_marocaine_266250/article.phtml?tab=texte.

[3] O. M'Hirit, M. Benzyane, F. Benchekroun, S. M. El Yousfi, and M. Bendaanoun, *L'Arganier. Une espèce fruitière à usages multiples*, P. Mardaga, Ed., Sprimont, Belgique, 1998.

[4] D. Rivera and J. B. Ruiz, "Argania spinosa L (Skeel) Sapotaceae. Subespontánea en la Península Ibérica," *Anales del Jardín Botánico de Madrid*, vol. 44, no. 1, p. 173, 1978.

[5] A. Benabid, *Flore et écosytèmes du Maroc: évaluation et préservation de la biodiversité*, Ibis Press, Paris, France, 2000.

[6] C. Orwa, A. Mutua, R. Kindt, R. Jamnadass, and S. Anthony, "Agroforestree Database: a tree reference and selection guide version 4.0," 28 December 2015, http://www.worldagroforestry.org/sites/treedbs/treedatabases.asp.

[7] Z. Charrouf and D. Guillaume, "Argan oil: occurrence, composition and impact on human health," *European Journal of Lipid Science and Technology*, vol. 110, no. 7, pp. 632–636, 2008.

[8] N. El Benni and S. Reviron, *Geographical Indications: Review of Seven Case-Studies World Wide; NCCR Trade Regulation*, Working Paper No 2009/15, Swiss National Centre of Competence in Research, 2009, http://phase1.nccr-trade.org/images/stories/publications/IP5/GI_Case-studies_2009.pdf.

[9] ITIS, "Catalogue of Life: 2014 Annual Checklist," 26 November 2015, http://www.catalogueoflife.org/annual-checklist/2014/details/species/id/9714131.

[10] JP. Peltier, G. Carlier, and A. El Aboudi, "Evolution journalière de l'état hydrique des feuilles d'arganier (Argania spinosa L. Skeels.) sous bioclimat aride à forte influence océanique (plaine du Souss, Maroc)," *Acta Oecol*, vol. 11, no. 5, pp. 643–668, 1990.

[11] Z. Charrouf and D. Guillaume, "Argan Oil," *Alternative Medicine Review*, vol. 16, pp. 275–279, 2011, http://altmedrev.com/publications/16/3/275.pdf.

[12] Y. le Polain de Waroux and E. F. Lambin, "Monitoring degradation in arid and semi-arid forests and woodlands: The case of the argan woodlands (Morocco)," *Applied Geography*, vol. 32, no. 2, pp. 777–786, 2012.

[13] FAO, "Ecocrop," (1993-2007) Available at: http://ecocrop.fao.org/ecocrop/srv/en/home Accessed 26 June 2015.

[14] F. Bani-Aameur and J. Sipple-Michmerhuizen, "Germination and seedling survival of Argan (Argania spinosa) under experimental saline conditions," *Journal of Arid Environments*, vol. 49, no. 3, pp. 533–540, 2001.

[15] Y. Mizrahi and A. Nerd, "New Crops as a Possible Solution to the Troubled Israeli Export Market," in *Progress in New Crops*, J. Janick and J. E. Simon, Eds., pp. 56–64, ASH Press, Alexandria, VA, USA, 1996.

[16] Z. Bouzoubaâ and A. El Mousadik, "Temperature drought and salt effect on Arganiaspinosa (L.) Skeels seed germination," *Acta Botanica Gallica*, vol. 150, no. 3, pp. 321–330, 2 June 2016.

[17] A. Zahidi and F. Bani-Aameur, "Germination des amandes d'arganier (Argania spinosa (L.) Skeels): effet du génotype, de la date de semis et de l'année de récolte," *Annales de la Recherche Forestiere au Maroc*, vol. 30, pp. 2–16, 1997.

[18] M. Zunzunegui, F. Ain-Lhout, J. Jáuregui et al., "Fruit production under different environmental and management conditions of argan, Argania spinosa (L.)," *Journal of Arid Environments*, vol. 74, no. 10, pp. 1138–1145, 2010.

[19] Z. Bouzoubaâ and A. El Mousadik, "Effet de la température, du déficit hydrique et de la salinité sur la germination de l'Arganier, Argania spinosa (L.) Skeels," *Acta Botanica Gallica*, vol. 150, no. 3, pp. 321–330, 2003.

[20] H. N. Le Houerou, *The desert and arid zones of northern Africa*, M. Evenari and D. W. Goodal, Eds., vol. 74 of *Hot deserts and arid shrublands*, Elsevier, Amsterdam, Netherlands, 1986.

[21] R. Michalet, *Une approche synthétique biopédoclimatique des montagnes méditerranéennes: exemple du Maroc septentrional*, Doctorat université Joseph Fourier, Grenoble I, France, 28 May 2016, http://cat.inist.fr/?aModele=afficheN&cpsidt=148671.

[22] A. Benabid, *Bref aperçu sur la zonation altitudinale de la végétation climatique du Maroc. Ecología Mediterranea VIII*, 26 June 2016, http://ecologia-mediterranea.univ-avignon.fr/uploads/media/Ecologia_mediterranea_1982-8_1-2.pdf.

[23] S. M. El Yousfi and F. Benchekroun, "La dégradation forestière dans le sud marocain, exemple de l'arganeraie d'Admine (Souss) entre," *Annales de la Recherche Forestiere au Maroc*, vol. 26, pp. 1969–1986, 1992.

[24] M. R. Tarrier and M. Benzyane, *L'arganeraie marocaine se meurt: problématique et bioindication. Revue Sécheresse en ligne*, 28 April 2016, http://www.secheresse.info/article.php3?id_article=228.

[25] P. Quézel and F. Médail, *Écologie et biogéographie des forêts du Bassin méditerranéen*, Collection Environnement, Elsevier, Paris, France, 2003.

[26] *Discover life (N/D)*, Argania spinosa, 26 March 2016, http://www.discoverlife.org/mp/20m?kind=Argania+spinosa.

[27] GBIF, *Global Biodiversity Information Facility (N/D)*, Argania spinosa, 26 August 2015, http://www.gbif.org/occurrence/search?taxon_key=5333832&HAS_COORDINATE=true&HAS_GEOSPATIAL_ISSUE=false&offset=20.

[28] A. Homrani Bakali, *Essai d'installation et de suivi de comportement de l'arganier (argania spinosa) dans le contexte aride de Tafilalet*, Actes du Premier Congrès International de l' Arganier, Décembre 2011, http://www.inra.org.ma/Docs/actes-arganier/arganier309318.pdf.

[29] D. Enneking, "Argan (Argania spinosa)," *Australian New Crops Newsletter*, vol. 10, no. 1-2, 1998.

21

Introducing Natural Farming in Black Pepper (*Piper nigrum* L.) Cultivation

Kevin Muyang Tawie Sulok,[1] **Osumanu Haruna Ahmed**,[2,3,4] **Choy Yuen Khew**,[1] **and Jarroop Augustine Mercer Zehnder**[1]

[1] *Research and Development Division, Malaysian Pepper Board, 93916 Kuching, Sarawak, Malaysia*
[2] *Department of Crop Science, Faculty of Agriculture and Food Sciences, Universiti Putra Malaysia (UPM), Bintulu Campus, 97008 Bintulu, Sarawak, Malaysia*
[3] *Institute of Tropical Agriculture and Food Security (ITAFoS), Universiti Putra Malaysia (UPM), 43400 Serdang, Selangor, Malaysia*
[4] *Institute of Tropical Forestry and Forest Products (INTROP), Universiti Putra Malaysia (UPM), 43400 Serdang, Selangor, Malaysia*

Correspondence should be addressed to Kevin Muyang Tawie Sulok; kevinmuyang@mpb.gov.my

Academic Editor: Maria Serrano

This paper reviews the role of Natural Farming as an ecological farming method to produce organically grown food of safe and high quality and at the same time improve soil quality and soil health. Currently, there is a dearth of information on the effects of Natural Farming approach on black pepper farms particularly in Sarawak, Malaysia. Previous studies on other crops had indicated positive outcome using the Natural Farming method. Thus, this paper discusses the essential role of effective microorganisms in Natural Farming and their potential in pepper cultivation. Through the action of effective microorganisms, this approach should be able to transform a degraded soil ecosystem into one that is fertile and has high nutrients availability. The mixed culture of effective microorganisms applied must be mutually compatible and coexist with one another to ensure its favorable establishment and interaction in the soil. Therefore, it is anticipated that introducing Natural Farming in black pepper cultivation can enhance the predominance of effective microorganisms in the soil, which in turn could lead to promising growth and yield of the crop.

1. Introduction

Current trends in pepper cultivation, fertilizers, and wide-scale applications of broad-spectrum organophosphate pesticides could result in a degrading ecological environment besides shortening the lifespan of pepper vines [1]. Therefore, as part of the National Commodity Policy of Malaysia heads towards an environment-friendly industry, introducing Natural Farming in pepper cultivation will be part of an effort by the Malaysian Pepper Board, Malaysia, to support green development. Moreover, as the global demand for organic farm produce increases, there is a need to develop an agricultural practice which is sound and sustainable for the betterment of both the environment and humans living in it.

Natural Farming is a term used to describe an ecological farming approach to produce organic based food crops [2, 3] (Talaat et al., 2017). It is a natural agriculture alternative which promotes lower production cost and at the same time is able to achieve product of high quality and yield with lower or without the usage of inorganic fertilizers and pesticides [4]. As the approach is closely related to soil fertility, introducing Natural Farming to enhance the predominance of beneficial and effective microorganisms can help to improve and maintain soil biological, chemical, and physical properties (Fukuoka, 2009) [5].

The role of using beneficial microorganisms in Natural Farming is in increase because many people are being aware of the hazards associated with consuming products with chemical substances. Nevertheless, enhancing food security through this approach in Malaysia is still a not well-established option due to lack of technical knowledge in the required subject. The desired effects for applying Natural Farming fertilizers to soils can vary, at least initially. In some soils, a single application may be enough to produce the expected results, whereas for other soils, repeated applications may be ineffective [6]. The reason for this is that, in some soils, it takes longer time for the introduced microorganisms to adapt to a new set of ecological and environmental conditions and to become well established as a stable, effective, and predominant part of the indigenous soil microflora. Megir and Paulus [1] observed that in most black pepper farms, intense use of farming chemicals causes the ecosystem to be degraded and therefore not suitable for newly introduced microbial communities to live in. Megir and Paulus [1] added that these microbes may not be able to withstand the intolerable pH, moisture, and temperature conditions of degraded soils. Hence, the proper and regular addition of organic amendments is often an important part of any strategy to ensure the survival of newly introduced microorganisms [5]. To this end, there is strong belief that Natural Farming can ensure very productive pepper farming, clean environment, increase profits of pepper growers, and at the same time make healthy and fresh organic produces affordable for everyone. Thus, this paper discusses the essential role of effective microorganisms in Natural Farming and their potential in pepper cultivation.

2. Fertilization Practice in Pepper

Pepper has high demand for nutrients [7]. It requires large quantities of nutrients to maintain significant growth and yield. It has been estimated that the nutrient uptake of a mature stand of pepper amounts to 202 kg N, 13 kg P, 156 kg K, 18 kg Mg, and 68 kg Ca per hectare per year [8]. Studies by the Malaysian Pepper Board [9] and Yap [8] estimated pepper yield by conventional practice was approximately 6 tons per hectare per year, whereas that of the yield associated with organic farming practice resulted in approximately 5 tons per hectare per year [8]. In Malaysia, the costs of buying compound inorganic fertilizers for pepper vines of different maturity stage are shown in Table 1. Mature vines or pepper aged three years and above requires more chemical or inorganic fertilizers than immature vines; therefore, more cost is involved. Hence, it is important to note that naturally cultivated pepper should be able to reduce the dependency on expensive inorganic fertilizers. Natural Farming approach must be able to convert a degraded soil ecosystem full of harmful microbes to one that is productive and contains useful microorganisms which in turn could contribute to high nutrients availability.

3. Sustainable Agriculture through Natural Farming

The term Natural Farming or nature farming is not simply farming without chemical fertilizers and pesticides, but rather it is organic farming with the added dimension of exploiting beneficial microorganisms to enhance soil quality and soil health [10, 11]. Sustainable agriculture can be defined as a set of practices that conserve resources and the environment without compromising human needs. In line with the concept of sustainable agriculture and development, Natural Farming approach qualifies as one of its main pillars (Talaat et al., 2017). Sun et al. [11] stated that intimate understanding of the local ecology is necessary for successful agriculture, and it may be important to extend this knowledge to the smallest of lifeforms.

Table 1: Cost of fertilization associated with pepper cultivation.

Pepper vine	Cost per hectare per year
Immature	USD 1,670
Mature	USD 2,131

4. Soil Microflora Associated with Natural Farming

Microorganisms are effective only when they are presented with suitable and optimum conditions for metabolizing their substrates including available water, oxygen (depending on whether the microorganisms are obligate aerobes or facultative anaerobes), pH, and temperature of their environment. The important consideration here is the careful selection of a mixed culture of compatible, effective microorganisms properly cultured and provided with acceptable organic substrates [12, 13].

In Natural Farming, these beneficial microorganisms can be grown to high populations and then applied to soils that also have a large stable population of beneficial microorganisms, especially facultative anaerobic bacteria [14]. The desired effects appear only after they are established, become dominant, and remain stable and active in the soil. It has been difficult to study the time required for the inoculated microorganisms to be effective due to variable conditions in soils including unanticipated changes in soil pH, temperature, and moisture [1]. However, it will have required at least a year for a conventional black pepper farm to be fully established as a naturally organic farm with wealth of inoculated beneficial microorganisms in soils [2]. Park and DuPonte [2] added that repeated monthly applications, especially during the first cropping season, can markedly facilitate early establishment of the introduced beneficial microorganisms.

Van Bruggen and Finckh [15] also reported that even when a beneficial microorganism is isolated from a soil, cultured in the laboratory, and reinoculated into the same soil at a very high population, it is immediately subjected to competitive and antagonistic effects from the indigenous soil microflora and its numbers soon decline. Studies by Van Der Heijden et al. [16] and Rashid et al. [17] mentioned that one of the essential requirements for the inoculated microorganisms to be well established was the introduction of mixed cultures instead of a single inoculum. Zimmerman (2008) and Thies et al. [18] found that amending soil with biochar can also enhance the effectiveness of beneficial or effective microorganisms as it provides conducive environment for the microbes to live in. Thies et al. [18] added that the tiny perforations in the biochar provide shelter for the inoculated microorganisms to house themselves. Thus, the possibility of shifting the "microbiological equilibrium" of a soil and controlling it to favor the growth, yield, and health of crops is greater if mixed cultures of beneficial microorganisms introduced are physiologically and ecologically compatible. If so, then it is also highly probable that they will exercise considerable control over the indigenous soil microflora which, in due course, would likely be transformed into or replaced by a "new" soil microflora [16].

5. Effective Microorganism (EM)

Effective Microorganism (EM), generally in liquid form, contains a variety of lactic acid bacteria, yeasts, and phototrophic bacteria that can be applied as inoculants to increase the microbial diversity of soil ecosystem [12, 19–21]. Talaat et al. [22] observed that EM contains selected species of microorganisms including predominant populations of lactic acid bacteria and yeasts, smaller numbers of photosynthetic bacteria, actinomycetes, and other types of organisms. All of these microorganisms are mutually compatible with one another and can coexist in liquid culture. Keymer and Lankau [14] added that usually EM shows best results in situations where the natural balance of microorganisms has been severely disrupted or where agricultural farming inputs are in short supply. In situations where natural microorganism populations are reasonably intact, or where a balanced supply of inputs is available, the addition of EM does not make significant difference.

Zuraihah et al. [21] noted that the exact mechanism of how EM acts and interacts in the soil-plant ecosystem is not known. However, there is evidence that supports a number of theories concerning the action of EM. Further studies done by Van Bruggen and Finckh [15] reported the beneficial effects of EM which include suppression of soil-borne pathogens, increased decomposition rate of organic waste, increased availability of mineralized nutrients to plants, enhancement of microbial activities, increased nitrogen fixation, and reduced requirement of chemical fertilizers, pesticides, and fungicides. Shin et al. [5] added that EM helps to speed up the natural composting process, without many of the side effects of foul odours and pests. Figure 1 shows the function of EM in the soil-plant ecosystem which was illustrated by Higa [19].

Sharma et al. [23] reported an increase in humus content of a soil amended with organic materials and inoculated with EM. Although there was little difference in the chemical analyses of the soil among treatments, EM enhanced or improved the growth of plants over those of the control and fertilizer-treated plants. The beneficial effect of EM was attributed to the utilization of plant root exudates and the solubilization and mineralization of certain soil nutrients into plant available forms [23]. In other experiments, pure cultures of lactic acid bacteria were unable to promote significant differences in the humus content or chemical content of the soil.

6. Biochar as Soil Amendment

Biochar is produced with the intent to be applied to soil as a means of improving soil productivity, carbon (C) storage, or filtration of percolating soil water [24]. Biochar is produced by thermal decomposition of organic material under limited supply of oxygen (O_2) and at relatively low temperatures (<700°C) [25]. The potential of biochar to improve plant growth is related to changes in the physicochemical properties of soils that lead to overall increases in microbial activity [26, 27]. When conditioning soils, it is possible to enhance the effectiveness of beneficial or effective microorganisms by adding biochar, or a similar material with tiny perforations, in which the microbes can house themselves (Zimmerman, 2008) [18].

Agegnehu et al. [28] listed some extensive scientific evidence demonstrating that additions of low application rates of biochar into bedding plant container media in the greenhouse, or as amendments to field soils, improve plant growth and yields significantly, regardless of nutrient supply. For example, Vaccari et al. [29] reported increases in the rates of germination, growth, and yields of tomato plants grown in a range of concentrations of a soil commercial container medium amended with biochar. Dunlop et al. [30] demonstrated improvements in the germination and growth of petunias, marigolds, bell pepper, and tomatoes in response to similar charcoal activated carbon amendments into bedding plant container media.

7. Composts as Soil Amendment

Beneficial or effective microorganisms (EM) including yeast, mold fungi, lactic acid bacteria (LAB), photosynthetic bacteria, and actinomycetes [19] microbial groups are responsible for decomposing a variety of organic materials for use in agriculture [31]. These microorganisms are very common in composts [31]. Composts inoculated with EM improve yields and nutrient uptake significantly for a variety of crops including wheat [11], soybean [6], rice [23], and cotton [32] more than plants grown with noninoculated composts. Talaat et al. [22] stated that improved productivity in crops treated with EM-inoculated composts compared with untreated composts is related to hastened decomposition of organic compounds into plant available nutrients. Sharma et al. [23] added that EM compost is a good source of nutrients for crops, which can provide favorable conditions for the growth of crops, promoting the mobilization of insoluble nutrients and activating the beneficial microorganisms in soils.

Significant effects of composts on soils depend on the type and rate of compost, soil C, pH, cation exchange capacity (CEC), and other components of soil fertility. The application of composts to soils can alter soil organic matter status [33, 34] as this is associated with the release of nutrients such as N through mineralization [35]. This process improves both the growth and yield of the plant. Besides decreasing soil bulk density, Agegnehu et al. [28] added that application of compost can increase soil pH, soil organic content (SOC), N content, available P, K, Ca, Mg, Na, and S. Compost addition to soil has also been shown to increase yield of crops similarly to, or beyond the effects of, mineral fertilizer application [36]. However, annual application of high doses of compost had an inhibitory effect on enzyme activity and crop yields [37].

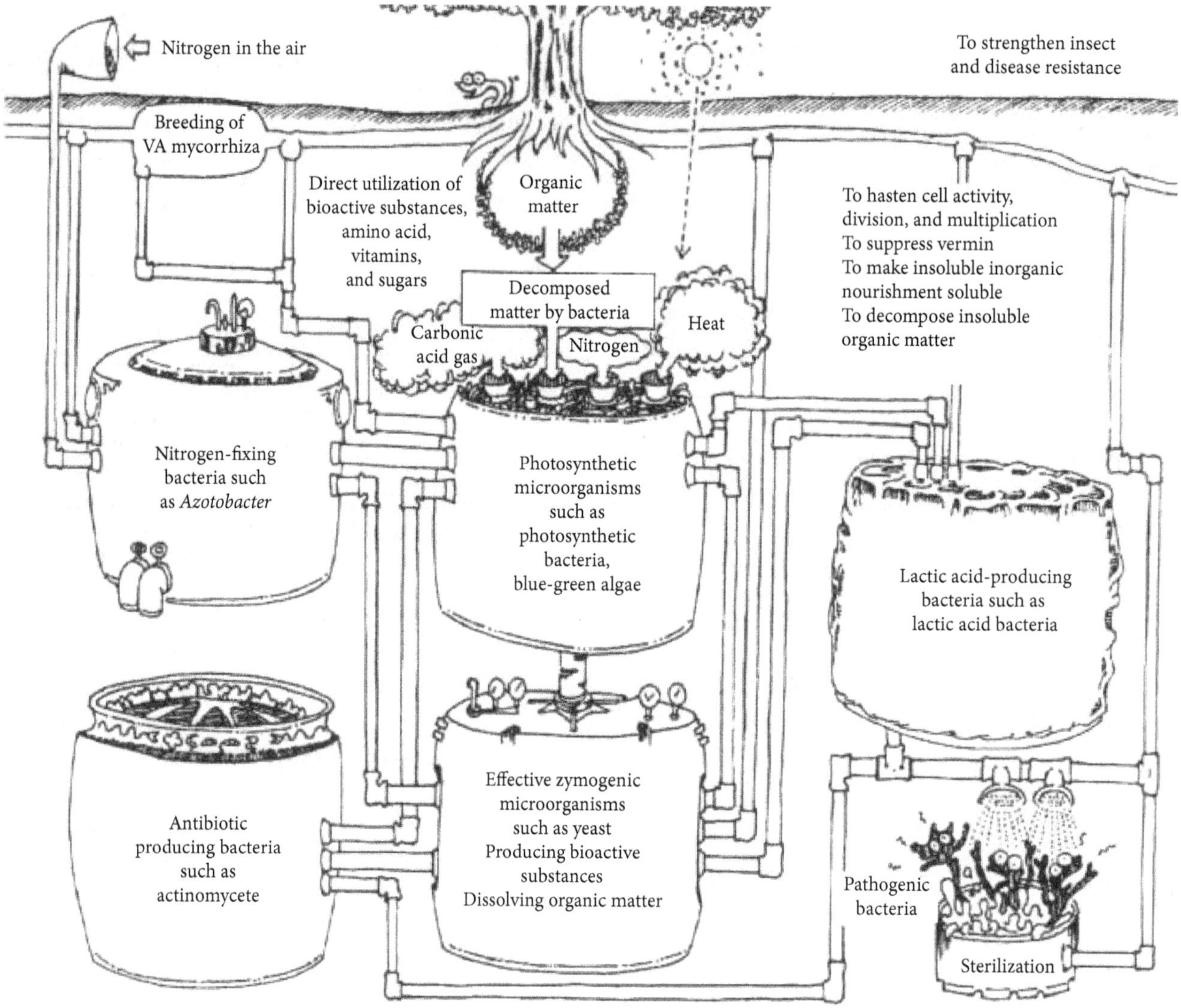

FIGURE 1: Functions of effective microorganisms in the soil in "Microorganisms for Agriculture and Environmental Preservation", Higa [19], Nou-bun Kyo (in Japanese).

8. Role of Natural Farming to Improve Pepper Cultivation

Smith et al. [13] opined that the foundation of Natural Farming production lies in the health of the soil. A fertile soil provides essential nutrients to the growing crop and helps support a diverse and active biotic community. The soil in Natural Farming production is often referred to as living soil. There is a wealth of microorganisms including bacteria, fungi, algae, and protozoa as well as macrofauna such as earthworms, snails, slugs, insects, spiders, centipede, and millipede living in it. Fungi such as mycorrhizae and bacteria *(Rhizobia)*, *Azotobacter*, and earthworms are known to contribute to the fertility of the soil.

Pepper is a nutrient demanding crop, especially during fruit setting. For optimal growth and yield, it requires sufficient amounts of macronutrients and micronutrients. A pepper grower who practices Natural Farming must be able to recognise nutrient deficiency symptoms so that remedial measures are timely taken [1, 11, 14]. Hunt et al. [26] reported that fermented plant juice (FPJ) and fermented fruit juice (FFJ) from organic materials incorporated with biochar can provide fertile soil for the vines to grow well and bear fruit. All crop residues and farm wastes available on the farm including branches and leaves from pepper vines can be recycled, so that soil fertility is restored and maintained. It is the process of transforming organic materials of plant origin into humus. The FPJ and FFJ can act as fertilizers as well as a soil conditioners [6]. Although a study by Paulus and Anyi Wan [38] showed that pepper yield obtained from organic farming was lower compared with that of the conventional practice of using chemical based farming input, on the production cost per yield basis, it provides higher benefit-cost ratio of 1.67 compared with that of conventional practice with benefit-cost ratio of 1.44.

9. Conclusion

The proper selection and dosage of Natural Farming liquid fertilizers should be studied and adopted to improve nutrient efficiency. If the microorganisms comprising the mixed culture can coexist and are physiologically compatible and mutually complementary, and if the initial inoculum density is sufficiently high, there is a high probability that these microorganisms will become established in the soil and will be effective as an associative group, whereby such positive interactions would continue. Therefore, it is important to set up and properly sustain pepper cultivation practices under Natural Farming to reduce investment on chemical based inputs and at the same time lower the environmental hazard caused by pesticides, weedicides, fungicides, and inorganic fertilizers.

Acknowledgments

The authors acknowledge the support of the Research and Development Division, Malaysian Pepper Board, Sarawak, Malaysia, and Universiti Putra Malaysia for this review paper.

References

[1] G. Megir and A. D. Paulus, *Pepper Production Technology in Malaysia*, L. K. Fong and S. S. Liang, Eds., Malaysian Pepper Board, Sarawak, Malaysia, 2011.

[2] H. Park and M. W. DuPonte, "How to cultivate Indigenous Microorganisms," *Biotechnology*, vol. 9, pp. 1–7, 2008.

[3] S. A. Miller, D. M. Ikeda Weinert, E. Weinert Jr. et al., *Fermented Plant Juice*, University of Hawai'I, College of Tropical Agriculture and Human Resources, Honolulu, Hawaii, 2013.

[4] H. A. Hamed, Y. A. Moustafa, and S. M. Abdel-Aziz, "In vivo efficacy of lactic acid bacteria in biological control against Fusarium oxysporum for protection of tomato plant," *Life Science Journal*, vol. 8, no. 4, pp. 462–468, 2011.

[5] K. Shin, G. van Diepen, W. Blok, and A. H. C. van Bruggen, "Variability of Effective Micro-organisms (EM) in bokashi and soil and effects on soil-borne plant pathogens," *Crop Protection*, vol. 99, pp. 168–176, 2017.

[6] A. Javaid and R. Bajwa, "Field evaluation of effective microorganisms (EM) application for growth, nodulation, and nutrition of mung bean," *Turkish Journal of Agriculture and Forestry*, vol. 35, no. 4, pp. 443–452, 2011.

[7] C. Zu, Z. Li, J. Yang et al., "Acid soil is associated with reduced Yield, root growth and nutrient uptake in black pepper (*Piper nigrum* L.)," *Agricultural Sciences*, vol. 05, no. 05, pp. 466–473, 2014.

[8] C. A. Yap, "Determination of nutrient uptake characteristics of black pepper (*Piper nigrum* L.)," *Journal of Agricultural Science and Technology*, vol. 6, pp. 86–89, 2012.

[9] Malaysian Pepper Board, *Laporan Kajian Verifikasi Hasil Lada*, Malaysian Pepper Board, Sarawak, Malaysia, 2017.

[10] Z. Zuraini, G. Sanjay, and M. Noresah, "Effective Microorganism (EM) technology for water quality restoration and potential for sustainable water resources and management," in *Proceedings of the International Congress on Environmental Modelling and Software Modelling for Environments Sake, Fifth Biennial Meeting Held between 5th–8th*, Ontario, Canada, 2010.

[11] P.-F. Sun, W.-T. Fang, L.-Y. Shin, J.-Y. Wei, S.-F. Fu, and J.-Y. Chou, "Indole-3-acetic acid-producing yeasts in the phyllosphere of the carnivorous plant Drosera indica L," *PLoS ONE*, vol. 9, no. 12, Article ID e114196, 2014.

[12] C. T. Lee, M. N. Ismail, F. Razali, I. I. Muhamad, M. R. Sarmidi, and A. K. Khamis, "Application of effective microorganisms on soil and maize," *Journal of Chemical and Natural Resources Engineering*, pp. 1–13, 2008.

[13] D. L. Smith, S. Subramanian, J. R. Lamont, and M. Bywater-Ekegård, "Signaling in the phytomicrobiome: Breadth and potential," *Frontiers in Plant Science*, vol. 6, no. 709, 2015.

[14] D. P. Keymer and R. A. Lankau, "Disruption of plant–soil–microbial relationships influences plant growth," *Journal of Ecology*, vol. 105, no. 3, pp. 816–827, 2017.

[15] A. H. C. Van Bruggen and M. R. Finckh, "Plant Diseases and Management Approaches in Organic Farming Systems," *Annual Review of Phytopathology*, vol. 54, pp. 25–54, 2016.

[16] M. G. A. Van Der Heijden, R. D. Bardgett, and N. M. van Straalen, "The unseen majority: soil microbes as drivers of plant diversity and productivity in terrestrial ecosystems," *Ecology Letters*, vol. 11, no. 3, pp. 296–310, 2008.

[17] M. I. Rashid, L. H. Mujawar, T. Shahzad, T. Almeelbi, I. M. I. Ismail, and M. Oves, "Bacteria and fungi can contribute to nutrients bioavailability and aggregate formation in degraded soils," *Microbiological Research*, vol. 183, pp. 26–41, 2016.

[18] J. E. Thies, M. C. Rillig, and E. R. Graber, "Biochar effects on the abundance, activity and diversity of the soil biota," in *Biochar for Environmental Management: Science, Technology and Implementation*, pp. 327–388, 2015.

[19] T. Higa, "34, Nou-bun Kyo," in *Microorganisms for Agriculture and Environmental Preservation*, p. 33, In Japanese, 1991.

[20] A. Ghasemzadeh and H. Z. E. Jaafar, "Effect of CO_2 enrichment on synthesis of some primary and secondary metabolites in ginger (*Zingiber officinale* Roscoe)," *International Journal of Molecular Sciences*, vol. 12, no. 2, pp. 1101–1114, 2011.

[21] I. I. Zuraihah, Z. Aini, and M. Faridah, "Effects of IMO and EM application on soil nutrients, microbial population and crop yield," *Journal of Tropical Agriculture and Food Science*, vol. 40, no. 2, pp. 257–263, 2012.

[22] N. B. Talaat, A. E. Ghoniem, M. T. Abdelhamid, and B. T. Shawky, "Effective microorganisms improve growth performance, alter nutrients acquisition and induce compatible solutes accumulation in common bean (Phaseolus vulgaris L.) plants subjected to salinity stress," *Plant Growth Regulation*, vol. 75, no. 1, pp. 281–295, 2015.

[23] A. Sharma, T. N. Saha, A. Arora, R. Shah, and L. Nain, "Efficient Microorganism Compost Benefits Plant Growth and Improves Soil Health in Calendula and Marigold," *Horticultural Plant Journal*, vol. 3, no. 2, pp. 67–72, 2017.

[24] M. I. Bird, C. M. Wurster, P. H. de Paula Silva, N. A. Paul, and R. de Nys, "Algal biochar: Effects and applications," *GCB Bioenergy*, vol. 4, no. 1, pp. 61–69, 2012.

[25] M. I. Bird, "Test procedures for biochar analysis in soils," in *Biochar for Environmental Management: Science, Technology and Implementation*, J. Lehmann and S. Joseph, Eds., pp. 677–714, Routledge, London, UK, 2nd edition, 2015.

[26] J. Hunt, M. DuPonte, D. Sato, and A. Kawabata, "The basics of Biochar: a natural soil amendment," *Soil and Crop Management*, vol. 30, pp. 1–6, 2010.

[27] P. Blackwell, S. Joseph, P. Munroe et al., "Influences of Biochar and Biochar-Mineral Complex on Mycorrhizal Colonisation and Nutrition of Wheat and Sorghum," *Pedosphere*, vol. 25, no. 5, pp. 686–695, 2015.

[28] G. Agegnehu, P. N. Nelson, and M. I. Bird, "Crop yield, plant nutrient uptake and soil physicochemical properties under organic soil amendments and nitrogen fertilization on Nitisols," *Soil & Tillage Research*, vol. 160, pp. 1–13, 2016.

[29] F. P. Vaccari, A. Maienza, F. Miglietta et al., "Biochar stimulates plant growth but not fruit yield of processing tomato in a fertile soil," *Agriculture, Ecosystems & Environment*, vol. 207, pp. 163–170, 2015.

[30] S. J. Dunlop, M. C. Arbestain, P. A. Bishop, and J. J. Wargent, "Closing the loop: Use of biochar produced from tomato crop green waste as a substrate for soilless, hydroponic tomato production," *HortScience*, vol. 50, no. 10, pp. 1572–1581, 2015.

[31] P. Partanen, J. Hultman, L. Paulin, P. Auvinen, and M. Romantschuk, "Bacterial diversity at different stages of the composting process," *BMC Microbiology*, vol. 10, article 94, 2010.

[32] A. Khaliq, M. K. Abbasi, and T. Hussain, "Effects of integrated use of organic and inorganic nutrient sources with effective microorganisms (EM) on seed cotton yield in Pakistan," *Bioresource Technology*, vol. 97, no. 8, pp. 967–972, 2006.

[33] D. Fischer and B. Glaser, *Synergisms between Compost and Biochar for Sustainable Soil Amelioration, Management of Organic Waste*, S. E. Kumar, Ed., InTech, Shanghai, China, 2012.

[34] H. Schulz and B. Glaser, "Effects of biochar compared to organic and inorganic fertilizers on soil quality and plant growth in a greenhouse experiment," *Journal of Plant Nutrition and Soil Science*, vol. 175, no. 3, pp. 410–422, 2012.

[35] J. E. Sanchez, T. C. Willson, K. Kizilkaya, E. Parker, and R. R. Harwood, "Enhancing the mineralizable nitrogen pool through substrate diversity in long term cropping systems," *Soil Science Society of America Journal*, vol. 65, no. 5, pp. 1442–1447, 2001.

[36] J. Kimpinski, C. E. Gallant, R. Henry, J. A. Macleod, J. B. Sanderson, and A. V. Sturz, "Effect of Compost and Manure Soil Amendments on Nematodes and on Yields of Potato and Barley: A 7-Year Study," *Journal of Nematology*, vol. 35, no. 3, pp. 289–293, 2003.

[37] I. Marcote, T. Hernández, C. García, and A. Polo, "Influence of one or two successive annual applications of organic fertilisers on the enzyme activity of a soil under barley cultivation," *Bioresource Technology*, vol. 79, no. 2, pp. 147–154, 2001.

[38] A. D. Paulus and L. Anyi Wan, *Pepper Production Technology in Malaysia*, L. K. Fong and S. S. Liang, Eds., Malaysian Pepper Board, Sarawak, Malaysia, 2011.

Application of the Flux-Variance Technique for Evapotranspiration Estimates in Three Types of Agricultural Structures

Ori Ahiman,[1] Yonatan Mekhmandarov,[1] Moran Pirkner,[1] and Josef Tanny[1,2]

[1]*Institute of Soil, Water and Environmental Sciences, Agricultural Research Organization, The Volcani Center, P.O. Box 15159, Rishon LeZion 7528809, Israel*
[2]*HIT-Holon Institute of Technology, P.O. Box 305, Holon 58102, Israel*

Correspondence should be addressed to Josef Tanny; tanai@volcani.agri.gov.il

Academic Editor: Paolo Inglese

Irrigation of protected crops requires sound knowledge of evapotranspiration. Previous studies have established that the eddy-covariance (EC) technique is suitable for whole canopy evapotranspiration measurements in large agricultural screenhouses. Nevertheless, the eddy-covariance technique remains difficult to apply in the farm due to costs, operational complexity, and postprocessing of data, thereby inviting alternative techniques to be developed. The subject of this paper is the evaluation of a turbulent transport technique, the flux variance (FV), whose instrumentation needs and operational demands are not as elaborate as the EC, to estimate evapotranspiration within large agricultural structures. Measurements were carried out in three types of agricultural structures: (i) a banana plantation in a light-shading (8%) screenhouse (S1), (ii) a pepper crop in an insect-proof (50-mesh) screenhouse (S2), and (iii) a tomato crop in a naturally ventilated greenhouse with a plastic roof and 50-mesh screened sidewalls (S3). Quality control analysis of the EC data showed that turbulence development and flow stationarity conditions in the three structures were suitable for flux measurements. However, within the insect-proof screenhouse (below the screen) and the plastic-covered greenhouse, R^2 of the energy balance closure was poor; hence, the alternative simple method could not be used. Results showed that the FV technique was suitable for reliable estimates of ET in shading and insect-proof screenhouses with R^2 of the regressions between FV latent heat flux and latent heat flux deduced from energy balance closure of 0.99 and 0.92 during validation for S1 and S2, respectively.

1. Introduction

In recent years, the area of vegetables and orchards grown in protected cultivation systems is constantly increasing. These include, among other structures, naturally ventilated greenhouses [1], insect-proof screenhouses [2], and shading screenhouses [3]. These structures are naturally ventilated and hence have significant interaction with the external environment. The advantages and limitations of such protected cultivation systems are well documented in the literature [4, 5].

Protected crops are exposed to microclimatic conditions that are significantly different from those in the open field. Hence, the interaction between protected crops and their microenvironment has been the topic of much research during the past years (e.g., [6–8]). One effect of covering the crops is in modifying the exchange of energy, mass, and momentum between the plants and their environment. This modification may affect the evapotranspiration; that is, the water vapor flux from the canopy to the atmosphere, which, in turn, will affect the irrigation demands. The possibility of water saving through reduced transpiration and irrigation demands initiated a number of research studies focused on evapotranspiration measurements and estimates, mainly in screenhouses [3, 9–13].

The most common method for direct measurements of evapotranspiration and other scalar fluxes is the eddy covariance [14]. The method was originally developed and mostly used for flux measurements over open surfaces like forests, natural, or agricultural fields and open water bodies. Due to its high capabilities in reliable measurements of whole canopy evapotranspiration, in recent years, its performance was also examined in protected environments like screenhouses.

Results obtained in various screenhouses [2, 3, 9, 12, 13] illustrated the reliability of the eddy-covariance technique within such protected environments. In all these studies, the EC system was deployed above the plants and below the screen, at a height which is smaller than twice the canopy height. Although conditions at such a height apparently do not meet the common requirements for flux measurements [15], turbulence analysis and flux results supported the use of the EC method at such heights within screenhouses. For example, Tanny et al. [3] evaluated the suitability of the eddy-covariance technique to directly measure evapotranspiration in a large banana screenhouse with almost mature plants. Results were promising: they found 94% closure of the energy balance and daily evapotranspiration values, in agreement with the irrigation applied by the grower. Even though their EC system was deployed relatively close to the canopy top, the spectral energy density decayed with the frequency in a rate close to −5/3, suggesting that turbulence properties resembled the flow in the inertial subrange of steady-boundary layers. Tanny et al. [12] extended these results by measuring turbulent fluxes simultaneously with two EC systems installed at two heights above the crop and below the screen within a large banana screenhouse. Similar friction velocities were measured at the two levels, validating the constant-flux layer assumption within the air gap between the canopy top and the horizontal screen.

Due to the high cost of sensors and complex operation and data analysis, the EC method is inaccessible for day-to-day use by growers as a tool for irrigation management. To assist growers in improving irrigation management, a family of simplified methods was developed in recent years that are capable of indirectly estimating the canopy sensible heat flux and extracting evapotranspiration as a residual of the energy balance closure. One method of this family is the flux-variance (FV) method, which is derived from the MOST principle that any scalar variance normalized by the scalar flux depends on atmospheric stability only. Using this theory, it can be shown that, under unstable conditions, the sensible heat flux is proportional to $\sigma_T^{3/2}$, where σ_T is the standard deviation of air temperature measured at high frequency (~10 Hz) above the canopy. Hence, the method can be applied using fast response single-point measurement of air temperature [16] and auxiliary, relatively simple measurements of net radiation and soil heat flux.

The flux-variance (FV) method, which is the topic of the present paper, has been applied in several studies in open fields. There are three technical aspects involved with applying the FV method: measurement height, sampling frequency, and the available fetch. *Measurement height*: no study in the literature identified the optimal height where the measurement of sensible heat flux, H, is in best agreement with a reference value. Most literature studies show measurements that were done at a single height in the surface layer, larger than 1.1 h_c (e.g., [16]), that is, within and above the roughness sublayer which is about 2 h_c. *Measurement frequency*: most studies reported sampling frequencies between 0.1 and 20 Hz [17–19]; however, no report compared the method's performance at different frequencies to identify the optimal one. *Fetch*: literature studies were conducted under height/fetch ratio in the range 1 : 90–1 : 200 [17, 18, 20].

Several studies examined the value of C_T, the similarity constant associated with the FV method (see Section 2) and the correlation between EC and FV sensible heat fluxes. The value of C_T was in the range 0.9–1.1, and the coefficient of correlation with EC sensible heat flux was in the range 0.72–0.98 [17–20].

The main goal of the present study was to examine the FV technique for crops cultivated in three modified environments that are very common in regions of mild winter climates like the Mediterranean basin [4]. The structures examined are a tomato greenhouse with impermeable plastic roof and screened sidewall openings for natural ventilation, an insect-proof screenhouses with dense net that blocks insect invasion, in which pepper was grown, and a banana screenhouse that protects the crop from hail, high wind speed, and supraoptimal solar radiation. The ultimate goal is to provide guidelines on the optimal use of the FV technique in estimating ET for irrigation management in such structures. Hence, sensible heat flux estimates using FV are used for extracting evapotranspiration from the energy balance closure, and results are compared with ET measurements.

2. Theory

The detailed theory of the flux-variance method is given by Wesson et al. [16]. This section provides only a brief outline with major equations.

The Monin-Obukhov similarity theory (MOST) implies that any nondimensional turbulence statistics depends on the atmospheric stability only, $\zeta = (z-d)/L$, where z is the measurement height, d is the zero-plane displacement height, and L is the Obukhov length defined as follows:

$$L = -\frac{u_*^3 T}{kg\langle w'T'\rangle}, \quad (1)$$

where $u_* = (\langle u'w'\rangle^2 + \langle v'w'\rangle^2)^{1/4}$ is the friction velocity, u', v', and w' are fluctuations in longitudinal, transversal, and vertical velocity components, respectively, T is air temperature, k is von-Karman's constant, g is the gravitational acceleration, and $\langle w'T'\rangle$ is the covariance between vertical velocity and temperature fluctuations which represents the mean kinematic sensible heat flux. Based on MOST, the air temperature standard deviation can be expressed as follows:

$$\frac{\sigma_T}{T_*} = f(-\zeta), \quad (2)$$

where T_* is the temperature scaling parameter given by $T_* = \langle w'T'\rangle/u_*$. As shown by Albertson et al. [21] and

Wesson et al. [16] under unstable conditions, the temperature standard deviation, σ_T, can be approximated by the following equation:

$$\frac{\sigma_T}{T_*} = C_T(-\zeta)^{-1/3}, \tag{3}$$

where $C_T = 0.99$ is a similarity constant [22]. From (2) and (3), it can be shown that

$$H_{FV} = \rho c_p \langle w'T' \rangle = \rho c_p \left(\frac{\sigma_T}{C_T}\right)^{3/2} \left(\frac{kg(z-d)}{T}\right)^{1/2}, \tag{4}$$

where ρ is the air density and c_p is the air specific heat at constant pressure. Hence, the sensible heat flux can be estimated by calculating the temperature standard deviation obtained from a single-point measurement of air temperature at high-sampling frequency. Note that (4) can only predict positive sensible heat fluxes under unstable conditions since $\sigma_T \geq 0$. An approximate expression for stable conditions was also suggested [16], but analysis of such conditions was outside the scope of the present study.

To estimate evapotranspiration, the energy balance closure equation is used as $LE + H = Rn - G$, where LE is the latent heat flux (evapotranspiration), H is the sensible heat flux, Rn is the net radiation, and G is the soil heat flux. The FV latent heat flux is extracted by $LE_{FV} = Rn - G - H_{FV}$, where H_{FV} is the flux variance sensible heat flux calculated by (4). Since in the present study, the measured energy balance was not perfectly closed (see Section 4), a "closed" LE_{EB} was estimated, derived by forcing the energy balance closure [23] using the eddy-covariance sensible heat flux, $LE_{EB} = Rn - G - H_{EC}$. Finally, the flux-variance latent heat flux, LE_{FV}, is validated against the "closed" latent heat flux, LE_{EB}.

3. Materials and Methods

3.1. Sites, Crops, and Structures. The study is based on three comprehensive field campaigns carried out in three different agricultural structures each with a different crop. Details on each of the campaigns are given below (see also [24] for details on S1 and S2).

3.1.1. Campaign S1: Banana, 8% Shading Screen. The screenhouse was located at 32°42′N; 34°57′E, 20 m AMSL, on the Mediterranean coastline in western Israel near the Carmel mountain. The calibration period was 7.08.2011–29.08.2011 (DOY 219–241), during which 21 days were analyzed. Screenhouse dimensions were $250 \times 450\,m^2$ and 5.5 ± 0.1 m high, with the longer side oriented nearly east-west (Figure 1). Screenhouse cover was a woven screen, with nominal shading of 8% (manufacturer's data), and a rectangular hole of 2.3 mm × 3 mm, made of clear, round polyethylene monofilaments 0.3 mm in diameter (Polysack Plastic Industries Inc., Israel). Banana, *Grand Nain* AAA, was planted during May 2011, in groups of four, separated 4.5 m between rows and 3.5 m between plants in a row. During the experiment, plant height was 4.3 m and Leaf area index (LAI) was 1.4. Plants were irrigated following

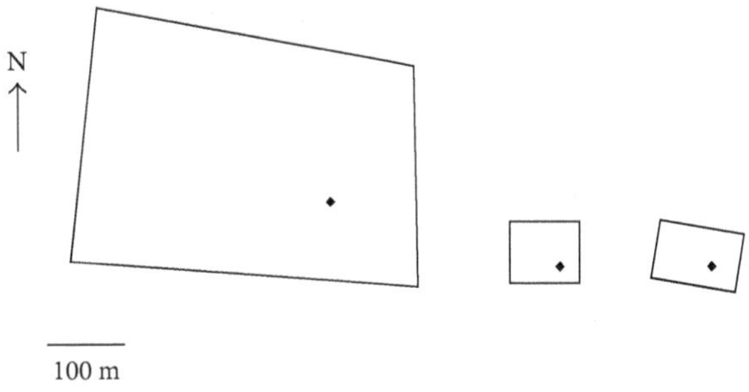

FIGURE 1: A schematic top-view of the three structures examined. The diamond symbol represents the location of the eddy-covariance and flux-variance sensors. From left to right: S1, banana screenhouse; S2, pepper screenhouse; S3, tomato greenhouse.

regional recommendations for screenhouse banana. Soil comprised 44% clay, 26% sand, and 30% silt. Dry and volumetric soil heat capacities were $C_d = 0.86\ J\cdot g^{-1}\cdot K^{-1}$ and $C_V = 1.57\ MJ\cdot m^{-3}\cdot K^{-1}$ [25].

3.1.2. Campaign S2: Pepper, 50-Mesh Insect-Proof Screenhouse. The screenhouse was located at 31°48′N; 35°28′E, −380 m AMSL, in the Jordan Valley of eastern Israel. The calibration period was 13.05.2013–15.06.2013 (DOY 133–166), during which 34 days were analyzed. Screenhouse dimensions were $100 \times 110\,m^2$ and 3.8 ± 0.1 m in height, with the longer side oriented east-west (Figure 1). Screenhouse was covered by a white insect-proof 50-mesh screen woven of round polyethylene monofilaments, 0.23 mm in diameter, with hole size of 0.46 mm (Ginegar Inc., Israel). Pepper 1715 (Hish-Till, Inc. Israel) was planted on the 15th of September 2012, upon rows 0.60 m wide, separated 1 m apart, with plant density of 3.5 plants·m^{-2}. During measurements, plant height was constant at 1.6 ± 0.1 m. Plants were planted in compost channels, which were dug into the local Reg soil, and irrigated following regional recommendations. The Reg soil comprised 50% clay, 25% sand, and 25% silt. The compost comprised 90% clay, 5% sand, and 5% silt [26]. Thus, the dry and volumetric soil heat capacities were $C_d = 0.9\ J\cdot g^{-1}\cdot K^{-1}$ and $C_V = 1.72\ MJ\cdot m^{-3}\cdot K^{-1}$ [25].

3.1.3. Campaign S3: Tomato, Naturally Ventilated Greenhouse. The greenhouse was located at 31°44′N; 35°27′E, −400 m AMSL in the western shore of the Dead Sea, near Kibbutz Kalia. The calibration period was 20.01.2014–23.03.2014 (DOY 20–83), during which 62 days were analyzed. The structure was a 12-span naturally ventilated greenhouse with dimensions of $125 \times 80\,m^2$ (Figure 1) and 5.5 m height arched gable, equipped with an insect-proof, 50-mesh net on the sidewalls, and impermeable 0.5 mm thick polyethylene plastic roof. Tomato (LA–JULIE) (A. B. Zeraim, Inc. Israel) was planted on the 6th of September 2013, upon rows 0.60 m wide, separated 1.8 m apart, with plant density of 0.9 plants·m^{-2}. Plant height during the measurements period varied between 1.3 and 1.8 m, depending on the trellis management and plant pruning. Plants were irrigated

Table 1: Instruments, their heights, and normalized heights in the 3 campaigns.

Measurement	Instrument	Model and manufacturer	Campaign S1; z (m) (Z_S)	S2a (**b**); z (m) (Z_S)	S3; z (m) (Z_S)
3D wind components, sonic temperature	3D sonic anemometer	CSAT3 Campbell Scientific	4.5 (0.48)	5 (**3**) (1.62)	2.5 (0.58)
H_2O concentration	IRGA	LI-7500 LI-COR	4.5 (0.48)	5 (**3**) (1.62)	2.5 (0.58)
Net radiation	Net radiometer	Q*7.1 REBS	4.9 (0.69)	5 (**3**) (1.24)	2.3 (0.54)
Soil heat flux	Thermopile	HF-3.1 REBS	−0.08 (NA)	−0.08 (NA)	−0.08 (NA)
Soil temperature	Thermocouple	T type, 24 gauge	−0.02 and −0.06 (NA)	−0.02 and −0.06 (NA)	−0.02 and −0.06 (NA)

following regional recommendations for greenhouse tomato. Before plantation, the farmer added compost to the soil (60 $m^3 \cdot Ha^{-1}$). Soil comprised 39% clay, 29% sand, 29% silt, and 3% organic matter. Hence, dry and volumetric soil heat capacities were $C_d = 0.88$ $J \cdot g^{-1} \cdot K^{-1}$ and $C_V =$ 2.05 $MJ \cdot m^{-3} \cdot K^{-1}$ [25].

3.2. Sensors and Data Acquisition. Table 1 presents the instruments, their heights above soil, z, and normalized height, Z_S, in the three campaigns. For S2, the table shows the heights of EC system and net radiometer above (S2a) and below (S2b) the screen. Figure 1 depicts the geometry of each structure and the position of the EC tower and FV miniature thermocouples in each of the structures.

Miniature thermocouples were placed on the same tower of the EC system, at 9 heights ranging from 1 to 5.8 m and from 1.9 to 5.4 m, in S1 and S2, respectively ($-1.34 \leq Z_S \leq 1.15$ and $0.14 \leq Z_S \leq 1.83$, resp.), and at 5 heights ranging from 1.5 to 3.8 m in S3 ($0.35 \leq Z_S \leq 0.89$). All high-frequency data were recorded on a CR3000 data logger (Campbell Scientific Inc., USA).

3.2.1. Eddy Covariance. In S1, the eddy-covariance (EC) system was deployed below the screen at a height of 4.5 m on one of the screenhouse-supporting poles (Figure 1). This position allowed a minimum fetch of about 200 m for the prevailing wind direction in the site. The height of the EC system was between 2.1 and 2.5 m above plant zero-plane displacement height, as estimated according to Stanhill [27] assuming neutral stability. Hence, the height/fetch ratio of the EC system was 1/95–1/80, suitable for surface flux measurements based on footprint models developed for open canopies [28, 29]. A similar EC setup in screenhouses that are similar to S1 had already been reported to be valid by Tanny et al. [3, 12]. In S2, the EC system was deployed on the southeast side of the screenhouse at a height of 5 m, which was 1.3 m above the 50-mesh screen (campaign S2a). In a screenhouse similar to S2, the zero-plane displacement of a canopy-screenhouse system was, for neutral stability, about 0.8 of the screen height [30]. In S2, the prevailing wind was north-northwest, and resulting fetch was about 100 m. Thus, the height/fetch ratio at S2 was about 1/50, which is larger than in S1 but still suitable for surface flux measurements according to simplified footprint models [28]. In S2, an internal EC system was also deployed below the screen (designated as S2b) and above the plants, at a height of 3 m above the ground. Three additional ultrasonic anemometers (CSAT3) were deployed in S2 at heights 1.15, 1.7, and 2.5 m above the ground and below the screen.

In S3, the EC system was deployed on the southeast region of the greenhouse at a height of 2.5 m, which was 2 m below the gable. The zero-plane displacement height of a canopy-greenhouse system was estimated, for neutral stability, to be about 0.67 of the plant height. The dominant wind was north-northwest where the fetch was about 100 m. Thus, the height/fetch ratio at S3 was 1/75–1/60, which is still suitable for surface flux measurement according to simplified footprint models [28].

Data regarding footprint distribution in the three campaigns are given in Table 4 (Section 4.1), which describes data quality analysis.

3.2.2. Flux Variance. For FV analysis, air temperature was measured at 10 Hz by fine-wire miniature thermocouples. Radiation load on thermocouples can introduce small errors; however, due to the small size of the thermocouple junctions and low-radiation load (due to the screen or plastic cover), the errors are assumed negligible [31].

In S1, T-type fTcs (fine-wire thermocouples) were self-constructed in our lab using fine wires of 50 or 76 μm in diameter. When installed in the field, the junction and additional 0.02 m of wire were exposed to the air. In S2 and S3, commercial T-type fTcs (model COCO-002, Omega Engineering Inc., UK), 50 μm in diameter, were installed. Length of exposed wire varied between 0.01 and 0.02 m. In all campaigns, the TC junction was directed upwind and extension wires of 24 gauge SLE thermocouples were used to connect the fTcs to the data logger.

3.2.3. Additional Measurements. In all campaigns, additional measurements (Table 1) were conducted to enable energy balance closure and data analysis. These included net radiation, soil heat flux, and soil temperature. Soil heat flux and storage were measured and calculated following Tanny et al. [3]. Data of these sensors were recorded on a CR23X data logger (Campbell Scientific Inc., USA).

3.3. Data Processing. Eddy-covariance calculations of sensible and latent heat fluxes were done using the EddyPro software (LI-COR, Inc.) which incorporates all necessary raw data filtering and corrections. Quality control of the EC data was implemented using the common criteria for testing the developed turbulence and steady-state conditions [15].

The developed turbulence criterion is quantified by the value of the integral turbulence characteristic, ITCσ [15]:

$$\mathrm{ITC}\sigma = \left| \frac{(\sigma_x/x_*)_{\mathrm{model}} - (\sigma_x/x_*)_{\mathrm{measurement}}}{(\sigma_x/x_*)_{\mathrm{model}}} \right|, \quad (5)$$

where σ_x is the standard deviation of the variable x and x_* is its dynamical parameter (e.g., u_* for the velocity). The steady-state test is estimated by

$$\mathrm{RN}x = \left| \frac{\left[\overline{w'x'}\right]_{\mathrm{SI}} - \left[\overline{w'x'}\right]_{\mathrm{WI}}}{\left[\overline{w'x'}\right]_{\mathrm{WI}}} \right|, \quad (6)$$

where x' is the scalar of interest, the subscript SI represents the covariance calculated during a short time interval (e.g., 5 min), and the subscript WI is the covariance calculated during the whole averaging interval (30 min).

Using this approach each half hour is categorized according to one of three quality levels from high to low. Highest quality data (classes 1–3) can be used for fundamental research and development of parametrizations (ITCσ, RNx < 0.5), intermediate quality data (classes 4–6) can be used for general use like routine flux measurements (0.51 < ITCσ, RNx < 2.5), and lowest quality data (classes 7–9) can be used for orientation only or rejected [15]. Since in the present study the EC data were used as a reference for flux measurements of sensible and latent heat fluxes, data up to class 6, representing the intermediate category of "routine flux measurements," were chosen for analysis.

For the FV data analysis, sensible heat flux was calculated by (4) based on the half-hourly standard deviation of the temperature measured by each fTc. All measurements by the miniature fTcs were conducted at 10 Hz; nevertheless, data were also analyzed at lower frequencies of 1, 2, and 5 Hz. Lower frequency analysis was done by skipping the corresponding data at the temperature time series. This analysis was performed to examine the possibility of measurements at sampling rates lower than 10 Hz, with the goal of using low-cost data acquisition systems in future application of the FV technique.

Data of each campaign were divided into two periods: calibration and validation, as described in Table 2. This division was chosen so as to simulate future application of the FV method, where simultaneous FV and EC data from the initial period are used for calibrating the system, while, during the consecutive period, the FV is operated independently. Hence, during calibration the corresponding value of C_T was extracted from regressions of H_{FV} from (4), against direct eddy-covariance measurements of sensible heat flux, H_{EC}. The value of C_T thus obtained was then used along with σ_T from the temperature measurements in the validation phase for estimating H_{FV} using (4).

Data were also categorized according to the atmospheric stability conditions within each of the structures using the value of L, the Obukhov length scale (1), to indicate the stability level. The criteria based on Webb [32], summarized in Table 3, were adopted in the present analysis.

TABLE 2: Calibration and validation periods of the three campaigns.

Experiment	Calibration period (DOY)	Validation period (DOY)
S1 (2011)	219–230 (11 days)	231–241 (10 days)
S2 (2013)	133–145 (13 days)	146–166 (21 days)
S3 (2014)	20–49 (29 days)	50–83 (33 days)

TABLE 3: Stability regimes and corresponding values of L.

Stability	L
Stable	$0 < L < 200$
Neutral	$\lvert L \rvert > 200$
Unstable	$-200 < L < 0$

4. Results

4.1. Quality Control of EC Data. Figure 2 shows the percentage of EC data points within each range of quality levels: 1–3 (research), 4–6 (flux data), and 7–9 (orientation). Figures 2(a) and 2(b) show the results for the steady-state test of the covariances, $w'T'$ and $w'q'$, that are proportional to the sensible and latent heat fluxes, respectively. Figures 2(c) and 2(d) show the results of developed turbulence tests for the horizontal and vertical velocity components, u and w, and Figures 2(e) and 2(f) for the temperature and absolute humidity, T and q, respectively. In the analysis, data from S2 were presented separately for measurements below (S2b) and above (S2a) the screen. Figure 2 shows that, for most variables, the percentage of data points with flags 7–9 is relatively low. This indicates that most data points in the three structures met the quality criteria of developed turbulence and steady state.

Table 4 summarizes the data quality results. Data points under neutral and stable conditions were excluded from further analysis in this paper. Hence, results in subsequent sections of this paper are based on data that met the three quality criteria: unstable conditions, quality flag <7, and footprint peak within the structure under study.

One of the basic assumptions of the eddy-covariance technique is that measurements are done within the constant-flux layer, namely, the vertical flux of the measured scalar is constant with height. This assumption was examined in campaign S2 where 4 ultrasonic anemometers were installed below the screen and one above it. Figure 3 shows the vertical distribution of mean friction velocity, which represents the vertical flux of horizontal momentum, measured during S2. Figure 3 clearly shows that under the screen and above the plants, at z (Z_S) = 1.7 (0.05), 2.45 (0.39), and 3 m (0.64), the friction velocity is nearly constant with height, supporting the constant-flux layer assumption. Above the screen (Z_S = 1.55), the friction velocity is much higher. This is presumably due to the screen being a sink of momentum due to drag. A similar change in friction velocity across a shading screen was observed by Tanny et al. [13].

4.2. Energy Balance Closure. Figure 4 presents the regressions of half-hourly values of consumed energy ($LE + H$) versus available energy ($Rn - G$) in each of the campaigns. Equations

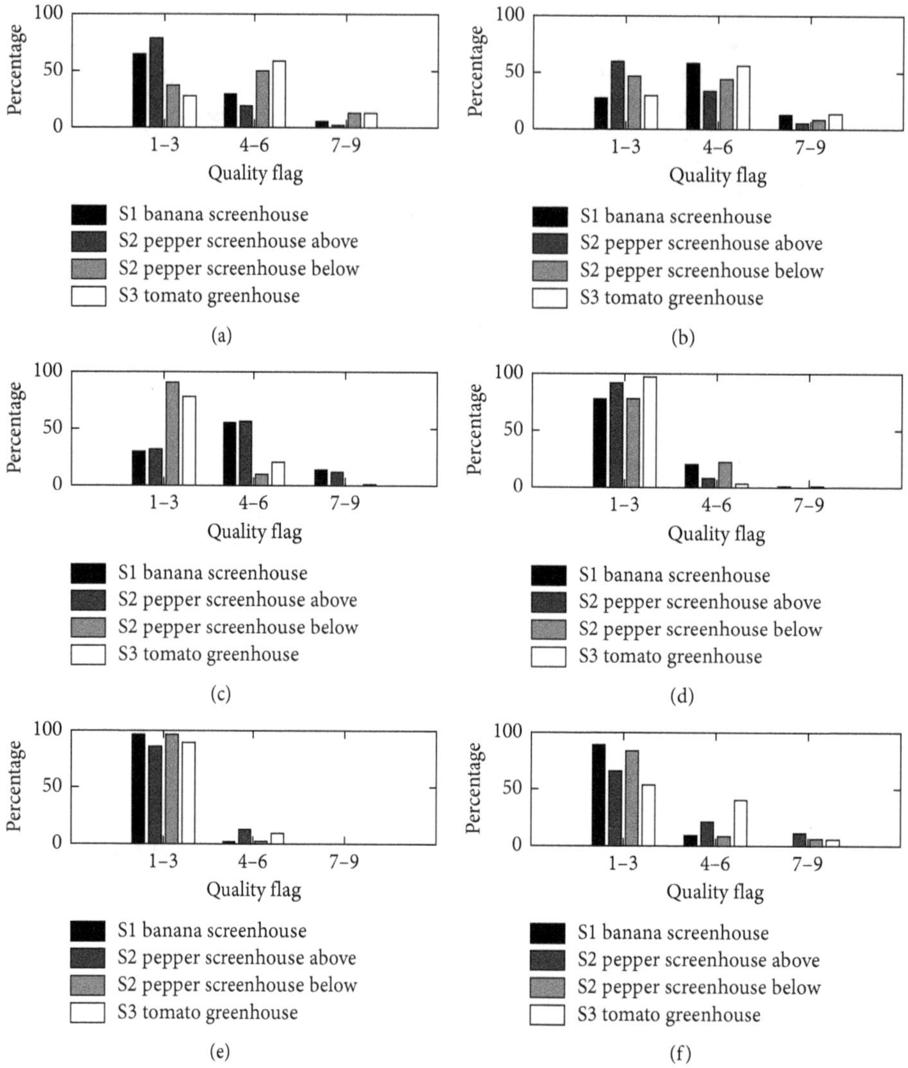

FIGURE 2: Results of quality control tests: the percentage of data points within each range of quality category: 1–3 (research), 4–6 (flux data), and 7–9 (orientation). (a) Steady-state test: Cov $(w'T')$; (b) steady-state test: Cov $(w'q')$; (c)ITC$\sigma(u)$; (d)ITC$\sigma(w)$; (e)ITC$\sigma(T)$; (f)ITC$\sigma(q)$.

TABLE 4: Summary of data quality control.

	Year	Total data points	Stable	Neutral	Unstable	Flags 1–6	Footprint peak within structure (%)
S1	2011	1103	405	68	630	501	97
S2 above	2013	1171	338	165	668	573	95
S2 below	2013	2347	1795	38	514	396	99
S3	2014	4645	1822	219	2604	2509	100

Columns from left to right represent campaign code, year, total number of available data points, number of data points under stable conditions, number of data points under neutral conditions, number of data points under unstable conditions, number of data points with quality flags 1–6, and percentage of data points whose flux footprint peak is within the structure.

(7)–(10) below present the linear fits for the energy balance closure analyses in Figure 4.

The results show that, for S1, energy balance slope was low (0.41); however, the high R^2 indicates close correlate between consumed $(LE+H)$ and available $(Rn-G)$ energy. For S2a (pepper screenhouse above the screen), both slope and R^2 were reasonable. For S2b (pepper screenhouse below the screen) and S3, low closure slopes and very low values of R^2 were obtained.

$$\text{S1}: \quad LE+H = 0.41(Rn-G), \quad R^2 = 0.91, \tag{7}$$

$$\text{S2a}: \quad LE+H = 0.76(Rn-G), \quad R^2 = 0.66, \tag{8}$$

$$\text{S2b}: \quad LE+H = 0.53(Rn-G), \quad R^2 = 0.17, \tag{9}$$

$$\text{S3}: \quad LE+H = 0.39(Rn-G), \quad R^2 = 0.29. \tag{10}$$

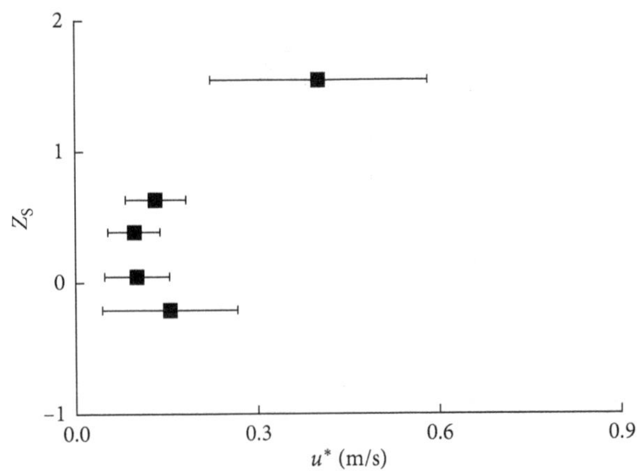

FIGURE 3: Vertical distribution of the half-hourly friction velocity averaged during campaign S2. Horizontal lines represent the standard error. Canopy height is $Z_S = 0$. Screen height is $Z_S = 1$.

Due to the poor R^2 values of the energy balance closure of S2b and S3, the subsequent analyses of this paper were conducted only for campaigns S1 and S2a.

4.3. Sensible Heat Flux: FV Method

4.3.1. FV Analysis Based on Temperature Signal of the Ultrasonic Anemometer. The FV method was initially applied by analyzing the temperature signal measured by the ultrasonic anemometers (CSAT3, Campbell Scientific Inc., USA). The sensible heat flux was estimated by (4) where during the calibration, the constant C_T was estimated from the slope of the regression between H_{FV} and H_{EC}. For validation, the constant C_T obtained during calibration was used to estimate H_{FV} that was eventually compared with H_{EC}. Figure 5 shows the regressions obtained during calibration and validation in the two campaigns, S1 and S2a.

The statistical variables associated with the regressions in Figure 5 are given in Table 5. The results in both Figure 5 and Table 5 show that good correlation was obtained between the two fluxes in campaigns S1 and S2a.

4.3.2. FV Analysis Based on Temperature Signal of the Miniature Thermocouples. In order to develop a simple measurement system for estimating the sensible heat flux by the FV technique, analysis was carried out for the data of the thermocouples installed at the different heights. For each sensor, a regression was calculated between H_{FV} and H_{EC}, and an optimal C_T value was determined, which guaranteed a regression slope closest to one. Figure 6 shows vertical distributions of R^2 and C_T as a function of normalized height during calibration and validation in the two campaigns.

The results in Figure 6 show that, in S1 (Figure 6(a)), the value of R^2 decreased with height from within the canopy and up to the canopy top, and then, it increases towards the screen level. A slight reduction in R^2 is observed above the screen. In S2 (Figure 6(c)), R^2 was nearly constant just above the canopy and increased sharply near the screen. Maximum R^2 during calibration was obtained in S1 at $z = 6$ m ($Z_S = 1.41$) and in S2 at $z = 5$ m ($Z_S = 1.54$). In both campaigns, the C_T value generally increased with height and reached a maximum value just above the screen.

Table 6 shows statistical variables and C_T values for the calibration and validation periods in campaigns S1 and S2a, for the miniature thermocouple that provided the best R^2 in each campaign, as deduced from Figure 6.

Farmers, who will apply the FV method, are not expected to use high frequency (10 Hz) measurements that require expensive data loggers, as done in this study. More probably, simple data acquisition systems with lower sampling frequency would be more practical for such purposes. Therefore, an attempt was made to examine the performance of the FV technique under sampling frequencies lower than 10 Hz. Hence, the standard deviation of the temperature signal was analyzed at frequencies of 5, 2, 1, 0.5, 0.1, and 0.0166 Hz. The analysis was done for the fTc at $z = 5$ m height ($Z_S = 1.54$) in S2a which provided the best R^2 at 10 Hz data analysis (Figure 6). Figure 7 shows relations between characteristics of regressions between H_{FV} and H_{EC}, when H_{FV} is calculated by each of the lower frequencies, to that obtained at 10 Hz.

It is shown that ratios between all regression parameters, namely, R^2 (Figure 7(a)), slope (Figure 7(b)), and intercept (Figure 7(c)) are unaffected by reducing the data analysis frequency down to 0.5 Hz (i.e., one sample every 2 s!). This result implies that data acquisition systems of relatively low sampling rates (and presumably lower price) can be used in future implementation of the FV technique for day-to-day use by farmers.

4.4. Latent Heat Flux: FV Method. The flux-variance latent heat flux was estimated in each campaign, S1 and S2a, assuming a closed energy balance closure and using H_{FV}. Figure 8 presents the regressions between latent heat flux estimated by the FV technique, LE_{FV}, and LE_{EB} deduced by forcing energy balance closure using H_{EC}. The corresponding coefficients of the regressions are given in Table 7. Results show high capability of the FV method in estimating the latent heat flux in the two campaigns with deviations of up to 8% (calibration period of S2a) between measured and estimated latent heat flux.

5. Discussion

This is the first study that examines the applicability of the flux-variance method in protected environments like screenhouses and greenhouses. The flux-variance method is based on MOST (Section 2), which assumes steady state and well-developed turbulence conditions. In campaign S1 (the banana screenhouse), best performance of the FV method, with largest R^2, was realized within the canopy and above the screen (Figure 6(a)). The performance was lowest just below the canopy top. In a banana plantation, the leaf area density is low near the ground, increases with height, and then decreases again at plants' top [8]. The regions of high R^2 (Figure 6(a)) are commensurate with regions where leaf area

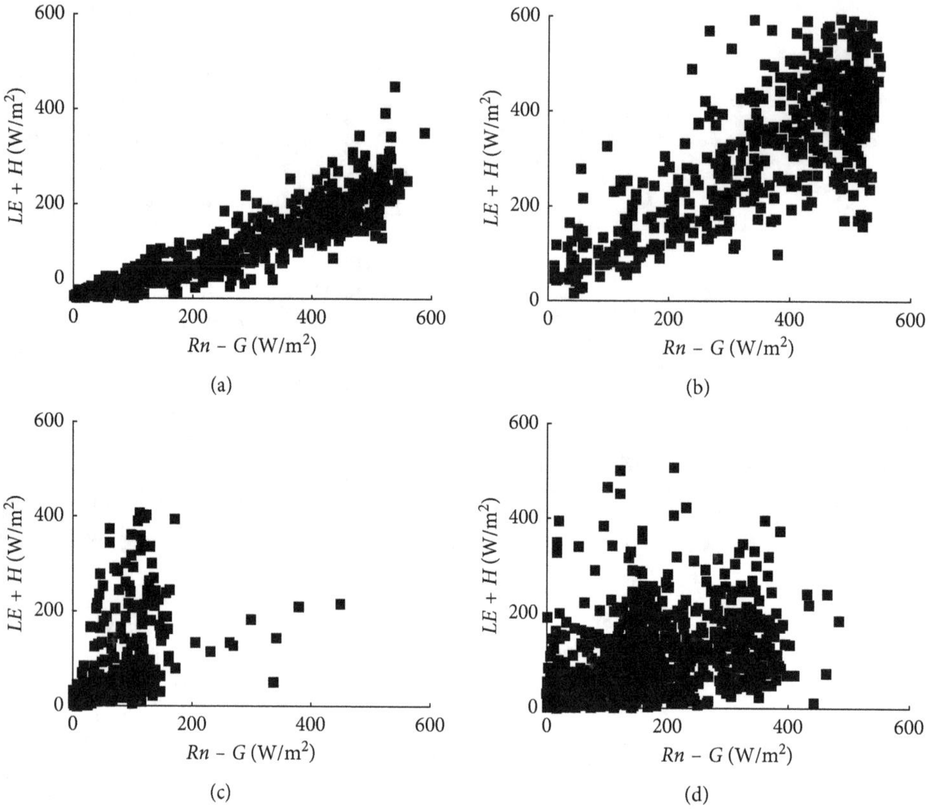

FIGURE 4: Energy balance closure analyses for the three campaigns. Each symbol represents a measured half-hourly data point. (a) S1: banana screenhouse; (b) S2: pepper screenhouse (above screen); (c) S2: pepper screenhouse (below screen); (d) S3: tomato greenhouse.

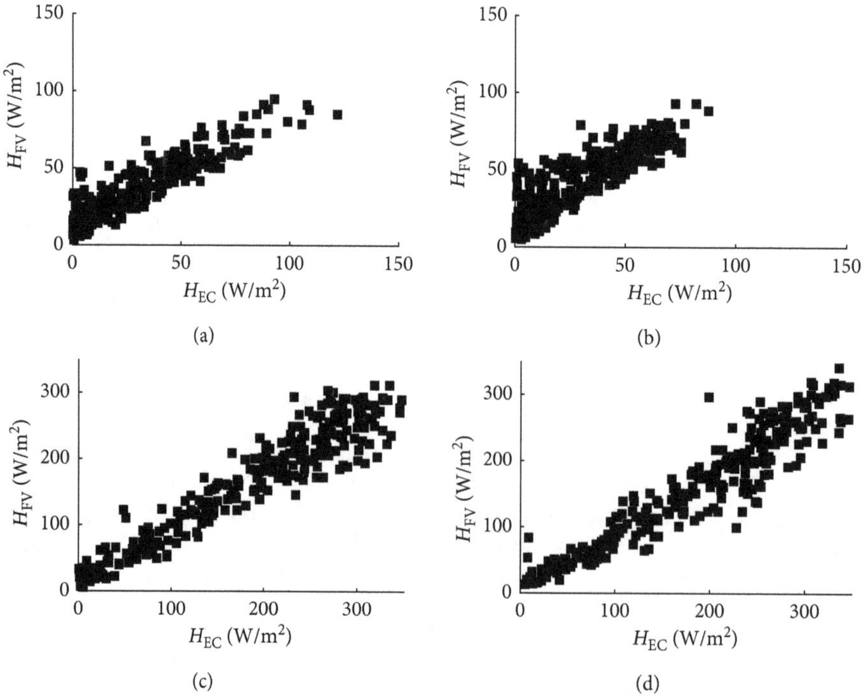

FIGURE 5: Sensible heat flux measured by EC versus that estimated by FV for calibration and validation during the two campaigns. FV data are calculated based on the temperature signal of the ultrasonic anemometer. S1 banana screenhouse: (a) calibration; (b) validation. S2 pepper screenhouse (above screen): (c) calibration; (d) validation.

TABLE 5: Statistical variables and C_T value for the regressions between sensible heat flux measured by the EC and estimated by FV.

		DOY	C_T	Slope, a	RMSE ($W{\cdot}m^{-2}$)	R^2
S1 banana 2011	Calibration	219–230	1.98	0.99	12.68	0.70
	Validation	231–241	1.98	1.03	13.91	0.85
S2a pepper 2013	Calibration	133–145	1.51	0.98	23.45	0.88
	Validation	146–166	1.51	1.01	24.18	0.91

FV data are based on the temperature signal of the ultrasonic anemometer. The results correspond to the graphs in Figure 5.

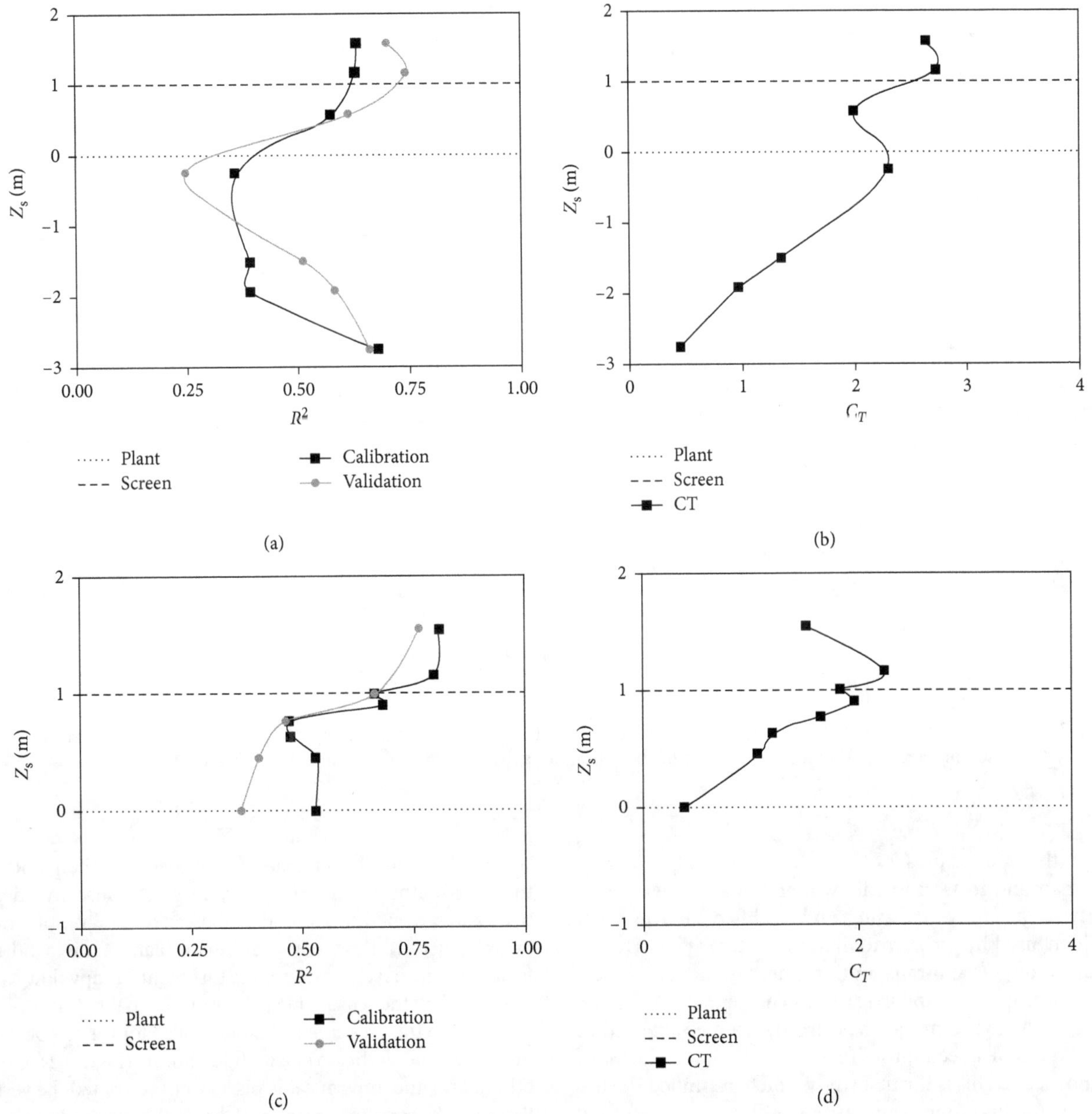

FIGURE 6: Vertical distributions of R^2 during calibration (black line) and validation (grey line) in the two campaigns, S1 (a) and S2 (c), respectively. (b) and (d) are the corresponding distributions of C_T for the validation conditions. In S2, calibration and validation were against reference measurements of H_{EC} above the screen. $Z_S = 0$, canopy top; $Z_S = 1$ screen height.

density is low, presumably since, in these regions, MOST is less violated by the high density of canopy elements. In the pepper screenhouse (Figure 6(c)), R^2 which represents the performance of the FV method increases with height from plants' tops towards the screen and reaches its maximum above the screen. This distribution may also be related to the fact that MOST validity is higher above the screen where conditions of well-developed turbulence are more favorable.

TABLE 6: Statistical variables and C_T value for the regressions between sensible heat fluxes measured by the EC and FV methods.

		DOY	C_T	Slope, a	RMSE ($W \cdot m^{-2}$)	R^2
S1 banana 2011	Calibration	219–230	2.64	0.99	8.38	0.82
	Validation	231–241	2.64	1.02	8.56	0.85
S2a pepper 2013	Calibration	133–145	1.52	1.01	26.78	0.9
	Validation	146–166	1.52	1.03	31.92	0.87

Results are presented for the miniature thermocouple positioned at the height that provided the best performance in each campaign.

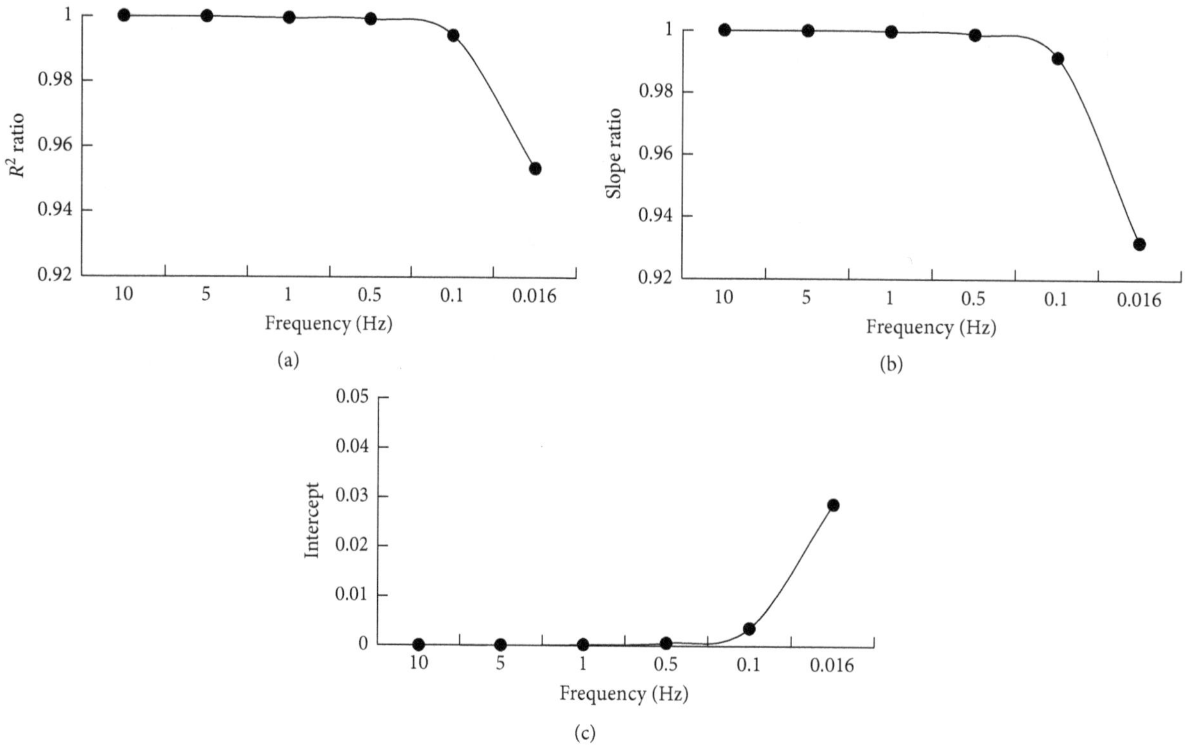

FIGURE 7: Ratio between the regression R^2, slope, and the difference between intercept, between H_{FV} and H_{EC}, of data analysis at each frequency to those obtained at 10 Hz. (a) Coefficient of determination; (b) slope; (c) Y-axis intercept difference.

The three structures examined in this study are very popular among growers in mild winter climates where water scarcity is constantly increasing, and modified environments are widely used by growers to increase water saving. Hence, whole canopy evapotranspiration measurements are essential for improving the irrigation management. Although the use of the EC technique has already been established in various types of screenhouses (e.g., [2, 3, 13]), to the best of our knowledge, this is the first study which examined the use of the EC method to measure turbulent fluxes in a naturally ventilated greenhouse (S3) with an impermeable plastic roof. Therefore, one of the goals of this study was to determine and compare the quality of data used for the EC analysis in the different structures.

Screenhouses allow relatively intense interaction between inside and outside through the permeable roof, as was already presented by Tanny et al. [3] for a banana crop under a light-shading screen. On the contrary, the greenhouse has an impermeable roof, which totally blocks any such interaction, as well as the vertical exchange of mass between the plants and the outside atmosphere. However, the greenhouse investigated in S3 had screened sidewalls that allowed natural ventilation by horizontal wind flow over the tomato plants. This wind flow facilitated the development of the turbulent canopy flow as was illustrated by the reasonable results of the quality tests of ITCσ and RNx in the greenhouse (Table 4), also for the temperature, humidity, and vertical velocity fluctuations (Figure 2). On the other hand, the impermeable plastic roof restricted the vertical flux of momentum, mass, and heat, thereby inducing significant advective effects in the greenhouse, which significantly deteriorated the energy balance closure and R^2 (10). In a similar manner, the dense insect-proof screen in campaign S2b also restricted the vertical fluxes, which resulted with a poor energy balance closure slope and R^2 values (9).

This observation suggests that, in greenhouses, the validity of the two common data quality criteria of developed turbulence and steady state is insufficient to guarantee reliable EC flux measurements. Conditions of developed

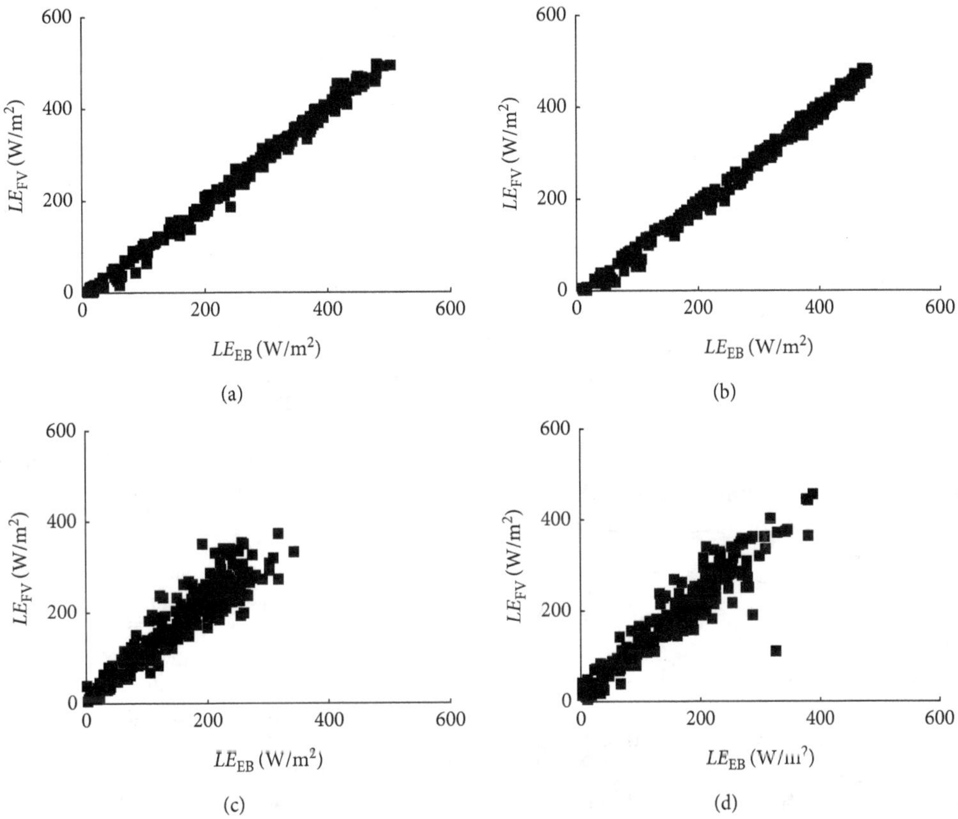

FIGURE 8: The latent heat flux estimated based on FV technique, LE_{FV}, versus the one estimated by forcing the energy balance closure, LE_{EB}. S1 banana screenhouse: (a) calibration; (b) validation. S2 pepper screenhouse (above screen): (c) calibration; (d) validation. Statistical data of these graphs are presented in Table 7.

TABLE 7: Statistical variables for the regressions between latent heat fluxes deduced from energy balance and estimated by the FV method (Figure 8).

		DOY	Slope, a	Intercept, b	RMSE (W·m^{-2})	R^2
S1 banana 2011	Calibration	219–230	1.02	−12.96	10.32	0.99
	Validation	231–241	1.01	−16.09	8.96	0.99
S2a pepper 2013	Calibration	133–145	1.08	9.02	32.39	0.93
	Validation	146–166	1.07	11.67	33.64	0.92

Results are presented for the thermocouple positioned at the measurement height that provided the best performance in each campaign.

turbulent flow and steady state can prevail in naturally ventilated greenhouses even when the scalars are eventually advected horizontally due to the impermeable roof of the greenhouse. Detailed advection measurements require the deployment of several EC systems at different downwind locations, a task that was outside the scope of the present study.

6. Conclusions

A field experiment was carried out in 3 different naturally ventilated agricultural structures, two screenhouses, and a greenhouse to study the suitability of a simple turbulent transport technique, namely, the flux variance, in estimating sensible and latent heat fluxes. The major conclusions of this study are as follows: (i) Quality control analysis of the EC method showed that conditions in the shading and insect-proof screenhouses (above the screen) were reasonable for flux measurements. However, in the plastic-covered greenhouse and in the insect-proof screenhouse below the screen, energy balance closure and its R^2 were poor. Therefore, the FV method could not be employed in these campaigns; (ii) The FV technique was suitable for reliable estimates of H and LE in shading and insect-proof screenhouses, provided the miniature temperature sensors are positioned above the screen; and (iii) data analysis frequency of the temperature signal used in the FV analysis could be reduced down to 0.5 Hz with no effect on the statistical parameters of the regressions between sensible heat flux measured by EC and estimated by FV. Hence, for the future application of this method, simple and low-cost temperature sensors and data acquisition systems can be used effectively.

Nomenclature

C_T: A similarity constant (–)
c_p: Air specific heat at constant pressure ($J \cdot kg^{-1} \cdot K^{-1}$)
d: Zero-plane displacement height (m)
g: Gravitational acceleration ($m \cdot s^{-2}$)
h_c: Canopy height (m)
h_s: Roof height, either screen (S1 and S2) or plastic (S3) (m)
H: Sensible heat flux ($W \cdot m^{-2}$)
H_{EC}: Sensible heat flux measured by eddy covariance ($W \cdot m^{-2}$)
H_{FV}: Sensible heat flux estimated by flux variance ($W \cdot m^{-2}$)
k: von-Karman's constant (–) (=0.41)
L: Obukhov length scale (m) (1)
LE: Latent heat flux ($W \cdot m^{-2}$)
LE_{FV}: Latent heat flux deduced from flux variance ($W \cdot m^{-2}$)
LE_{EB}: Latent heat flux deduced by forcing the energy balance closure ($W \cdot m^{-2}$)
T: Air temperature (°C)
T_*: Temperature-scaling parameter, $T_* = \langle w'T' \rangle / u_*$ (°C)
T': Air temperature fluctuations (°C)
U: Horizontal mean air velocity ($m \cdot s^{-1}$)
u_*: Friction velocity ($m \cdot s^{-1}$)
u', w': Horizontal and vertical velocity fluctuations ($m \cdot s^{-1}$)
Z: Measurement height (m)
Z_S: $(z\text{-}h_c)/(h_s\text{-}h_c)$ normalized measurement height (–).
ζ: Atmospheric stability parameter, $(z-d)/L$ (–)
ρ: Air density ($kg \cdot m^{-3}$)
σ_T: Standard deviation of air temperature (°C)
EC: Eddy covariance
fTc: Fine-wire thermocouple
FV: Flux variance
S1: Measurements campaign in the banana screenhouse with 8% shading screen
S2: Measurements campaign in the pepper screenhouse covered with a 50-mesh insect-proof screen
S2a: Measurements above the screen during S2
S2b: Measurements below the screen during S2
S3: Measurements campaign in the tomato greenhouse with a plastic roof.

Acknowledgments

The authors thank Mr. Victor Lukyanov for technical help. The authors also thank Professor Gabriel Katul for his helpful suggestions. This research was supported by Research Grant Award no. IS-4374-11C from BARD, the United States–Israel Binational Agricultural Research and Development Fund, and the Chief Scientist of the Israeli Ministry of Agriculture under Grant no. 304-0444.

References

[1] M. Teitel, O. Liran, J. Tanny, and M. Barak, "Wind driven ventilation of a mono-span greenhouse with a rose crop and continuous screened side vents and its effect on flow patterns and microclimate," *Biosystems Engineering*, vol. 101, no. 1, pp. 111–122, 2008.

[2] M. Moeller, J. Tanny, Y. Li, and S. Cohen, "Measuring and predicting evapotranspiration in an insect-proof screenhouse," *Agricultural and Forest Meteorology*, vol. 127, no. 1-2, pp. 35–51, 2004.

[3] J. Tanny, L. Haijun, and S. Cohen, "Airflow characteristics, energy balance and eddy covariance measurements in a banana screenhouse," *Agricultural and Forest Meteorology*, vol. 139, no. 1-2, pp. 105–118, 2006.

[4] C. von Zabeltitz, *Integrated Greenhouse Systems for Mild Climates: Climate Conditions, Design, Construction, Maintenance, Climate Control*, Springer-Verlag, Berlin, Heidelberg, Germany, 2011.

[5] N. Gruda and J. Tanny, "Protected crops-recent advances, innovative technologies and future challenges," *Acta Horticulturae*, vol. 1107, pp. 271–277, 2015.

[6] E. Kitta, A. D. Baille, N. Katsoulas, N. Rigakis, and M. M. Gonzalez-Real, "Effects of cover optical properties on screenhouse radiative environment and sweet pepper productivity," *Biosystems Engineering*, vol. 122, pp. 115–126, 2014.

[7] C. Kittas, N. Katsoulas, N. Rigakis, T. Bartzanas, and E. Kitta, "Effects on microclimate, crop production and quality of a tomato crop grown under shade nets," *Journal of Horticultural Science and Biotechnology*, vol. 87, no. 1, pp. 7–12, 2012.

[8] M. B. Siqueira, G. G. Katul, and J. Tanny, "The effect of the screen on the mass, momentum, and energy exchange rates of a uniform crop situated in an extensive screenhouse," *Boundary-Layer Meteorology*, vol. 142, no. 3, pp. 339–363, 2012.

[9] U. Dicken, S. Cohen, and J. Tanny, "Effect of plant development on turbulent fluxes of a screenhouse banana plantation," *Irrigation Science*, vol. 31, no. 4, pp. 701–713, 2013.

[10] E. Kitta, A. Baille, N. Katsoulas, and N. Rigakis, "Predicting reference evapotranspiration for screenhouse-grown crops," *Agricultural Water Management*, vol. 143, pp. 122–130, 2014.

[11] M. Pirkner, U. Dicken, and J. Tanny, "Penman-Monteith approaches for estimating crop evapotranspiration in screenhouses-a case study with table-grape," *International Journal of Biometeorology*, vol. 58, no. 5, pp. 725–737, 2014.

[12] J. Tanny, U. Dicken, and S. Cohen, "Vertical variation in turbulence statistics and energy balance in a banana screenhouse," *Biosystems Engineering*, vol. 106, no. 2, pp. 175–187, 2010.

[13] J. Tanny, V. Lukyanov, M. Neiman, S. Cohen, M. Teitel, and I. Seginer, "Energy balance and partitioning and vertical profiles of turbulence characteristics during initial growth of a banana plantation in a screenhouse," *Agricultural and Forest Meteorology*, vol. 256-257, pp. 53–60, 2018.

[14] M. Aubinet, T. Vesala, and D. Papale, *Eddy Covariance: a Practical Guide to Measurement and Data Analysis*, Springer, Dordrecht, Netherlands, 2012.

[15] T. Foken, R. Leuning, S. R. Oncley, M. Mauder, and M. Aubinet, "Corrections and data quality control," in *Eddy Covariance*, pp. 85–131, Springer, Dordrecht, Netherlands, 2012.

[16] K. H. Wesson, G. Katul, and C.-T. Lai, "Sensible heat flux estimation by flux variance and half-order time derivative methods," *Water Resources Research*, vol. 37, no. 9, pp. 2333–2343, 2001.

[17] H. A. R. De Bruin, W. Kohsiek, and B. J. J. M. Van Den Hurk, "A verification of some methods to determine the fluxes of momentum, sensible heat, and water vapour using standard deviation and structure parameter of scalar meteorological

quantities," *Boundary-Layer Meteorology*, vol. 63, no. 3, pp. 231–257, 1993.

[18] C. I. Hsieh, L. Mei-Chun, H. Yue-Joe, and C. Tsang-Jung, "Estimation of sensible heat, water vapor, and CO2 fluxes using the flux-variance method," *International Journal of Biometeorology*, vol. 52, no. 6, pp. 521–533, 2008.

[19] C.-I. Hsieh, G. G. Katul, J. Schieldge, J. Sigmon, and K. R. Knoerr, "Estimation of momentum and heat fluxes using dissipation and flux-variance methods in the unstable surface layer," *Water Resources Research*, vol. 32, no. 8, pp. 2453–2462, 1996.

[20] G. G. Katul and C. I. Cheng, "A note on the flux-variance similarity relationships for heat and water vapour in the unstable atmospheric surface layer," *Boundary-Layer Meteorology*, vol. 90, no. 2, pp. 327–338, 1999.

[21] J. D. Albertson, M. B. Parlange, G. G. Katul, C.-R. Chu, H. Stricker, and S. Tyler, "Sensible heat flux from arid regions: a simple flux-variance method," *Water Resources Research*, vol. 31, no. 4, pp. 969–973, 1995.

[22] J. C. Wyngaard, O. R. Cot, and Y. Izumi, "Local free convection, similarity, and the budgets of shear stress and heat flux," *Journal of the Atmospheric Sciences*, vol. 28, no. 7, pp. 1171–1182, 1971.

[23] T. E. Twine, W. P. Kustas, J. M. Norman et al., "Correcting eddy-covariance flux underestimates over a grassland," *Agricultural and Forest Meteorology*, vol. 103, no. 3, pp. 279–300, 2000.

[24] Y. Mekhmandarov, M. Pirkner, O. Achiman, and J. Tanny, "Application of the surface renewal technique in two types of screenhouses: sensible heat flux estimates and turbulence characteristics," *Agricultural and Forest Meteorology*, vol. 203, pp. 229–242, 2015.

[25] D. A. DeVries, "Thermal Properties of soils," in *Physics of Plant Environment*, W. R. van Wijk, Ed., pp. 210–235, North-Holland Publishing Co., Amsterdam, Netherlands, 1963.

[26] S. Ravikovitch, *Soils of Israel, Formation and Characteristics of the Soils in Israel*, (in Hebrew), Hakibbutz Hameuchad, Tel-viv, Israel,1981.

[27] G. Stanhill, "A simple instrument for the field measurement of turbulent diffusion flux," *Journal of Applied Meteorology*, vol. 8, no. 4, pp. 509–513, 1969.

[28] C. I. Hsieh, G. Katul, and T. W. Chi, "An approximate analytical model for footprint estimation of scalar fluxes in thermally stratified atmospheric flows," *Advances in Water Resources*, vol. 23, no. 7, pp. 765–772, 2000.

[29] H. P. Schmid, "Experimental design for flux measurements: matching scales of observations and fluxes," *Agricultural and Forest Meteorology*, vol. 87, no. 2-3, pp. 179–200, 1997.

[30] J. Tanny, M. Moeller, and S. Cohen, "Aerodynamic properties of boundary layers along screens," *Biosystems Engineering*, vol. 102, no. 2, pp. 171–179, 2009.

[31] E. Erell, V. Leal, and E. Maldonado, "Measurement of air temperature in the presence of a large radiant flux: an assessment of passively ventilated thermometer screens," *Boundary-Layer Meteorology*, vol. 114, no. 1, pp. 205–231, 2005.

[32] E. K. Webb, "Profile relationships: the log-linear range, and extension to strong stability," *Quarterly Journal of the Royal Meteorological Society*, vol. 96, no. 407, pp. 67–90, 1970.

23

The Ideotype for Seed Size: A Model Examining the Relationship between Seed Size and Actual Yield in Pea

Jamin Smitchger and Norman F. Weeden

Department of Plant Sciences and Plant Pathology, Montana State University, Bozeman, MT 59717, USA

Correspondence should be addressed to Norman F. Weeden; nweeden@montana.edu

Academic Editor: Othmane Merah

Seed size plays a large role in determining productivity of large seeded legumes. In many large seeded legumes such as pea and bean, actual yield, defined here as grain yield at harvest minus the weight of seed planted, is often a better measure of actual productivity than grain yield at harvest, because the weight of planted seed varies with seed size. In many grain legumes, the weight of planted seed can be equal to 10% of the total grain yield, and minimizing the weight of planted seed could significantly impact actual yield. This study produced an algorithm to examine the relationship between seed size, yield, and actual yield in silico. The output of this algorithm predicted the ideotype for seed size in peas to be much lower (12.5 g./100 seeds) than the seed size of nearly all commercial varieties, indicating that efficiency in pea cropping systems could be increased by reducing seed size. Modifications to the algorithm would allow the prediction of the ideal seed size in other legumes. The algorithm predicts that there is likely very little correlation between seed size and grain yield, although larger seeded legumes will likely have a higher harvest index. Plant breeders can use the ideotype predicted by this function to create varieties of peas and other large seeded legumes that have higher actual yield. The ideotype for seed size was defined in pea.

1. Introduction

In dryland production regions, pea yield is mainly limited by summer heat and availability of soil water. Although pea has poor yield stability in dryland cropping systems [1], future yield potential can be estimated based on prior yields, allowing the breeding of advanced cultivars that can thrive in a specific region. Averaged across 2001-2016, the yield potential in Montana, the largest pea production region in the US, is 1735 kg/ha [2]. Because environmental conditions are the limiting factor, maximizing economic returns within the constraints of a specific environment is a major breeding objective.

Actual yield as defined here is the weight of seed harvested minus the weight of seed planted. Particularly in large seeded legumes in low yielding environments, yield *per se* is not a perfect measurement of economic productivity. Given two varieties of peas with the same yield potential (1735 kg/ha) but different seed sizes, where variety A has a 100 seed weight of 15 g and variety B has a 100 seed weight of 30 g, variety A would have 124 kg/ha higher actual yield than variety B at current seeding rates of 80 seeds/m^2. Actual yield is a better measure of productivity than the common measurement of grain yield at harvest because it accounts for seed cost. Given seeding rates for pea of 170 kg/ha and an average crop yield of 1735 kg/ha), the ratio between harvested grain and seeds planted (harvest ratio) is approximately 10:1. Based on current seeding rates and yields, actual crop yield declines 0.46% with every 1 g increase in hundred seed weight. Chickpeas, beans, and other large seeded legumes also have a similar harvest ratio, especially in many dryland cropping systems. This ratio is low compared to most other crops. Camelina and canola, for example, have a harvest ratio of ~500:1, and wheat has a harvest ratio of ~40:1 (assuming a seeding rate of 78.5 kg/ha and 50-bushel grain yield).

Yield potential is the genetically determined ability of a crop to generate optimal yield in a given growth environment [3]. Yield potential is thought to be partially determined by seed size, and numerous studies have tried to understand the relationship between seed size and yield in pea and other pulses with contradictory results [1, 4–9]. A lack of correlation between seed size and yield was found in grass pea under drought conditions [5]. In chickpea, a positive correlation was found between seed size and yield, but no effect was seen in lentil [4]. In one study in pea, a strong correlation was found between seed size and yield [6], and Krajewski et al. [8] found that seed size was positively correlated with yield, explaining 20% of the variation in yield in pea. However, peduncle number explained nearly 50% of the variation in yield, indicating that seed size is not the most important factor in determining seed yield. Other research has also shown a correlation between the number of fertile nodes and seed yield rather than seed weight and seed yield [10]. Timmerman-Vaughn et al. [1] indicated that seed size is negatively correlated with seed yield, contradicting the study by Krajewski et al. [8], and Irzykowska and Wolko [7] indicated that QTLs for seed size and yield were not colocated. Gambin and Borras [11] concluded that crop yield within a species was more related to variations in seed number than in seed weight. Hence, there appears to be no consensus regarding the relationship, and the ideotype as defined by Donald [12] for seed size in pea has not been defined.

The theoretical effect of seed size on grain yield is positive because larger seeded peas have a higher harvest index than smaller seeded peas (when all other factors are held constant). There may be an association between low pod wall proportion and high seed yield [13], but seed weight did not seem to affect yield in this study. Pate et al. [3] in a previous study in lupin estimated that for every 100 units of carbon imported from the parent plant into pods or seeds, 52 are incorporated into seeds, 37 are into the pod, and the remaining 11 units are lost as CO_2, indicating that up to 41.5% of photosynthate may end up in pod material. However, 96% of the nitrogen in the mature fruit is incorporated into seeds, with 16% of that being remobilized from pod material, indicating that seeds may be more nutrient rich than pods. Pate et al. [3] indicated that approximately 7% of carbon is photosynthesized from the pod itself. Flynn and Pate [14] indicated that approximately an eighth of the carbon in seeds (2 mg/ photoperiod after 10 days after anthesis) is synthesized from pods and that many of the nutrients in pod tissues are transported to the seed prior to senescence.

When compared to other legumes, peas have a much lower proportion of pod walls to percent pod biomass, and it is estimated that this percentage is around 13% [13]. A hypothetical plant that could produce 100 g of seed could produce 400 seeds that weighed a 0.25 g or 100 seeds that weighed 1 g [15]. However, in reality the amount of photosynthate required to create 1 kg of small seeded peas is higher than that required to create 1 kg of large seeded peas because small seeded peas require more pod surface area per unit of mass and more flowering nodes in order to produce the same amount of yield. Therefore, smaller seeded varieties may have an inherent theoretical disadvantage in achieving high grain yields. The consideration of the plasticity of yield components is important when determining the trade-offs between seed size and seed number [15]. Assuming that pod wall thickness is the same in large seeded and small seeded genotypes and that pod length is the same, we can determine the relative amount of photosynthate required to produce pods and seeds of a certain size. Pod material is photosynthetic so it may not be a major burden for the plant to produce, but pods are a sink during their formation [14]. Grain yield is associated with harvest index, which is the ratio of grain yield to plant biomass. This source-sink relationship in peas is dependent on seed size, pod length, plant height, leaf size, stem width, stem wall thickness, leaf thickness, stipule size, and other factors. Here we develop a mathematical model to examine the effect of pod length and seed size on the proportion of photosynthate available for partitioning into seed and pod material. All other parameters of harvest index were assumed to be fixed, and only pod size and seed size were varied.

2. Materials and Methods

The functions to predict the ideal pod length and ideal seed size were created in Excel spreadsheets. All of the equations were derived from commonly used mathematical concepts.

2.1. A Function to Predict the Ideal Pod Length. The formula for the surface area of a cylinder ($SA=2\pi r^2 + 2\pi rh$) allows the prediction of total pod surface area per plant when pod length is varied. Yield, seed size, spacing of seeds in the pod, and all other parameters were held constant. It is assumed that pea pods have the basic shape of a cylinder. Seed radius (r in the formula) is held constant and only pod length is varied. Thus, seed number is fixed, and the number of pods per plant will vary with pod length. For example, if a 4 cm pod holds 5 seeds and the yield potential is fixed at 100 seeds in the model, the plant will produce 20 pods. The surface area of the twenty pods based on the formula for the surface area of a cylinder is 188.4 cm^2. Changing the pod length to 8 cm will increase the number of seeds per pod, and only 10 pods will be needed to have the same yield potential. The surface area of the 10 pods is 172.7 cm^2. The pod surface area was compared between long-podded and short-podded pea to determine the ideotype for pod size.

2.2. A Function to Predict the Ideal Seed Size. The core of the ideal seed size model is the equation for actual yield (yield per acre, seed planted). A number of equations (see below) are used to increase the theoretical accuracy of the model by predicting changes in harvest index based on seed size. The algorithm is based on the idea that increasing harvest index will result in increased yield, a trend which is commonly seen in other crops [16]. This algorithm did not account for variation in other factors besides seed size.

2.3. Assumptions of the Function to Predict the Ideal Seed Size

(i) The density of a dry pea seed is 1.176 g/cm^3, which was empirically determined in this study from a single large sample of dry pea seed.

(ii) The density of a dry pea pod is approximately $1.05 g/cm^3$, which was empirically determined from a single sample of pea pods.

(iii) Seed weight (which also influences pod size and pod number when the yield potential is held constant) is the only characteristic of a plant that varies.

(iv) The amount of photosynthate available to produce either pods or seeds is fixed at a specified yield potential dependent on specific environmental factors.

(v) Formulas for the surface area and volume of a cylinder are a good approximation of the shape of a pea pod. This model does not account for the portion of the tip and base of the pod not occupied by seed.

Several other assumptions will be identified and discussed in the Results and Discussion.

The function to determine the ideal seed size uses the following equations: each one of the parameters is a column title in the Excel spreadsheet found in supplementary data and explains the underlying calculations occurring.

Seed planted per acre: ((seeding rate/m^2)/(sq. feet per m^2))∗(square feet in an acre).

Seed cost multiplier (cost of seed planted per lb/ value of harvested grain per lb). This accounts for the difference in value between seed grown by the farmer and seed purchased by the farmer.

Mass per seed (Mass): 100 seed weight/100. This can be varied or held constant at a specific value in order to determine the ideal seed weight.

Volume of a seed: (Mass)/(seed density). Seed density is approximately $1.176 g/cm^3$.

Radius of a seed (r): the radius of a seed is found by using the formula for the volume of a sphere (volume = $4/3\pi r^3$) and solving for the radius. Therefore $r = \{(\text{Volume})/((4/3) * \pi)\}^{1/3}$.

Seed diameter (Seed Diam): 2 times the radius.

Pod length: this can be varied in the model or held constant at a specific value. Pod length is generally about 6 cm in peas, but this varies depending on the variety. This parameter affects seeds per pod, and pod number.

Correction for space between seeds in a pod: this accounts for the fact that there is often space between the seeds in a dry pod. The multiplier, which represents the actual space taken up by the seed divided by the dry seed diameter, accounts for the pod wall material required to produce a seed if the seed actually requires more space in the pod than its dry seed diameter. It should be noted that the multiplier increases pod material by decreasing the number of seeds per pod. Therefore, more pods must be produced if there is more space between seeds. Seed yield will likely decrease if more photosynthate is partitioned into pod material versus seed material.

Pod width: seed diameter + 2 x pod wall thickness. Pod wall thickness was determined empirically to be approximately 0.194 mm on average.

Photosynthate: this is the total amount of sugars and other nutrients available for partitioning into seeds, seed testa, and pods (in units of grams per plant). This parameter is held constant at an arbitrarily determined value. The algorithm assumes that pea plant sizes, leaf number, and other factors are the same. Only seed size and a few other yield parameters are varied. The percentage of photosynthate dedicated to seeds or pods will vary depending on seed size and pod number. It is quite often that pea plants have enough energy to create about 20 g of pod wall material, seeds, and seed testa, based on empirical data. Photosynthate (g) = (mass per seed∗((length of pod/ radius of seed)∗pod number)) + (pod number $*(\pi r_1^2 h - \pi r^2 h)*$pod density in g/cm2). In this case h=pod length, rl = pod width, and r = seed diameter. This equation can be partitioned into two components where (mass per seed∗((length of pod/ radius of seed)∗pod number)) is the seed yield component and (pod number $*(\pi r_1^2 h - \pi r^2 h)*$pod density in g/cm2) is the pod weight component. The percentage of photosynthate being directed toward the seed or pods for any seed weight can be determined by dividing the seed yield component or pod weight component by the total weight of photosynthate.

Pod number based on amount of photosynthate produced (Pod #): this the number of pods per plant. It is important to note that, to produce the same amount of photosynthate per plant, pod number must vary when seed size varies. Pod number can be determined with the following formula. Pod # = photosynthate/ {(mass per seed∗(h /seed diameter))+($(\pi r_1^2 h - \pi r^2 h)*$pod density)}. In this case h=pod length, r_1=half of the pod width, and r=seed radius. Photosynthate is held constant at an arbitrary value (10 g), and the weight of photosynthate is divided by the weight of a pod at a given seed size.

Amount of photosynthate allocated to seed: this is number of grams of photosynthate that is allocated to the seed and not allocated to the pods. The amount of photosynthate allocated to the seed is dependent on seed size and pod length. The equation for the amount of photosynthate allocated to seed = Mass∗(Pod L/Seed Diam)∗Pod#.

% Allocated to seed: (photosynthate allocated to seed/photosynthate)∗100.

Expected yield potential: this will vary depending on the environment and location. According to the National Agricultural Statistics Service the average yield in Montana during the 15-year period between 2001 and 2016 was 1735 kg/ha.

Expected weight of yield components: this is the total weight of pods, seed, and seed testa produced by a specific plant. The relative proportion of pods, seeds, and seed testa is dependent on seed diameter, and the total amount of pod material is required to achieve specific yield potential decreases with increases in seed size. All other factors are held constant. In general, the weight of seed is 85% of the weight of the total weight of all the yield components, so the expected yield in Montana (1735 kg/ha) is divided by .85. The expected weight of yield components is held constant in the seed size model, but the number is dependent on expected yield.

Yield at harvest: the expected yield at harvest of seed given a specific seed weight. Yield at harvest = (% Allocated to seed/100) ∗Expected weight of yield components. Expected yield at harvest increases with increasing seed weight because less photosynthate is going to pod weight.

Actual seed yield: yield at harvest minus the weight of seed planted. This is essentially parabolic. It reaches a high point at the ideal seed weight and then starts to decline.

Volume of seed that is occupied by the testa: This is part of the volume of a seed that is composed of the seed coat. The seed coat (testa) is the outer surface of the seed, and it is often stripped from the seed during processing. It was empirically determined in this study that the seed testa has an average thickness of 0.0125 cm. The volume can be found by determining the volume of a seed and subtracting the volume of the seed without its testa. Volume of testa= $4/3\pi r_1^3 - 4/3\pi r^3$. In this case r_1 = seed radius, and r = seed radius minus testa thickness.

Mass of testa: volume of testa∗seed density. It is possible that the testa may be more or less dense than a whole seed, but it sinks in water, indicating that it has a density over 1 g/cm^3.

Actual seed yield minus testa: seed mass - mass of testa.

Number of g of seed: the number of grams of photosynthate dedicated to the seed without its testa.

Percent of photosynthate going to seed without its testa: (number of g of seed/photosynthate) ∗100.

Yield minus the weight of seed testa: expected weight of yield components∗Percent of photosynthate going to seed without its testa.

Weight of seeds planted with varying 100 seed weights: (total seeds per acre∗Mass per seed) ∗the number of grams per pound (there are 453.59237 g per lb).

3. Results and Discussion

Our mathematical algorithm (results shown in Figure 1) was created to predict the ideotype for pod length. This function shows the relationship between pod surface area per plant and pod length.

It should be noted that increasing pod length also increases the number of seeds per pod, reducing the number of pods needed per plant to reach a given yield potential. Increases in harvest index are the major mechanism by which increases in pod length increase theoretical yield. The function for pod length predicts longer pods are a more efficient use of plant material than shorter pods. Shorter pods with fewer seeds per pod require more pod ends to reach the same yield potential as a longer podded variety. As a result, the surface area of pod material for a given yield potential declines with increased pod length. However, the plant may not always have enough photosynthate to create longer pods. In peas, average pod length is approximately 6 cm based on data collected in this study, but this number will differ in various germplasm. The ideal pod length would need to be determined empirically, but this model indicates that the ideotype of pea is a plant with longer pods since it would be a more efficient use of plant material. However, the current average length of pods (6 cm) is already nearly ideal. Therefore, pod length should not be an important target for breeders, although increasing pod length in short-podded lentil and chickpea would theoretically increase harvest index.

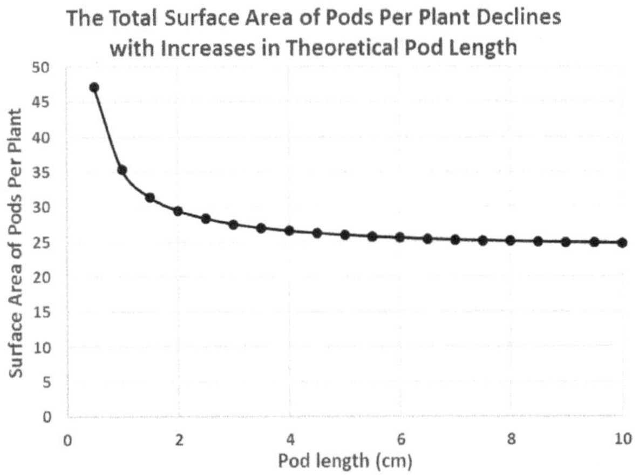

Figure 1: Pod length and pod surface area (assuming a seed yield per plant of 15 seeds and a 0.5 cm seed diameter).

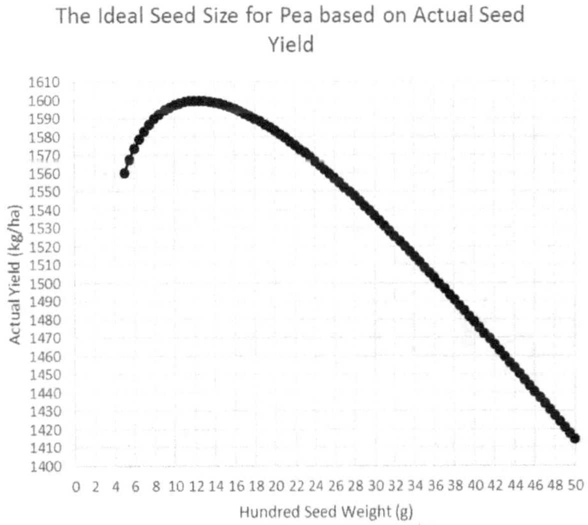

Figure 2: The ideal seed size for pea in Montana. The ideal seed size is at the vertex of the curve.

The function for seed size was created in a Microsoft Excel spreadsheet (the equation predicting the ideal pod length is on the first sheet of the spreadsheet). The Excel spreadsheet containing the algorithm predicting the ideal seed size in pea can be viewed electronically at https://zenodo.org/record/1189038#.WudMIaQvyUk and is also available in supplementary data. The DOI for the file is 10.5281/zenodo.831165. If trouble is encountered opening the spreadsheet, the spreadsheet is also available by contacting the author. Once the spreadsheet is opened, yield potential, seed cost, pod length, or seeding density can be changed to a single value in all the cells in any green column and the effect on ideal seed size can be visualized. This function was used to predict an ideal 100 seed weight based on a yield potential of 1735 kg/ha and a seeding rate of 80 seeds/m^2 (Figure 2). This weight was determined to be 12.5 g, but yield varies only slightly when a given seed weight is near the vertex of the curve. Therefore,

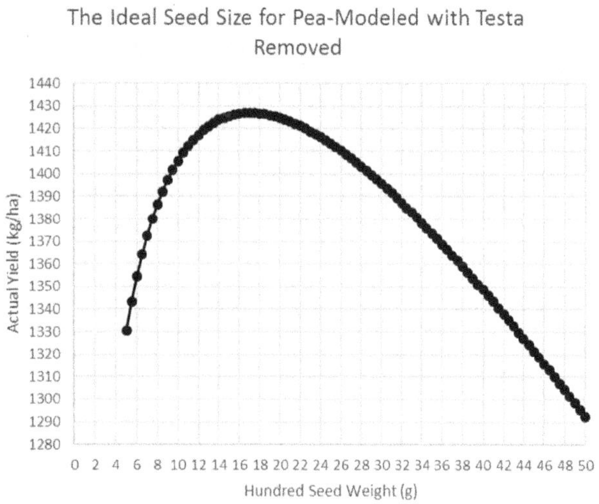

FIGURE 3: The ideal seed size for pea-modeled with seed testa removed. The ideal seed size is at the vertex of the curve.

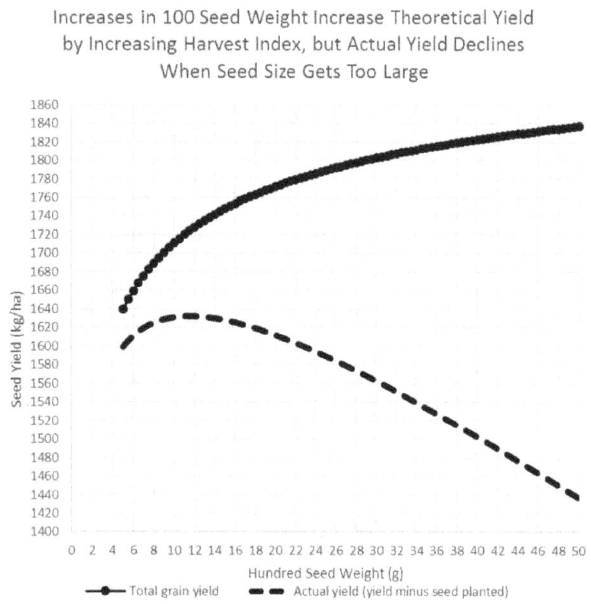

FIGURE 4: The contrast between the weight of the grain harvested and actual yield (grain harvested minus seed planted) as seed size increases.

the ideal 100 seed weight may be better estimated as a range from 8.5 to 16 g.

We also used the function to predict the ideal 100 seed weight with the testa removed (assuming no value for the seed testa) to be ~17.5 g based on a yield potential of 1548 lb/A and a seeding rate of 80 seeds/m^2 (Figure 3).

Finally, the relationship between 100 seed weight and theoretical (total) yield versus actual yield is presented in Figure 4. Theoretical yield, which is typically reported in agricultural summaries, continues to increase with 100 seed weight, but fails to indicate that a portion of that yield must be subtracted as planting inputs. The net actual yield reaches an apex and then declines after a certain seed size is passed.

Several assumptions involved with the overall algorithm need further discussion to clarify their effect on the function should the assumption prove inaccurate. The function is most sensitive to seeding rate and expected grain yield. Halving the seeding rate will increase the ideal 100 seed weight from 12.5 g. to 20 g. Doubling expected grain yield from 1735 to 3470 kg/ha will also increase the ideal 100 seed weight to 20 g. The values used in the function are based on years of data from Montana sites but can be easily modified to other locations with a history of growing pulses. The assumption that the amount of photosynthate available to produce either pods or seeds is equal may be debated because it is likely that less photosynthate is required to produce 1 g of pod versus 1 g of seed, seeds being enriched in nutrient content and young pods probably produce more photosynthate than green seeds. If such is the case, the ideal 100 seed weight would be lower than the seed weight predicted by the function. Also, no account is made for the weight of the peduncle. The peduncle generally is short and light, and it was empirically determined from a small dataset to weigh approximately 10% of the weight of the pods and seed. Each peduncle can support 2 pods. Generally, only 1-2 peduncles are produced per plant in Montana. If the peduncles contribute significantly to photosynthate then the ideal seed size would be smaller than the seed size predicted by the function. The assumption that the number of pods per plant and seed number can vary at no cost to the plant is probably inaccurate. As only 1-2 pods are produced per node, more nodes must be produced in order to generate additional pods and seed. If producing more nodes significantly drains photosynthate, then the ideal seed size would be larger than predicted by our algorithm.

Some of the parameters used in the function such as seed density, pod density, pod wall thickness, testa thickness, and testa density could be refined because it is likely that these values vary considerably in different genotypes or legume species. However, a change in seed density does not affect the algorithm, and changes in pod density, testa thickness, and testa density also have a rather weak effect on the overall calculations, indicating the function is robust. Doubling pod wall thickness to 0.08 mm would increase the ideal seed size to 18 g, but empirical evidence indicates that wall thickness used in the function is a good estimate. Huyghe [13] estimated the proportion of pod wall biomass for each pod to be approximately 0.13, and this study indicates that the proportion of pod wall biomass for each pod is approximately 0.15. By using the estimate of pod wall thickness in this model (0.194 mm), the formula for the volume of a cylinder, and the pod wall density, it was determined that the estimate for pod wall biomass per pod produced by the model was only 1% lower than the empirically determined data. The space between seeds in each pod was also estimated based on the proportion of pod wall biomass measured in this study. Based on the estimates of pod wall biomass, it is estimated that the ratio between space occupied by the seed and dry seed diameter is approximately 1.1. While this estimate could be better defined in future studies, increasing this ratio beyond 1.1 will result in an allocation of more photosynthate to pod wall biomass than was empirically determined and previously reported.

Finally, our algorithm assumes that the purchase of seed and the use of saved seed have an equal cost to the grower. Typically, growers feel that using saved seed is more economical, but processors often require growers to purchase seed from a specific supplier at a greater cost. If the cost of seed is higher than the cost of the harvested product, the ideal seed size would be smaller than the seed size predicted by this function. The ideal seed size varies depending on the yield potential of a given environment. Smaller expected yields indicate that a smaller seed size is ideal, but higher yielding areas require peas with larger seed sizes to achieve their highest actual yield because harvest index increases with increases in seed size. Processors also may demand larger seed sizes. Larger seed sizes could also be more attractive to consumers. When using the algorithm to determine the economic impacts of seed weight, these factors will deserve more attention.

The algorithm initially assumes that the only variable is seed size, but it does not account for the fact that the seed testa is often peeled from the seed after harvest. The function could be altered to account for the weight of the testa because the weight of the testa per unit of mass will vary depending on the size of the seed. It was determined empirically in this study that the average testa thickness is approximately 0.125 mm in pea (this estimate was based on a small dataset). A function accounting for testa thickness indicates that the ideal hundred seed weight is 17.5 g at the average yield potential in Montana. The increase in ideal seed weight when the testa is removed is due to the fact that larger seeds have a smaller surface area per unit of mass than smaller seeds. It is assumed in the model that seed testa has no value.

4. Conclusions

The ideal 100 seed weight predicted by this algorithm (~12.5 g.) is much lower than the seed weight of current cultivars grown in Montana [17], indicating that efficiency could be increased by decreasing seed size. Based on the output of this function, selecting for individuals with larger seed sizes will increase yield due to increased harvest index, but it will also increase the cost of seed. The point of diminishing returns occurs at approximately 12.5 g. Pea breeders should correct yield results for variations in seed size by using actual yield rather than yield, assuming that there is no premium or requirement for a specific seed size. Additionally, extension publications in peas and other pulses should be reporting actual yield per acre rather than grain yield per acre, when no premium for larger seed exists. Processors who remove the pea seed testa during processing should offer premium prices for larger seeded peas, if the testa is not utilized in other processes. However, for many of the end uses of the dry pea crop such as pea flour and split peas, seed size is likely irrelevant and efficiency could be increased by growing a small seeded variety. Since seed costs are often higher than the cost of harvested seed, it is likely that the ideal seed size is lower than the seed size predicted by this model. This algorithm could be modified to predict the ideal seed size under a wide range of conditions.

The harvest ratio describes the overall efficiency of planting a crop, and a higher harvest ratio is generally beneficial. There are several mechanisms whereby harvest ratio could be increased. Firstly, seeding rate could be decreased. However, yield is often highly dependent on seeding rate. Secondly, seed size could be decreased through breeding, which would reduce the weight of seed planted. Thirdly, the seeding rate could remain the same, but varietal differences in optimum seeding rate could be exploited. It has been shown that branching varieties of peas achieve optimum yield potential at lower seeding rates [18], and the branching habit appears to decrease lodging susceptibility (data not shown). Branched varieties also may be more competitive with weeds [19], which are a consistent problem in pulse crops.

The ideotype for seed size in peas and other large seeded legumes has not been previously defined. It should be noted that a similar analysis would likely be valid in other crops with a low harvest ratio, such as chickpea and bean, if no premium for seed size exists. Particularly in chickpea, consumer preferences are likely more important than efficiency of production; however, defining the ideotype for seed size in peas and other large seeded legumes may allow crop breeders and agronomists to increase efficiency of cropping systems for end uses where seed size is of no importance.

Acknowledgments

The senior author would like to thank the MSU Graduate School for funding a Ph.D. Completion Award on his behalf. Funding was also provided by the Northern Pulse Growers Association, AGT Foods USA, and the USDA Hatch Act.

References

[1] G. M. Timmerman-Vaughan, A. Mills, C. Whitfield et al., "Linkage mapping of QTL for seed yield, yield components, and developmental traits in pea," *Crop Science*, vol. 45, no. 4, pp. 1336–1344, 2005.

[2] "NASS, "Pea," National Agricultural Statistics Service," 2017, http://www.nass.usda.gov.

[3] J. S. Pate, P. J. Sharkey, and C. A. Atkins, "Nutrition of a Developing Legume Fruit: Functional Economy in Terms of Carbon, Nitrogen, Water," *Plant Physiology*, vol. 59, no. 3, pp. 506–510, 1977.

[4] B. T. Biger, "The effect of seed size on yield and yield components of chickpea and lentil," *African Journal of Biotechnology*, vol. 8, no. 8, pp. 1482–1487, 2009.

[5] M. Gusmao, K. H. M. Siddique, K. Flower, H. Nesbitt, and E. J. Veneklaas, "Water Deficit during the Reproductive Period of Grass Pea (Lathyrus sativus L.) Reduced Grain Yield but Maintained Seed Size," *Journal of Agronomy and Crop Science*, vol. 198, no. 6, pp. 430–441, 2012.

[6] D. J. Bing and Q. Liu, "Investigation of relationships of yield, seed size, seed protein and starch content and development of varieties with improved protein content of field pea (Pisum sativum L.)," *Canadian Journal of Plant Science*, vol. 91, pp. 381-381, 2011.

[7] L. Irzykowska and B. Wolko, "Interval mapping of QTLs controlling yield-related traits and seed protein content in Pisum sativum," *Journal of Applied Genetics*, vol. 45, no. 3, pp. 297–306, 2004.

[8] P. Krajewski, J. Bocianowski, M. Gawłowska et al., "QTL for yield components and protein content: A multienvironment study of two pea (Pisum sativum L.) populations," *Euphytica*, vol. 183, no. 3, pp. 323–336, 2012.

[9] G. M. Timmerman-Vaughan, J. A. McCallum, T. J. Frew, N. F. Weeden, and A. C. Russell, "Linkage mapping of quantitative trait loci controlling seed weight in pea (Pisum sativum L.)," *Theoretical and Applied Genetics*, vol. 93, no. 3, pp. 431–439, 1996.

[10] V. Kosev and A. Mikić, "Short communication. Assessing relationships between seed yield components in spring-sown field pea (Pisum sativum L.) cultivars in bulgaria by correlation and path analysis," *Spanish Journal of Agricultural Research*, vol. 10, no. 4, pp. 1075–1080, 2012.

[11] B. L. Gambín and L. Borrás, "Resource distribution and the trade-off between seed number and seed weight: A comparison across crop species," *Annals of Applied Biology*, vol. 156, no. 1, pp. 91–102, 2010.

[12] C. M. Donald, "A barley breeding programme based on an ideotype," *The Journal of Agricultural Science*, vol. 93, no. 02, p. 261, 1979.

[13] C. Huyghe, "Genetics and genetic modifications of plant architecture in grain legumes: A review," *Agronomy for Sustainable Development*, vol. 18, no. 5-6, pp. 383–411, 1998.

[14] A. M. Flinn and J. S. Pate, "A quantitative study of carbon transfer from pod and subtending leaf to the ripening seeds of the field pea (Pisum arvense L.)," *Journal of Experimental Botany*, vol. 21, no. 1, pp. 71–82, 1970.

[15] V. O. Sadras, "Evolutionary aspects of the trade-off between seed size and number in crops," *Field Crops Research*, vol. 100, no. 2-3, pp. 125–138, 2007.

[16] M. Reynolds, J. Foulkes, R. Furbank et al., "Achieving yield gains in wheat," *Plant, Cell & Environment*, vol. 35, no. 10, pp. 1799–1823, 2012.

[17] Y. Mohammed and C. Chen, "2016 Montana Cool-Season Spring Pulse Variety Evaluation Annual Report, Montana Agricultural Experiment Stations 2016 Annual Report," 2016.

[18] J. M. Spies, T. Warkentin, and S. J. Shirtliffe, "Basal branching in field pea cultivars and yield-density relationships," *Canadian Journal of Plant Science*, vol. 90, no. 5, pp. 679–690, 2010.

[19] J. M. Spies, T. D. Warkentin, and S. J. Shirtliffe, "Variation in field pea (pisum sativum) cultivars for basal branching and weed competition," *Weed Science*, vol. 59, no. 2, pp. 218–223, 2011.

24

Optimization of Protocols for In Vitro Regeneration of Sugarcane *(Saccharum officinarum)*

Shakra Jamil,[1] Rahil Shahzad,[1] Ghulam Mohyuddin Talha,[2] Ghazala Sakhawat,[1] Sajid-ur-Rahman,[1] Razia Sultana,[2] and Muhammad Zaffar Iqbal[1]

[1] *Agricultural Biotechnology Research Institute, AARI, Faisalabad, Pakistan*
[2] *Department of Plant Breeding and Genetics, University of Agriculture, Faisalabad, Pakistan*

Correspondence should be addressed to Rahil Shahzad; rahilshahzad91@gmail.com

Academic Editor: Siva P. Kumpatla

Sugarcane contributes 60–70% of annual sugar production in the world. Somaclonal variation has potential to enhance genetic variation present within a species. Present study was done to optimize an in vitro propagation protocol for sugarcane. The experiments included four varieties, 9 callus induction media, 27 regeneration media, and 9 root induction media under two-factor factorial CRD. Data were recorded on callus induction, embryogenic callus formation, shoot elongation (cm), root induction, and plant regeneration. Statistically significant differences existed between genotypes and treatments for callus induction (%), embryogenic callus formation (%), shoot elongation (cm), root induction, and plant regeneration (%). All parameters showed dependency on genotypes, culture media, and their interaction. Highest callus induction (95%) embryogenic callus formation (95%) was observed in callus induction media 5. Highest plantlet regeneration (98.9%) capacity was observed in regeneration media 11 whereas maximum shoot elongation (12.13 cm) and root induction (8.32) were observed in rooting media 4. G_1 showed best response for all traits and vice versa for G_4. Hence it was concluded that G_1, callus induction media 5, regeneration media 11, and rooting media 4 are the best conditions for in vitro propagation of sugarcane.

1. Introduction

Sugarcane is a strong and tall crop. It belongs to perennial grasses. It has high sugar content. Commonly grown sugarcane cultivar is *Saccharum officinarum* which belongs to family Poaceae. It is an important industrial crop in tropical and subtropical regions of the world. Properties such as an efficient photosynthetic machinery and efficient biomass production make it an excellent target for industrial processing and a valuable alternative for animal feeding and the production of byproducts. It is also a prime candidate as a fuel crop (production of ethanol). Byproducts from sugarcane include molasses, stock feed, alcoholic drinks, bagasse, and cane wax [1].

Sugarcane provides major raw material for sugar producing industries [2]. It ranks among the world's top ten food crops and contributes 60–70% of annual sugar production in the world. Nowadays it has appeared as an important bioenergy field crop [3]. Pakistan ranks 5th in sugarcane producing countries with respect to area under cultivation and holds 15th position with respect to sugar production [4]. Sugarcane crop contributes 3.2 percent as value addition in agriculture and 0.7% in GDP [5].

With the uses of conventional, modern breeding, and advanced biotechnological approaches, sugarcane is becoming model crop for the future. It also promises to fulfill food, fiber, feed, and biofuel requirements of quickly growing world population. To meet the increasing sugar demand and to make sugarcane a beneficial crop to farmers it is required to advance the quality and quantity of sugarcane and sugar production per unit area [6, 7]. The other major challenges to be addressed by the sugarcane breeders include biotic stresses (pests, diseases, and weed competition), abiotic stresses (drought, salinity, cold, aluminum toxicity, and compact soils), and plant architecture including stalk number, height, leaf angle, and plant vigor [8–11].

In vitro propagation and other in vitro techniques are in common use in agriculture and horticulture for solving various problems related to plant and mass propagation of crop plants [12]. In vitro propagation of sugarcane is a model system to investigate sugar uptake [13], sugar metabolism [14], disease control [15], improvement of yield components [16], and weed control via herbicide tolerance [17]. Use of plant tissue culture is a preferable technique to solve the agricultural problems over conventional methods, as plants are grown in controlled environment in tissue culture.

It has been observed that, in many plant regeneration systems, callus induction is initiated by 2, 4-D at different concentrations [18]. Also the success of the sugarcane tissue culture is also dependent on sugarcane genotypes [19], culturing medium and use of growth enhancer & regulators [20]. Therefore the present study was designed with four sugarcane cultivars developed by Sugarcane Research Institute, AARI, Faisalabad, to check the effect of cultivar, different media compositions on the callus induction, embryogenic callus formation, and plant regeneration. In addition, we have also carried out optimization of protocol for in vitro propagation of sugarcane.

2. Material and Methods

The present research work was done in the Genetically Modified Organism Testing Laboratory, Agricultural Biotechnology Research Institute, AARI Faisalabad, during 2016.

2.1. Media Preparation for Callus Induction, Callus Regeneration, and Plantlet Regeneration. Different types of callus induction and regeneration media with different concentrations of 2, 4-D, BAP, and NAA were prepared (Table 1). The pH of the media was adjusted to 5.8 with 0.1 N NaOH and 0.1 N HCl. Phytagel at 2.5 g/L was added to solidify the media. Media were then autoclaved for 20 min at 121°C.

2.2. Explant Selection, Preparation, and Culturing for Callus Induction. The four sugarcane *(Saccharum* sp.*)* varieties, namely, CPF-246 (G_1), CPF-247 (G_2), CPF-248 (G_3), and CPF-249 (G_4), were obtained from the Sugarcane Research Institute, Ayub Agricultural Research Institute, Faisalabad. Stalk pieces were sown in pots containing soil in greenhouse. Pots were irrigated every two days with tap water. Sugarcane plants were kept in the same conditions approximately for six months. The plant material used for callus inductions was innermost layer of the leaves. The explants were surface sterilized for 10 min in ten percent bleach, followed by three rinses with sterile distilled water. Then plant part was put on sterile filter paper for drying purpose. Innermost layer of the leaf was obtained and was cut into 5 mm pieces which were separately placed on aseptic media. Two explants per test tubes were cultivated and cultures were kept in the dark at 25± 1°C. Callus induction percentage was determined after four weeks. Media were refreshed in each callus after every second week.

2.3. Identification of Embryonic and Nonembryogenic Callus. Difference between nonembryogenic and embryogenic callus was achieved on the basis of callus physical characteristic [21, 22]. Embryogenic callus is of smooth trait, compact, and described by its white-cream colour and its nodular-like arrangement. But nonembryogenic callus is of wet trait, translucent, or of brownish colour. After four weeks of culture, the number of embryogenic calli was noted for all treatment combinations.

2.4. Plantlet Regeneration, Shoot Elongation, Root Formation, and Acclimatization of the Regenerated Plants. Embryogenic calli were transferred in Petri dishes containing regeneration media (MS modified). Different types (27) of media prepared with different combinations of sucrose, BAP, and NAA keeping the MS salt in constant amount (e.g., 4.6 g/L) were used for plantlet regeneration. Cultures were placed in growth room at 25 ± 1°C under 16 h light (1000 lux). Regeneration of plants from each of the calli was recorded consecutively for 4 weeks. Data were recorded on number of regenerated shoots for four weeks and media were changed after every second week. Regenerated plantlets were then shifted on different combinations of (NAA) root induction media (Table 1). Shoot elongation and rooting induction were recorded for four weeks. Plants with well-developed roots were transferred to disposable cups filled with soil, under high humidity (>90%), by covering the plants with plastic wraps after cutting their leaves [23]. Plants were kept in green house at 25 ± 1°C under 16 h light for hardening. Later on plants were shifted to field. Plants were said to be acclimatized when new leaves and roots were formed in the field conditions.

2.5. Statistical Analysis. Data regarding number of induced callus, the number of embryogenic callus, plantlet regeneration, shoot elongation, and root induction were subjected to analysis of variance according to Steel et al. [24].

3. Result and Discussion

An efficient system for in vitro propagation of sugarcane has been optimized and results obtained in the above experiments are summarized here.

3.1. Callus Induction. For callus induction 9 different media formulations with varying concentrations of the sucrose and 2, 4-D (Table 1) along with four genotypes CPF-246, CPF-247, CPF-248, and CPF-249 were used. Genotypes, media, and their interaction significantly differed for callus induction ($p < 0.05$) Table 2. The results indicated that the callus induction varied from 40% to 95% (Figure 1). These high callus induction percentages explained the high capacity of the tested sugarcane genotypes to induce callus. Genotype 1 and 3 have the highest callus induction percentages (95% and 90%, resp.) with white creamy callus (Figure 5(a)), while genotype 4 has the lowest callus induction (40%). Genotype 3 showed an intermediate (70%) callus induction. Similar results were also reported by Gandonou et al. [19] who studied the effect of genotypes on callus induction in sugarcane and reported high callus induction (95%) in Clone III.

Highest callus induction was observed with media 5 (MS salt 4.6 g/L, 2, 4-D 2.0 mg/L, and sucrose 30 g/L) and

Table 1: Different callus inductions and regenerations media compositions.

Sr. number	Callus induction media			Regeneration media				Rooting media		
	MS salt (g/L)	Sucrose (g/L)	2, 4-D (mg/L)	MS salt (g/l)	Sucrose (g/l)	NAA (mg/l)	BAP (mg/l)	MS salt (g/l)	Sucrose (g/l)	NAA (mg/l)
1	4.60	20.00	1.00	4.60	20.00	0.50	1.00	4.60	20.00	0.50
2	4.60	30.00	1.00	4.60	20.00	0.50	2.00	4.60	20.00	1.00
3	4.60	40.00	1.00	4.60	20.00	0.50	3.00	4.60	20.00	1.50
4	4.60	20.00	2.00	4.60	20.00	1.00	1.00	**4.60**	**30.00**	**0.50**
5	**4.60**	**30.00**	**2.00**	4.60	20.00	1.00	2.00	4.60	30.00	1.00
6	4.60	40.00	2.00	4.60	20.00	1.00	3.00	4.60	30.00	1.50
7	4.60	20.00	3.00	4.60	20.00	1.50	1.00	4.60	40.00	0.50
8	4.60	30.00	3.00	4.60	20.00	1.50	2.00	4.60	40.00	1.00
9	4.60	40.00	3.00	4.60	20.00	1.50	3.00	4.60	40.00	1.50
10				4.60	30.00	0.50	1.00			
11				**4.60**	**30.00**	**0.50**	**2.00**			
12				4.60	30.00	0.50	3.00			
13				4.60	30.00	1.00	1.00			
14				4.60	30.00	1.00	2.00			
15				4.60	30.00	1.00	3.00			
16				4.60	30.00	1.50	1.00			
17				4.60	30.00	1.50	2.00			
18				4.60	30.00	1.50	3.00			
19				4.60	40.00	0.50	1.00			
20				4.60	40.00	0.50	2.00			
21				4.60	40.00	0.50	3.00			
22				4.60	40.00	1.00	1.00			
23				4.60	40.00	1.00	2.00			
24				4.60	40.00	1.00	3.00			
25				4.60	40.00	1.50	1.00			
26				4.60	40.00	1.50	2.00			
27				4.60	40.00	1.50	3.00			

5 ml iron (prepared by adding 1000 mg/L myoinositol, 500 mg/L nicotinic acid, 500 mg/L pyridoxine, and 100 mg/L thiamine) and 5 ml vitamin (prepared by adding 0.430 g iron and volume made up to 1 L) solution were also added in the above media.

Table 2: Mean squares for callus induction (%), embryogenic callus formation (%) and plantlet regeneration (%) along with their significance.

Callus induction experiment						Regeneration experiment			Plant elongation and rooting experiment					
Callus induction %			Embryogenic callus % for first week			Plant regeneration % for first week			Shoot elongation (cm)			Root induction		
SOV	DF	MS	SOV	DF	MS	SOV	DF	MS	SOV	DF	MS	SOV	DF	MS
G	3	343.04**	G	3	403.7**	G	3	1005.8**	G	3	130.3**	G	3	43.91**
M	8	199.19**	M	8	234.1**	M	26	164.2**	M	8	39.21**	M	8	29.41**
G * M	24	44.78**	G * M	24	46.72**	G * M	78	42.03**	G * M	24	37.76**	G * M	24	20.53**
Error	72	21.41	Error	72	1.99	Error	216	21.85	Error	72	39.92	Error	72	60.31

** indicates that $P < 0.01$ and results are highly significant.

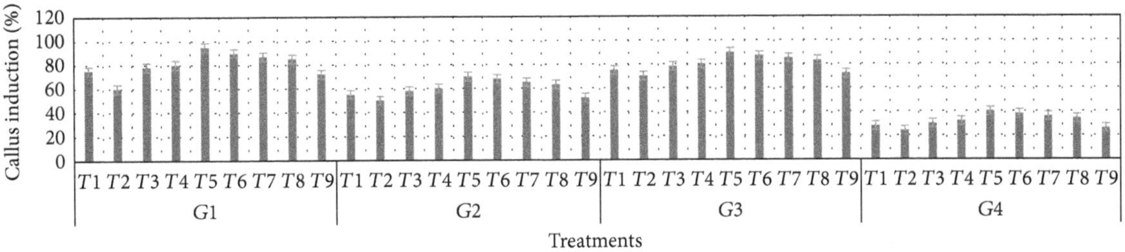

FIGURE 1: Effect of different callus induction media on callus induction (%) in sugarcane. G_1–G_4 are CPF-246, CPF-247, CPF-248, and CPF-249, respectively, and T_1–T_9 are different media compositions in accordance with callus induction media portion of Table 1.

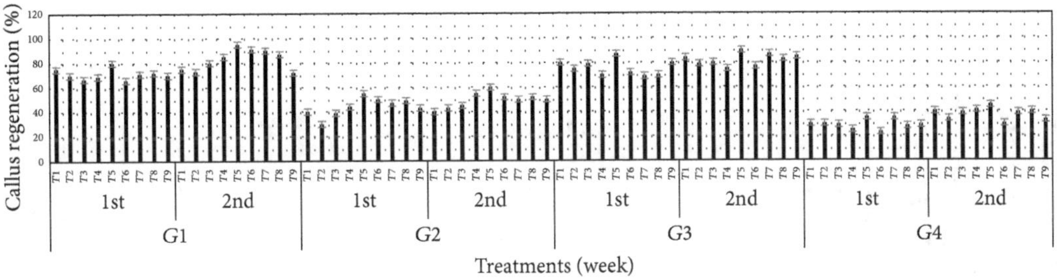

FIGURE 2: Effect of different plantlet regeneration media on embryogenic callus formation (%) in sugarcane. 1st to 2nd week indicates the reading for week 1 and week 2, respectively. G_1–G_4 are CPF-246, CPF-247, CPF-248, and CPF-249, respectively, and T_1–T_{27} are different media compositions in the regeneration media portion of Table 1.

lowest callus induction was observed with callus induction media 2 (MS salt 4.6 g/L, 2, 4-D 1 mg/L, and sucrose 30 g/L). These results showed that callus induction ability depends on genotype, culture media, and their interaction. Similar results were also observed by Gandonou et al. [19] and Khan et al. [2]. Begum et al. [25] observed that 3.5 mg/l of 2, 4-D gave maximum callus induction in Bangladeshi sugarcane variety Nagabari. Islam et al. [26] demonstrated that 0.5–5.0 mg/l of 2, 4-D gave maximum callus induction from the sugarcane leaf tissues supplemented with MS media.

3.2. Embryogenic Callus. Embryogenic callus shows the ability of the callus to regenerate plant from one cell or few numbers of cells. Difference between embryogenic and nonembryogenic callus was achieved on the basis of callus physical appearance as reported by [19]. In addition to these two previous types, we observed an intermediary type with a nonembryogenic tissue covered by an embryogenic tissue. This type of callus was also observed previously for sugarcane [27] and sorghum [28]. For embryogenic calli percentage determination, we classified the intermediary type as embryogenic because, in further subcultures, the embryogenic tissue grows faster than nonembryogenic tissue.

Embryogenic callus formation was recorded for two weeks and was higher in the second week as compared to week 1 irrespective of the genotypes and treatments. Statistical analysis showed significant differences between genotype, treatments, and their interaction for embryogenic callus formation (Table 2). The highest embryogenic callus was observed in G_1 (95%) followed by G_3 (90%) in 2nd week. The trend was similar for the genotype in the first week. The lowest embryogenic callus was observed in G_4 (23, 35%), respectively, in the first and second week. Maximum embryogenic callus formation was found in media 5 for second week (95, 60, 90, and 45%) and first week (80, 55, 87, and 35%), respectively, for G_1 to G_4 (Figure 2). Previously similar results were reported by Verma et al. [29] in crocus and Gandonou et al. [19] in sugarcane. Molina et al. [30] also reported that somatic embryogenesis in coffee plant is affected by genotype and month of leaf collection. Somatic embryogenesis varied from 4.8 to 72.4% across genotypes in their experiment.

3.3. Plant Regeneration. Embryogenic callus was grown on regeneration media. Media 11 showed best results for plantlet regeneration. The number of plantlets regenerated per callus was also very high (Figure 3). High regeneration ability was found in G_1 followed by G_3 whereas G_2 and G_4 showed low regeneration ability. Maximum and minimum regeneration ability were observed in G_1 (91.3, 93.3, 94.3, and 98.9%) and G_4 (60.9, 62.8, 65.7, and 69.9%), respectively, from 1st week to the 4th week. Regeneration ability of the remaining two genotypes was in between these two (Figure 3). Cultivars, having high regeneration ability, are useful in tissue culture programs. These results illustrated that plant regeneration capacity is a cultivar dependent trait and it is likely that this trait is being controlled by one or a block of genes (Figure 3). These results are in line with the previous results of Birsin and Özgen [31] in Triticosecale who reported that Mikham 2002 and Presto cultivars showed higher regeneration ability than the other genotypes.

Different types of regeneration media with varying concentrations of sucrose, BAP, and NAA were used to optimize media for plantlet regeneration. Best media combination for plant regeneration was media 11 (MS salt 4.6 g/L, sucrose

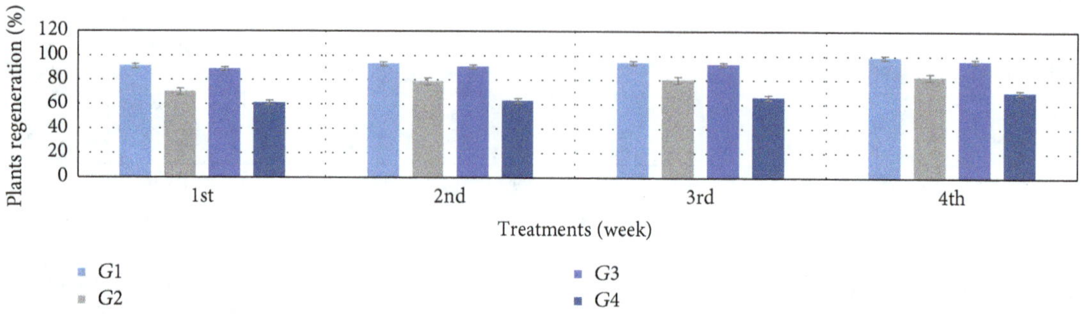

FIGURE 3: Effect of regeneration media 11 on plants regeneration (%) in sugarcane. 1st to 4th week indicate the reading for each week whereas G_1–G_4 are CPF-246, CPF-247, CPF-248, and CPF-249, respectively.

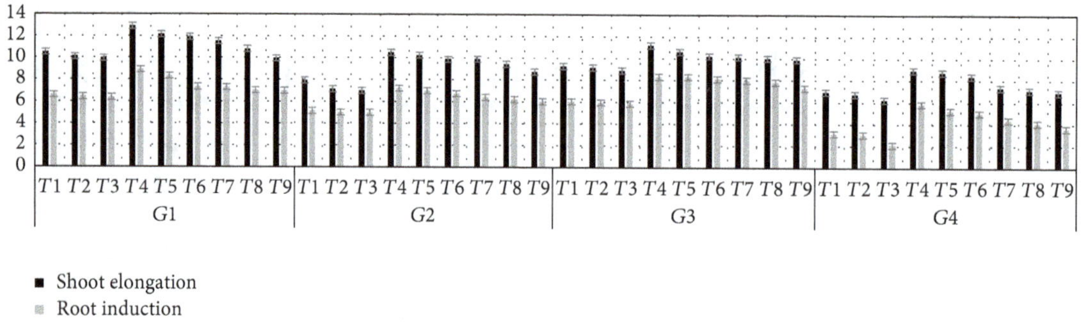

FIGURE 4: Effect of different rooting media on shoot elongation and root induction. G_1–G_4 are CPF-246, CPF-247, CPF-248, and CPF-249, respectively, and T_1–T_9 are different media compositions in the rooting media portion of Table 1.

30 g/L, BAP 2.0 mg/L, and NAA 0.50 mg/L). Plant regeneration was comparatively low in the rest of the media tested. The results for the best treatment are given in Figure 3 from 1st week to 4th week and best regenerated plants in Figure 5(b). However the genotypic response for plantlet regeneration was the same as was found for callus induction and embryogenic callus formation; that is, maximum plantlet was observed in G_1 followed by G_3 and lowest plantlet was observed in the cultures of G_4. The results are also in line with the Verma [29] in crocus. They reported that using NAA and BAP together gave better regeneration. However Ather et al. [32] reported that only BAP @ 1 mg/L alone was found optimum and showed maximum shoot emergence from sugarcane callus.

3.4. Shoot Elongation and Root Induction. Auxins (NAA) are involved in several crucial morphogenetic and biochemical responses, that is, promotion of root induction and shoot elongation in plant axial portions (stem and flower pedicels) [33]. Effect of naphthalene acetic acid on shoot elongation and root induction in sugarcane was evaluated in this study. Nine different compositions of rooting media with varying levels of NAA and sucrose keeping MS salt constant (4.6 g/L) were used (Table 1). All the treatments showed significant differences for shoot elongation and root induction (Table 2). Highest shoot length (12.9 cm, Figure 5(c)) and maximum root induction (8.91 roots, Figures 5(d)–5(f)) were observed with rooting media 4 and G_4 (Figure 4). From these results it was concluded that NAA at 0.5 mg/L is optimum level for shoot elongation. Lal and Sing [34] also reported that NAA is the ideal growth regulator for induction of rooting in the sugarcane. However Khan et al. [2] reported that instead of NAA, IBA is ideal growth regulator for root induction in sugarcane.

4. Conclusion

Our findings showed that callus induction capacity, embryogenic callus response, plantlet regeneration, shoot elongation, and root induction in sugarcane are influenced by genotype and culture media. Growth regulator concentrations used for callus induction also differently affected plant regeneration in the sugarcane varieties. Furthermore, the four genotypes described in this study reacted well to in vitro culture especially to embryogenic callus formation and plantlet regeneration. Maximum callus induction, embryogenic callus formation, plantlet regeneration, shoot elongation, and root induction were observed in G_1 followed by G_3. Further, we have also observed that the media compositions were equally important and, out of the media tested in this study, we have identified the best media for callus induction, plantlet regeneration, and shoot elongation and rooting. The study will fill the knowledge gap and will be helpful in deciding the suitable cultivar for micropropagation studies in sugarcane as well as other crops. A complete protocol for the in vitro propagation of sugarcane has been optimized which will also help researchers interested in micropropagation of sugarcane. It is further suggested that the plethora of genes involved in

Figure 5: Pictorial presentations of the (a) in vitro callus induction, (b) regenerated callus, (c) regenerated shoots, (d) mature shoot, (e) rooted plants, and (f) potted plant under best treatment of callus induction media (media 5) and regeneration media (media 11).

specific response of cultivar to tissue culture media needs be explored for better understanding of the phenomena.

Conflicts of Interest

The authors declare that they have no conflicts of interest.

Acknowledgments

The authors are thankful to Genetically Modified Organisms Testing Lab, Agricultural Biotechnology Research Institute, and Sugarcane Research Institute, AARI Faisalabad, for the funding of this project and providing plant material, respectively.

References

[1] M. Gallo-Meagher, R. G. English, and A. Abouzid, "Thidiazuron stimulates shoot regeneration of sugarcane embryogenic callus," *In Vitro Cellular and Developmental Biology - Plant*, vol. 36, no. 1, pp. 37–40, 2000.

[2] I. A. Khan, A. Khatri, M. A. Siddiqui, G. S. Nizamani, and S. Raza, "Performance of promising sugarcane clone for yield and quality traits in different ecological zones of Sindh," *Pakistan Journal of Botany*, vol. 36, no. 1, pp. 83–92, 2004.

[3] A. J. Waclawovsky, P. M. Sato, C. G. Lembke, P. H. Moore, and G. M. Souza, "Sugarcane for bioenergy production: an assessment of yield and regulation of sucrose content," *Plant Biotechnology Journal*, vol. 8, no. 3, pp. 263–276, 2010.

[4] A. Bashir, S. Haq, M. Abbas, M. A. Munir, and A. Afzal, "Impact of sugarcane mills development activities on cane production in Punjab," *Pakistan J. Agri. Res*, vol. 25, pp. 21–27, 2012.

[5] Economic Survey of Pakistan, "Impact of Social Media Applications on Small Business Entrepreneurs," in *Management and Economics Research Journal*, p. 20, Finance and Economic Affairs Division, Ministry of Finance, Govt. of Pakistan, Islamabad, Pakistan, 2012.

[6] Qureshi, M. Asghar, and S. Afghan, *Sugarcane cultivation in*, Pakistan. Sugar Book Pub. Pakistan Society of Sugar Technologist, 2005.

[7] M. Y. Arain, R. N. Panhwar, N. Gujar et al., "Evaluation of new candidate sugarcane varieties for some qualitative and quantitative traits under Thatta agro-climatic conditions," *Journal of Animal and Plant Sciences*, vol. 21, no. 2, pp. 226–230, 2011.

[8] L. C. C. Carvalho, "Evolution of Sugarcane Industry in the State of São Paulo," in *Sugar Cane Bioethanol: R and D for productivity and sustainability*, L. A. B. Cortez, Ed., vol. 3, Blucher, São Paulo, Brazile, 16 edition, 2010.

[9] S. Creste, L. R. Pinto, M. A. Xavier, and M. G. A. Landell, "Sugarcane Breeding Method and Genetic Mapping," in *Sugar Cane Bioethanol: R and D for productivity and sustainability*, L. A. B. Cortez, Ed., pp. 353–357, Blucher, São Paulo, Brazile, 2010.

[10] P. H. Moore, F. C. Botha, R. T. Furbank, and C. P. L. Grof, "Potential for Overcoming Physico-biochemical Limits to Sucrose Accumulation," in *Proceedings of the Intensive Sugarcane Production: Meeting the Challenges beyond, 2000, CAB International*, B. A. Keating and J. R. Wilson, Eds., pp. 141–156, Wallingford, UK, 2000.

[11] A. Khatri, I. A. Khan, M. A. Javed et al., "Studies on callusing and regeneration potential of indigenous and exotic sugarcane clones," *Asian Journal of Plant Sciences*, vol. 1, no. 1, pp. 41–43, 2002.

[12] H. Schmauder, "E. F. George and P. D. Sherrington, Plant Propagation by Tissue Culture — Handbook and Directory of Commercial Laboratories. VIII + 709 S., 12 Abb., 40 Tab. Eversley,

Basingstoke 1984. Exegetics Limited. £ 65.00. ISBN: 0-9509325-0-7," *Journal of Basic Microbiology*, vol. 25, no. 7, pp. 475-475, 1985.

[13] A. Maretzki and M. Thom, "Membrane transport of sugars in cell suspensions of sugarcane: I. Evidence for sites and specificity," *Plant Physiology*, vol. 49, no. 2, pp. 177–182, 1972.

[14] R. Wendler, R. Veith, J. Dancer, M. Stitt, and E. Komor, "Sucrose storage in cell suspension cultures of Saccharum sp. (sugarcane) is regulated by a cycle of synthesis and degradation," *Planta*, vol. 183, no. 1, pp. 31–39, 1991.

[15] Y. Parmessur, S. Aljanabi, S. Saumtally, and A. Dookun-Saumtally, "Sugarcane yellow leaf virus and sugarcane yellows phytoplasma: Elimination by tissue culture," *Plant Pathology*, vol. 51, no. 5, pp. 561–566, 2002.

[16] J. W. Hoy, K. P. Bischoff, S. B. Milligan, and K. A. Gravois, "Effect of tissue culture explant source on sugarcane yield components," *Euphytica*, vol. 129, no. 2, pp. 237–240, 2003.

[17] I. A. Nasir, B. Tabassum, Z. Qamar et al., "Herbicide-tolerant sugarcane (Saccharum officinarum L.) plants: An unconventional method of weed removal," *Turkish Journal of Biology*, vol. 38, no. 4, pp. 439–449, 2014.

[18] G. Zahra Jahangir, I. Ahmad Nasir, R. Ahmad Sial, M. Aslam Javid, and T. Husnain, "Various Hormonal Supplementations Activate Sugarcane Regeneration In-Vitro," *Journal of Agricultural Science*, vol. 2, no. 4, 2010.

[19] C. Gandonou, T. Errabii, J. Abrini, M. Idaomar, F. Chibi, and N. Skali Senhaji, "Effect of genotype on callus induction and plant regeneration from leaf explants of sugarcane (Saccharum sp.)," *African Journal of Biotechnology*, vol. 4, no. 11, pp. 1250–1255, 2005.

[20] V. Saharan, R. C. Yadav, N. R. Yadav, and B. P. Chapagain, "High frequency plant regeneration from desiccated calli of indica rice (Oryza Sativa l.)," *African Journal of Biotechnology*, vol. 3, no. 5, pp. 256–259, 2004.

[21] M. W. Nabors, J. W. Heyser, T. A. Dykes, and K. J. DeMott, "Long-duration, high-frequency plant regeneration from cereal tissue cultures," *Planta*, vol. 157, no. 5, pp. 385–391, 1983.

[22] A. Pellegrineschi, R. M. Brito, S. McLean, and D. Hoisington, "Effect of 2,4-dichlorophenoxyacetic acid and NaCl on the establishment of callus and plant regeneration in durum and bread wheat," *Plant Cell, Tissue and Organ Culture*, vol. 77, no. 3, pp. 245–250, 2004.

[23] P. Paulet and C. Glaszmann, "Les biotechnologies ensoutien α la diffusion variΘtale chez la canne a sucre," *Agriculture et development*, vol. 2, p. 1, 1994.

[24] R. G. Steel, J. H. Torrie, and D. A. Dickey, *Principles and procedures of statistic: a biometrical approach*, 1980.

[25] S. Begum, L. Hakim, and M. A. Azam, "Efficient regeneration of plants from leaf base callus in sugarcane," *Plant Tissue Cult*, p. 5, 1995.

[26] A. S. Islam, H. A. Begum, and M. M. Haque, "Regeneration of Saccharumofficinarum for disease resistant Varieties," in *Proc. Int. Cong. Plant Tissue and Cell Culture*, vol. 5, pp. 709-710, 709-710, 5, 1982.

[27] E. Guiderdoni, "L'embryogenèse somatique des explants foliaires de canne à sucre (*Saccharumsp.*) cultivés in vitro," in *L Agronomie Tropicale*, vol. 41, pp. 50–57, I- Initiation des cultures. L' Agronomie Tropicale, 1986.

[28] C. MacKinnon, G. Gunderson, and M. W. Nabors, "Plant regeneration by somatic embryogenesis from callus cultures of sweet sorghum," *Plant Cell Reports*, vol. 5, no. 5, pp. 349–351, 1986.

[29] S. K. Verma, A. K. Das, G. S. Cingoz, E. Uslu, and E. Gurel, "Influence of nutrient media on callus induction, somatic embryogenesis and plant regeneration in selected Turkish crocus species," *Biotechnology Reports*, vol. 10, pp. 66–74, 2016.

[30] D. M. Molina, M. E. Aponte, H. Cortina, and G. Moreno, "The effect of genotype and explant age on somatic embryogenesis of coffee," *Plant Cell, Tissue and Organ Culture*, vol. 71, no. 2, pp. 117–123, 2002.

[31] M. A. Birsin and M. Özgen, "A comparison of callus induction and plant regeneration from different embryo explants of triticale (x Triticosecale Wittmack)," *Cellular and Molecular Biology Letters*, vol. 9, no. 2, pp. 353–361, 2004.

[32] A. Ather, S. Khan, A. Rehman, and M. Nazir, "Optimization of the protocols for callus induction, regeneration and acclimatization of sugarcane CV. Thatta-10," *Pakistan Journal of Botany*, vol. 41, no. 2, pp. 815–820, 2009.

[33] B. Kaviani, A. A. Hesar, A. Tarang, S. B. Zanjani, D. Hashemabadi, and M. H. Ansari, "Effect of kinetin (Kn) and naphthalene acetic acid (NAA) on the micropropagation of Matthiolaincana using shoot tips, and callus induction and root formation on the leaf explants," *Afr. J. Agri. Res*, vol. 8, no. 30, pp. 4134–4139, 2013.

[34] N. Lal and N. H. Sing, "Rapid clonal multiplication of sugarcane through tissue culture," *Plant Tissue Cult*, vol. 4, pp. 1–7, 1994.

25

Comparative Analysis of METRIC Model and Atmometer Methods for Estimating Actual Evapotranspiration

Arturo Reyes-González,[1] Jeppe Kjaersgaard,[1,2] Todd Trooien,[1] Christopher Hay,[3] and Laurent Ahiablame[1]

[1]Department of Agricultural and Biosystems Engineering, South Dakota State University, Brookings, SD 57007, USA
[2]Minnesota Department of Agriculture, Saint Paul, MN, USA
[3]Iowa Soybean Association, Ankeny, IA, USA

Correspondence should be addressed to Arturo Reyes-González; reyes.arturo@inifap.gob.mx

Academic Editor: Othmane Merah

Accurate estimation of crop evapotranspiration (ET) is a key factor in agricultural water management including irrigated agriculture. The objective of this study was to compare ET estimated from the satellite-based remote sensing METRIC model to in situ atmometer readings. Atmometer readings were recorded from three sites in eastern South Dakota every morning between 8:15 and 8:30 AM for the duration of the 2016 growing season. Seven corresponding clear sky images from Landsat 7 and Landsat 8 (Path 29, Row 29) were processed and used for comparison. Three corn fields in three sites were used to compare actual evapotranspiration (ET_a). The results showed a good relationship between ET_a estimated by the METRIC model (ET_a-METRIC) and ET_a estimated with atmometer (ET_a-atm) (r^2 = 0.87, index of agreement of 0.84, and RMSE = 0.65 mm day^{-1}). However, ET_a-atm values were consistently lower than ET_a-METRIC values. The differences in daily ET_a between the two methods increase with high wind speed values (>4 m s^{-1}). Results from this study are useful for improving irrigation water management at local and field scales.

1. Introduction

With increasing demands placed on freshwater resources worldwide, it is necessary to accurately estimate crop water consumption efficiently. Crop water use information is needed for a range of applications, including improving agricultural water management, irrigation and crop selection, water resource planning, water rights management, and water regulations [1–3]. Irrigated agriculture produces 40% of global food and fiber supply from 20% of the world's croplands [4]. In arid areas, up to 90% of all water withdrawals may be for irrigation purposes [5]. With increasing population and water uses a scarce water supply is put under additional pressure and other water users relying on the same water supply may experience insufficient water allocations. At the same time, a reduction of irrigation water supply may result in loss of production and, ultimately, threatened food security. There is an opportunity, however, to optimize the management of water in agricultural production systems, and the accurate estimation of evapotranspiration (ET) is critical in that regard.

ET is the loss of water from the land surface to the atmosphere through two processes, namely, evaporation (E) from soil and water surfaces and transpiration (T) from vegetative surfaces [6, 7]. ET rates are affected by weather conditions such as solar radiation, air temperature, wind speed and air vapor pressure deficit, and plant and soil characteristics conditions [8, 9]. Different methods, direct and indirect, exist to estimate ET. Direct methods include weighing lysimeters and soil water balance estimations, while indirect methods include pan evaporation, atmometer, Bowen Ratio Energy Balance System (BREBS), Eddy Covariance (EC), scintillometer, sap flow, and remote sensing [1]. An attractive property of satellite-based remote sensing ET estimates using Landsat imagery is its coverage on a field by field basis at a regional scale [1, 10, 11].

Several models have been developed to estimate ET using remote sensing. One of them is the Mapping EvapoTranspiration at High Resolution using Internalized Calibration (METRIC) model. METRIC utilizes the innovative Surface Energy Balance Algorithm for Land (SEBAL) method for estimating sensible heat flux. METRIC uses the near surface to air temperature gradient (dT) for each pixel within an image based on a regression relationship between the dT and radiometric surface temperature (T_s) of two anchor pixels. The anchor pixels represent the conditions of an agricultural field with full vegetation cover and maximum crop ET (cold condition) and a bare agricultural field with no vegetation cover (hot condition) [10, 12].

One of the advantages of the METRIC model compared to previous surface energy balance-based models for use in arid areas is that it utilizes reference evapotranspiration (ET_r) for estimating actual evapotranspiration (ET_a) at the cold pixel condition [10]. Because ET_r is based on ground-based meteorological measurements and because ET_r is calibrated to account for atmospheric conditions common in arid and semiarid conditions, such as horizontal advection, the METRIC model is particularly useful for ET_a estimations under arid or semiarid conditions [10].

Previous studies have compared METRIC model outputs to other methods of ET_a estimation such as weighing lysimeter [10], soil water balance [13], Bowen Ratio Energy Balance System (BREBS) [14–17], Eddy Covariance (EC), for example, [18–21], Large Aperture Scintillometer (LAS) [22], and the METRIC-MODIS method [23]. These studies showed from moderate to strong relationships between observed and METRIC-estimated ET_a, indicating that the METRIC model is a useful tool for estimating accurate ET_a at local and field scales. In addition, the METRIC model has been compared with other models such as water balance model [24], trapezoid interpolation model (TIM) [25, 26], two-source energy balance model (TSEB) [27, 28], SIMDualKc model [29], and the Landsat-MODIS fusion model [30]. However, comparison of METRIC model outputs to an atmometer for ET_a estimation has not yet been attempted.

An atmometer is a device that measures the amount of water evaporated from wet porous surface to the atmosphere [31]. Atmometers are simple and inexpensive devices, which consist of a ceramic evaporation plate (Bellani plate) covered by a green canvas, mounted on top of a cylindrical water reservoir, to provide a visual interpretation of atmospheric demand for pulling water out of the vegetation and the soil [32, 33]. The standard model with number 54 green canvas is recommended for measuring alfalfa ET_r similar to the alfalfa-based Penman-Monteith ET_r, while number 30 green canvas is designed to simulate grass ET_r similar to the grass-based Penman-Monteith ET_0 [32, 34].

Research demonstrated that ET_r estimated with atmometers was moderately correlated ($r^2 = \pm 0.70$) with weighing lysimeters values [35, 36], strongly correlated ($r^2 = 0.90$) with pan evaporation values [37, 38], and strongly correlated ($r^2 = 0.92$) with agrometeorological data values, for example, [39–43].

While METRIC is a well-recognized and widely used method to estimate ET, as noted by you, it is a complex and time intensive method that requires an experienced user to be run. For practical applications of crop water use estimations such as for irrigation scheduling, other in situ methods may be more appropriate. This is where atmometers may serve a purpose since they are simple and intuitive to use. They may be less accurate compared to other methods, but for applications where slightly higher uncertainty is acceptable, atmometers may be a good option. The objective of this study was to compare ET_a estimated from satellite-based remote sensing METRIC model to ET_a estimated with atmometers in corn fields in eastern South Dakota.

2. Material and Methods

2.1. Study Area. The study was carried out at three sites in eastern South Dakota at Brookings (44°19′N, 96°46′W), Volga (44°18′N, 96°55′W), and Oak Lake (44°30′N, 96°31′W) at elevations 500, 497, and 574 m above sea level, respectively. Three corn fields near to each atmometer (within 5 km) were selected and considered to estimate ET_a (Figure 1). The population density was approximately 78,000 plants ha^{-1} in all fields. Corn fields in Brookings and Volga had 0–2% slope, while the Oak Lake had 2–6% slope [44]. All fields used in this study are in corn-soybean crop rotation system. All corn fields emerged in early May and harvested in late October. The average annual precipitation is 533 mm, of which 3/4 typically falls during the growing season April through October. The mean annual maximum temperature is 12.3°C, minimum 0.3°C, and mean 6.3°C. The climate of the study area is classified as moist subhumid according to the Thornthwaite climate classification system [45].

2.2. Landsat Images. We used seven clear sky images from Landsat 7 Enhanced Thematic Mapper Plus (ETM+) and Landsat 8 Operational Land Imager (OLI) and Thermal Infrared Sensor (TIRS) (Path 29, Row 29), Table 1. The images were downloaded from the United States Geological Survey (USGS) EROS Datacenter (http://glovis.usgs.gov) [46]. The images were selected based on the temporal coverage and cloud-free conditions. Images with cloud located > 10 km from all study sites were considered acceptable. The images were processed using the METRIC model running in ERDAS Imagine Software environment [47]. The time of satellite overpass of both Landsats ranged from 11:11 to 11:14 AM, local time (Table 1). Landsat 7 and Landsat 8 have a pixel resolution of 30 m by 30 m in the shortwave bands and 60 m by 60 m and 100 m by 100 m in the thermal band, respectively.

The wedge-shaped gaps appearing within the Landsat 7 images as a result of the SLC-off issue were removed using the Imagine built-in focal analysis tool. During the process, the gap is filled iteratively based on information from nearby pixels. The gap filling is completed prior to image processing (http://landsat.usgs.gov/gap-filling-landsat-7-slc-single-scenes-using-erdas-imagine-TM) [48]. An example of the process is shown in Figure 2.

TABLE 1: Day of year (DOY), selected acquisition dates, Landsat satellite, path/row, and overpass time during the 2016 corn growing season at standard time, used for ET_a estimations.

DOY	Acquisition dates	Satellite	Path/row	Overpass time (local)
154	06/02/16	Landsat 8	29/29	11:11:03 AM
178	06/26/16	Landsat 7	29/29	11:13:56 AM
194	07/12/16	Landsat 7	29/29	11:13:55 AM
202	07/20/16	Landsat 8	29/29	11:11:21 AM
218	08/05/16	Landsat 8	29/29	11:11:24 AM
234	08/21/16	Landsat 8	29/29	11:11:30 AM
258	09/14/16	Landsat 7	29/29	11:14:05 AM

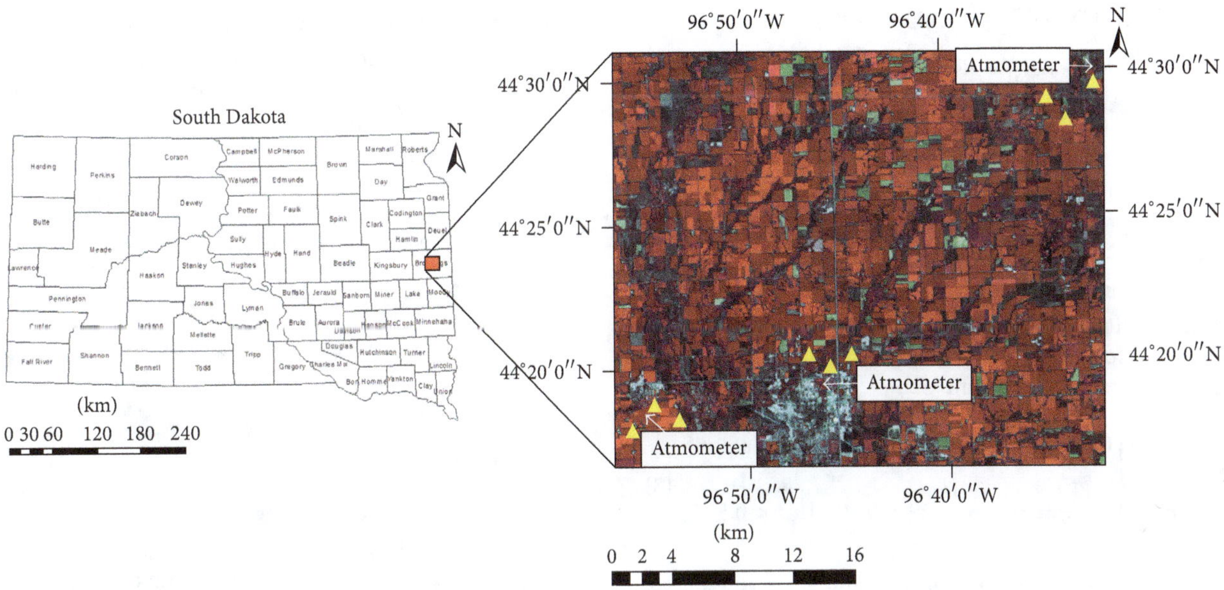

(a) (b)

FIGURE 1: South Dakota with county boundaries: the red rectangle shows the study area in eastern South Dakota (a). Landsat 7 with false color composite (bands 4, 3, and 2) indicates the atmometer locations and the nine yellow triangles show corn field sites (b).

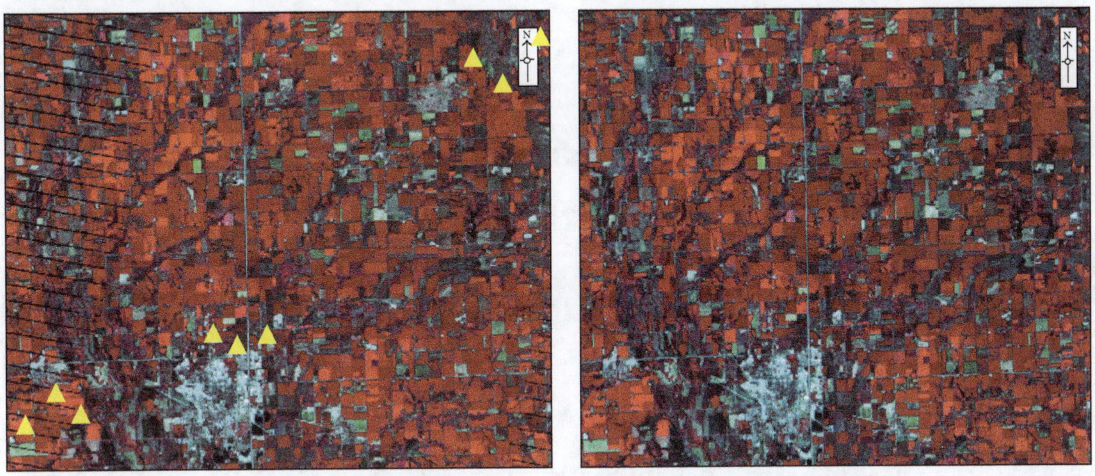

(a) (b)

FIGURE 2: Stripes removed from Landsat 7 image. Original image with nine yellow triangles that indicate corn field sites (a) and final image without stripes, where SLC-off image is filled after employing the focal analysis tool two times (b).

2.3. METRIC Model. ET_a estimations use the METRIC model approach as described by Allen et al. [1, 6].

METRIC model is a remote sensing image processing model that computes instantaneous ET values as a residual of the surface energy balance equation [1, 6, 49, 50]:

$$\mathrm{LE} = R_n - G - H, \tag{1}$$

where LE is the latent heat flux (W m^{-2}), or ET (mm day^{-1}), R_n is the net radiation (W m^{-2}), G is the soil heat flux (W m^{-2}), and H is the sensible heat flux (W m^{-2}).

Net radiation (R_n) is calculated using surface reflectance and surface temperature (T_s) derived by satellite imagery. R_n is the difference between incoming shortwave radiation and outgoing longwave radiation computed as

$$R_n = R_{S\downarrow} - \alpha R_{S\downarrow} + R_{L\downarrow} - R_{L\uparrow} - (1 - \varepsilon_o) R_{L\downarrow}, \tag{2}$$

where $R_{S\downarrow}$ is the incoming shortwave radiation (W m^{-2}) (solar radiation), α is surface albedo (dimensionless), $R_{L\downarrow}$ is the incoming longwave radiation (W m^{-2}), $R_{L\uparrow}$ is the outgoing longwave radiation (W m^{-2}), and ε_o is the surface thermal emissivity (dimensionless). $(1-\varepsilon_o)R_{L\downarrow}$ is the fraction of incoming longwave radiation reflected from the surface.

Soil heat flux (G) is the magnitude of the heat flux stored or released into the soil. G was computed using the following equations described by Tasumi [51]:

$$\frac{G}{R_n} = \begin{cases} 0.05 + 0.18e^{-0.521\mathrm{LAI}} & \mathrm{LAI} \geq 0.5 \\ \dfrac{1.80\,(T_s - 273.16)}{R_n} + 0.084 & \mathrm{LAI} < 0.5. \end{cases} \tag{3}$$

Sensible heat flux (H) was determined using the aerodynamic based heat transfer equation as follows:

$$H = \rho_{\mathrm{air}} C_p \frac{dT}{r_{\mathrm{ah}}}, \tag{4}$$

where ρ_{air} is the air density (kg m^{-3}), C_p is the air specific heat (1004 J kg^{-1} K^{-1}), dT is the temperature difference between two heights z_1 (0.1 m) and z_2 (2 m), and r_{ah} is the aerodynamic resistance to heat transfer (s m^{-1}).

For the H estimations, the METRIC model uses the CIMEC (Calibration using Inverse Modeling of Extreme Conditions) procedure described by Bastiaanssen et al. [52] and Allen et al. [6] to calibrate the near surface to air temperature difference for each pixel within an image based on a regression relationship between dT and T_s of two anchor pixels (hot and cold). The advantage of the CIMEC approach within the METRIC model reduces possible impacts of biases in estimation of aerodynamic stability correction and surface roughness [6].

In this study, hot and cold pixels were selected for each image in agricultural fields near to the weather stations (<15 km). The hot pixel was selected in a bare agricultural field with no vegetation cover, while the cold pixel was selected in an agricultural field with full vegetation cover. An adjustment factor is used to adjust the ET estimation to reflect the actual evaporation amount at the cold and hot pixels. A value of 1.05 is typically used for the cold pixel but may be adjusted down in situations where the no field with vegetation cover is present within the image. For the hot pixel, a value of 0.05 is typically used for situations where the top 10–15 cm of the soil is dry and no rainfall has occurred for several weeks. The bare soil evaporation model by Allen et al. [8] was used to establish the appropriate adjustment factor higher than 0.05 to account for residual soil evaporation from antecedent rainfall.

Based on LE values, the instantaneous values of ET were computed for each pixel as

$$\mathrm{ET_{inst}} = 3600 \frac{\mathrm{LE}}{\lambda \rho w}, \tag{5}$$

where $\mathrm{ET_{inst}}$ is the hourly instantaneous ET (mm h^{-1}), 3600 is used to convert to hours, LE is the latent heat flux (W m^{-2}) consumed by ET, ρw is the density of water (1000 kg m^{-3}), and λ is the latent heat of evaporation (j kg^{-1}), which is computed as

$$\lambda = \left(2.501 - 0.00236\,(T_s - 273.15) \times 10^6\right). \tag{6}$$

The reference ET fraction ($\mathrm{ET_rF}$) or crop coefficient (K_c) was calculated based on $\mathrm{ET_{inst}}$ for each pixel and $\mathrm{ET_r}$ was obtained from local weather data.

$$\mathrm{ET_rF} = \frac{\mathrm{ET_{inst}}}{\mathrm{ET_r}}. \tag{7}$$

Daily values of ET ($\mathrm{ET_{24}}$) (mm day^{-1}) for each pixel were calculated as follows:

$$\mathrm{ET_{24}} = \mathrm{ET_rF} \times \mathrm{ET_r24}, \tag{8}$$

where $\mathrm{ET_rF}$ is the reference ET fraction, $\mathrm{ET_r24}$ is the cumulative alfalfa reference for the day (mm day^{-1}), and $\mathrm{ET_{24}}$ is the actual evapotranspiration for the entire 24-hour period (mm day^{-1}).

Monthly and seasonal $\mathrm{ET_a}$ are calculated by interpolation daily values of $\mathrm{ET_rF}$ between images and multiplying by $\mathrm{ET_r}$ for each day and then integrated over the specific month [6]. The interpolation values of $\mathrm{ET_rF}$ are made using a linear interpolation or a curvilinear interpolation function such as a spline function [53]. According to Allen et al. [6] one cloud-free satellite image per month is enough to develop $\mathrm{ET_rF}$ curves for seasonal $\mathrm{ET_a}$ estimations.

2.4. Meteorological Data. Hourly weather data was used for the internal calibration of the METRIC model. The weather observations were taken from the automatic Brookings, Volga, and Oak Lake stations. All weather stations are located in Brookings County, SD. The weather stations at Brookings and Oak Lake sites are surrounded by grass, whereas the weather station at Volga is surrounded by corn fields.

The $\mathrm{ET_r}$ values were calculated using weather dataset using the Penman-Monteith equation [8, 54] as follows:

$$\mathrm{ET_{ref}} = \frac{0.408\Delta\,(R_n - G) + \gamma\,(C_n/\,(T + 273))\,u_2\,(e_s - e_a)}{\Delta + \gamma\,(1 + C_d u_2)}, \tag{9}$$

(a) (b)

FIGURE 3: Atmometer mounted in wooden post damaged by mice (a) and mounted on a metal post (b) at Volga site.

where ET_{ref} is the alfalfa reference (mm day^{-1}), Δ is the slope pressure versus air temperature curve (kPa°C^{-1}), R_n is the net radiation at the crop surface (MJ m^{-2} day^{-1}), G is the soil heat flux at the soil surface (MJ m^{-2} day^{-1}), T is the mean air temperature at 1.5 to 2.5 m height (°C), u_2 is the mean daily wind speed at 2 m height (m s^{-1}), e_s is the saturation vapor pressure of the air (kPa), e_a is the actual vapor pressure of the air (kPa), γ is the psychrometric constant (0.0671 kPa°C^{-1}), $e_s - e_a$ is the vapor pressure deficit (kPa), C_n is the numerator constant (1600 K mm s^3 Mg^{-1} day^{-1}), C_d is the denominator constant (0.38 s m^{-1}) for alfalfa reference, and 0.408 is the coefficient constant (m^2 mm MJ^{-1}).

All weather data were subjected to quality control (QC) prior to being used in any calculations as suggested by Allen et al. [8] and ASCE-EWRI [54]. Hourly QC included the following weather variables such as solar radiation, air temperature (maximum and minimum), wind speed, and air vapor pressure deficit. Therefore, accurate estimations of ET_a depend on the quality weather data.

2.5. Atmometers. Three atmometers were used to measure daily ET_r. One automated atmometer Model E (ETgage Company, Loveland, Colorado, USA) was placed adjacent to the Oak Lake weather station. The automated atmometer Model E was connected to the automated Oak Lake weather station controlled by a CR1000 datalogger (Campbell Scientific, CSI, Logan, UT, USA), where the evaporated data were recorded every 5 minutes. Two manual atmometers were located adjacent to the Brookings and to the Volga weather stations, respectively. The target corn field was located within 5 km from the weather stations. Manual atmometers were manually recorded every morning between 8:15 and 8:30 AM at the Brookings and Volga sites, respectively. The evaporated water from the green canvas in manual atmometers was measured as the difference between the observed water levels on consecutive days [55].

All atmometers were covered with a number 54 green canvas which simulate evaporation rates of alfalfa reference crop. The atmometers were installed on a vertical wooden post using metal brackets and with the top of the ceramic evaporation surface 1.0 m above the ground surface. The atmometers were surrounded by grass at the Brookings and Oak Lake sites, while at the Volga site the atmometer was surrounded by rainfed corn fields (<5 m) in all directions. Due to rodent damage to the canvas, the wooden post at Volga site was replaced by a metal rod in early August (DOY 217) as shown in Figure 3 which prevented further damage and resulting loss of ET_r data.

The atmometer observation period was from May 17, 2016 (DOY 138), to September 18, 2016 (DOY 262); during this period the atmometers were refilled two times with distilled water. Distilled water was used in the atmometer reservoir to prevent accumulation of solutes in and on the top of the plate that can decrease the porosity of the plate and affect the evaporation rates [34].

2.6. Development of Crop Coefficient (K_c) Curves. K_c curves were developed for each corn field at three sites based on the alfalfa reference crop coefficient using appendix E method from Manual 70 [3]. This method uses percent of time from planting to effective cover and days after effective cover to harvest for K_c calculations. In our study, the effective cover for corn fields occurred around 55 days after emergence (DAE). Thus the effective cover was used such as a reference point to calculate local K_c values. Local K_c values for different corn growth stages were calculated and adjusted using DAE. According to Irmak et al. [56] the DAE is more accurate because it ignores the period prior to emergence and is more closely related to the corn growing period, from emergence until physiological maturity.

The K_c curves generated in this study for different corn fields (Figure 6) were multiplied by the ET_r obtained from atmometers to estimate ET_a (ET_a-atm) and compare it with ET_a estimated with the METRIC model (ET_a-METRIC).

2.7. Statistical Analysis between ET_a-METRIC and ET_a-Atm. Statistical comparison between ET_a-METRIC and ET_a-atm was established using a simple linear regression. For the simple regression the model was $y = a + bx$, where y is ET_a-atm and x is ET_a-METRIC. Other statistical evaluations such as mean bias error (MBE) (see (10)), root mean square error (RMSE) (see (11)), coefficient of determination (r^2) (see (12)),

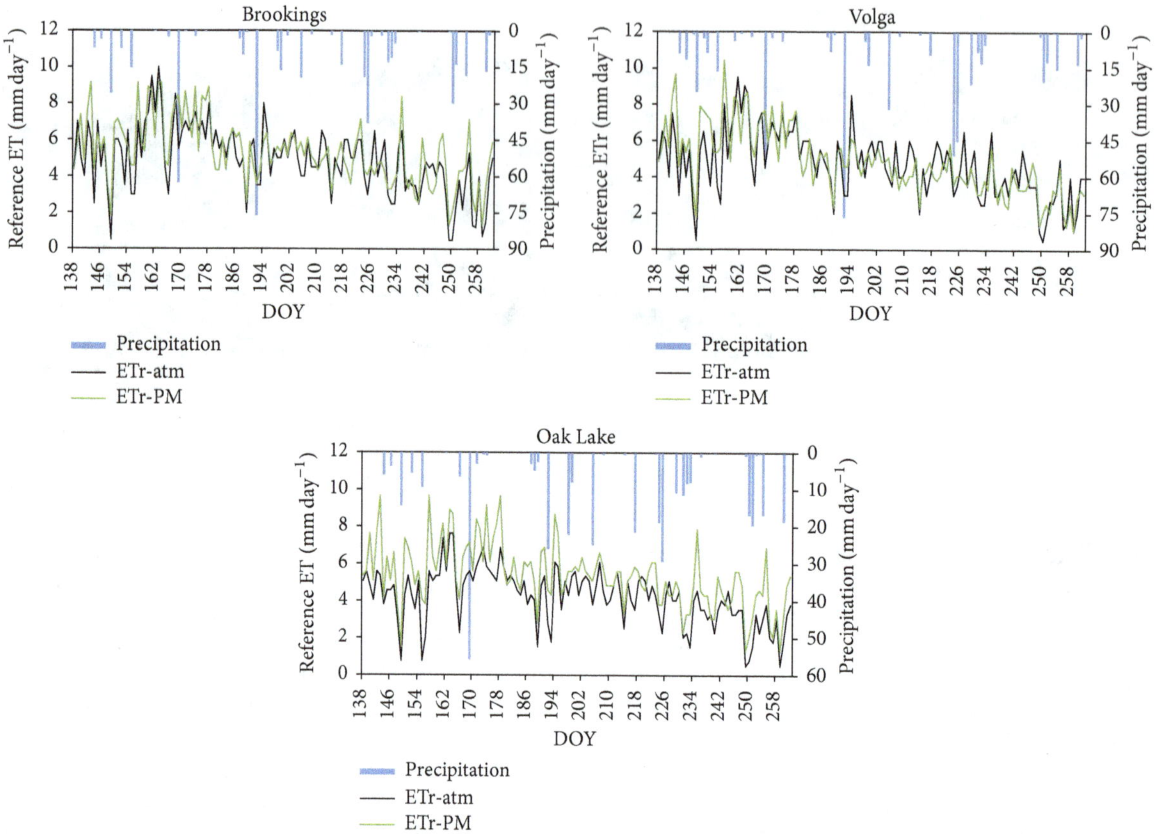

FIGURE 4: Daily ET_r and precipitation for three different sites during the 2016 growing season in eastern South Dakota.

and Willmott index of agreement "d" (see (13)) were used to determine agreement between ET_a-atm and ET_a-METRIC [57].

$$\text{MBE} = \frac{1}{n}\sum_{i=1}^{n}(x_i - y_i) \quad (10)$$

$$\text{RMSE} = \sqrt{\frac{1}{n}\sum_{i=1}^{n}(x_i - y_i)^2} \quad (11)$$

$$r^2 = \frac{\sum_{i=1}^{n}(x_i - \overline{x})(x_i - \overline{y})}{\sqrt{\sum_{i=1}^{n}(x_i - \overline{x})^2 \sum_{i=1}^{n}(y_i - \overline{y})^2}} \quad (12)$$

$$d = 1 - \frac{\sum_{i=1}^{n}(x_i - y_i)^2}{\sum_{i=1}^{n}(|x_i - \overline{x}| + |y_i - \overline{x}|)^2}, \quad (13)$$

where n is the observation number, x_i is the estimated value with the METRIC model, y_i is estimated value using atmometer, and the bars above the variables indicate averages.

3. Results and Discussion

3.1. Precipitation and Reference Evapotranspiration (ET_r). The cumulative precipitation values for the growing period were 450 mm, 497 mm, and 380 mm for Brookings, Volga, and Oak Lake, respectively. In 2016 the cumulative precipitation for the three sites was greater than the average rain (±360 mm) that typically falls during the growing season (April-October). The precipitation events had good distribution during the corn growing season due to the major events occurring in development stage, vegetation stage (V5) (June, 17), and tassel stage (VT) (July, 10) (Figure 4).

Daily values of ET_r from atmometers (ET_r-atm) varied from 0.5 to 10, 0.5 to 9.5, and 0.5 to 7.6 mm day^{-1} for Brookings, Volga, and Oak Lake, respectively (Figure 4). ET_r from Penman-Monteith equation (ET_r-PM) varied from 1.3 to 9.1 mm day^{-1} for Brookings, 1.0 to 10.4 mm day^{-1} for Volga, and 1.3 to 9.6 mm day^{-1} for Oak Lake (Figure 4). The highest ET_r-atm values recorded in the three sites were in early June (DOY 161) and the lowest values were in early September (DOY 249). The highest ET_r-PM values registered in the three sites were in early June (DOY 157) and the lowest values were in middle of September (DOY 259). Even so, moderate correlations between ET_r-PM values and ET_r-atm values in the three sites were found with r^2 of 0.64, 0.59, and 0.67 for Brookings, Volga, and Oak Lake sites, respectively (Figure 5).

3.2. Development of Crop Coefficient (K_c) Curves. The trends of K_c for each corn field at three different sites during the growing season are shown in Figure 6. The K_c curves showed similar tendencies for all corn fields, where K_c values

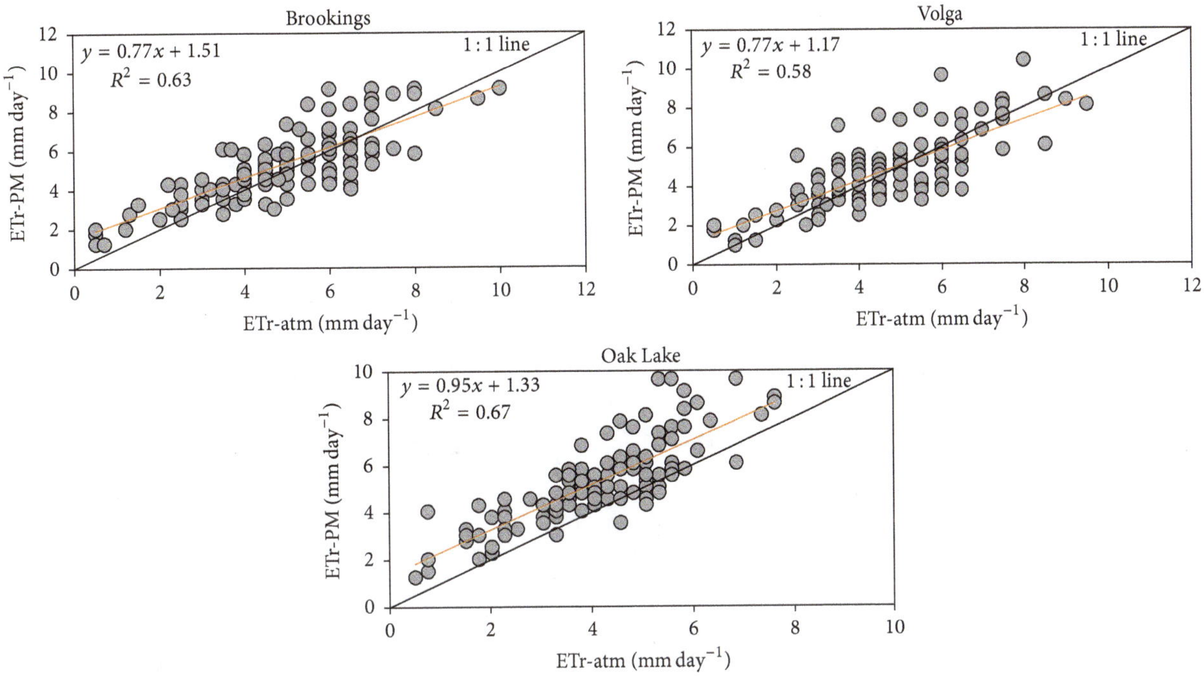

FIGURE 5: Relationship between ET_r-PM values and ET_r-atm values at the Brookings, Volga, and Oak Lake sites throughout the 2016 corn growing season.

increased from initial stage (vegetation stage (V3)) to mid-season stage (VT). In this period the K_c values increase as a function of time between 10% of crop cover to 100% of effective cover. In the mid-season the K_c remains constant (K_c = 1.0), while in the late season the K_c values gradually decreased indicating the crop senescence. At the end of the season (reproductive stage (R6)) the K_c values are low again (K_c = ±0.6).

The K_c curves depend of vegetation index, soil water content, weather conditions, crop variety, and growing degree days [8, 56, 58, 59]. In this study, K_c curves for corn fields presented little variability because rainfall events, emergence days, and air temperature were almost homogenous in our study area.

The maximum K_c values observed in this study were similar to the Kc_r (from alfalfa reference) values reported by Djaman and Irmak [58], who reported maximum Kc_r values from 50 to 70 DAE in corn with rainfed treatment. Also, Wright [53] found maximum Kc_r values at the 100% of effective full cover for a corn field. However, our K_c values are different from those reported by other researchers [60, 61]. They found the peak Kc_r values (±1.0) from late July to early August (±70 DAE) for corn fields planted in south central Nebraska.

3.3. ET_a Maps and Variation of ET_a throughout Growing Season. Spatial and temporal distribution of ET_a maps during the growing season were generated by the METRIC model for Brookings, Volga, and Oak Lake (data not showed). However, Figure 7 shows two ET_a maps for Brookings site, one ET_a map displays high ET_a-METRIC value (dark blue color) in a corn field (white rectangle) at mid-season (DOY 194), and another ET_a map shows low ET_a-METRIC value (light green color) at the end of the season (DOY 258). Generally, high ET rates are related to high crop water demands, which normally occurred in the mid-season period, while low ET rates occurred in early and late growing season when the crop has low vegetation and leaf senescence, respectively. Daily ET_a maps helped to explain the variability of crop water requirements throughout the growing season.

The ET_a maps developed by the METRIC model in this study were similar to other ET maps generated by the METRIC model and reported by Chávez et al. [13], Santos et al. [24], Folhes et al. [18], Droogers et al. [62], Healey et al. [15], Zhang et al. [20], and Carrillo-Rojas et al. [63], where they reported the spatial and temporal distribution of daily ET_a for different crops including corn. In other situations, Chávez et al. [13] reported maximum ET_a values (14.1 mm day^{-1}) due to high wind speed values (7.0 m s^{-1}) at the time of satellite overpass in corn field in Texas High Plains, USA.

Figure 8 shows the variation of ET_a-METRIC values throughout the growing season at three sites. These results are in agreement with [13, 20, 62]; they reported high ET values at mid-season stage and low ET values at the end of the growing season in agricultural crops using the METRIC model.

3.4. ET_a Correlations between the METRIC Model and Atmometer. In the METRIC model ET_a was taken from ten randomly selected pixels in three corn fields at three locations for each satellite image, while for the atmometer method ET_a was the result of multiplying ET_r (from atmometer) by K_c values generated in this study for each corn field.

The linear relationship between ET_a-METRIC values and ET_a-atmometer (ET_a-atm) values in three fields at three sites

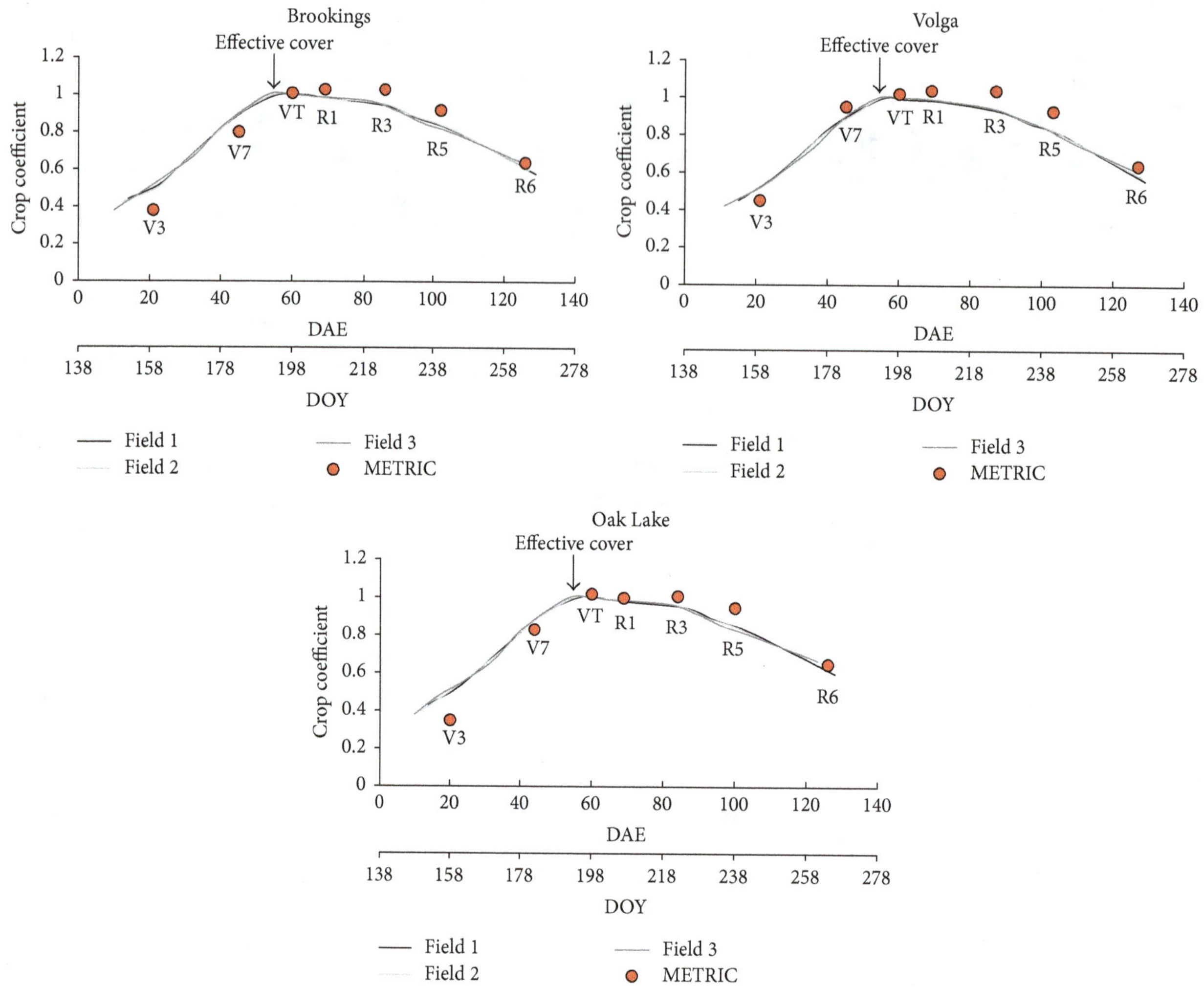

FIGURE 6: Crop coefficient curves based on the alfalfa reference crop coefficient in three fields at the Brookings, Volga, and Oak Lake sites. The red circles indicate the average K_c (ET_rF) values derived from the METRIC model.

TABLE 2: Regression coefficients between ET_a-METRIC values and ET_a-atm values for three corn fields at three sites.

Corn field	Brookings			Volga			Oak Lake		
	Slope	Intercept	r^2	Slope	Intercept	r^2	Slope	Intercept	r^2
Field 1	0.73	1.28	0.92	0.90	0.61	0.79	0.65	0.73	0.93
Field 2	0.72	1.20	0.91	0.97	0.32	0.81	0.66	0.75	0.88
Field 3	0.78	1.14	0.87	0.98	0.33	0.82	0.64	0.86	0.88

is shown in Figure 9. Brookings and Volga sites demonstrated good distribution of points around the 1:1 line, whereas Oak Lake shows that the points were distributed below the 1:1 line; this means that ET_a-METRIC values were higher than ET_a-atm values except on DOY 154. Even so, strong relationships were observed for Brookings and Oak Lake, and a good relationship was observed for Volga (Table 2). In addition, the sum for all corn fields, the ET_a-METRIC values, and the ET_a-atm values correlated well (Table 3).

In general, the difference between ET_a-METRIC (5.36 mm day^{-1}) and ET_a-atm (4.95 mm day^{-1}) at three sites was approximately 8%. The coefficient of determination (r^2) and index of agreement ("d") were 0.87 and 0.84, respectively. The corresponding MBE was 0.41 mm day^{-1} and RMSE was 0.65 mm day^{-1} (Table 3). According to RMSE value, this can be acceptable assuming an average daily ET_a-METRIC value of 5.36 mm day^{-1} and average daily ET_a-atm value of 4.95 mm day^{-1}. The METRIC model has been tested with other methods for ET_a estimation. In this study METRIC was compared to atmometers, where the relationship between them was good (r^2 = 0.87) indicating that the METRIC model is a useful tool to estimate ET_a at field scale. It should

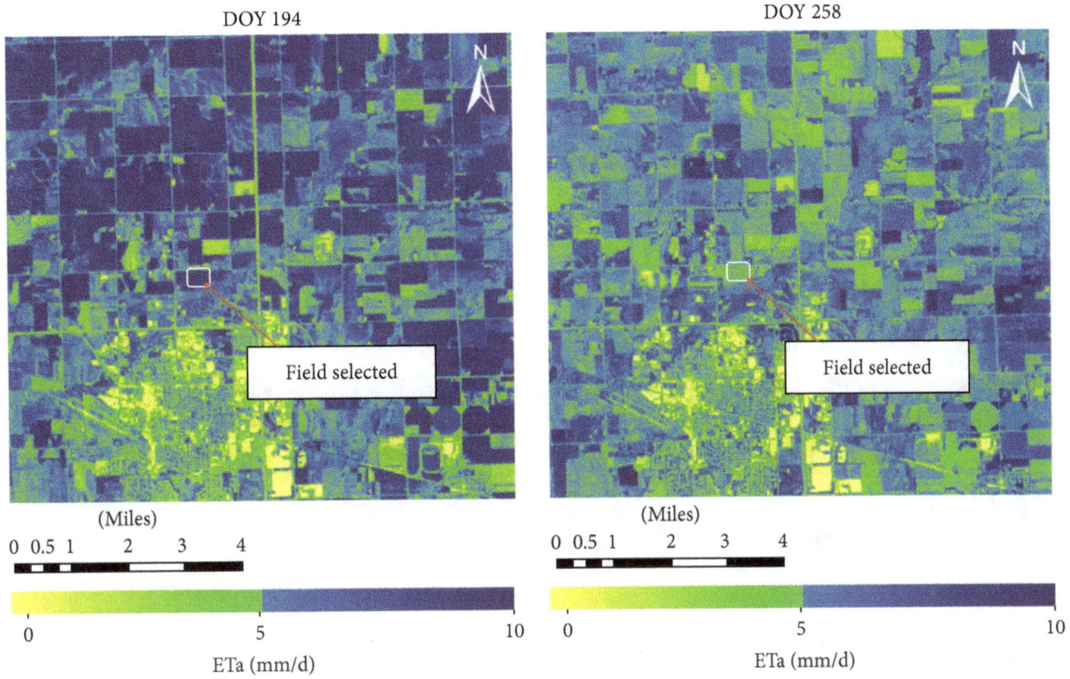

FIGURE 7: ET_a maps developed by the METRIC model. White small rectangles show the corn field selected on DOY 194 and DOY 258. These dates showed high and low ET_a-METRIC values throughout the 2016 growing season at the Brookings site.

TABLE 3: Statistics comparing between ET_a-METRIC and ET_a-atm at the Brookings, Volga, and Oka Lake sites.

Site	ET_a-METRIC (mm day^{-1})	ET_a-atm (mm day^{-1})	MBE (mm day^{-1})	RMSE (mm day^{-1})	r^2	"d"
Brookings	5.71	5.44	0.27	0.56	0.89	0.91
Volga	4.88	5.07	−0.19	0.91	0.81	0.89
Oak Lake	5.50	4.35	1.15	0.48	0.90	0.73
Average	5.36	4.95	0.41	0.65	0.87	0.84

be noted that the comparison between METRIC and the atmometers was done for cloud-free conditions. While daily ETrF values are generated from METRIC enabling the construction of daily ET estimates throughout the growing season, those values were not part of this analysis.

Similar results in r^2 (±0.86) were reported by Healey et al. [15], Morton et al. [64], Gordillo et al. [19], French et al. [28], and Liebert et al. [21], who compared ET_a estimated with the METRIC model and ET_a measured with Bowen Ratio Energy Balance System (BREBS) and Eddy Covariance (EC) methods. Similar results (r^2 = ±0.85) were found in ET_r measured with atmometers by other researchers [32, 65], although different ET_r results (low r^2 = ±0.70) were reported by Chen and Robinson [66] and Lamine et al. [41]. All these authors compared ET_r measured using atmometer covered with a No 54 green canvas (alfalfa reference) with ET_r estimated using agrometeorological data under different weather conditions.

3.5. ET_a Differences between the METRIC Model and Atmometer. The difference between the daily ET_a estimated with the METRIC model (ET_a-METRIC) and ET_a estimated by atmometer (ET_a-atm) is presented in Figure 10. Negative values indicated that the ET_a-METRIC estimates are lower than ET_a-atm, while positive values indicated that the ET_a-METRIC estimates exceed ET_a-atm.

In Brookings, the daily ET_a difference ranged from −0.95 to 1.32 mm day^{-1}, found in field 1 (DOY 154) (V3) and field 2 (DOY 202) (R1), respectively. The more negative values were presented early in the growing season (DOY 154) (V3) due to high K_c value (0.51) used with atmometer method compared to the low K_c value (0.38) used in the METRIC model method. In addition, on DOY 178 (V7) corn field 3 shows negative value (−0.4 mm day^{-1}); this is attributed to hailstorm, which occurred on DOY 169 (V5) nine days before that satellite image overpass. The high positive values (DOY 178, 194, and 202) were related to high wind speed values (>4 m s^{-1}) at the time of satellite overpass (red columns in Figure 12). On DOY 218, 234, and 258 the difference between ET_a-METRIC and ET_a-atm was small (±0.5 mm day^{-1}).

At the Volga site, the daily ET_a difference varied between −1.93 and 1.33 mm day^{-1}. These values were found in field 1 for DOY 194 (VT) and for DOY 178 (V7), respectively. The higher positive values were during the development stage

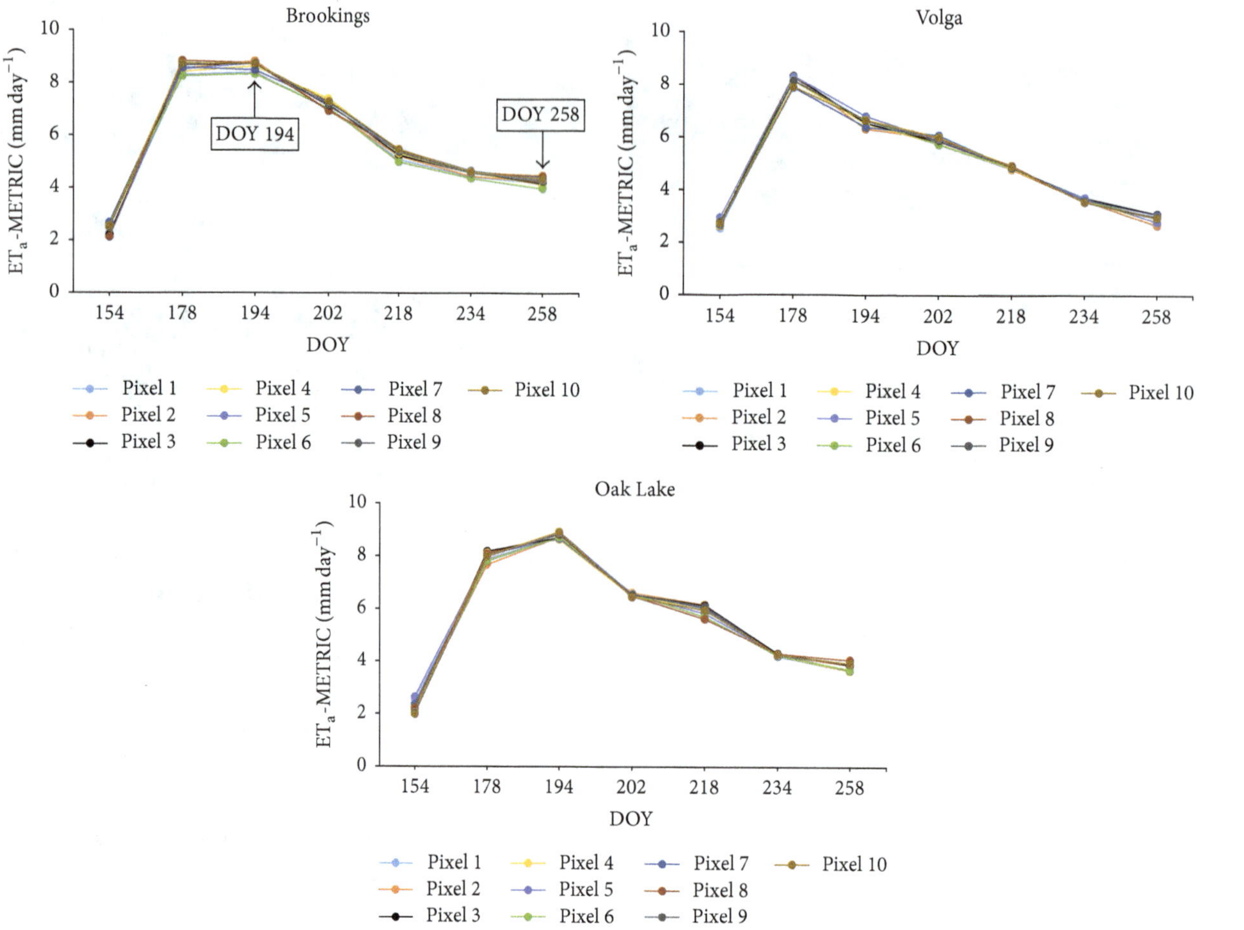

FIGURE 8: Evolution of ET_a-METRIC values at three sites in eastern South Dakota. Ten random pixels were selected within a field in each site. The values from those same pixels were observed during the 2016 growing season.

(DOY 178) (V7) when the corn was 1 m tall. The lower negative values were during the mid-season (DOY 194) (VT) when the crop was 2 m tall. This discrepancy was due to not only the wind speed values but also ET_r recorded in atmometer on DOY 194 (VT). The ET_r recorded in this date was one of the highest values registered during the corn growing season (Figure 4, Volga). After DOY 202 (R1) the difference between ET_a-METRIC and ET_a-atm was minimal (<0.6 mm day^{-1}), because daily average wind speed values were less than 0.8 m s^{-1} and ET_r-PM were around of 10% lower than the ET_r-atm. It is important to mention that at the Volga site the weather station was surrounded by corn fields 2 m tall and the wind speed is reduced by the corn height. So, when low wind speed is used to estimate ET_r using the P-M equation, the resulting ET_r are too low. For this reason the wind speed did not affect the ET_a difference between ET_a-METRIC and ET_a-atm from mid-season to late-season period.

At the Oak Lake site, the daily ET_a difference ranged from −0.62 to 2.61 mm day^{-1}, reported for field 3 (DOY 154) (V3) and for field 1 (DOY 194) (VT), respectively. The negative values were found on DOY 154 (V3) for the three fields; this is attributed to that fact that the ET_a-atm was calculated using K_c equal to 0.5, while ET_a-METRIC used K_c (ET_rF) equal to 0.35, indicating an overestimation of 30% with ET_a-atm method. The higher positive (overestimated) values were observed on DOY 194 (VT) (2.61 mm day^{-1}). At the Oak Lake site, the ET_a-METRIC values tend to be higher than the ET_a-atm values in almost all corn growing season. This noticeable difference is due to the high wind speed values registered throughout the growing season. These high values of wind speed may be attributed to the elevation of the weather station (574 m above sea level), which is 13% higher than Brookings and Volga elevations. Also, at the Oak Lake site the weather station and automated atmometer were located in smooth hill. On the other hand, we observed different ET_a values on DOY 178 (V7) between corn fields. This difference is attributed to hailstorm, which affected the canopy cover in the fields 2 and 3 (Figure 10, Oak Lake).

Similar ET_a differences were found by Choi et al. [26], who reported ET difference between −2.2 and 2.5 mm day^{-1} for different land cover types using METRIC model and trapezoid interpolation model (TIM). They found high discrepancy in ET due to low values of elevation; also they reported that as elevation increases the TIM model slightly overestimates the METRIC ET. On the contrary, low ET

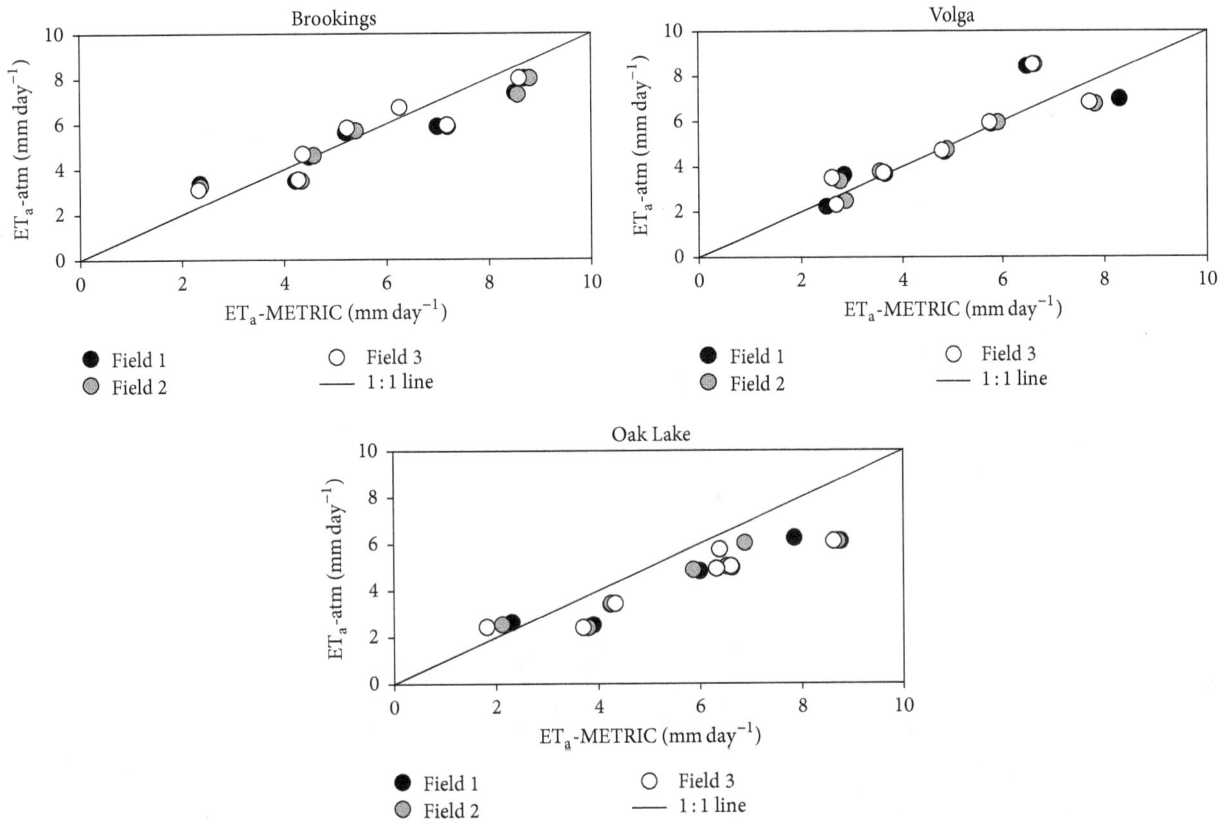

FIGURE 9: Relationship between ET$_a$-METRIC and ET$_a$-atm at three sites and nine corn fields in eastern South Dakota for the 2016 growing season. The black line represents the 1 : 1 line.

difference (±1.0 mm day^{-1}) was reported by Chávez et al. [13], who compare ET estimated by METRIC and ET derived from soil water balance in irrigation corn in Texas. Also, Healey et al. [15] and Hankerson et al. [16] reported that ET difference ranged from −1.0 to 1.0 mm day^{-1} between the METRIC model and BREBS method in different crops.

In our study, daily ET$_a$ differences were attributed to high wind speed values (>4 m s^{-1}) at time of satellite overpass. Generally, as the wind speed increases, the ET$_a$ difference increases (Figure 11). Figure 11 shows that at the Volga site (columns with dark gray color) the difference between ET$_a$-METRIC and ET$_a$-atm was minimal after mid-season stage; however at the Oak Lake site (black columns) the difference was major due to high wind speed.

3.6. Hourly Wind Speed at Three Sites. Hourly average wind speed (m s^{-1}) for each overpass date at the Brookings, Volga, and Oak Lake sites is shown in Figure 12. The wind speed recorded at Volga is similar to Brookings and Oak Lake early in the season but then is quite a bit lower later during the growing season. This is because Volga weather station is too close to the corn fields, and the wind speed is reduced by the height of corn crop. On the contrary, at the Oak Lake site higher average wind speed values were found, especially at time of satellite overpass (red column in Figure 12). Also, the Oak Lake weather station recorded the maximum wind speed values throughout the season.

In Maui Island, USA, for example, Anderson et al. [67] and Zhang et al. [20] reported high variation in evapotranspiration due to high wind speeds values in sugarcane fields. Our results are similar to results reported by Westerhoff [68], who found that as the wind speed increases the ET$_a$ values slightly increase. In addition, Gleason et al. [65] reported ET underestimation with high wind speed conditions. Mokhtari et al. [69] reported that as the wind speed increases ET$_a$ decreases; also they concluded that the METRIC-based ET is highly sensitive to surface temperature but less sensitive to wind speed values.

Based on the results from our study, ET$_r$ values from atmometer need to be adjusted during the windy days. The adjustment factors (average ratio of ET$_r$-atm to ET$_r$-METRIC) for Brookings, Volga, and Oak Lake were 0.83, 0.87, and 0.68, respectively. These adjustment factors can be used to adjust the ET$_r$-atm values to get close estimates to the ET$_r$-PM values on windy days (>4 m s^{-1}). These adjustment factors are necessary to correct ET$_r$-atm values to obtain accurate ET$_a$ estimations. On the other hand, ET$_r$ measured with atmometer can be used by the METRIC model instead of ET$_r$ derived from weather data for estimating ET$_a$.

4. Conclusions

The objective was to compare ET$_a$ estimated from satellite-based remote sensing METRIC model to ET$_a$ estimated with atmometers.

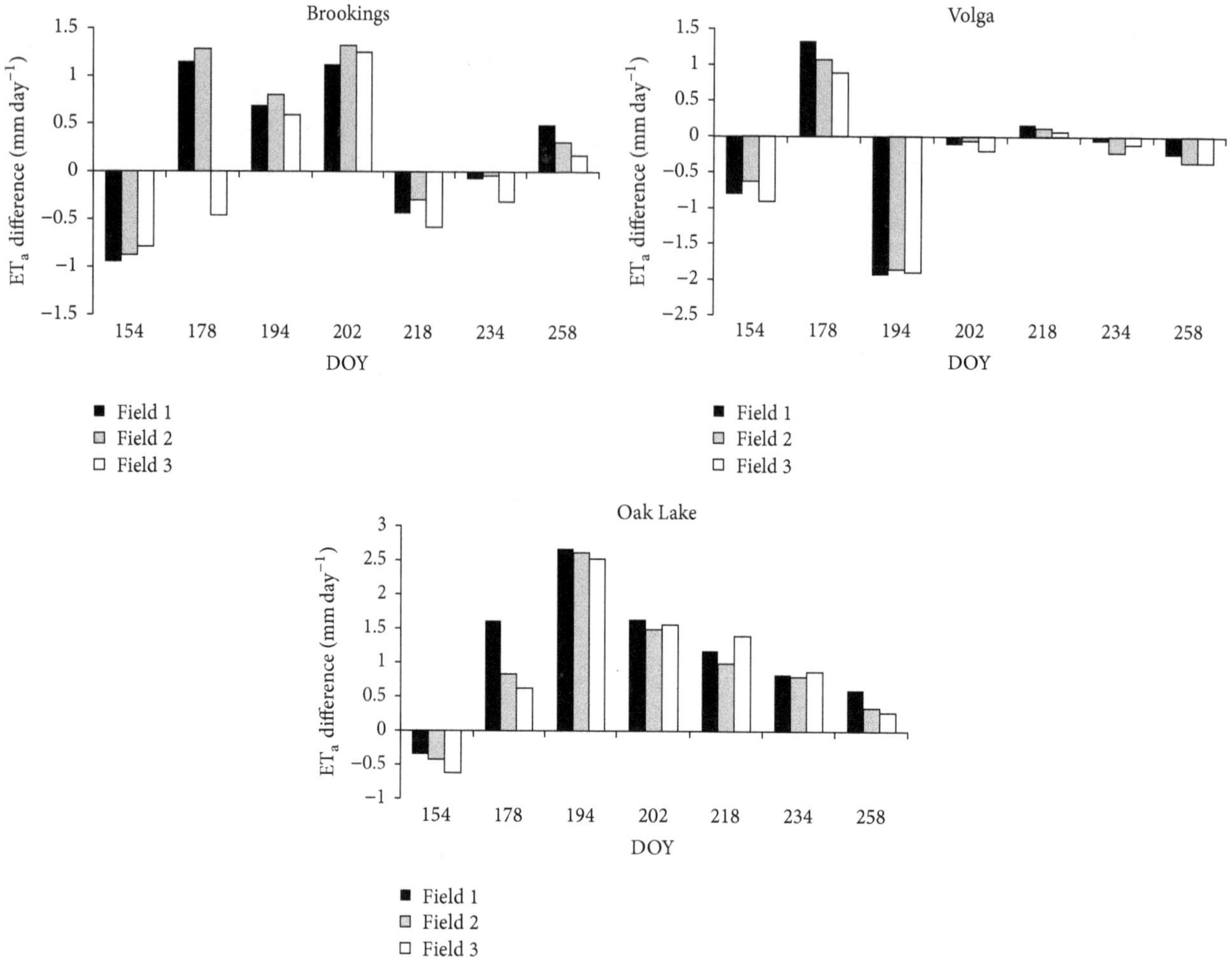

FIGURE 10: Daily ET_a difference between ET_a-METRIC and ET_a-atm at three different sites throughout the 2016 corn growing season.

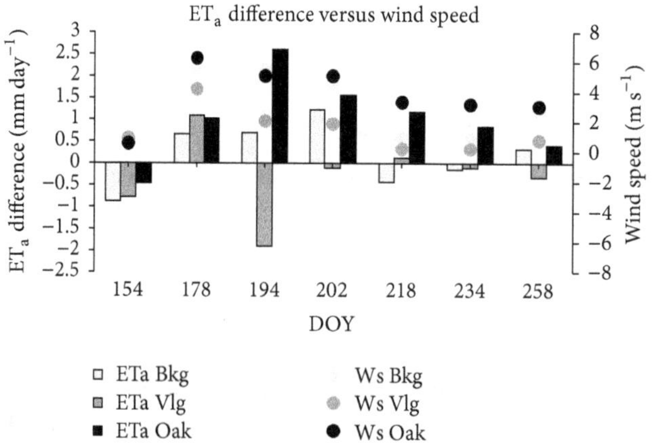

FIGURE 11: Comparison between average ET_a difference and 3-hour average wind speed values at time of satellite overpass at three sites in eastern South Dakota. The columns indicate ET_a difference values and the circles indicate wind speed values.

Spatial and temporal ET_a values were observed at different crop growth stages using the METRIC model. The ET_a values derived from METRIC were higher than ET_a values estimated with atmometers. However, the ET_a-METRIC and ET_a-atm correlated well with coefficient of determination (r^2) of 0.87 and index of agreement ("d") of 0.84. The main difference in ET_a estimation with both methods was attributed to high wind speed values (>4 m s^{-1}) at the time of satellite image overpass. According to our results, the ET_r-atm values need to be

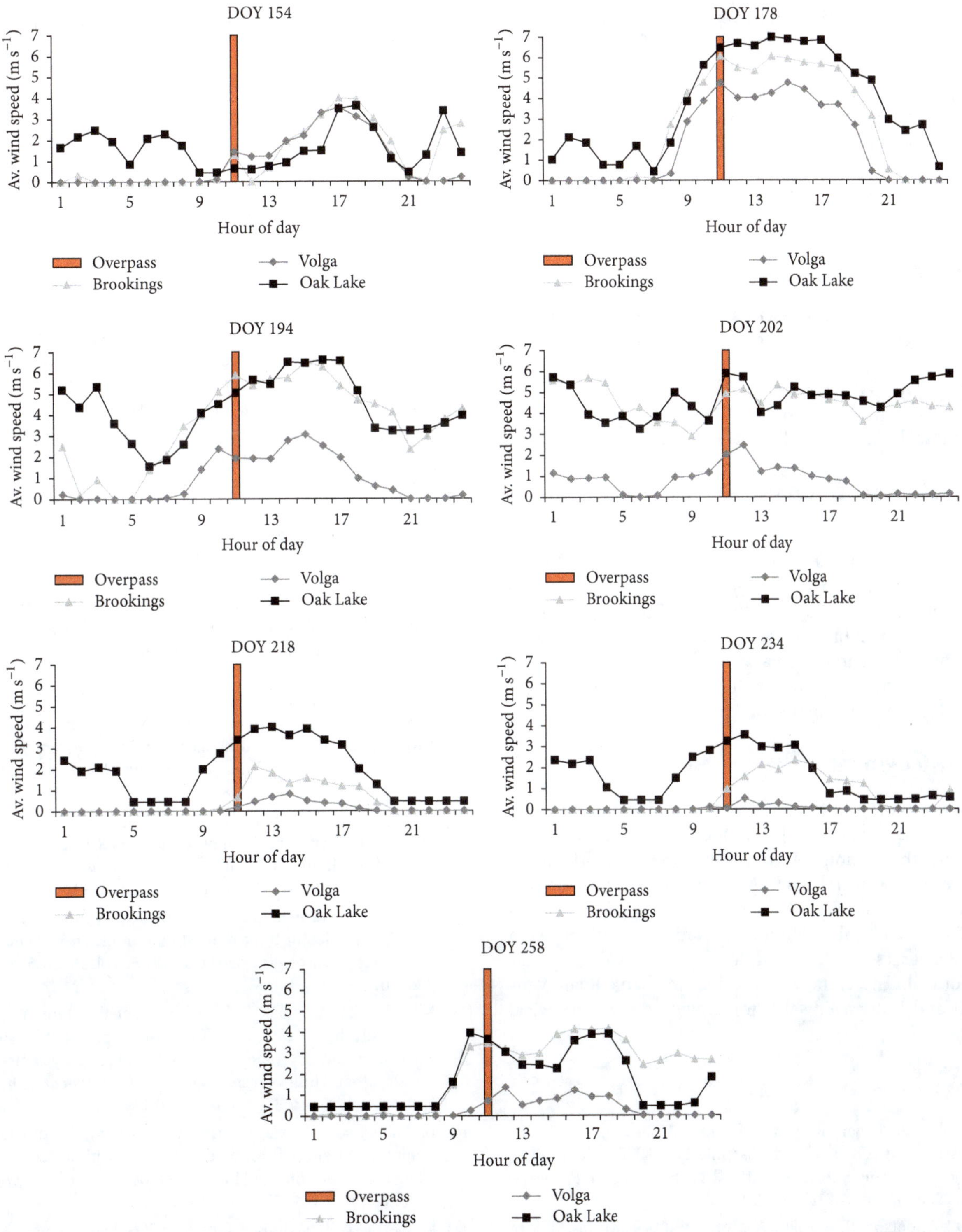

FIGURE 12: Hourly average wind speed values at three sites in eastern South Dakota. The red columns denote the time of satellite overpass (±11:12 AM).

adjusted during the windy days to get close estimates to the ET_r-PM values to get accurate ET_a estimations. In this study adjusted ET_r-atm values were not validated.

Based on our results the METRIC model was sensible to wind speed values at the time of satellite overpass. A limitation of satellite derived ET is that ET_a is estimated only for clear sky days, while ET_a from atmometer is estimated for any weather condition. However, a limitation of ET_a from atmometer is that ET_a is estimated only at field scale or few miles (±20) from atmometer location, whereas with remote sensing ET_a is estimated at field, local, and regional scales.

In conclusion the results of this study can be used by researchers and producers to estimate actual evapotranspiration and improve irrigation water management at regional and field scales using satellite-based remote sensing METRIC model method and at field scale using atmometer method.

Disclosure

This article is a part of a doctoral thesis: http://openprairie.sdstate.edu/cgi/viewcontent.cgi?article=2785&context=etd.

Conflicts of Interest

The authors declare no conflicts of interest.

Authors' Contributions

Arturo Reyes-González, Todd Trooien, and Jeppe Kjaersgaard conceived and designed the research. Arturo Reyes-González and Jeppe Kjaersgaard analyzed satellite images and ET_a graphs. Arturo Reyes-González, Todd Trooien, and Jeppe Kjaersgaard contributed to the data analysis and discussion of the results. Laurent Ahiablame and Christopher Hay provided a critical revision of the manuscript. All authors contributed to reviewing the final paper.

Acknowledgments

The first author would like to acknowledge the National Council for Science and Technology of México (CONACYT) and the National Institute of Forestry, Agriculture, and Livestock Research (INIFAP) for funding his doctoral scholarship. Additional funding is provided by the South Dakota Agricultural Experiment Station, South Dakota Water Resources Institute, and South Dakota Corn Utilization Council. The authors would like to thank Ryan Vanderleest and Nathan Edwards for providing meteorological datasets.

References

[1] R. Allen, A. Irmak, R. Trezza, J. M. H. Hendrickx, W. Bastiaanssen, and J. Kjaersgaard, "Satellite-based ET estimation in agriculture using SEBAL and METRIC," *Hydrological Processes*, vol. 25, no. 26, pp. 4011–4027, 2011.

[2] M. C. Anderson, R. G. Allen, A. Morse, and W. P. Kustas, "Use of Landsat thermal imagery in monitoring evapotranspiration and managing water resources," *Remote Sensing of Environment*, vol. 122, pp. 50–65, 2012.

[3] M. E. Jensen and R. G. Allen, "Evaporation, evapotranspiration, and irrigation water requirements," in *ASCE Manuals and Reports on Engineering Practice*, vol. 70, p. 744, 2nd edition, 2016.

[4] P. S. Thenkabail, M. A. Hanjra, V. Dheeravath, and M. Gumma, "A holistic view of global croplands and their water use for ensuring global food security in the 21st century through advanced remote sensing and non-remote sensing approaches," *Remote Sensing*, vol. 2, no. 1, pp. 211–261, 2010.

[5] M. G. Bos, R. A. L. Kselik, R. G. Allen, and D. J. Molden, *Water Requirements for Irrigation and the Environment*, Springer Science and Business Media, Dordrecht, Netherlands, 2009.

[6] R. G. Allen, M. Tasumi, A. Morse et al., "Satellite-based energy balance for mapping evapotranspiration with internalized calibration (METRIC)—applications," *Journal of Irrigation and Drainage Engineering*, vol. 133, no. 4, pp. 395–406, 2007.

[7] P. H. Gowda, J. L. Chavez, P. D. Colaizzi, S. R. Evett, T. A. Howell, and J. A. Tolk, "ET mapping for agricultural water management: present status and challenges," *Irrigation Science*, vol. 26, no. 3, pp. 223–237, 2008.

[8] R. G. Allen, L. S. Pereira, D. Raes, and M. Smith, "Crop evapotranspiration: guide-lines for computing crop requirements," in *Irrigation and Drainage Paper*, vol. 56, FAO, Rome, Italy, 1998.

[9] B. A. George, B. R. S. Reddy, N. S. Raghuwanshi, and W. W. Wallender, "Decision support system for estimating reference evapotranspiration," *Journal of Irrigation and Drainage Engineering*, vol. 128, no. 1, pp. 1–10, 2002.

[10] R. Allen, M. Tasumi, and R. Trezza, "Satellite-based energy balance for mapping evapotranspiration with internalized calibration (METRIC)—model," *Journal of Irrigation and Drainage Engineering*, vol. 133, no. 4, pp. 380–394, 2007.

[11] J. Kjaersgaard, R. Allen, and A. Irmak, "Improved methods for estimating monthly and growing season ET using METRIC applied to moderate resolution satellite imagery," *Hydrological Processes*, vol. 25, no. 26, pp. 4028–4036, 2011.

[12] J. Kjaersgaard, R. G. Allen, C. Robinson et al., "Computation of evapotranspiration in parts of South Platte and North Platte River basin using Landsat Imagery," in *Proceedings of the Pecora 17-The Future of Land Imaging...Going Operational*, Denver, Colo, USA, November 2008.

[13] J. L. Chávez, P. H. Gowda, S. R. Evett, P. D. Colaizzi, T. A. Howell, and T. Marek, "An application METRIC for ET mapping in the Texas high plains," in *Proceedings of the ASABE Annual International Meeting*, Minneapolis, Minn, USA, June 2007.

[14] R. K. Singh and A. Irmak, "Treatment of anchor pixels in the METRIC model for improved estimation of sensible and latent heat fluxes," *Hydrological Sciences Journal*, vol. 56, no. 5, pp. 895–906, 2011.

[15] N. C. Healey, A. Irmak, T. J. Arkebauer et al., "Remote sensing and in situ-based estimates of evapotranspiration for subirrigated meadow, dry valley, and upland dune ecosystems in the semi-arid sand hills of Nebraska, USA," *Irrigation and Drainage Systems*, vol. 25, no. 3, pp. 151–178, 2011.

[16] B. Hankerson, J. Kjaersgaard, and C. Hay, "Estimation of evapotranspiration from fields with and without cover crops using remote sensing and in situ methods," *Remote Sensing*, vol. 4, no. 12, pp. 3796–3812, 2012.

[17] M. Carrasco-Benavides, S. Ortega-Farías, L. O. Lagos, J. Kleissl, L. Morales-Salinas, and A. Kilic, "Parameterization of the satellite-based model (METRIC) for the estimation of instantaneous surface energy balance components over a drip-irrigated vineyard," *Remote Sensing*, vol. 6, no. 11, pp. 11342–11371, 2014.

[18] M. T. Folhes, C. D. Rennó, and J. V. Soares, "Remote sensing for irrigation water management in the semi-arid Northeast of Brazil," *Agricultural Water Management*, vol. 96, no. 10, pp. 1398–1408, 2009.

[19] V. M. Gordillo, H. M. Flores, L. Tijerina, and R. Arteaga, "Estimation of evapotranspiration using energy balance and

satellite images," *Revista Mexicana de Ciencias Agrícolas*, vol. 5, no. 1, pp. 143–155, 2014 (Spanish).

[20] H. Zhang, R. G. Anderson, and D. Wang, "Satellite-based crop coefficient and regional water use estimates for Hawaiian sugarcane," *Field Crops Research*, vol. 180, pp. 143–154, 2015.

[21] R. Liebert, J. Huntington, C. Morton, S. Sueki, and K. Acharya, "Reduced evapotranspiration from leaf beetle induced tamarisk defoliation in the Lower Virgin River using satellite-based energy balance," *Ecohydrology*, vol. 9, no. 1, pp. 179–193, 2016.

[22] M. Mkhwanazi, J. L. Chávez, and E. H. Rambikur, "Comparison of Large Aperture Scintillometer and satellite-based energy balance models in sensible heat flux and crop evapotranspiration determination," *International Journal of Remote Sensing Applications*, vol. 2, no. 1, pp. 24–30, 2012.

[23] R. Trezza, R. G. Allen, and M. Tasumi, "Estimation of actual evapotranspiration along the Middle Rio Grande of New Mexico using MODIS and landsat imagery with the METRIC model," *Remote Sensing*, vol. 5, no. 10, pp. 5397–5423, 2013.

[24] C. Santos, I. J. Lorite, M. Tasumi, R. G. Allen, and E. Fereres, "Integrating satellite-based evapotranspiration with simulation models for irrigation management at the scheme level," *Irrigation Science*, vol. 26, no. 3, pp. 277–288, 2008.

[25] M. Choi, W. P. Kustas, M. C. Anderson, R. G. Allen, F. Li, and J. H. Kjaersgaard, "An intercomparison of three remote sensing-based surface energy balance algorithms over a corn and soybean production region (Iowa, U.S.) during SMACEX," *Agricultural and Forest Meteorology*, vol. 149, no. 12, pp. 2082–2097, 2009.

[26] M. Choi, T. W. Kim, M. Park, and S. J. Kim, "Evapotranspiration estimation using the Landsat-5 Thematic Mapper image over the Gyungan watershed in Korea," *International Journal of Remote Sensing*, vol. 32, no. 15, pp. 4327–4341, 2011.

[27] M. P. Gonzalez-Dugo, C. M. U. Neale, L. Mateos et al., "A comparison of operational remote sensing-based models for estimating crop evapotranspiration," *Agricultural and Forest Meteorology*, vol. 149, no. 11, pp. 1843–1853, 2009.

[28] A. N. French, D. J. Hunsaker, and K. R. Thorp, "Remote sensing of evapotranspiration over cotton using the TSEB and METRIC energy balance models," *Remote Sensing of Environment*, vol. 158, pp. 281–294, 2015.

[29] T. A. Paço, I. Pôças, M. Cunha et al., "Evapotranspiration and crop coefficients for a super intensive olive orchard. An application of SIMDualKc and METRIC models using ground and satellite observations," *Journal of Hydrology*, vol. 519, pp. 2067–2080, 2014.

[30] N. Bhattarai, L. J. Quackenbush, M. Dougherty, and L. J. Marzen, "A simple Landsat–MODIS fusion approach for monitoring seasonal evapotranspiration at 30 m spatial resolution," *International Journal of Remote Sensing*, vol. 36, no. 1, pp. 115–143, 2015.

[31] I. Broner and R. A. P. Law, "Evaluation of a modified atmometer for estimating reference ET," *Irrigation Science*, vol. 12, no. 1, pp. 21–26, 1991.

[32] M. Alam and T. P. Trooien, "Estimating reference evapotranspiration with an atmometer," *Applied Engineering in Agriculture*, vol. 17, no. 2, pp. 153–158, 2001.

[33] V. Magliulo, R. D'Andria, and G. Rana, "Use of the modified atmometer to estimate reference evapotranspiration in Mediterranean environments," *Agricultural Water Management*, vol. 63, no. 1, pp. 1–14, 2003.

[34] S. Irmak, M. D. Dukes, and J. M. Jacobs, "Using modified Bellani plate evapotranspiration gauges to estimate short canopy reference evapotranspiration," *Journal of Irrigation and Drainage Engineering*, vol. 131, no. 2, pp. 164–175, 2005.

[35] J. C. Mendonça, E. F. Sousa, S. Bernardo, G. P. Dias, and S. Grippa, "Comparação entre métodos de estimativa da evapotranspiração de referência (ETo) na região Norte Fluminense, RJ," *Revista Brasileira de Engenharia Agrícola e Ambiental*, vol. 7, no. 2, pp. 275–279, 2003 (Portuguese).

[36] M. P. Casanova, I. Messing, A. Joel, and A. M. Cañete, "Methods to estimate lettuce evapotranspiration in greenhouse conditions in the central zone of Chile," *Chilean Journal of Agricultural Research*, vol. 69, pp. 60–70, 2009.

[37] W. L. Pelton, "Evaporation from atmometers and pans," *Canadian Journal of Plant Science*, vol. 44, no. 5, pp. 397–404, 1964.

[38] G. J. Kidron, "Measurements of evaporation with a novel mini atmometer in the Negev," *Weather*, vol. 60, no. 9, pp. 268–272, 2005.

[39] J. W. Knox, J. A. Rodriguez-Diaz, and T. M. Hess, "Estimating evapotranspiration by using atmometers for irrigation scheduling in a humid environment," *Journal of Irrigation and Drainage Engineering*, vol. 137, no. 11, pp. 685–691, 2011.

[40] S. Taghvaeian, J. L. Chávez, W. C. Bausch, K. C. DeJonge, and T. J. Trout, "Minimizing instrumentation requirement for estimating crop water stress index and transpiration of maize," *Irrigation Science*, vol. 32, no. 1, pp. 53–65, 2014.

[41] D. Lamine, B. Ansoumana, and D. Dior, "Use of atmometers to estimate reference evapotranspiration in Arkansas," *African Journal of Agricultural Research*, vol. 10, no. 48, pp. 4376–4383, 2015.

[42] K. W. Peterson, D. J. Bremer, and J. D. Fry, "Evaluation of atmometers within urban home lawn microclimates," *Crop Science*, vol. 55, no. 5, pp. 2359–2367, 2015.

[43] A. Reyes-González, T. Trooien, C. Hay, and J. Kjaersgaard, "Comparison of reference evapotranspiration estimates by automated weather station and measured an atmometer," in *Proceedings of the 2016 Western South Dakota Hydrology Conference*, Rapid City, SD, USA, April 2016.

[44] Soil Survey Staff, Natural Resources Conservation Service, Unite States Department of Agriculture. Web soil survey, October 2016, http://websoilsurvey.nrcs.usda.gov.

[45] B. D. Keim, "The lasting scientific impact of the thornthwaite water-balance model," *Geographical Review*, vol. 100, no. 3, pp. 295–300, 2010.

[46] USGS-GLOVIS, "Global visual viewer," 2011, http://glovis.usgs.gov.

[47] J. Kjaersgaard and R. G. Allen, "Remote sensing technology to produce consumptive water use maps for the Nebraska Panhandle," Tech. Rep. 88, University of Nebraska, Lincoln, Neb, USA, 2010.

[48] USGS-Landsat, 2016, http://landsat.usgs.gov/gap-filling-landsat-7-slc-single-scenes-using-erdas-imagine-TM.

[49] W. G. M. Bastiaanssen, E. J. M. Noordman, H. Pelgrum, G. Davids, B. P. Thoreson, and R. G. Allen, "SEBAL model with remotely sensed data to improve water-resources management under actual field conditions," *Journal of Irrigation and Drainage Engineering*, vol. 131, no. 1, pp. 85–93, 2005.

[50] M. Tasumi, R. G. Allen, R. Trezza, and J. L. Wright, "Satellite-based energy balance to assess within-population variance of crop coefficient curves," *Journal of Irrigation and Drainage Engineering*, vol. 131, no. 1, pp. 94–109, 2005.

[51] M. Tasumi, *Progress in operational estimation of regional evapotranspiration using satellite imagery [Ph.D. thesis]*, University of Idaho, Moscow, Idaho, USA, 2003.

[52] W. G. M. Bastiaanssen, M. Menenti, R. A. Feddes, and A. A. M. Holtslag, "A remote sensing surface energy balance algorithm for land (SEBAL): 1. Formulation," *Journal of Hydrology*, vol. 212, no. 1–4, pp. 198–212, 1998.

[53] J. L. Wright, "New evapotranspiration crop coefficients," *Journal of the Irrigation & Drainage Division*, vol. 108, pp. 57–74, 1982.

[54] ASCE-EWRI, "The ASCE Standardized Reference Evapotranspiration Equation," Report of the ASCE-EWRI Task Committee on Standardization of Reference Evapotranspiration, ASCE, Reston, Va, USA, 2005.

[55] P. Gavilán and F. Castillo-Llanque, "Estimating reference evapotranspiration with atmometers in a semiarid environment," *Agricultural Water Management*, vol. 96, no. 3, pp. 465–472, 2009.

[56] S. Irmak, L. O. Odhiambo, J. E. Specht, and K. Djaman, "Hourly and daily single and basal evapotranspiration crop coefficients as a function of growing degree days, days after emergence, leaf area index, fractional green canopy cover, and plant phenology for soybean," *Transactions of the ASABE*, vol. 56, no. 5, pp. 1785–1803, 2013.

[57] C. J. Willmott, "On the validation of models," *Physical Geography*, vol. 2, no. 2, pp. 184–194, 1981.

[58] K. Djaman and S. Irmak, "Actual crop evapotranspiration and alfalfa- and grass-reference crop coefficients of maize under full and limited irrigation and rainfed conditions," *Journal of Irrigation and Drainage Engineering*, vol. 139, no. 6, pp. 433–446, 2013.

[59] A. Reyes-González, T. Trooien, J. Kjaersgaard, C. Hay, and D. G. Reta-Sánchez, "Development of crop coefficients using remote sensing-based vegetation index and growing degree days," in *Proceedings of the ASABE Annual International Meeting*, Orlando, Fla, USA, July 2016.

[60] A. Irmak and S. Irmak, "Reference and crop evapotranspiration in South Central Nebraska. II: Measurement and estimation of actual evapotranspiration for corn," *Journal of Irrigation and Drainage Engineering*, vol. 134, no. 6, pp. 700–715, 2008.

[61] R. K. Singh and A. Irmak, "Estimation of crop coefficients using satellite remote sensing," *Journal of Irrigation and Drainage Engineering*, vol. 135, no. 5, pp. 597–608, 2009.

[62] P. Droogers, W. W. Immerzeel, and I. J. Lorite, "Estimating actual irrigation application by remotely sensed evapotranspiration observations," *Agricultural Water Management*, vol. 97, no. 9, pp. 1351–1359, 2010.

[63] G. Carrillo-Rojas, B. Silva, M. Córdova, R. Célleri, and J. Bendix, "Dynamic mapping of evapotranspiration using an energy balance-based model over an andean páramo catchment of southern ecuador," *Remote Sensing*, vol. 8, no. 2, article 160, 2016.

[64] C. G. Morton, J. L. Huntington, G. M. Pohll, R. G. Allen, K. C. Mcgwire, and S. D. Bassett, "Assessing calibration uncertainty and automation for estimating evapotranspiration from agricultural areas using METRIC," *JAWRA Journal of the American Water Resources Association*, vol. 49, no. 3, pp. 549–562, 2013.

[65] D. J. Gleason, A. A. Andales, T. A. Bauder, and J. L. Chávez, "Performance of atmometers in estimating reference evapotranspiration in a semi-arid environment," *Agricultural Water Management*, vol. 130, pp. 27–35, 2013.

[66] F. Chen and P. J. Robinson, "Estimating reference crop evapotranspiration with ETgages," *Journal of Irrigation and Drainage Engineering*, vol. 135, no. 3, pp. 335–342, 2009.

[67] R. G. Anderson, D. Wang, R. Tirado-Corbalá, H. Zhang, and J. E. Ayars, "Divergence of actual and reference evapotranspiration observations for irrigated sugarcane with windy tropical conditions," *Hydrology and Earth System Sciences*, vol. 19, no. 1, pp. 583–599, 2015.

[68] R. S. Westerhoff, "Using uncertainty of Penman and Penman-Monteith methods in combined satellite and ground-based evapotranspiration estimates," *Remote Sensing of Environment*, vol. 169, pp. 102–112, 2015.

[69] M. H. Mokhtari, B. Ahmad, H. Hoveidi, and I. Busu, "Sensitivity analysis of METRIC-based evapotranspiration algorithm," *International Journal of Environmental Research*, vol. 7, no. 2, pp. 407–422, 2013.

Permissions

All chapters in this book were first published in IJA, by Hindawi Publishing Corporation; hereby published with permission under the Creative Commons Attribution License or equivalent. Every chapter published in this book has been scrutinized by our experts. Their significance has been extensively debated. The topics covered herein carry significant findings which will fuel the growth of the discipline. They may even be implemented as practical applications or may be referred to as a beginning point for another development.

The contributors of this book come from diverse backgrounds, making this book a truly international effort. This book will bring forth new frontiers with its revolutionizing research information and detailed analysis of the nascent developments around the world.

We would like to thank all the contributing authors for lending their expertise to make the book truly unique. They have played a crucial role in the development of this book. Without their invaluable contributions this book wouldn't have been possible. They have made vital efforts to compile up to date information on the varied aspects of this subject to make this book a valuable addition to the collection of many professionals and students.

This book was conceptualized with the vision of imparting up-to-date information and advanced data in this field. To ensure the same, a matchless editorial board was set up. Every individual on the board went through rigorous rounds of assessment to prove their worth. After which they invested a large part of their time researching and compiling the most relevant data for our readers.

The editorial board has been involved in producing this book since its inception. They have spent rigorous hours researching and exploring the diverse topics which have resulted in the successful publishing of this book. They have passed on their knowledge of decades through this book. To expedite this challenging task, the publisher supported the team at every step. A small team of assistant editors was also appointed to further simplify the editing procedure and attain best results for the readers.

Apart from the editorial board, the designing team has also invested a significant amount of their time in understanding the subject and creating the most relevant covers. They scrutinized every image to scout for the most suitable representation of the subject and create an appropriate cover for the book.

The publishing team has been an ardent support to the editorial, designing and production team. Their endless efforts to recruit the best for this project, has resulted in the accomplishment of this book. They are a veteran in the field of academics and their pool of knowledge is as vast as their experience in printing. Their expertise and guidance has proved useful at every step. Their uncompromising quality standards have made this book an exceptional effort. Their encouragement from time to time has been an inspiration for everyone.

The publisher and the editorial board hope that this book will prove to be a valuable piece of knowledge for researchers, students, practitioners and scholars across the globe.

List of Contributors

Stephen D. Merrill, Mark A. Liebig and John D. Hendrickson
Northern Great Plains Research Laboratory, Agricultural Research Service, U.S. Department of Agriculture, 1701 10th Ave. SW, Mandan, ND 58554, USA

Abbey F. Wick
Department of Soil Science,Walster Hall, North Dakota State University, Department 7680, Fargo, ND 58108-6050, USA

Olga S. Walsh and Sanaz Shafian
Department of Plant Sciences, Southwest Research and Extension Center, University of Idaho, Parma, ID 83660, USA

Robin J. Christiaens
Private Enterprise, Roundup, MT 59072, USA

Pangirayi Tongoona, Agyemang Danquah, Samuel Offei and Eric Danquah
West Africa Centre for Crop Improvement, University of Ghana, PMB LG 30, Legon, Ghana

Ndiaga Cisse
Centre d'Etudes Régional pour l'Amélioration de l'Adaptation àla Sècheresse, BP 3320,Thies, Senegal

Ghislain Kanfany and Amadou Fofana
Centre National de Recherches Agronomiques de Bambey, BP 211, Bambey, Senegal

Dipesh KC and Bradley S. Osbon
Louisiana State University AgCenter, Hill Farm Research Station, Homer, LA 71040, USA

Michael C. Tyree
Department of Biology, Indiana University of Pennsylvania, Indiana, PA 15705, USA

Michael A. Blazier and Wilson G. Hood
School of Forestry, Louisiana Tech University, Ruston, LA 71272, USA

Mary Anne Sword Sayer
USDA Forest Service, Southern Research Station, Pineville, LA 71360, USA

Richard Moise Alansou Dieme and Papa Madiallacke Diedhiou
Université Gaston Berger, BP 211 Saint-Louis, Senegal

Issa Faye
Centre National de Recherches Agronomiques de Bambey, Institut Sénégalais de Recherches Agricoles, BP 211 Bambey, Senegal

Yedomon Ange Bovys Zoclanclounon and Daniel Fonceka
Centre d'Etudes Régional pour l'Amélioration de l'Adaptation à la Sécheresse, Route de Khombole, BP 3320 Thiès, Senegal

Ousmane Ndoye
Conseil Ouest et centre Africain pour la Recherche et le Développement Agricoles, Avenue Bourguiba, BP 48 Dakar RP, Senegal

S. Ganya
Ministry of Agriculture, Mashonaland West Province, Zimbabwe

E. Svotwa
Senior Lecturer, Department of Crop Science and Postharvest Technology, Chinhoyi University of Technology, Chinhoyi, Zimbabwe

R. D. Katsaruware
Senior Lecturer, Department of Agriculture, Zimbabwe Open University, Mashonaland West Province, Zimbabwe

Zachary D. Lancaster and Jason K. Norsworthy
Department of Crop, Soil, and Environmental Sciences, University of Arkansas, Fayetteville, AR 72701, USA

Robert C. Scott
Department of Crop, Soil, and Environmental Sciences, University of Arkansas, Lonoke, AR 72086, USA

Asmiaty Sahur, Ambo Ala and Elkawakib Syam'un
Department of Agronomy, Faculty of Agriculture, Hasanuddin University, Makassar, Indonesia

Baharuddin Patandjengi
Research Center (PKP), Hasanuddin University, Makassar, South Sulawesi 90245, Indonesia

Lord Abbey, Nana Annan, Samuel Kwaku Asiedu, Ebenezer Oluwaseun Esan and Ekene Mark-Anthony Iheshiulo
Faculty of Agriculture, Department of Plant, Food, and Environmental Sciences,Dalhousie University, Truro, NS, Canada B2N5E3

Imtiaz Hussain and Akhter Ali
CIMMYT, CSI, NARC, Park Road, Islamabad, Pakistan

Ansaar Ahmed
CIMMYT, Cereal Crop Research Institute, Nowshera, Pakistan

Hafiz Nasrullah
Agronomic Research Station, Bahawalpur, Pakistan

Badar ud Din Khokhar
Wheat Research Institute Sindh, Sakrand, Pakistan

Shahid Iqbal
Agriculture Research Institute, DI Khan, Pakistan

Azhar Mahmood Aulakh
Arid Zone Research Institute, Bhakkar, Pakistan

Atta ullah Khan
Arid Zone Research Institute, Umerkot, Pakistan

Jamil Akhter
National Rural Support Program, Islamabad, Pakistan

Gulzar Ahmed
Cereal Crop Research Institute, Nowshera, Pakistan

Dylan N. Dillaway, Michael C. Tyree and John K. Jackson
Work Performed: School of Forestry, Louisiana Tech University, Ruston, LA 71272, USA

Dylan N. Dillaway
Center for Natural Resources Mgmt. & Protection, Unity College, Unity, ME 04988, USA

Michael C. Tyree
Department of Biology, Indiana University of Pennsylvania, Indiana, PA 15705-1001, USA

Albert Kobina Mensah
Department of Soil Science and Soil Ecology, Institute of Geography, Ruhr-Universitaet Bochum, Universitätsstraße 150, 44801 Bochum, Germany

Kwame Agyei Frimpong
Department of Soil Science, School of Agriculture, University of Cape Coast, Cape Coast, Ghana

Aliakbar Enghiad
Business Administration, California State University, Northridge, 18111 Nordhoff Street, Northridge, CA 91330, USA

Danielle Ufer, Amanda M. Countryman and Dawn D. Thilmany
Department of Agricultural and Resource Economics, Colorado State University, 1200 Center Ave. Mall, Fort Collins, CO 80523-1172, USA

Nader Soltani, Lynette R. Brown and Peter H. Sikkema
University of Guelph Ridgetown Campus, Ridgetown, ON N0P 2C0, Canada

Bouchaib Khadari
AGAP, University Montpellier, CIRAD, INRA, Montpellier Sup Agro, Montpellier, France

Bouchaib Khadari
Conservatoire Botanique National Méditerranéen de Porquerolles (CBNMed), UMR AGAP, Montpellier, France

Ahmed El Bakkali
INRA, UR Amélioration des Plantes et Conservation des Ressources Phytogénétiques, Meknés, Morocco

Daniel Nyoki and Patrick A. Ndakidemi
School of Life Science and Bio-Engineering, The Nelson Mandela African Institution of Science and Technology, Arusha, Tanzania
Centre for Research, Agricultural Advancement, Teaching Excellence and Sustainability (CREATES) in Food and Nutrition Security, The Nelson Mandela African Institution of Science and Technology, Arusha, Tanzania

Venera S. Kamburova, Elena V. Nikitina, Shukhrat E. Shermatov, Zabardast T. Buriev and Ibrokhim Y. Abdurakhmonov
Center of Genomics and Bioinformatics, Academy of Sciences of the Republic of Uzbekistan, University Street-2, Qibray Region, 111215 Tashkent, Uzbekistan

Siva P. Kumpatla and Zabardast T. Buriev
Dow Agro Sciences LLC, Indianapolis, IN 46268, USA

Chandrakanth Emani and Zabardast T. Buriev
Western Kentucky University-Owensboro, Owensboro, KY 42303, USA

W. James Grichar
Texas A&M Agri Life Research and Extension Center, Corpus Christi, TX 78406, USA

Jack J. Rose
Sesaco Corporation, Austin, TX 78741, USA

Peter A. Dotray
Texas A&M Agri Life Research and Extension Center, Lubbock, TX 79403, USA

Todd A. Baughman
Institute for Agricultural Biosciences, Oklahoma State University, Ardmore, OK 73402, USA

D. Ray Langham
Sesame Research, LLC, San Antonio, TX 78217, USA

Kaisa Werner and Muthu Bagavathiannan
Texas A & M Agri Life Research, College Station, TX 77843, USA

Ali Mansouri and Bachir Oudjehih
Agronomy Department, El Hadj Lakhdar University, 5000 Batna, Algeria

Abdelkader Benbelkacem
Plant Breeding and Biotechnology Division, National Agronomic Research Institute of Algeria (INRAA), 25000 Constantine, Algeria

Zine El Abidine Fellahi
Department of Agronomy, Faculty of Natural, Life and Earth Sciences and the Universe, University of Mohamed El Bachir El Ibrahimi, 34030 Bordj Bou Arreridj, Algeria

Hamenna Bouzerzour
Valorization of Natural Biological Resources Laboratory, University of Setif, 19000 Setif, Algeria

Silvia L. Falasca
Climate and Water Institute, INTA, Las Cabãnas y Los Reseros, s/n, Castelar, Buenos Aires, Argentina

Sandra Pitta-Alvarez
Laboratorio de Cultivo Experimental de Plantas y Microalgas, Departamento de Biodiversidad y Biología Experimental, Facultad de Ciencias Exactas y Naturales, Universidad de Buenos Aires, Buenos Aires, Argentina
Instituto de Micología y Botánica (INMIBO), Universidad de Buenos Aires-CONICET, Buenos Aires, Argentina

Ana Ulberich
Centro de Investigaciones y Estudios Ambientales (CINEA), Facultad de Ciencias Humanas (FCH), Universidad Nacional del Centro de la Provincia de Buenos Aires (UNICEN), Campus Universitario, Paraje Arroyo Seco, s/n, Tandil, Argentina

Kevin Muyang Tawie Sulok, Choy Yuen Khew and Jarroop Augustine Mercer Zehnder
Research and Development Division, Malaysian Pepper Board, 93916 Kuching, Sarawak, Malaysia

Osumanu Haruna Ahmed
Department of Crop Science, Faculty of Agriculture and Food Sciences, Universiti Putra Malaysia (UPM), Bintulu Campus, 97008 Bintulu, Sarawak, Malaysia
Institute of Tropical Agriculture and Food Security (ITAFoS), Universiti Putra Malaysia (UPM), 43400 Serdang, Selangor, Malaysia
Institute of Tropical Forestry and Forest Products (INTROP), Universiti Putra Malaysia (UPM), 43400 Serdang, Selangor, Malaysia

Ori Ahiman, Yonatan Mekhmandarov, Moran Pirkner and Josef Tanny
Institute of Soil, Water and Environmental Sciences, Agricultural Research Organization, The Volcani Center, Rishon LeZion 7528809, Israel

Josef Tanny
HIT-Holon Institute of Technology, Holon 58102, Israel

Jamin Smitchger and Norman F. Weeden
Department of Plant Sciences and Plant Pathology, Montana State University, Bozeman, MT 59717, USA

Index

R

S

T

U

V

W

Z